RENEWALS 458-4574
DATE DUE

Robert B. Heimann
Plasma Spray Coating

Further Reading

J. Friedrich

The Plasma Chemistry of Polymer Surfaces
Advanced Techniques for Surface Design

2009
ISBN: 978-3-527-31853-7

W. E. S. Unger (Ed.)

Surface Chemical Analysis
Characterization Techniques for Plasma-deposited Organic Films

2009
ISBN: 978-3-527-31851-3

K. E. Schneider, V. Belashchenko, M. Dratwinski, S. Siegmann, A. Zagorski

Thermal Spraying for Power Generation Components

2006
ISBN: 978-3-527-31337-2

Robert B. Heimann

Plasma Spray Coating

Principles and Applications

Second, Completely Revised and Enlarged Edition

WILEY-VCH

WILEY-VCH Verlag GmbH & Co. KGaA

The Author

Prof. Dr. Robert B. Heimann
Oceangate Consulting
Questenbergweg 48
34346 Hann. Münden
Germany

All books published by Wiley-VCH are carefully produced. Nevertheless, authors, editors, and publisher do not warrant the information contained in these books, including this book, to be free of errors. Readers are advised to keep in mind that statements, data, illustrations, procedural details or other items may inadvertently be inaccurate.

Library of Congress Card No.: applied for

British Library Cataloguing-in-Publication Data
A catalogue record for this book is available from the British Library.

Bibliographic information published by the Deutsche Nationalbibliothek
Die Deutsche Nationalbibliothek lists this publication in the Deutsche Nationalbibliografie; detailed bibliographic data are available in the Internet at http://dnb.d-nb.de.

© 2008 WILEY-VCH Verlag GmbH & Co. KGaA, Weinheim

All rights reserved (including those of translation into other languages). No part of this book may be reproduced in any form – by photoprinting, microfilm, or any other means – nor transmitted or translated into a machine language without written permission from the publishers. Registered names, trademarks, etc. used in this book, even when not specifically marked as such, are not to be considered unprotected by law.

Typesetting   Thomson Digital, Noida, India
Printing   betz-druck GmbH, Darmstadt
Binding   Litges & Dopf Buchbinderei GmbH, Heppenheim

Printed in the Federal Republic of Germany
Printed on acid-free paper

ISBN: 978-3-527-32050-9

Contents

Preface *XIII*
Preface to the First Edition *XV*
Synopsis *XVII*

1 **Introduction** *1*
1.1 Coatings in the Industrial Environment *2*
1.1.1 Market Position *2*
1.2 Survey of Surface Coating Techniques *3*
1.3 Brief History of Thermal Spraying *9*
1.4 Synergistic Nature of Coatings *12*
1.5 Applications of Thermally Sprayed Coatings *13*
References *15*

2 **Principles of Thermal Spraying** *17*
2.1 Characterization of Flame vs. Plasma Spraying *21*
2.2 Concept of Energy Transfer Processes *22*
2.3 Unique Features of the Plasma Spray Process *22*
References *24*

3 **The First Energy Transfer Process: Electron–Gas Interactions** *25*
3.1 The Plasma State *25*
3.1.1 Characteristic Plasma Parameters *26*
3.1.1.1 Langmuir Plasma Frequency *26*
3.1.1.2 Debye Screening Length *27*
3.1.1.3 Landau Length *27*
3.1.1.4 Collision Path Length *27*
3.1.1.5 Collision Frequency *29*
3.1.2 Classification of Plasmas *29*
3.1.2.1 Low Density Plasmas *30*

Plasma Spray Coating: Principles and Applications. Robert B. Heimann
Copyright © 2008 WILEY-VCH Verlag GmbH & Co. KGaA, Weinheim
ISBN: 978-3-527-32050-9

3.1.2.2	Medium Density Plasmas	30
3.1.2.3	High Density Plasmas	32
3.1.3	Equilibrium and Nonequilibrium Plasmas	32
3.1.4	Maxwellian Distribution of Plasma Energies	33
3.1.5	Equilibrium Composition of Plasma Gases (Phase Diagrams)	34
3.2	Plasma Generation	37
3.2.1	Plasma Generation by Application of Heat	37
3.2.2	Plasma Generation by Compression	39
3.2.2.1	z-Pinch	39
3.2.2.2	Θ-Pinch	39
3.2.2.3	Plasma Focus	39
3.2.3	Plasma Generation by Radiation	39
3.2.4	Plasma Generation by Electric Currents (Gas Discharges)	40
3.2.4.1	Glow Discharges	41
3.2.4.2	Arc Discharges	43
3.2.5	Structure of the Arc Column	44
3.2.5.1	Positive Column	44
3.2.5.2	The Cathode Fall Region	44
3.2.5.3	The Anode Region	47
3.3	Design of Plasmatrons	48
3.3.1	Arc Discharge Generators and their Applications	50
3.3.1.1	Electrode-supported Plasmas	50
3.3.1.2	Electrode-less Plasmas	54
3.3.1.3	Hybrid Devices	57
3.3.2	Stabilization of Plasma Arcs	57
3.3.2.1	Wall-stabilized Arcs	59
3.3.2.2	Convection-stabilized Arcs	59
3.3.2.3	Electrode-stabilized Arcs	60
3.3.2.4	Other Stabilization Methods	60
3.3.3	Temperature and Velocity Distributions in a Plasma Jet	61
3.3.3.1	Turbulent Jets	61
3.3.3.2	Quasi-laminar Jets	63
3.4	Plasma Diagnostics: Temperature, Enthalpy and Velocity Measurements	65
3.4.1	Temperature Measurements	66
3.4.1.1	Spectroscopic Methods	66
3.4.1.2	Two-wavelengths Pyrometry	68
3.4.1.3	Chromatic Monitoring	69
3.4.2	Velocity Measurements	70
3.4.2.1	Enthalpy Probe and Pitot Tube Techniques	70
3.4.2.2	Laser Doppler Anemometry (LDA)	72
3.4.2.3	Other Methods	75
	References	76

4	**The Second Energy Transfer Process: Plasma–Particle Interactions**	*79*
4.1	Injection of Powders *79*	
4.2	Characteristics of Feed Materials *80*	
4.2.1	Solid Wires, Rods and Filled Wires *80*	
4.2.2	Powders *81*	
4.2.2.1	Atomization *84*	
4.2.2.2	Fusion and Crushing *84*	
4.2.2.3	Compositing *85*	
4.2.2.4	Agglomeration *85*	
4.3	Momentum Transfer *86*	
4.3.1	Connected Energy Transmission *86*	
4.3.2	Transfer of Plasma Velocity to Particles *86*	
4.3.3	Surface Ablation of Particles *87*	
4.4	Heat Transfer *87*	
4.4.1	Heat Transfer under Low Loading Conditions *87*	
4.4.2	Exact Solution of Heat Transfer Equations *93*	
4.4.2.1	Particle Heating without Evaporation *93*	
4.4.2.2	Particle Heating with Evaporation *94*	
4.4.2.3	Evaporation Time of a Particle *97*	
4.4.3	Heat Transfer under Dense Loading Conditions *99*	
4.4.4	Heat Transfer Catastrophy *99*	
4.4.5	Energy Economy *102*	
4.5	Particle Diagnostics: Velocity, Temperature and Number Densities *103*	
4.5.1	Particle Velocity Determination *103*	
4.5.2	Particle Temperature Determination *107*	
4.5.3	Particle Number Density Determination *107*	
	References *109*	
5	**The Third Energy Transfer Process: Particle–Substrate Interactions**	*111*
5.1	Basic Considerations *111*	
5.2	Estimation of Particle Number Density *114*	
5.3	Momentum Transfer from Particles to Substrate *116*	
5.4	Heat Transfer from Particles to Substrate *126*	
5.4.1	Generalized Heat Transfer Equation *126*	
5.4.2	Heat Transfer from Coating to Substrate *128*	
5.5	Crystallinity of Coatings *132*	
5.6	Fractal Properties of Surfaces *135*	
5.6.1	Box Counting Method *137*	
5.6.2	Density Correlation Function *138*	
5.6.3	Mass Correlation Function *139*	
5.6.4	Slit Island Analysis (SIA) *139*	
5.6.5	Fracture Profile Analysis (FPA) *139*	
5.6.6	Scale-sensitive Fractal Analysis *140*	

5.7	Residual Stresses	*143*
5.7.1	Blind Hole Test	*145*
5.7.2	X-ray Diffraction Measurements ($\sin^2\psi$-technique)	*145*
5.7.3	Curvature Monitoring Technique (Almen-type test)	*151*
5.7.4	Photoluminescence Piezospectroscopy	*155*
5.7.5	Neutron Diffraction	*157*
	References	*159*
6	**Modeling and Numerical Simulation**	*163*
6.1	Principal Aspects of Modeling	*163*
6.2	Modeling of Plasma Properties	*164*
6.3	Modeling of the Plasma–Particle Interaction	*166*
6.3.1	Modeling of Heat Transfer	*166*
6.3.1.1	Conservation Equations	*166*
6.3.1.2	Experimental Validation of Modeling under Dense Loading Conditions	*167*
6.3.1.3	Modeling of Heat Transfer in Two-fluid Interfacial Flow	*169*
6.3.2	Modeling of Momentum Transfer	*169*
6.3.2.1	Modeling of the Drag Coefficient	*170*
6.3.3	Modeling of Particle Dispersion	*172*
6.4	Modeling of Particle–Substrate Interaction	*172*
6.4.1	3D Simulation of Coating Microstructure	*173*
6.4.2	Modeling of Splat Shapes	*177*
6.4.3	Modeling of Residual Coating Stresses	*180*
6.4.4	Modeling of Thermal Conductivity	*185*
	References	*188*
7	**Solutions to Industrial Problems I: Structural Coatings**	*191*
7.1	Carbide Coatings	*192*
7.1.1	Pure Carbides	*192*
7.1.2	Cemented Carbides	*193*
7.1.2.1	Tungsten Carbide-based Coatings	*194*
7.1.2.2	Titanium Carbide-based Coatings	*205*
7.1.2.3	Chromium Carbide-based Coatings	*209*
7.1.2.4	Other Hard Carbide Coatings	*212*
7.2	Nitride Coatings	*213*
7.2.1	Titanium Nitride-based Coatings	*214*
7.2.2	Silicon Nitride-based Coatings	*214*
7.3	Oxide Coatings	*216*
7.3.1	Alumina-based Coatings	*217*
7.3.2	Chromium Oxide-based Coatings	*220*
7.3.3	Other Oxide Coatings	*221*
7.4	Metallic Coatings	*221*
7.4.1	Refractory Metal Coatings	*221*
7.4.2	Superalloy Coatings	*223*

7.5	Diamond Coatings *224*
	References *226*

8	**Solutions to Industrial Problems II: Functional Coatings** *233*
8.1	Thermal (TBC) and Chemical (CBC) Barrier Coatings *233*
8.1.1	Partially Yttria-stabilized Zirconia Coatings (Y-PSZ) *233*
8.1.2	Stress Development and Control *239*
8.1.3	Sealing of As-sprayed Surfaces *244*
8.1.3.1	Laser Surface Remelting of Y-PSZ Coatings *244*
8.1.4	Other Thermal and Chemical Barrier Coatings *249*
8.2	Superconducting and Electrocatalytic Coatings *253*
8.2.1	HT-Superconducting Coatings *253*
8.2.2	Electrocatalytic Coatings for Solid Oxide Fuel Cells (SOFC) *254*
8.3	Photocatalytic Coatings *263*
8.4	Coatings with High Friction Coefficient *267*
	References *270*

9	**Solutions to Industrial Problems III: Bioceramic Coatings** *277*
9.1	Classification of Biomaterials and Mechanism of Bone Bonding *277*
9.2	Properties of Bioceramic Coatings *278*
9.3	Plasma Spraying of Osseoconductive Hydroxyapatite *284*
9.3.1	Phase Composition *285*
9.3.2	Parametric Study of HAp Coating Properties *287*
9.4	Bioinert Bond Coats *289*
9.4.1	Composition of Bioinert Bond Coats *290*
9.4.1.1	Calcium Silicate Bond Coats *290*
9.4.1.2	Zirconia Bond Coats *291*
9.4.1.3	Zirconia/Titania Bond Coats *292*
9.4.1.4	Titania Bond Coats and TiO_2/HAp Composite Coatings *293*
9.5	Other Thermal Coating Techniques *294*
9.6	Outlook *297*
	References *298*

10	**Quality Control and Coating Diagnostic Procedures** *303*
10.1	Quality Implementation *303*
10.1.1	Total Quality Management (TQM) *303*
10.1.1.1	Quality Tools *304*
10.1.1.2	Quality Philosophy *304*
10.1.1.3	Management Style *305*
10.1.2	Qualification Procedures *306*
10.2	Characterization and Test Procedures *307*
10.2.1	Powder Characterization *307*
10.2.2	Microstructure of Coatings *309*
10.2.2.1	Splat Configuration *310*

10.2.3	Mechanical Properties *312*
10.2.3.1	Adhesion of Coatings and Determination of Bond Strength *312*
10.2.3.2	Macro- and Microhardness Tests *328*
10.2.3.3	Fracture Toughness *332*
10.2.3.4	Porosity of Coatings *335*
10.2.4	Tribological Properties *338*
10.2.4.1	Simulation of Basic Wear Mechanisms *339*
10.2.4.2	Surface Roughness of Coatings *347*
10.2.5	Chemical Properties *348*
10.2.5.1	Thermally Induced Chemical Changes *348*
10.2.5.2	Tests of Chemical Corrosion Resistance *351*
10.2.5.3	Burner Rig Test *354*
	References *355*

11 Design of Novel Coatings *361*

11.1	Property-Based Approaches *361*
11.1.1	Coating Design Based on Chemical Bonding *361*
11.1.2	Design of Novel Advanced Layered Coatings *363*
11.1.2.1	Gradient Layers *363*
11.1.2.2	Multilayers *364*
11.1.3	Coating Design Based on Materials Informatics *365*
11.1.4	Process Mapping *365*
11.2	Stochastic Approaches *367*
11.2.1	Statistical Design of Experiments (SDE) *367*
11.2.1.1	The Experimental Environment and Its Evolution *367*
11.2.1.2	Screening Designs *368*
11.2.1.3	Response Surface (RSM) Designs *369*
11.2.1.4	Theoretical Models *369*
11.2.1.5	Anatomy of Screening Designs *369*
11.2.1.6	Anatomy of Factorial Designs *371*
11.2.1.7	Box–Behnken Designs *373*
11.2.1.8	Designs of Higher Dimensionality *374*
11.2.1.9	Neyer D-optimal Designs *374*
11.2.2	Optimization of Coating Properties: Case Studies *375*
11.2.2.1	Plackett–Burman (Taguchi) Screening Designs *375*
11.2.2.2	Full Factorial Designs *376*
11.2.2.3	Fractional Factorial Designs *380*
11.2.3	Artificial Neuronal Network Analysis *392*
11.2.4	Fuzzy Logic Control *396*
11.3	Future Developments *398*
	References *403*

Appendix A: Dimensionless Groups *407*

A.1	Momentum Transfer *407*
A.2	Heat Transfer *407*

A.3	Mass Transfer	*408*
A.4	Materials Constants	*408*

Appendix B: Calculation of Temperature Profiles of Coatings *409*
B.1	Heat Conduction Equations	*409*
B.2	Solutions of the Equations	*410*
B.2.1	Substrate Temperature Profile	*410*
B.2.2	Deposit Temperature Profile	*410*
B.3	Real Temperature Profiles	*412*
	References	*412*

Appendix C: Calculation of Factor Effects for a Fractional Factorial Design 2^{8-4} *413*

References *416*

Index *117*

Preface

Much work has been accomplished in the field of plasma spraying since this treatise first appeared in 1996. Since then the author and his research group at the Department of Mineralogy, Technische Universität Bergakademie Freiberg have conducted a series of studies on biomedical coatings, hard coatings based on TiC, protective mullite coatings for space-bound structures, photocatalytic coating based on titania and novel silicon nitride-based coatings for a variety of applications. This work has been prominently included in the new edition.

New substantial achievements in the field of modeling and numerical simulation of plasma spraying have prompted the author to collect the information scattered throughout the text of the first edition to form a new Chapter 6 by adding recent work in this area that is supposed to reflect the move from a more pragmatic, while still experimentally-based, approach towards theoretical approaches to support development of coatings purposely designed for ever-broadening areas of industrial applications. I owe much of this information to Professors Javad Mostaghimi (University of Toronto) and Sanjay Sampath (State University of New York, Stony Brook). Knowledge-based stochastic approaches to coating design such as artificial neuronal network analysis and fuzzy logic control have been included in the text and examples of their analytical power are given.

Information on suspension plasma spraying (SPS) and thermal plasma chemical vapor deposition (TPCVD) has been added to the text as well as several case studies to elucidate the advantages of these novel techniques. Also, a more in-depth account has been given on the explosively developing field of bioceramic coatings for implants designed to augment or replace damaged or missing tissue and body parts. Some results of high-velocity oxyfuel (HVOF) and cold dynamic gas (CDG) spraying, while not subject of this treatise, have been included occasionally to contrast the advantages or disadvantages of these techniques with those of atmospheric plasma spraying *per se*.

Furthermore, some misconceptions have been removed, typographical errors omitted, and a host of recent references, and new or revised figures added. However, the vast amount of information published during the last few decades in the field of thermal spraying cannot be accounted for in their totality. Alas, the resigned comment by the German poet Johann Wolfgang von Goethe relating to his autobiography "Out of my Life: Poetry and Truth" applies: "Such (...) work will never be finished; one has to declare it finished when one has done the utmost in terms of time and circumstances".

I am thankful to Gabriele whose love, deep understanding and enthusiastic support have been my guiding lights through many a dark day in the past years. Wiley-VCH Weinheim provided much support and encouragement during preparation of the text of this book. I am particularly indebted to Dr. Rainer Münz for his advice and constant help.

And yes, my musical taste has not changed since. I am still much devoted to the ancient Italian masters and their inspiring music.

Robert Heimann
Hann. Münden, November 2007

Preface to the First Edition

Thermal spraying encompasses a variety of apparently simple surface engineering processes by which solid material (wire, rods, particles) are rapidly heated by a plasma jet or a combustion flame, melted and propelled against the substrate to be coated. Rapid solidification of the molten particles at the substrate surface builds up, splat by splat, a layer whose various functions include protection against wear, erosion, corrosion, and thermal or chemical degradation but also impart special electrical, optical, magnetic or decorative properties to the substrate/coating system. Also, thick coatings are applied in many industrial areas to restore or attain desired workpiece dimensions and specifications.

The text has been written with the theoretical and practical requirements in mind of students of materials engineering and materials science. It emerged from a nucleus that contained the topics presented to classes of Master students of the Materials Engineering Programme at the School of Energy and Materials, King Mongkut's Institute of Technology Thonburi, Bangkok, Thailand between 1991 and 1995 as well as to students of Technical (Applied) Mineralogy at Freiberg University of Mining and Technology since 1993. The author has also gleaned experience in plasma spray technology during his work from 1987–1988 as the head of the Industrial Products and Materials Section of the Industrial Technologies Department of the Alberta Research Council, Edmonton, Alberta, Canada, and from 1988–1993 as the manager of the Institutional and International Programs Group of the Manufacturing Technologies Department of the same organization.

It is nearly impossible to consider the entire body of literature on the subject of thermal spraying. Therefore, instead of an exhaustive coverage of applications of plasma-sprayed coatings only typical examples and case studies will be given that illustrate the various physical processes and phenomena occurring within the realm of this technology.

Many colleagues and friends helped with the production of this text. I owe thanks to Professor Dr. Dr. h.c. Walter Heywang, formerly director of Corporate Research and Development of Siemens AG in Munich for suggesting the idea of this book. I am much indebted to Mr. Liang Huguong, Dr. Ulrich Kreher and Mr. Dirk Kurtenbach who prepared diligently the numerous diagrams and graphs. The critical comments of my graduate students and Professor Jürgen Niklas, Institute of Experimental

Plasma Spray Coating: Principles and Applications. Robert B. Heimann
Copyright © 2008 WILEY-VCH Verlag GmbH & Co. KGaA, Weinheim
ISBN: 978-3-527-32050-9

Physics, Freiberg University of Mining and Technology were most welcome. My wife Giesela patiently endured my idiosyncrasies and irritations during some phases of the preparation of this text, and suffered through many lonely weekends. Last but not least, the Verlag Chemie Weinheim, represented by Frau Dr. Ute Anton, supported and encouraged me in the endeavor to complete the manuscript in time. In this I failed miserably, though. Special thanks are due to Tommaso Albinoni and Antonio Vivaldi.

Robert Heimann
Freiberg, March 1996

Synopsis

Atmospheric Plasma Spraying (APS) – A Brief Account of the Underlying Physics

This synopsis provides a concise summary of the basic physics of plasma spraying, aimed to initiate the novice in the underlying principles of one of the most versatile materials processing technologies. The different aspects of the connected energy transfer processes and the interaction of powder particles with the plasma jet and the substrate to be coated will be treated in much more detail in the remainder of the book.

Plasma spraying is a rapid solidification technology during which material introduced into a plasma jet is melted and propelled against a surface to be coated. This technology is versatile: any thermally reasonably stable metallic, ceramic or even polymeric material with a well-defined melting point can be coated onto nearly any surface. However, in practice many limitations persist related to high coating porosity, insufficient adhesion to the substrate, occurrence of residual coating stresses, and line-of-sight technology.

Plasma spraying can be conveniently described as a connected energy transfer process, starting with the transfer of energy from an electric potential field to a suitable gas forming a plasma by ionization, proceeding with the transfer of thermal energy and impulse (heat and momentum transfer) from the plasma to the injected powder particles, and concluding with the transfer of thermal and kinetic energy from the molten or semimolten particles to the substrate to be coated.

The mode of injection of powder particles into the plasma jet depends on the grain size, the melting temperature and the thermal stability of the powder material. In general, injection can be done perpendicularly to the jet at its point of exit from the anode nozzle of the plasmatron (plasma 'torch'), in upstream or downstream mode at an angle to the jet axis, directly into the nozzle, or coaxially through a bore in the cathode. Upstream injection is used when increased residence time of the powder particles in the jet is required, that is when spraying high refractory materials. Downstream injection protects a powder with a low melting point from decomposition and vaporization, respectively.

Plasma Spray Coating: Principles and Applications. Robert B. Heimann
Copyright © 2008 WILEY-VCH Verlag GmbH & Co. KGaA, Weinheim
ISBN: 978-3-527-32050-9

The plasma originates from ionization in an electric potential field of a suitable gas, preferentially argon or nitrogen. Hence a plasma by definition consists of positively charged ions and electrons, but also neutral gas atoms and photons. Moving charges within the plasma column induce a magnetic field *B* perpendicular to the direction of the electric field characterized by the current *j*. The vector cross-product of the current and the magnetic field, $[j \times B]$ is the magneto-hydrodynamic Lorentz force whose vector is mutually perpendicular to *j* and *B*. Hence an inward moving force is created that constricts the plasma jet by the so-called *magnetic* or *z-pinch* (Cap, 1984; Goldston and Rutherford, 1997). In addition to the magnetic pinch there is a *thermal pinch* that stems from reduction of the conductivity of the plasma gas at the cooled inner wall of the anode nozzle leading to an increase in current density at the center of the jet. Consequently the charged plasma tends to concentrate along the central axis of the plasmatron thereby confining the jet. As the result of these two effects the pressure in the plasma core increases drastically and the jet is blown out of the anode nozzle of the plasmatron with supersonic velocity.

A portion of this supersonic velocity will be transferred to the injected powder particles, that is, the powder particles will gain acceleration from the plasma jet by momentum transfer. The particle acceleration dV_p/dt is proportional to the viscous drag coefficient C_D and the velocity gradient between the gas velocity and the particle velocity, $V_g - V_p$, and inversely proportional to the particle diameter d_p and the particle density ρ_p as expressed by the Basset–Boussinesq–Oseen (BBO) equation of motion:

$$dV_p/dt = [3C_D \times \rho_g / 4d_p \times \rho_p] \times (V_g - V_p)|V_g - V_p|,$$

where $C_D = 2[F_D/A_p]/[\rho_g \times u_R^2]$ with F_D Stokesian drag force, A_p cross-sectional area of the particle, ρ_g gas density, and $u_R = V_g - V_p$. Since C_D is also inversely proportional to the Reynolds number **Re** the numerical value of the latter determines the flow regime in the plasma jet, that is, laminar or turbulent flow. Most atmospheric plasmas jets are turbulent, that is, characterized by a nonsteady flow field around the immersed particles and thus a rapid change of the particle Reynolds number with time. Together with the problem of arc root fluctuation this turbulence introduces nonlinearity to the process (see below).

A large part of the (electric) energy spent on ionization of the plasma gas will be recovered by recombination in the form of heat, and the powder particles accelerated by momentum transfer along a trajectory in the jet will be heated by the hot plasma. The amount of heat a particle acquires can be approximated by the balance of the amounts of heat gained by convective energy transfer, $Q_C = h A (T_\infty - T_s)$ and of heat lost by radiative energy transfer, $Q_R = \sigma \cdot \varepsilon \cdot A \cdot (T_s^4 - T_a^4)$ with h = convective heat transfer coefficient, A = surface area of the particle, T_∞ = free-stream plasma temperature, T_s = particle surface temperature, T_a = temperature of the surrounding atmosphere, σ = Stefan–Boltzmann coefficient, and ε = particle emissivity. The heat transfer coefficient between a fluid and a (spherical) particle is frequently expressed by the Ranz–Marshall equation (Ranz and Marshall, 1952) as a function of the nondimensional Nusselt number, **Nu**:

$$\mathbf{Nu} = 2.0 + b\mathbf{Re}^m \mathbf{Pr}^n$$

where **Re** = Reynolds number and **Pr** = Prandtl number. The coefficient b and the exponents n and m vary widely depending on the plasma conditions. In the original Ranz–Marshall equation m and n were set to be 0.5 and 0.33, respectively. To completely melt a (refractory) particle a certain degree of superheating is required, that is, the temperature of the particle has to be raised sufficiently high beyond the melting temperature to account for limited heat conduction within the particle, radiative energy losses and other more complex effects stemming from the nonlinearity of the process. Also, the viscosity of the liquid droplet has to be low enough to flow easily on impact with the substrate and hence facilitate proper bonding with its roughened surface. The degree of superheating depends on the Biot number, **Bi** that in this application is defined as the ratio of the thermal conductivity of the plasma gas, k_g to the thermal conductivity of the particle k_p, **Bi** = k_g/k_p. For a particle to have a uniform temperature **Bi** < 0.01 holds. The Biot number can be adjusted by selecting appropriate auxiliary plasma gases such as hydrogen or helium with increased thermal conductivities and specific heats that will increase the heat transfer rates from plasma gas to particles.

Finally, the more or less liquid droplet will impact the surface to be coated and, given a low enough viscosity, splash across the already deposited and frozen splats whose roughness determine to a large extent the solidification kinetics as well as the size and morphology of the newly arriving particles (Ghafouri-Azar *et al.*, 2004; Raessi *et al.*, 2005). The electron micrograph (Figure S.1) shows the surface of an APS hydroxyapatite coating with the typical overlapping particle splats as well as some apparently incompletely melted quasi-spherical particles that cling loosely to the surface.

The selection of the proper intrinsic (plasma power, argon gas flow rate, auxiliary gas flow rate, powder carrier gas flow rate *etc.*) and extrinsic (spray distance, powder feed rate, powder grain size, particle morphology, surface roughness *etc.*) plasma parameters is crucial for sufficient powder particle heating, flow and surface wetting on impact, and hence development of the desired coating porosity and adhesion to the substrate. However, overheating leads to an 'exploded' splat configuration with

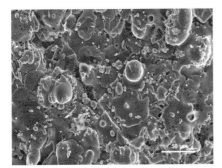

Figure S.1 Characteristic surface features of a typical APS hydroxyapatite coating with well-developed, overlapping particle splats and some loosely adhering incompletely melted particles.

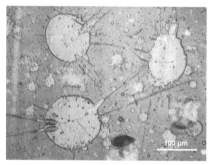

Figure S.2 Frozen-in-time traces of superheated 'exploded' alumina splats with material ejected on impact with a solid surface (Heimann, 1991).

increased microporosity, high residual stresses, and consequently reduced adhesion strength of the coatings as shown in Figure S.2.

The heat transfer from the molten particle to the solid substrate follows the generalized heat transfer relation expressed by the simplified Fourier equation

$$\text{div grad } \Theta - (1/a)(\partial\Theta/\partial t) = 0,$$

where Θ = temperature, a = thermal diffusivity, and t = time. The kinetic energy acquired by the particle will cause the splat to deform on impact thereby creating a shock wave moving with supersonic speed into the solid substrate. The ratio of deformation (flattening or spreading ratio) $\xi = D/d$ (D = splat diameter, d = original particle diameter during flight) depends on many parameters. Under simplifying assumptions the flattening ratio can be estimated using the semiquantitative Madejski splat-quench solidification model $\xi = D/d = A \times (\rho \times v/\mu)^z \approx \mathbf{Re}^z$, with $\mathbf{Re}$ = Reynolds number, ρ = density, v = impact velocity, and μ = viscosity of the melt. Many values for the pre-exponential coefficient A and the exponent z have been proposed in the literature, for example A = 1.2941, z = 0.2 (Madejski, 1976); A = 0.5, z = 0.25 (Pasandideh-Fard and Mostaghimi, 1996); A = 0.82, z = 0.2 (Watanabe et al., 1992); A = 0.5, z = 0.25 (Hadjiconstantinou, 1999) or more recently, A = 0.925, z = 0.2 (Dyshlovenko et al., 2006).

While in a dilute situation, with low-plasma loading conditions, particle movement, momentum and heat transfer can be considered a ballistic process without particle interaction, dense loading conditions appear to drastically change both the flow and temperature fields in a plasma jet. As such dense loading conditions are in practice selected for economic reasons during industrial plasma spraying, the impact of high powder feed rates on melting and spreading behavior of the particles must be considered. With increasing powder feed rates both momentum and temperature of the plasma will decrease for three reasons. First, the heat extracted from the plasma to melt a large mass of powder causes substantial local cooling of the plasma. Second, the evaporated fraction of powder can drastically alter the thermophysical, thermodynamic and transport properties of the plasma gas. Third, small powder particles with high optical emissivities evaporate easily and radiate away substantial amounts of plasma energy.

Larger rigid particles or the solidified part of an impacting particle can shatter on collision with the substrate surface owing to differential pressures across the particle as the leading and the trailing faces are subject to different dynamic pressures $p_{dyn} \approx C_D \times \rho_p \times v^2/2$. When the dynamic pressure exceeds the yield strength of the solid material fragmentation occurs. However, fragmentation is not confined to particle impact. Using fractal analysis it has been concluded that fragmentation may already occur in-flight along the particle trajectory when large thermal stresses develop owing to limited thermal conductivity within the particle (Reisel and Heimann, 2004).

Selection of the correct numerical non-dimensional parameters **Re**, **Nu**, **Pr**, **We**, **St** and **Bi** allows the process of plasma spraying to be modeled making somewhat simplified assumptions that have to satisfy the four general plasma conservation equations, that is, the continuity equation of mass conservation, the nonlinear Navier–Stokes equation of momentum conservation, the energy conservation equation and the species conservation equation as well as the Maxwell electromagnetic field equations (see for example Proulx *et al.*, 1985; Mostaghimi *et al.*, 2003; Ghafouri-Azar *et al.*, 2003; Raessi and Mostaghimi, 2005; also Dyshlovenko *et al.*, 2006).

It is a fact well known to the plasma spraying community that the properties can vary widely from coating to coating even though the spray parameters have supposedly been set within narrow ranges using sophisticated microprocessor-controlled metering devices and stringent quality control measures including intelligent statistical process control (iSPC), Taguchi methodology and the like. As it turns out, infinitesimally small changes of the input parameters will cause large, and in general nondeterministic changes of the output parameters and hence the properties of the coatings and their performance in service. Such behavior signals nonlinearity that arises from electromagnetic and magneto-hydrodynamic turbulences that affect the local magnetic field strength *B* and the electric current density *j*, and hence the Lorentz force [*j* × *B*]. The Lorentz force fluctuates rapidly with frequencies that are on the order of the residence time of the particles in the jet. In lockstep with the Lorentz force the plasma compression, that is, the *z* pinch changes and hence the rate with which turbulent eddies of cool air surrounding the plasma column are being entrained by the pumping action of the plasma. This alters the temperature distribution within the turbulent plasma jet dramatically and instantaneously so that the local thermal equilibrium breaks down on a scale that is small compared to the overall volume of the plasma. Then the system enters the realm of a heat transfer catastrophy of co-dimension 2, that is, a Riemann–Hugoniot (cusp) catastrophy (Thom, 1975). Since such nondeterministic behavior cannot be properly controlled by even the most stringent quality control measures plasma spraying is still by and large an experimental technique based on trial-and-error methodology and thus relies heavily on experience and expert knowledge.

References

F. Cap, *Einführung in die Plasmaphysik I. Theoretische Grundlagen*. Vieweg + Sohn: Wiesbaden. 1984.

S. Dyshlovenko, L. Pawlowski, B. Pateyron, I. Smurov, J.H. Harding, *Surf. Coat. Technol.*, 2006, **200** (12/13), 3757.

R. Ghafouri-Azar, J. Mostaghimi, S. Chandra, M. Charmchi, *J. Thermal Spray Technol.*, 2003, **12** (1), 53.

R. Ghafouri-Azar, J. Mostaghimi, S. Chandra, *Int. J. Comput. Fluid Dynamics*, 2004, **18** (2), 133.

R.J. Goldston, P.H. Rutherford, *Introduction to Plasma Physics*. Inst. of Physics: Bristol. 1997.

N.G. Hadjiconstantinou, in: Proc. IMECE'99 (Intern. Mech. Eng. Congress and Exposition), Nov 14–19, 1999, Nashville, USA.

R.B. Heimann, *Process. Adv. Mater.*, 1991, **1**, 181.

J. Madejski, *Int. J. Heat Mass Transfer*, 1976, **19**, 1009.

J. Mostaghimi, S. Chandra, R. Ghafouri-Azar, A. Dolatabadi, *Surf. Coat. Technol.*, 2002, **163/164**, 1.

M. Pasandideh-Fard, J. Mostaghimi, *Plasma Chem. Plasma Proc.*, 1996, **16**, 83S.

P. Proulx, J. Mostaghimi, M.I. Boulos, *Int. J. Heat Mass Transfer*, 1985, **28**, 1327.

M. Raessi, J. Mostaghimi, *Num. Heat Transfer B*, 2005, **47**, 1.

M. Raessi, J. Mostaghimi, M. Bussmann, *Thin Solid Films*, 2005, **506/507**, 133.

W.E. Ranz, W.R. Marshall, *Chem. Eng. Prog.*, 1952, 18.

G. Reisel, R.B. Heimann, *Surf. Coat. Technol.*, 2004, **185**, 215.

R. Thom, *Structural Stability and Morphogenesis*. Benjamin: New York. 1975.

T. Watanabe, I. Kuribayashi, T. Honda, A., Kanazawa, *Chem. Eng. Sci.*, 1992, **47**, 3059.

1
Introduction

Thermal spraying has emerged as an important tool of increasingly sophisticated surface engineering technology. Research and development are increasing rapidly and many applications are being commercialized. An indication of this rapid development is the fact that over 80% of the advances made over the last 90 years were made in the last two decades, a truly unique corollary to Pareto's 80/20 rule! (Alf, 1988). Many exciting niches are now being filled for metallic and ceramic surface coatings that include such well established markets as aerospace, energy and consumer industries but also developing coating markets in the automotive, computer and telecommunication industries. The particularly important segment of biomedical coatings for implant is today served by a variety of coating technologies with atmospheric plasma spraying (APS) still the leading contender.

The goal of this text is to give materials engineering/materials science students, mechanical and chemical engineers, researchers in the field of physics, chemistry and materials science, and last but not least spray shop managers, supervisors and foremen an appreciation of the fundamental physical processes governing plasma spray technology; to provide familiarization with advantages and disadvantages of the technology compared with other surface coating techniques; to discuss basic equipment requirements and limitations; to present case studies and typical applications of plasma spray technology to solve industrial problems and in general to lay a foundation for future research and development work in this field.

The material covered will discuss the basic nature of the plasma state, plasma–particle interactions, heat and momentum transfer, particle–substrate interactions, analyses of the microstructure, adhesion strength and residual stresses of coatings, optimization of coatings by SDE (statistical design of experiments) and SPC (statistical process control) methods, modeling of the plasma spray process, an account of a novel fractal approach to coating properties and other nonlinear considerations. The fundamental physical processes underlying the plasma spray process have been treated in some detail since it was felt that other existing texts frequently neglect this topic. However, its knowledge is crucial to the understanding of the process and, most importantly will enable the materials and maintenance engineer to choose the most appropriate combination of materials, equipment and parameter selection to lay down coatings with high performance, new functional properties and improved service life.

It should be emphasized, however, that this text will in no way cover the totality of this fast developing field. In particular, limitations on space and proprietary information, and the wide variety of types of equipment and coatings as well as applications, prevented detailed treatment of some aspects of plasma spray technology. For those reasons, differences in the type of plasma spray systems used successfully in many applications have not been given much consideration. However, the following text will give brief references to the most pertinent aspects of plasma spray technology, and will enable the reader to build on this knowledge in order to perform generic research and development work.

Collaboration of universities and government research organizations with industry in the resource and manufacturing sectors will lead increasingly to strategic alliances that enable industry to produce in a more competitive and environmentally compatible manner within the framework of a global concept of sustained development. Process control, including three-dimensional (3D) modeling of complex plasma–particle–substrate interactions, on-line process diagnostics, and design and development of novel coatings with improved performance using knowledge-based analysis by artificial neuronal network and fuzzy logic control tools are areas rich in research needs and opportunities. Specialists in plasma processing including plasma spray technology will find an ever increasing rewarding field of endeavor.

1.1
Coatings in the Industrial Environment

1.1.1
Market Position

There is increasing interest worldwide in thermal spray coatings. The 1986 global sales for ceramic coatings (total sales: US$ 1.1 billion) were achieved predominantly in the construction industry (36%), metal fabrication industries (21%), military (12%) and other industries (31%) including chemical processing, internal combustion engines, petrochemical and metal producing industries. 39% of the ceramic coatings were produced by physical vapor deposition techniques (PVD), 26% by chemical vapor deposition (CVD), 23% by thermal spraying, and 12% by wet processing including sol-gel technique (Chan and Wachtman, 1987). At this time predictions showed that these markets were expected to triple to more than US$ 3 billion by the year 2010 with an annual average growth rate of 12%. The industrial segments with the largest predicted individual annual growth rates are engines (28%), marine equipment (18%), chemical processing (15%), military (11%) and construction (11%).

According to a study conducted by the Gorham Advanced Materials Institute 1990, the sales in advanced ceramics were predicted to top US$ 4 billion in 1995, 80% of which was supposed to be in ceramic coatings. While these predictions overestimated the actual growth rate by close to 100% (Gorham Advanced Materials

Institute, 1999), the real growth is still impressive. For example, the global thermal spray market for the year 2003 consisted of the following market segments: original equipment manufacturers and end users: US$ 1.4 billion; large coating service companies: US$ 0.8 billion; small coating service companies: US$ 0.6 billion; powder and equipment sales: US$ 0.7 billion. This adds up to a total of US$ 3.5 billion as estimated by Read 2003. This figure has increased to US$ 5 billion in 2004 of which 45% were obtained by atmospheric DC plasma spray technology (Fauchais et al., 2006).

A field of application with a large potential are bioceramic coatings based on plasma-sprayed hydroxyapatite and second-generation bioceramics. Such biocompatible coatings for bone implants have promising applications to solve health problems of an aging population (Hench, 1991). High-temperature superconductive and diamond coatings as well as electrode and electrolyte coatings for solid oxide fuels cells (SOFCs) are on the verge of making technological breakthroughs in the microelectronics and power industries (Heimann, 1991a; Müller et al., 2002).

Besides these high-technology applications a major market exists in the resource industries including oil and gas, mining, forestry, pulp and paper and agricultural industries, construction, manufacturing and electronics, automotive and aerospace industries. In particular, industries threatened by international competition, erosion of raw materials prices, and shifts to new materials and technologies must radically improve operational efficiency, industrial diversification and environmental compatibility to survive. An important component of this struggle are coatings that combat wear, erosion and corrosion found at all levels of operations in industry, that impart new functional properties, extent the service life of machinery, and contribute in general to the sustainable development required for the increasingly environmentally conscious world in the decades to come. Just to give an impression on the economic scale, in Germany alone, wear and corrosion of technical systems destroy annually the equivalent of about 4.5% of the GNP, corresponding to €35 billion (Henke et al., 2004).

The market development of ceramic coatings depends to a large extent on individual products that must, in terms of materials and technology alternatives, show a better performance than competing materials. The field of application of the product appears to be less important. Figure 1.1 shows that the onset of substantial market penetration in automotive applications was considerably earlier for ceramic coatings than for monolithic ceramics and ceramic composites (Reh, 1989; Maier, 1991). More recently, applications of thermally sprayed coatings designed to meet today's environmental requirements for nonferrous engine blocks, hybrid solutions for aluminum automotive engines and coatings for engine cylinder bores for both gasoline and diesel engines have been highlighted by Barbezat (2002, 2005) (see also Woydt et al., 2004).

1.2
Survey of Surface Coating Techniques

Before providing details on plasma spray technology and the wide field of applications of plasma spray-generated surface coatings, a short review will be given of

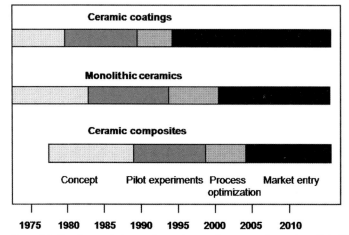

Figure 1.1 Development stages of ceramic parts for automotive engines (Reh, 1989; Maier, 1991).

various other surface coating techniques. For more details see, for example Bunshaw 1982, Sayer 1990 and Fauchais *et al.*, 2006.

The advantage of any coating technology in general lies in the fact that it marries two dissimilar materials to improve, in a synergistic way, the performance of the 'tandem' substrate/coating. Thereby mechanical strength and fracture toughness are provided by the substrate whereas the coating provides protection against environmental degradation processes including wear, corrosion, erosion, and biological, environmental and thermal attack.

There are three main advantages in using modern surface coatings technology.

- *Technical advantages:* Creation of 'new' materials (composites) with synergistic property enhancement, or completely new functional properties, for example electronic conductivity, piezo- or ferroelectric properties, electromagnetic shielding, bioconductivity and so on.

- *Economic advantages:* Expensive bulk materials such as stainless steel or Ni-based superalloys can be replaced by relatively thin overlays of a different material. These savings are enhanced by longer life of equipment and reduced downtime and shortened maintenance cycles.

- *Attitudinal advantages:* Materials engineers trained in metals handling need not be afraid to deal with 'new' materials with unfamiliar properties, specifications and performance such as ceramics, polymers or their composites. The thin ceramic coating added to a metal substrate just becomes a part of a familiar metal materials technology. However, increasingly protective coatings are considered 'prime reliant' components that are included *a priori* in the design, not as add-ons. This is happening predominantly in the aerospace and automotive industries where novel stringent quality requirements call for enhancement of both reliability and reproducibility of coatings. This necessitates a concerted, integrated interdisciplinary approach for each specific system of coating technology and material. Among other routes process mapping is a valiant vehicle to establish and verify correlations among intrinsic and extrinsic spray parameters and coating performance (for example Sampath *et al.*, 2003; Chapter 11.1.4).

Coatings are, of course, not new inventions. Since time immemorial, wood and metals have been painted with organic varnish or inorganic pigments to improve their esthetic appearance and enhance their environmental stability. Corrosion, wear and abrasion resistance of the substrate materials were significantly improved by such organic *paint coatings*, however, they did not adhere well nor survive high temperatures. The performance of traditional coatings has been extended by the use of chemically-cured paints in which components are mixed prior to application and polymerized by chemical interaction, that is, cross-linking. Epoxy resins, polyurethane and various polyester finishes show considerable resistance to alkalis and acids and also to a wide range of oils, greases and solvents.

Traditional enamels are glass-based coatings of inorganic composition applied in one or more layers that are designed to protect steel, cast iron or aluminum surfaces from corrosion. This technology has been extended today to manufacture thick film electronics in which metals or metal oxides are added to a fusible glass base to generate a range of thick film conductors, capacitors and other electronic components.

Chemical coatings are frequently applied by electroplating of metals such as copper and nickel. Nickel, for example, forms a highly adherent film for wear applications by an electroless plating process, and thus is applied to manufacture aerospace composites. A related technology is the anodization of aluminum by electrochemically induced growth of aluminum oxide in a bath of sulfuric or phosphoric acids. *Spray pyrolysis* involves chemical reactions at the surface of a heated substrate. Increasingly transparent conducting coatings of tin oxide, bismuth-manganese alloy or indium tin oxide (ITO) are used to coat glass windows for static control, radio frequency shielding and environmental temperature control.

Sol-gel coatings based on the pyrolysis of organometallic precursors such as metal alkoxides are used today in many applications. The process was originally developed for aluminum and zirconium oxide but now extends to a wide range of glasses including silicates and phosphates, and has recently been applied to complex ferroelectrics such as BT (barium titanate), PZT (lead zirconate titanate) and PLZT (lead lanthanum zirconium titanate). Sol-gel coatings enjoy a high compositional flexibility and provide ease of preparation at generally ambient temperature but because of the frequently expensive precursor materials their application is limited to high-value added devices, in particular in electronics. Problems still exist related to inherent porosity and frequently insufficient adhesion of such coatings.

Thin coatings produced by *chemical vapor deposition (CVD)* are widely employed in the semiconductor industry for large band-gap materials such as gallium arsenide, indium phosphide and other compound semiconductor materials. The technology uses vapor phase transport to grow epitaxial and highly structured thin films including insulating oxide films on single crystal silicon substrates. A related technology is the growth of thin crystalline diamond films by decomposition of methane or other hydrocarbons in a hydrogen (> 95%)–argon (< 5%) microwave plasma. Much activity is currently devoted to the improvement of the thickness and the crystallographic perfection of diamond thin films in Japan and the United States. Major potential industrial applications of such films can be found for protective coatings on compact discs, optical lenses, in particular such carried by low-earth orbit (LEO) spacecraft and substrates for ULSI (Ultra Large Scale Integration) devices.

Physical Vapor Deposition (PVD) technologies using evaporation, sputtering, laser ablation and ion bombardment are a mainstay of present-day surface engineering technology.

Evaporation is the simplest vacuum technique. The materials to be deposited on a substrate are melted and vaporized either on a resistively heated tungsten, tantalum or molybdenum boat, or by an electron beam. The method is suitable for many metals, some alloys, and compounds with a high thermal stability such as silica, yttria and calcium fluoride. While films deposited by evaporation are inferior to other vacuum techniques in terms of adhesion to the substrate and chemical purity, the excellent process control to generate optical films with well-defined thicknesses and indices of refraction has made this technique popular.

Sputtering methods deposit material by causing atoms to separate from a target by highly energetic ions from gas plasma or a separately exited ion beam, and to deposit them on a substrate. Magnetron sputtering uses confinement of the exiting plasma by a strong magnetic field that results in high deposition rates and good reproducibility. Large area sources are used to coat plastic foils and ceramic substrates for packaging materials in the food industry, as well as for metalizing plastic ornaments and automotive components such as bumpers, for architectural glass, and for multi-layer holographic coatings on identification and credit cards to prevent forgeries.

Molecular beam epitaxy (MBE) permits deposition of very thin layers of ultra-pure elements and compounds in ultrahigh vacuum (10^{-8} Pa) at extremely low deposition rates (10^{-3} to 10^{-1} µm min^{-1}). Typically the elements to form the coating are heated in quasi-Knudsen effusion cells from which a 'beam' of material emerges. The evaporated elements, for example gallium and arsenic to deposit high-quality crystalline or precisely-doped gallium arsenide layers on silicon substrates condense on the substrate wafer surface. Since the mean free paths of the evaporated molecules or atom clusters are comparatively long they do not interact with each other until they reach the surface. In modern equipment, computer controlled shutters in front of the evaporation cells are used to control the rate of evaporation and hence the thickness and stoichiometry of the deposited layers. Most importantly, the slow deposition rates allow the films to be deposited epitaxially, and component layers are built up atom by atom with unmatched precision, crystallographic orientation and purity. During deposition, Reflection High Energy Electron Diffraction (RHEED) is frequently used to monitor the growth of the layers. Such close control allows the creation of structures to confine electrons in space forming quasi one-dimensional quantum wells and even quasi zero-dimensional quantum dots.

Laser ablation is a modern technology that has been developed in particular for high temperature superconducting ceramics. A focused laser beam is used to vaporize the target material. Since this material comes from a highly localized region the ion flux reproduces the target composition faithfully. Even though the cost of the lasers is still high and the deposition rates are quite low, future developments may lead to much wider application of this technique if reproducible process control can be achieved.

The PVD methods mentioned so far all rely on a coating deposited on an existing substrate surface with given composition. However, *ion implantation* modifies the properties of the substrate itself. A beam of high energy ions can be created in an accelerator and brought in contact with a substrate surface. Thus corrosion and wear performance can be improved dramatically. For example, implantation of 18% chromium into steel results in an *in-situ* stainless steel with high corrosion resistance, the use of boron and phosphorus produces a glassy surface layer inhibiting pitting corrosion, and the implantation of titanium and subsequent carbon ions creates hard-phase titanium carbide precipitates. Likewise implantation of nitrogen produces order of magnitude improvement of the wear resistance of steel, vanadium alloys and even ceramics.

Modern high performance machinery, subject to extremes of temperature and mechanical stress, needs surface protection against high temperature corrosive media and mechanical wear and tear. For such coatings a highly versatile, low cost technique must be applied that can be performed with a minimum in equipment investment and does not require sophisticated training procedures for the operator. Such a technique has been found in *thermal spraying*. It uses partially or complete melting of a wire, rod or powder as it passes through a high temperature regime generated electrically by a gas plasma or by a combustion gas flame. The molten droplets impinge on the substrate and form the coating layer by layer. This technique is used widely to repair and resurface metallic surfaces but also, in recent years, to build up wear- and corrosion-resistant coatings, as well as chemical (CBCs) and thermal barrier coatings (TBCs) based on alumina or zirconia, in particular for applications in the aerospace and automotive industries. High temperature-erosion protection of boiler tubes and fire chambers of coal-fired power plants, corrosion protection of bars and wires in reinforced concrete in bridge construction, coating of the inner surface of cylinders of internal combustion engines, and application of bioceramic coatings to hip and dental implants are just a short list of ever increasing fields of service of thermally sprayed coatings.

In recent years much research effort was expended on *kinetic* or *cold gas dynamic spraying* (CGDS). The decisive advantage of CGDS over competing conventional techniques is its low deposition temperature (gas temperature between 350 and 600 °C) in conjunction with high pressure of the carrier gas that imparts high kinetic energy to the solid powder particles. The CGDS technique was invented by Papyrin and co-workers in Russia (see for example Alkhimov *et al.*, 1994) and has been subject to intense industrial interest ever since (Voyer *et al.*, 2003; Gärtner *et al.*, 2006). The advantages of the technique include low temperature powder processing, phase preservation, very little oxidation of both coating and substrate, high coating hardness due to formation of a cold-worked microstructure, elimination of solidification stresses and hence the possibility of depositing thicker coatings, coatings with low defect density, low heat input into the substrate hence reduced cooling requirements, as well as reduced or even eliminated need for fuel gases, electrical heating and sample masking. However, there are still some noticeable disadvantages including high gas flow rates of costly gases such as helium, limitation of coating materials to ductile metallic powders, and the fact that the technology is still mainly in its R&D stage with little coating performance and history data. Notwithstanding the materials limitations, attempts have been reported in the literature to deposit ceramic coatings by CGDS. In particular, titania for photocatalytic application was deposited by CGDS onto stainless steel (Li and Li, 2003; Li et al., 2004) and polymeric materials (Burlacov et al., 2007).

Promising developments are under way such as spray systems that combine kinetic with thermal spraying (van Steenkiste and Fuller, 2004). For example, two particle populations are injected simultaneously into a de Laval-type nozzle one of which is thermally softened and the other is not. This can be achieved by using a mixture of small and larger particles of the same material whereby the smaller particles will melt or thermally soften at a main gas temperature that is insufficient to soften the larger particles. Also, particles with different morphologies can be injected: irregularly shaped particles will reside longer in the hot gas stream thus being more easily softened than spherical particles that offer the gas stream less resistance and hence leave the hot region faster. Alternatively, different material combinations can be used such as aluminum and copper, or aluminum and silicon carbide.

Dynamic deposition at even lower temperature (room temperature) can be successfully used to lay down thin (1–10 μm) or thick (100–100 μm), dense (> 95%) and well-adhering (20–50 MPa) ceramic coatings (alumina, yttria, titania, PZT, AlN, MgB_2) on a variety of substrates including stainless steel, glass, silicon and polymers. The process is called aerosol deposition (AD) that relies on so-called room temperature impact consolidation (RTIC). An aerosol flow consisting of a carrier gas (nitrogen, helium) and a fine ceramic powder (0.1–1 μm) is ejected through a micro-orifice nozzle to acquire a velocity of up to 500 m s^{-1}, and made to impact onto the substrate in an evacuated deposition chamber (Akedo and Lebedev, 1999; Akedo et al., 2005; Akedo, 2006). The exact physics behind the process, however, has not been elucidated in much detail to date. While the temperature increase gained during impact is too low to cause widespread sintering in contrast to shock consolidation, fracture and plastic deformation of the ceramic powder particles may play a role. Also, generation by impact of clean and active surfaces that subsequently promote chemical reactions between the coating and substrate materials can be envisaged (Akedo, 2006).

In conclusion, advanced materials coatings are becoming more and more popular with materials engineers. The equipment, ease of application of coatings to complex surfaces, and the availability of tailored feedstock materials together with sophisticated process control and modern coating characterization techniques make novel surface coatings increasingly attractive. Thus it has been said that coatings technology will be the materials technology of the twentyfirst century.

1.3
Brief History of Thermal Spraying

After several patents in 1882 and 1899, in 1911 M.U. Schoop in Switzerland started to apply tin and lead coatings to metals surfaces by flame spraying to enhance corrosion performance (Schoop and Guenther, 1917). The field developed quickly and spread to other countries, and in 1926 a comprehensive book was published by T.H. Turner and N.F. Budgen on *Metal Spraying* (Turner and Budgen, 1926). This book was re-edited in 1963 under the title *Metal Spraying and the Flame Deposition of Ceramics and Plastics* to reflect the shift from metals to other materials.

Figure 1.2 gives an impression of the pace of development of the thermal spray technology (Smith, 2004). It illustrates those applications that have been either pushed or made possible by the technological progress (Heimann, 1991b) while highlighting advances being led by entrepreneurs or companies.

The curve follows a typical life-cycle curve: slow at first after inception by Schoop in 1911, and then increasing at a modest rate until the late 1950s. At this time the appearance of a variety of then modern plasmatrons boosted the development considerably. In particular, the D-gun coatings developed and applied by Union Carbide Corp. found a receptive market in the aerospace industry, and a large proportion of the subsequent technological growth was due to plasma-sprayed TBCs based on stabilized zirconia. The second growth spurt shown in the figure occurred in the 1980s by the invention of the VPS/LPPS and the Jet Kote™/HVOF techniques as well as the high temperature coatings for aerospace gas turbines associated therewith (Figure 1.3, Sivakumar and Mordike, 1989).

Future developments, undoubtedly highlighted by a further increase in the rate of technological innovation, will include improved on-line real-time feedback control with close-loop strategies, intelligent statistical process control (iSPC), design of new equipment and spray powders as well as 3D process modeling and improved understanding of the complex nonlinear physics underlying the plasma spray process. A recent summary of the development of DC plasma spraying (Fauchais *et al.*, 2006) indicates four main focus points of current developments:

- process on-line control requiring correlation of particle in-flight characteristic as well as substrate and coating temperatures to thermo-mechanical and other in-service coating properties;
- study of arc root fluctuations and their causes by experiments and 3D modeling, and relating them to structure and properties of coatings;
- study of splat formation and layering mechanisms and their effect on the properties of splat-substrate and splat-splat interfaces;
- development of novel spray techniques to produce nano-structured coatings, in particular using suspension and solution plasma spraying.

In the last three to four decades there has been increasing interest in coatings by the military and the commercial sector that led to a wealth of information. Journals totally dedicated to thermal spraying exist, for example the *Journal of Thermal Spray Technology*, but also journals devoted to the entire realm of surface engineering, for

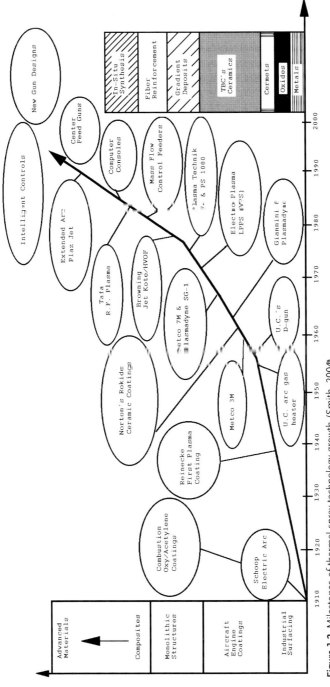

Figure 1.2 Milestones of thermal spray technology growth (Smith, 2004).

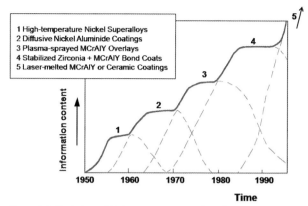

Figure 1.3 Evolution of high-temperature coatings for aerospace gas turbines (Sivakumar and Mordike, 1989).

example *Surface and Coating Technology, Thin Solid Films, Surface Technology, Acta Materialia, Materials Science and Engineering, Advanced Materials* and *Journal of Materials Science*. Biannual national and triannual international conferences present ongoing worldwide university and industry research and development efforts in a wide variety of thermal spray processes. Several monographs and handbooks dealing with plasma spraying are available, for example Matejka and Benko 1989, Barthelmess *et al.* 1992, Wachtman and Haber 1993, ASM Handbook 1994, Pawlowski 1995, Stern 1996, Davis 2004, d'Agostino *et al.* 2007 and more.

Activities are well under way to develop expert systems that integrate exhaustive databases with expert knowledge and practical experiences, being supported by ever more sophisticated 3D modeling and numerical simulation of the complex interaction of plasma, particles and substrate. For example, powerful software has been developed that allows the engineer to determine the best coatings for a given part or application, as well as information as to how and where the coatings are being used. These databases provide detailed information on over 1300 thermal spray applications and contain a listing of international suppliers. The materials covered are categorized into 12 groups including iron-based materials, nickel-based materials, cobalt-based materials and nonferrous metallic-based materials (Longo *et al.*, 1994).

To unravel the inherent complexity of the deposition process, and the influence of plasma parameters on coating properties traditionally a statistical design of experimentation (SDE) approach is used. However, insight into the complex interrelation of the parameters, intermediate sub-processes and their corresponding mechanisms is largely missing. Hence a concerted, integrated interdisciplinary approach is required that has been found in the development of so-called *process maps* (Sampath *et al.*, 2003). Details of a case study of Mo coatings on steel can be found in Chapter 11.1.4.

Process simulation plays an increasingly important role in estimating the interdependence of spraying parameters and desired coating properties (Knotek and Schnaut, 1993; Nylén *et al.*, 1999; Ahmed and Bergman, 2000; Sevostianov and Kachanov, 2000; Feng *et al.*, 2002; Williamson *et al.*, 2002; Ghafouri-Azar *et al.*, 2003;

Raessi and Mostaghimi, 2005) as well as ever more complex modeling and numerical simulation of the generation of plasma and its interaction with materials immersed in the plasma jet (see Chapter 6).

The main research and development activities today center around:

- improving powder feedstocks, spray application equipment, and process control through Statistical Design of Experiments (SDE), intelligent Statistical Process Control (iSPC), and Quality Function Deployment (QFD) techniques (see Chapter 11);

- designing new control devices, on-line real-time feedback looping, mass flow controllers, powder metering equipment, manipulators and robots (Borbeck, 1983; Barbezat, 2002, 2005);

- development of novel spray technologies, e.g. reactive plasma spraying (Pfender, 1999; see Chapter 7.1.2.2), suspension plasma spraying (SPS; Gitzhofer *et al.*, 1997; Müller, 2001; Fauchais *et al.*, 2005; see Chapter 8.2.2), thermal plasma chemical vapor deposition (TPCVD; Zhu *et al.*, 1991; Kolman *et al.*, 1998; Müller, 2001; see Chapter 8.2.2), reactive laser plasma coating (Schaaf *et al.*, 2005), high velocity pulsed plasma spraying (HVPPS; Witherspoon *et al.*, 2002), high velocity suspension flame spraying (HVSFS; Killinger *et al.*, 2006; Gadow *et al.*, 2006) and electromagnetically accelerated plasma spraying (EMAPS, Kitamura *et al.*, 2003; Usuba and Heimann, 2006; see Chapter 8.4);

- design and commissioning of new plasmatrons such as three-cathode systems (Triplex I and II; Sulzer Metco, Zierhut *et al.*, 1998), plasmatrons with axial injection (Axial III; Northwest Mettech Corp., Moreau *et al.*, 1995), plasmatrons with rotating heads for coating internal surfaces (Sulzer Metco, Barbezat, 2002), and r.f. plasmatrons for high power levels up to 200 kW (Tekna Corp., Boulos, 1997);

- effective innovation and technology transfer from research organizations to small and medium-sized enterprises (SMEs) (Riesenhuber, 1989);

- development of data bases and expert systems (Kern *et al.*, 1990) as well as process maps (Sampath *et al.*, 2003).

1.4
Synergistic Nature of Coatings

A metal substrate/ceramic coating system combines the mechanical strength of a metal with the environmental stability of a ceramic material. Typical metal properties exploited in industry are:

- creep strength;
- fatigue strength;
- flexural strength;
- ductility;

- high fracture toughness;
- high coefficient of thermal expansion;
- high heat conductivity;
- low porosity.

Typical ceramic properties are:

- high thermal stability;
- chemical stability;
- high hardness;
- low fracture toughness;
- low coefficient of thermal expansion;
- low heat conductivity;
- medium to high porosity.

The combination of these properties yields a superior 'composite' material. However, several aspects must be carefully controlled, such as the difference in the coefficients of thermal expansion of metal and ceramic which leads to undesirable stresses at the interface substrate/coating that impose the risk of cracking, spalling, and eventually delamination of the coating. Also, the generally high porosity of plasma-sprayed ceramics has to be dealt with, unless it is a desired property, by infiltrating the coating with another material, hot isostatically pressing, or by laser densification or other techniques (see Chapter 8.1.3). The low fracture toughness of the ceramic is also point of concern. Intense research is ongoing worldwide to develop ceramics with improved fracture toughness. For example, research is being pursued to thermally spray fiber-reinforced ceramics, such as silicon carbide/alumina composite coatings.

1.5
Applications of Thermally Sprayed Coatings

There are an ever increasing number of technical applications of plasma-sprayed metal and ceramic coatings. Many of such applications resulted from the demand of the users of machinery and equipment to protect their investment from wear, corrosion, erosion and thermal and chemical attack. Others result from the desire to impart new functional properties to conventional materials, such as high-temperature superconducting coatings, bioceramic coatings, diamond coatings and electrocatalytic coatings for solid oxide fuel cells. It is not possible to cover here all present and potential applications of thermally sprayed coatings. The listing below shows but a few industrial areas where coatings have been successfully used to solve performance problems. Practical solutions to a specific problem frequently follow a well-established sequence of events (Pawlowski, 1995):

- problem identification;
- specification of coating properties;
- suggestion of solutions including selection of materials to be sprayed, equipment and technique used and so on;

- application of coating;
- evaluation of results in terms of technical performance and economic viability.

Typical applications of metallic and ceramic coatings are:

- wear and erosion control of machinery parts and turbine vanes, shrouds and blades in coal-fired power generating stations;
- particle erosion control in boiler tubes and superheaters of coal-fired power plants;
- chemical barrier coatings for ethane steam cracking furnace tubes for coking and erosion protection;
- wear control and improvement of friction properties in a variety of machine parts, including pump plungers, valves, bearings, and calender and printing rolls;
- metal coatings for corrosion protection of engineering structures such as steel and concrete bridges in coastal regions;
- corrosion protection against liquid metals in extrusion dies, ladles and tundishes;
- corrosion protection of equipment for petrochemical and chemical plants and high-temperature corrosion in internal combustion engines;
- thermal and chemical barrier coatings for pistons and valves in adiabatic diesel engines and related machinery, as well as for aerospace gas turbine blades and combustor cans;
- coating of the inner surface of cylinders of internal combustion engines to facilitate the move from steel of light metal motor blocks;
- resurfacing of worn equipment, for example in railway transportation application, ship building and maintenance and mining tools and equipment;
- superalloy coatings for aerospace gas turbine vanes and shrouds to prevent hot gas erosion and corrosion;
- ceramic membranes for osmotic filtering and ultrafiltration;
- biomedical coatings for dental and bone implants with biocompatible and, in particular bioconductive properties based on hydroxyapatite and tricalcium phosphate, respectively;
- stabilized zirconia electrolyte and electrocatalytically active compound coatings for solid oxide fuel cells (SOFCs);
- high-temperature superconducting coatings for electromagnetic interference (EMI) shielding;
- abradable coatings and seals for clearance control in gas turbines;
- coatings to protect concrete floors from corrosive action of fruit juices and other agricultural products;
- sealing of concrete floors in dairy industrial operations;
- thick metal overlays for PVD sputter targets;
- photocatalytically active coatings for UV-activated cleaning devices for domestic and industrial waste water;
- high grip/high friction coatings for space-bound applications;
- diamond-like coatings for wear applications and for heat sinks in high-power electronic chips.

Some of these applications are dealt with in more detail in the case studies presented in Chapters 7 to 9.

References

R. d'Agostino, P. Favia, Y. Kawai, H. Ikegami, N. Sato, F. Arefi-Khonsari. *Advanced Plasma Technology*. Wiley-VCH: Berlin. ISBN: 3-527-40591-7. 2007.

I. Ahmed, T.L. Bergman, *J. Thermal Spray Technol.*, 2000, **9** (2), 215.

J. Akedo, *J. Am. Ceram. Soc.*, 2006, **89** (6), 1834.

J. Akedo, M. Lebedev, *Jpn. J. Appl. Phys.*, 1999, **38**, 5397.

J. Akedo, M. Lebedev, H. Sato, J. Park, *Jpn. J. Appl. Phys.*, 2005, **44**, 7072.

L.S. Alf, *Quality Improvement Using Statistical Process Control*, HDJ Publishers, New York, 1988.

P. Alkhimov, A.N. Papyrin, V.P. Dosarev, N.J. Nesterovich, M.M. Shuspanov, US Patent 5,302,414 (April 12, 1994).

ASM Handbook. Vol 05, *Surface Engineering*, ASM International: Materials Park, OH. ISBN: 0-87170-384-X. 1994.

G. Barbezat, *Automot. Technol.*, 2002, **2**, 47.

G. Barbezat, *Application of thermal spraying in automotive industry*, Proc. 2ièmes RIPT, Lille, France, 1–2 Décembre 2005, 231.

H. Barthelmess, H. Becker, U. Klimpel, *Plasmaspritzen*, DS Informationssysteme für die Schweisstechnik, Fachbibliothek Schweisstechnik, 53 series, ISBN: 3-871-55650-5. 1992.

K.D. Borbeck, in: Proc.10th ITSC, Essen 1983, DVS 80, 99.

M.I. Boulos, *J. High Temp. Mater. Process.*, 1997, **1**, 17.

R.F. Bunshaw, *Deposition Technologies for Thin Films and Coatings*, Noyes Publications, Park Ridge, N.J., 1982.

I. Burlacov, J. Jirkovsky, L. Kavan, R. Ballhorn, R.B. Heimann, *J. Photochem., Photobiol. A*, 2007, **187** (2–3), 285.

K. Chan, J.B. Wachtman, *Ceramic Industry*, 1987, **129**, 24.

J.R. Davis (ed.), *Handbook of Thermal Spray Technology*, ASM International: Materials Park, OH. ISBN: 9-780-87170-795-6. 2004.

P. Fauchais, V. Rat, C. Delbos, J.F. Coudert, T. Chartier, LK. Bianchi, *IEEE Trans. Plasma Sci.*, 2005, **33** (2), 920.

P. Fauchais, G. Montavon, M. Vardelle, J. Cedelle Surf. Coat. Technol., 2006, **201** 1908.

Z.G. Feng, M. Domaszewski, G. Montavon, C. Coddet, *J. Thermal Spray Technol.*, 2002, **11** (1), 62.

R. Gadow, F. Kern, A. Killinger, *Mater. Sci. Eng. B.*, doi:10.1016/j.mseb.2007.09.066.

F. Gärtner, T. Stoltenhoff, T. Schmidt, H. Kreye, *J. Thermal Spray Technol.*, 2006, **15** (2), 223.

R. Ghafouri-Azar, J. Mostaghimi, S. Chandra, M. Charmchi, *J. Thermal Spray Technol.*, 2003, **12** (1), 53.

F. Gitzhofer, E. Bouyer, M.I. Boulos, *Suspension Plasma Spray*, US Patent 5,609,921, 1997.

Gorham Advanced Materials Institute, *Thermal Spray Coatings*, Gorham, Maine, USA, 1990.

Gorham Advanced Materials Institute, *Thermal Spray Coatings*, Gorham, Maine, USA, 1999.

R.B. Heimann, *Process. Adv. Mat.* 1991a, **1**, 181.

R.B. Heimann, *Am. Ceram. Soc. Bull.*, 1991b, **70**, 1120.

L.L. Hench, *J. Am. Ceram. Soc.* 1991, **74**, 1487.

H. Henke, D. Adam, A. Köhler, R.B. Heimann, *Wear*, 2004, **256**, 81.

H. Kern, M. Fathi-Torbaghan, M. Stracke, in: Proc. TS'90, Essen 1990, DVS 130, 247.

A. Killinger, M. Kuhn, R. Gadow, *Surf. Coat. Technol.*, 2006, **201** (5), 1922.

J. Kitamura, S. Usuba, Y. Kakudate, H. Yokoi, K. Yamamoto, A. Tanaka, S. Fujiwara, *J. Thermal Spray Technol.*, 2003, **12** (1), 70.

O. Knotek, U. Schnaut, in: Proc.TS'93, Aachen 1993, DVS 152, 138.

D. Kolman, J. Heberlein, E. Pfender, *Plasma Chem. Plasma Process.*, 1998, **18** (1), 73.

C. J. Li, W.Y. Li, *Surf. Coat. Technol.*, 2003, **167**, 278.

C.J. Li, G.J. Yang, X.C. Huang, W.Y. Li, A. Ohmori, in: E. Lugscheider (ed.), Proc. Intern. Thermal Spray Conf., ITSC 2004, Osaka, Japan, DVS-Verlag, p. 315.

F.H. Longo, H. Herman, K. Kowalski, *The Thermal Spray Source Program*, MS Card. CRC Press, London, ISBN: 0-849-38639-X. 1994.

R. Maier (ed.), *Technische Keramik als Innovationsgrundlage für die Produkt- und Technologie-Entwicklung in NRW*. Ministerium für Wirtschaft, Mittelstand und Technologie des Landes Nordrhein-Westfalen, December 1991, p. 108.

D. Matejka, B. Benko, *Plasma Spraying of Metallic and Ceramic Materials*, Wiley: Chichester. ISBN: 0-471-91876-8. 1989.

C. Moreau, P. Gougeon, A. Burgess, D. Ross, in: S. Sampath, C.C. Berndt (eds.), *Advances in Thermal Spray: Science and Technology*, ASM Int., Materials Park, OH, USA, 1995, 41.

M. Müller, *Entwicklung elektrokatalytischer Oxidschichten mit kontrollierter Struktur und Dotierung aus flüssigen Prekursoren mittels thermischer Hochfrequenzplasmen*, Unpublished Ph.D. Thesis, Technische Universität Bergakademie Freiberg, Germany, June 2001.

M. Müller, E. Bouyer, M. v.Bradke, D.W. Branston, R.B. Heimann, R. Henne, G. Lins, G. Schiller, *Mater.-wiss.u.Werkstofftech.*, 2002, **33**, 322.

P. Nylén, J. Wigren, L. Pejryd, M.-O. Hansson, *J. Thermal Spray Technol.*, 1999, **8** (3), 393.

L. Pawlowski, *The Science and Engineering of Thermal Spray Coatings*. Wiley: Chichester, New York, Brisbane, Toronto, Singapore. 1995.

E. Pfender, *Plasma Chem. Plasma Proc.*, 1999, **19** (1), 1.

M. Raessi, J. Mostaghimi, *Numer. Heat Transfer B*, 2005, **47**, 1.

J. Read, ITSC, Keynote address, China Intern. Thermal Spray Conf., Dalian, China, Sept 22–25, 2003.

H. Reh, *Keram. Zeitschrift*, 1989, **41** (3), 176.

H. Riesenhuber, *Vakuum-Technik*, 1989, **38** (5/6), 119.

S. Sampath, X. Jiang, A. Kulkarni, J. Matejicek, D.L. Gilmore, R.A. Neiser, *Mater. Sci. Eng. A*, 2003, **348**, 54.

M. Sayer, *Can. Ceramics Quarterly*, 1990, **59** (1), 21.

P. Schaaf, M. Kahle, E. Carpene, *Surf. Coat. Technol.*, 2005, **200**, 608.

M.U. Schoop, H. Guenther, *Das Schoopsche Metallspritz-Verfahren*, Franckh-Verlag: Stuttgart, Germany, 1917.

I. Sevostianov, M. Kachanov, *Acta Mater.*, 2000, **48**, 1361.

R. Sivakumar, B.L. Mordike, *Surf. Coat. Technol.*, 1989, **37**, 139.

R.W. Smith, in: *Handbook of Thermal Spray Technology*, 06994G, ASM International: Materials Park, OH, 2004.

K.H. Stern (ed.), *Metallurgical and Ceramic Protective Coatings*, Chapman & Hall: London, ISBN: 0 121 51440-7. 1996.

T.H. Turner, N.E. Budgen, *Metal Spraying*, Charles Griffin: London, 1926.

S. Usuba, R.B. Heimann, *J. Thermal Spray Technol.*, 2006, **15** (3), 356.

T.H. van Steenkiste, B.K. Fuller, US Patent 7,108,893 2004.

J. Voyer, T. Stoltenhoff, H. Kreye, C. Moreau, B. Marple (eds.), *Thermal Spray 2003: Advancing the Science & Applying the Technology*, ASM International, Materials Park, OH, USA, 2003, p. 71.

J.B. Wachtman, R.A. Haber (eds.), *Ceramic Films and Coatings*, Noyes Publications: Park Ridge, N.J. 1993.

R.L. Williamson, J.R. Fincke, C.H. Chang, *J. Thermal Spray Technol.*, 2002, **11** (1), 107.

F.D. Witherspoon, D.W. Massey, R.W. Kincaid, G.C. Whichard, T.A. Mozhi, *J. Thermal Spray Technol.*, 2002, **11** (1), 119.

M. Woydt, N. Kelling, M. Buchmann, *Mat.-wiss.u.Werkstofftech.*, 2004, **35** (10/11), 824.

H. Zhu, Y.C. Lau, E. Pfender, *J. Appl. Phys.*, 1991, **69** (5), 3404.

J. Zierhut, P. Haslbeck, K.D. Landes, G. Barbezat, M. Müller, M. Schütz, in: C. Coddet (ed.), *Thermal Spray: Meeting the Challenges of the 21st Century*, ASM Int., Materials Park, OH, USA, 1998, 1375.

2
Principles of Thermal Spraying

Thermal spraying requires a device that creates a high temperature flame or plasma jet. In case of plasma, such a device is called a plasmatron,[1] i.e. a plasma generator powered by an arc or a high frequency discharge whose plasma is superposed by streaming gas. Metal, ceramic or even polymer powders are injected into the fluid medium, melted during the very short residence time in the flame or plasma jet, and propelled towards a target where the molten droplets solidify and build up a solid coating. Any material can be sprayed as long as it has a well defined melting point, and does not decompose or sublimate during melting.

Figure 2.1 shows the organization of thermal spraying methods (Lugscheider, 1992) divided into the two principal energy sources, *chemical energy* provided by the combusting gases that power the flame spray torches and *electric currents* providing energy for the plasmatrons. The plasma spray methods *per se* are shown in Figure 2.2. The different techniques are distinguished here by the type of the surrounding atmosphere, i.e. air for conventional air plasma spraying (APS) and high power plasma spraying (HPPS), as well as APS with an inert gas shroud to prevent oxidation of oxygen-sensitive powders, and plasma spray methods relying on special atmospheres, most importantly inert gas plasma spraying (IGS) and 'vacuum', i.e. low pressure plasma spraying (VPS and LPPS, respectively).

The principal difference between flame spraying and plasma spraying is given by the maximum temperatures achievable. In flame spraying the temperature is limited by the internal enthalpy of combustion of the gases (acetylene, propane, butane). Conventional oxyacetylene torches reach temperatures of around 3000 K; special devices such as Union Carbide Corporation's (now Praxair Surface Technology, Inc.) D-Gun™ and Jet Kote™ System are limited to about 3500 K, whereas the present limit of the Hypervelocity Oxyfuel Gun (HVOF) seems to be below 3200 K (Lugscheider, 1992). The advantage of the latter techniques lies in the exceptionally dense structure, high adhesion strength and rather low oxygen content of the coatings (Table 2.1).

---

[1] The terms plasma burner, plasma torch, plasma arc torch etc. should be avoided since they convey the inaccurate notion of something 'burning'. Also, the term plasma 'gun' should not be used for obvious reasons.

Plasma Spray Coating: Principles and Applications. Robert B. Heimann
Copyright © 2008 WILEY-VCH Verlag GmbH & Co. KGaA, Weinheim
ISBN: 978-3-527-32050-9

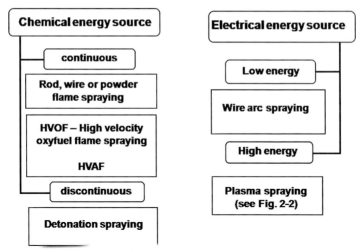

Figure 2.1 Thermal spray techniques divided by their principal energy sources (according to Lugscheider, 1992).

Using electric energy as the source to create the plasma provides for theoretically unlimited temperatures that are only controlled by the amount of energy input. The maximum power supplied to the plasmatron depends on the cross-sectional area of the power leads between a DC or AC transformer and the device. Electric arc wire-spraying reaches arc temperatures of close to 6000 K. Air plasma spraying (APS) temperatures are typically around 15 000 K depending on the type of the plasma gas used (argon, nitrogen, auxiliary hydrogen or helium) and the power input. Similar temperatures can be obtained by inductively-coupled radiofrequency plasmatrons and low-pressure plasma spraying (LPPS) devices.

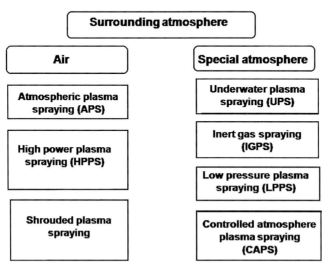

Figure 2.2 Plasma spraying techniques divided by their surrounding atmospheres (according to Lugscheider, 1992).

Table. 2.1 Comparison of different thermal spraying processes (adapted from Boulos et al., 1989; Thorpe, 1993; ASM, 2004)

Process	Gas flow rate (m³/h)	Temperature (°C)	Particle impact velocity (m/s)	Adhesive strength[a]	Cohesive strength	Oxide content (%)	Relative process cost	Spray rate (kg/h)	Power (kW)
Flame powder	10	2200	30	3	low	6	3	7	25–75
Flame wire	70	2800	180	4	medium	4	3	9	50–100
HVOF	30–60	3100	600–1100	8	very high	0.2	5	14	100–270
D gun	10	3900	>1000	8	very high	0.1	10	1	100–300
Wire arc	70	5500	240	6	high	<3	1	16	4–6
APS	4	>12,000	300–500	6	high	<1	5	5	30–80
LPPS	8	>10,000	250–600	9	very high	{ppm}	10	10	50–100

[a]1 (low) to 10 (high)

Thermal spraying is emerging as an active area of research and development. Parts of these worldwide activities involve the development of new techniques and associated devices. Modern applications include underwater plasma spraying (UPS, Lugscheider and Bugsel, 1988), laser-assisted spraying (LS, Petitbon *et al.*, 1989; Uchiyama *et al.*, 1992; Suutala *et al.*, 2006), high-frequency plasma spraying (ICPS, Boulos, 1997), reactive plasma spraying (Smith, 1993; Fauchais *et al.*, 1997; Pfender, 1999), suspension plasma spraying (SPS, Gitzhofer *et al.*, 1997), liquid precursor plasma spraying (PSLP, Karthikeyan *et al.*, 1998) high velocity pulsed plasma spraying (HVPPS, Witherspoon *et al.*, 2002) and thermal plasma chemical vapor deposition (TPCVD, Kolman *et al.*, 1998). So-called cold gas dynamic spraying (CGDS, Alkhimov *et al.*, 1994) is at present subject to intense research effort (Gärtner *et al.*, 2006; Burlacov *et al.*, 2007).

Modern equipment is so versatile that a polymeric substrate can be coated with metal, or a metal substrate with plastic, just by changing the plasma or flame spray parameters. (Pawlowski, 1995). As simple as it sounds there are highly sophisticated requirements for selecting and controlling numerous intrinsic and extrinsic plasma spray parameters. Some researchers say that there are many hundreds of parameters which can potentially influence the properties of the coatings.

For economic (time requirements) and principal reasons (interdependence of parameters) it is not possible to control all possible parameter variations. In fact, only eight to twelve parameters are routinely controlled at preset levels, using principles of Statistical Design of Experiments (SDE) and Statistical Process Control (SPC) (see Chapter 11). The most common control parameters are:

- power input;
- arc gas pressure (argon, nitrogen);
- auxiliary gas pressure (hydrogen, helium);
- powder gas pressure (argon);
- powder feed rate (mass flow controlled);
- grain size/shape;
- injection angle (orthogonal, downstream, upstream, axial);
- surface roughness;
- substrate heating or cooling;
- spray distance;
- spray divergence;
- spray atmosphere (air, low pressure, inert gas, water).

These parameters can control a variety of secondary parameters such as quench rate, residence time of particles in the plasma jet, gas composition of plasma jet, heat content and so on. Some of the more important parameters are shown in Figure 2.3 (Nicoll, 1984). For economical reasons, one of the most important secondary parameters is the deposition efficiency.

Figure 2.4 illustrates schematically the way in which specific variables affect deposition efficiency. Variables that can greatly influence efficiency include power input, arc gas flow and spray (stand-off) distance. Variables with moderate effects are the powder feed rate and the powder gas flow rate within limits. Finally, changes in

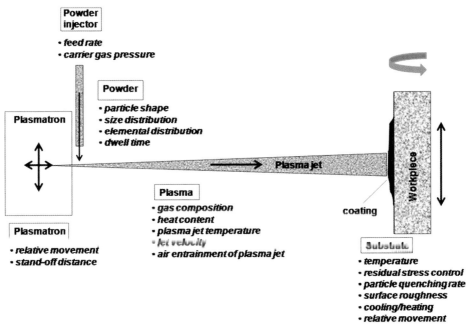

Figure 2.3 Main plasma spray parameters influencing coating properties.

the traverse rate have little or no effect on deposition efficiency and coating density (Marsh *et al.*, 1961; Ghafouri-Azar *et al.*, 2003).

2.1
Characterization of Flame vs. Plasma Spraying

The main difference between these two techniques is that of the temperature of the powder melting. The temperature of a combustion flame is limited by the enthalpy of the chemical reaction created by the combustion of gases such as acetylene or propane in the presence of oxygen. While the flame velocity can be boosted in modern equipment (such as High Velocity Oxyfuel technique, HVOF) to values near Mach 5, the temperatures achievable are limited to approximately 3300 K. On the other hand, plasma jets have temperatures limited only by the amount of electrical energy supplied which in turn is a function of the cross-section of the power leads. Temperatures as high as 15 000 K can be easily generated. Thus, flame spray including HVOF systems are used mainly for the spraying of materials with lower melting points such as metals or metal/ceramic (cermets) composites, for example tungsten carbide/cobalt or chromium carbide/nickel-chromium composites. The high flame velocity of the HVOF system leads to dense, well adhering coatings for a wide variety of applications in the resource and manufacturing industries. Thermal

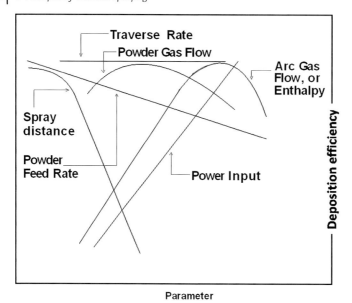

Figure 2.4 Dependence of deposition efficiency on various plasma spray parameters (Marsh et al., 1961).

spraying of ceramics with high melting points, such as alumina (2050 °C) or zirconia (2680 °C) is the true domain of plasma spray systems.

Table 2.1 shows a comparison of various flame and plasma arc spray techniques.

2.2
Concept of Energy Transfer Processes

The plasma spraying process can be conveniently described as a connected energy transfer process, starting with the transfer of electrical energy from an electrical potential field to the plasma gas (ionization and thus plasma heating), proceeding with the transfer of thermal energy and impulse (heat and momentum transfer) from the plasma to the injected powder particles and concluding with the transfer of thermal and kinetic energy from the particles to the substrate to be coated. Figure 2.5 shows the three stages of this connected energy transfer process together with fundamental constituent parts of the plasma spray system.

2.3
Unique Features of the Plasma Spray Process

The plasma pray process is characterized by a set of unique features, which are listed below (Herman et al., 1993).

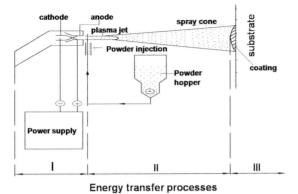

Figure 2.5 The three stages of the connected energy transfer process governing atmospheric plasma spraying.

- A wide range of materials, from metal to ceramics to polymers and any combination of them can be deposited.
- Mixed ceramics and alloys containing components with widely differing vapor pressures can be deposited without significant changes in coating composition.
- Homogeneous coatings with time-invariant changes in composition, i.e. without compositional changes across the thickness can be produced.
- Microstructures with fine, equiaxed grain but without columnar defects can be deposited in contrast to electron-beam deposition.
- Gradient coatings can be produced with the same type of equipment whereby the coating composition can be changed from that of a pure metal to that of a pure ceramics via continuously changing metal-ceramic mixtures.
- High deposition rates on the order of mm/s can be achieved with only modest investment in capital equipment.
- Free-standing thick forms can be sprayed in near-net shape fashion of pure and mixed ceramics.
- The process can be carried out in any conceivable environment such as air, under reduced pressure, inert gas, or even underwater.

Despite these apparently simple and straightforward characteristics it must be emphasized that the underlying physical principles are complex and in many cases nonlinear. Achieving a coating with the desired mechanical or functional properties requires great care and stringent control of many plasma spray parameters that in a generally synergistic way influence the coating properties and thus performance. With automated technology such as robotics and adaptive statistical process control as well as the incipient development of on-line feedback control reproducibility and consistent coating quality can be achieved. Details of these approaches can be found in Chapters 6 and 11.

References

P. Alkhimov, A.N. Papyrin, V.P. Kosarev, N.J. Nesterovich, M.M. Shuspanov, US Patent 5,302,414 (1994).

ASM International, *Handbook of Thermal Spray Technology*, ASM International: Materials Park, OH. 2004.

M.I. Boulos, P. Fauchais, E. Pfender, *Fundamentals of Materials Processing Using Thermal Spray Technology*, CUICAC Short Course, Edmonton, Alberta, Canada, October 17–18, 1989, p. 275.

M.L. Boulos, *High Temp. Mater. Process.*, 1997, **1**, 17.

I. Burlacov, J. Jirkovský, L. Kavan, R. Ballhorn, R.B. Heimann, *J. Photochem. Photobiol. A: Chemistry*, 2007, **187**, 285.

P. Fauchais, A. Vardelle, A. Denoirjean, *Surf.Coat.Technol.*, 1997, **97**, 66.

F. Gärtner, T. Stoltenhoff, H. Kreye, *J. Thermal Spray Technol.*, 2006, **15** (2), 223.

R. Ghafouri-Azar, J. Mostaghimi, S. Chandra, M. Charmchi, *J. Thermal Spray Technol.*, 2003, **12** (1), 53.

F. Gitzhofer, E. Bouyer, M.I. Boulos, *Suspension Plasma Spray*, US Patent 5,609,921, 1997.

H. Herman, C.C. Berndt, H. Wang, in: *Ceramic Films and Coatings*. J.B. Wachtman, R.A. Haber, (ed.), Noyes Publications: Park Ridge, N.J., 1993, Chapter 5, 131–188.

J. Karthikeyan, C.C. Berndt, S. Reddy, J.Y. Wang, A.H. King, H. Herman, *J. Am. Ceram. Soc.*, 1998, **81** (1), 121.

D. Kolman, J. Heberlein, E. Pfender, *Plasma Chem. Plasma Process.*, 1998, **18** (1), 73.

E. Lugscheider, *Technica*, 1992, **9**, 19.

E. Lugscheider, B. Bugsel, in: 1st Plasma-Technik Symposium, Lucerne, Switzerland, May 18–20, 1988, p. 55.

D.R. Marsh, N.E. Weare, D.L. Walker, *J. of Metals*, July 1961, 473.

A.R. Nicoll, *Thermal Spray*, CEI Course on High Temperature Materials and Coatings, Plasma-Technik AG, Wohlen, Switzerland, 1984.

L. Pawlowski, *The Science and Engineering of Thermal Spray Coatings*. John Wiley & Sons, Chichester, New York, Brisbane, Toronto, Singapore, 1995, Chapter 3, 28–50.

A. Petitbon, Guignot, U. Fischer, J.-M. Guillemot, *Mat.Sci.Eng.* 1989. **A121**, 545.

E. Pfender, *Plasma Chem. Plasma Process.*, 1999, **19** (1), 1.

R.W. Smith, *Powder Metall. Int.* 1993, **25** (1), 9.

J. Suutala, J. Tuominen, P. Vuoristo, *Surf. Coat. Technol.* 2006, **201** 1981.

M.L. Thorpe, *Adv. Mater. Process.*, 1993, **143** (5), 50.

F. Uchiyama *et al*, in: Proc. 13th ITSC, Orlando, Florida, May 28–June 5, 1992, p. 27.

F.R. Witherspoon, D.W. Massey, R.W. Kincaid, G.C.F. Whichard, T.A. Mozhi, *J. Thermal Spray Technol.* 2002, **11** (1), 119.

3
The First Energy Transfer Process: Electron–Gas Interactions

3.1
The Plasma State

The interaction of an electric current with a gas leads to dissociation and ionization. Hence plasmas consist of electrons, positively charged ions, neutral gas atoms and also high-energy (UV) photons.

The definition of the plasma state is as follows (Rutscher, 1982).

> Plasmas are quasi-neutral multiparticle systems characterized by gaseous or fluid mixtures of free electrons and ions, as well as neutral particles (atoms, molecules, radicals) with a high average kinetic energy of electrons or of all plasma components ($<\varepsilon> \approx 1$ eV...2 MeV per particle), and a considerable interaction of the charge carriers with the properties of the system.[1]

Since the states of matter can be defined by the mean kinetic energy $<\varepsilon>$ of the constituent particles, an energy threshold ε_n exists at the phase boundaries that marks the transition between states and can be treated as typical binding energies. A criterion of the existence of the nth state of matter is $\varepsilon_{n-1} \leq <\varepsilon> \leq \varepsilon$. With increasing $<\varepsilon>$, i.e. increasing temperature matter passes through the states n = 1...3 (solid, liquid, gaseous), and finally, through ionization of gas atoms, free electrical charge carriers appear that characterize the systems as the 4th state of matter. We have arrived at the plasma state.

While the plasma is electrically conductive, its overall charge is neutral. This quasi-neutrality is maintained by strong electric fields. The neutrality restoring electric field can be estimated as follows (Cap, 1984).

If a cubic centimeter of a plasma contains n^* electrons, then the total charge contained in a sphere of radius r is $Q = -(4/3)\pi r^3 n^* e$ ($-e$ is the Millikan elementary electron charge $= 1.6 \times 10^{-19}$ C $= 4.8 \times 10^{-10}$ ESU (electrostatic charge units)). Assuming a centro-symmetrical charge distribution the total charge Q is concentrated at the centerpoint of the sphere. This charge generates an electric field of strength

---

[1] In high energy and plasma physics usually photons of various energy levels are included as constituents of plasmas.

$E = Q/r^2$ at a distance r from the centerpoint. If $r = 1$ cm and $n^* = 10^{15}$ (thermal or 'hot' plasma, see below) then $E = -6 \times 10^8 \, \text{V cm}^{-1}$. One of the characteristics of the state of quasi-neutrality of plasmas is the fact that the number of positive and negative charge carriers per unit volume are exactly equal. If n_1 is the number and $+Ze$ the individual charge of the Z-times positively charged ions then the condition of quasi-neutrality is $n^* = Z \times n_1$ or $|n^* - Z \times n_1| \ll n^*$. Since the ions generate a field $+6 \times 10^8 \, \text{V cm}^{-1}$ the total electric field strength is zero in the state of quasi-neutrality. However, any slight deviation, for example by statistical density variations of a ratio $1:10^{-6}$, results in the occurrence of an electric field strength of $10^{-6} \times 6 \times 10^8 = 600 \, \text{V cm}^{-1}$ that instantaneously restores the quasi-neutrality. Only oscillations of very high frequency, for example a displacement current lead to a spatial separation of the centers of the positive and the negative charge clouds because the ions with their relatively high masses can no longer follow the high frequency oscillations.

3.1.1
Characteristic Plasma Parameters

Since the plasma state occupies an extremely wide range of densities, temperatures and magnetic field strengths it is convenient to define several characteristic plasma parameters (Cap, 1984; Goldston and Rutherford, 1997). Hereby microscopic and macroscopic parameters can be distinguished. The former refer to interactions among individual particles in a plasma, the latter to specific plasma properties in a continuum that can be described by the basic magneto-hydrodynamic (MHD) equations. As we will see, the electron density-temperature-magnetic field space can be subdivided, based on the magnitude of these microscopic parameters. The most important microscopic parameters are the Langmuir plasma frequency, ω_{Pl}, the Debye screening length, λ_D, the Landau length, l_L, the mean free path length, λ (collision length) and the collision frequency, ν_c.

3.1.1.1 Langmuir Plasma Frequency
From the one-dimensional radial equation of movement of electrons

$$m^*(d^2r/dt^2) = -(4/3)\pi n^* e^2 r \tag{3.1}$$

(m^* = mass of electron, n^* = number of electrons per cm^3) with the exponential relations $r(t) = r_0 \exp(i\omega t)$ it follows that the electrons in a plasma undergo harmonic oscillations with a frequency

$$\omega = [4\pi n^* e^2/3m^*]^{1/2}. \tag{3.2}$$

More accurate 3D calculations (see for example Cap, 1984) lead to the equation for the Langmuir frequency

$$\omega_{Pl} = [4\pi n^* e^2/m^*]^{1/2} \tag{3.3a}$$

or, in the SI system,

$$\omega_{Pl} = [n^* e^2/m^* \varepsilon_0]^{1/2}, \tag{3.3b}$$

where ε_0 = vacuum dielectric permittivity ($= 8.86 \times 10^{-12} \, \text{As Vm}^{-1}$).

3.1.1.2 Debye Screening Length

To obtain characteristic plasma properties the electric space charge effects must exceed those of the thermal movement of the carriers. Only then can collective interaction phenomena occur that distinguish a plasma from a normal gas or a cloud of charged individual particles. Simplified one-dimensional calculations yield

$$\text{div } E = (\delta E_x/\delta x) = 4\pi\rho^*, \quad \rho^* = -n^*e \tag{3.4a}$$

($E_x = E$, $E_y = E_z = 0$), and after integration

$$E = -4\pi n^* ex. \tag{3.4b}$$

The displacement of an electron by a distance λ_D in the electric field generated by the n^* electrons per cm^3 leads to an increase in energy defined by

$$\int eE \, dx = 2\pi n^* e^2 \lambda_D^2. \tag{3.5}$$

This energy increase equals the mean thermal energy per degree of freedom, kT, and from $2\pi n^* e^2 \lambda_D^2 = kT/2$ it follows

$$\lambda_D = [kT/4\pi e^2 n^*]^{1/2}. \tag{3.6}$$

This quantity is called the Debye screening length, and can further be expressed as

$$\lambda_D = 6.905 [T/n^*]^{1/2}. \tag{3.7}$$

Since λ_D can be defined as that distance from a point charge within which the potential of the charge decreases to the eth part as the result of space charge effects, it is also called the 'screening length' of the potential, $\varphi(r) = r^{-1} \exp[-r/\lambda_D]$.

3.1.1.3 Landau Length

That distance from a point charge at which the vacuum electrostatic energy equals the kinetic energy kT is called 'critical distance' or Landau length: $l_L = e^2/kT$. The necessary condition preventing a recombination of ions and electrons is $a > l_L$, i.e. the mean distance a between plasma particles must exceed the Landau length. Since on the other hand the distance of the electrostatic interaction of plasma particles is the Debye length λ_D, the relation $a > \lambda_D$ means that there is no cooperative interaction between particles, i.e. there is no plasma.

Assuming spherical and continuous charge distribution, the volume of interaction, the so-called *Debye sphere* is given by $(4/3)\pi\lambda_D^3$. The Debye sphere thus contains $(4/3)\pi\lambda_D^3 n^*$ electrons. In order to apply continuity conditions there is the requirement $(4/3)\pi\lambda_D^3 > 1$ or $\Lambda = (n^* \times \lambda_D^3)^{-1} \ll 1$ or $(n^*)^{-1/3} \ll \lambda_D$. If the 'plasma parameter' $\Lambda \ll 1$ a true plasma exists!

3.1.1.4 Collision Path Length

In electrically neutral gases only van der Waals forces play a role with short interaction distances. Thus collisions among two or more particles are generally negligible. In plasmas, however, electric forces occur whose interaction distances (Debye length) exceed those of the intermolecular van der Waals forces. All charged particles interact constantly with a cloud (Debye sphere) of surrounding particles. For a minimum particle density an average ('smeared out') electric field can be assumed, and calculations can be performed as if a particle is scattered only by this averaged field. Thus all elastic collisions among particles can be treated as binary collisions.

In general, plasma particles can collide inelastically and elastically. During *inelastic collisions* the total kinetic energy of a particle changes by transferring a portion of it to internal energy. Such inelastic collisions are responsible for excitation, ionization, recombination and charge transfer in plasmas. The hallmark of an *elastic collision* of particles is the conservation of its total kinetic energy. Such collisions occur predominantly in neutral gases at low temperatures.

The number of collisions per cm^3 and second is determined by $\sigma(v) \times n \times F$, where $\sigma(v)$ is the collision cross-section, n is the number of scattering particles per cm^3 and F is the number of particles per cm^3 and second that are being scattered. The mean free path length λ is the average path length traversed by a particle between two collisions. From the laws of kinetic gas theory it follows that $\lambda = [n \times \sigma(v)]^{-1} = kT/p \times \sigma(v)$, where p is the plasma pressure. For $p = 1$ atm and $T = 300$ K, λ is about 0.4 μm.

For elastic collisions the cross-section depends on the relative velocity ω of the particles and the scattering angle χ. However, assuming Coulomb forces limited by a finite value of λ_D, a mean free path length λ can be calculated independent of the scattering angle to yield

$$\lambda = [0.6 \times 10^6 \, T^2]/[n \ln(\lambda_D/l_L)] \approx (1.3 \times 10^4 \, T^2/n). \tag{3.8}$$

The term $\ln(\lambda_D/l_L)$ is called the *Coulomb logarithm* Λ_c. The particle velocity can be expressed through the temperature for a Maxwell–Boltzmann distribution as $m^*u^2 = kT$. For a thermal plasma ($T_e = T_h$; see Figure 3.1) it follows $\lambda_{Ion}/\lambda_{El} = m_{Ion}/m^*$, and

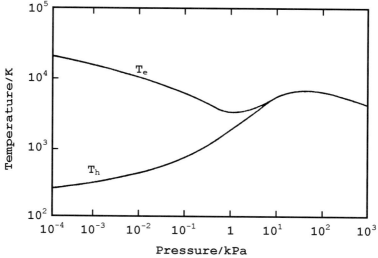

Figure 3.1 Electron (T_e) and heavy particle (T_h) temperatures as a function of plasma pressure (Boulos et al., 1989).

for a collision angle of 90° $\lambda_{90} = (4\pi l_L^2 n)^{-1}$. With $l_L = e^2/kT$ and $\lambda_D = (kT/4\pi e^2 n^*)^{1/2}$ we obtain

$$\lambda_D^2 = \lambda_{90} \, l_L. \tag{3.9}$$

Combination with the inequalities expressed above leads to the *definition of a plasma*:

$$l_L \ll \lambda_D, \; (n^*)^{-1/3} \ll \lambda_D, \; \lambda_D \ll \lambda_{90}. \tag{3.10}$$

Furthermore the condition $\Lambda \approx l_L/\lambda_D = \lambda_D/\lambda_{90} \ll 1$ must be fulfilled.

In addition to these conditions, another one must be fulfilled. To treat a plasma as a continuous medium all statistic fluctuations must be alleviated to obtain a Maxwellian distribution of particle velocities. This requires frequent collisions of particles. The characteristic distance of the plasma cloud, l must be large compared with the collision length λ in order to guarantee many collisions: $\lambda \ll l$.

3.1.1.5 Collision Frequency

The velocity distribution of particles within plasmas is given, under equilibrium conditions, by a Maxwell distribution. The mean velocity $\bar{u}$ is defined by the thermodynamic temperature

$$\bar{u} = (8kT/\pi m)^{1/2}; \quad \bar{u}^2 \approx 3kT/m. \tag{3.11}$$

Without applied strong magnetic fields a particle in a plasma moves between two collisions in a force-free fashion, i.e. with constant speed $\bar{u}$. The relaxation time τ between two collisions is $\tau = \lambda/\bar{u} \sim T\sqrt{T/n}$, and for the collision frequency v_c we obtain $v_c = \tau^{-1} = \bar{u}/\lambda$, or in terms of the collision cross-section, $v_c = n \times \bar{u} \times \sigma(v)$. Introducing the plasma parameter $\Lambda \approx l_L/\lambda_D$, and using the expressions for the Langmuir frequency ω_{Pl}, the Debye length λ_D and $kT \approx m\bar{u}^2$ it follows

$$\Lambda \approx \lambda_D/\lambda \approx v_c/\omega_{Pl}. \tag{3.12}$$

3.1.2
Classification of Plasmas

Depending on the electron density, it is possible to distinguish low pressure 'cold' (nonthermal) plasmas found in outer space with electron densities typically around 10^4 electrons m^{-3}, and high pressure 'thermal' plasmas with electron densities exceeding 10^{24} electrons m^{-3} (Figures 3.2 and 3.3). Thermal plasmas have pressures high enough to facilitate energy exchanges among light, fast moving and therefore energetic electrons and heavy, slowly moving charged ions. This energy exchange leads to efficient transfer of energy, increasing the plasma temperature and thus reaching equilibrium. In such plasmas the electron temperature T_e equals the heavy particle temperature T_h; the plasma is in general or local thermodynamic equilibrium, and can be described according to the laws of thermodynamics (Figure 3.1). The behavior of such a plasma is predictable. The energy extracted

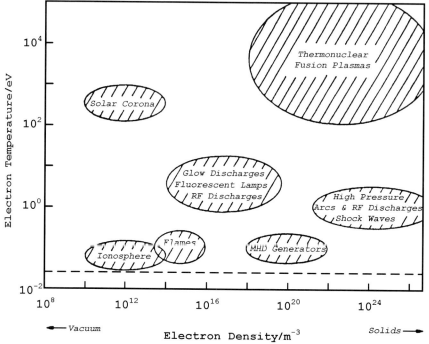

Figure 3.2 Classification of plasmas as a function of electron temperature and electron density (Boulos et al., 1989).

by the electrons from the electric field is transferred to enthalpy and, in turn, to an increase in temperature.

3.1.2.1 Low Density Plasmas
The variation of particle number per cm^3 (particle density) and temperature can be expressed by the plasma parameter Λ (see above). If $\Lambda \ll 1$, the plasma condition $\lambda \ll l$ is not guaranteed. In this case the number of collisions is low. If $\lambda > l$ then low density conditions prevail. Such a plasma is called a collision-free or Vlasov plasma (Figure 3.3). Its relaxation time is $\tau = l(m/2\,kT)^{1/2}$.

3.1.2.2 Medium Density Plasmas
With increasing particle density the mean free collision paths decrease. If $\lambda \approx l$, medium density plasmas can be stabilized depending on the strength of the magnetic field. In a homogeneous static magnetic field an electrical charge describes a circular path around the field lines. The radius of this path is called the radius of gyration (Larmor radius) and given by

$$r_L = (m\,u\,c/eB) \approx (8\,kTm)^{1/2}/eB\sqrt{\pi}, \tag{3.13}$$

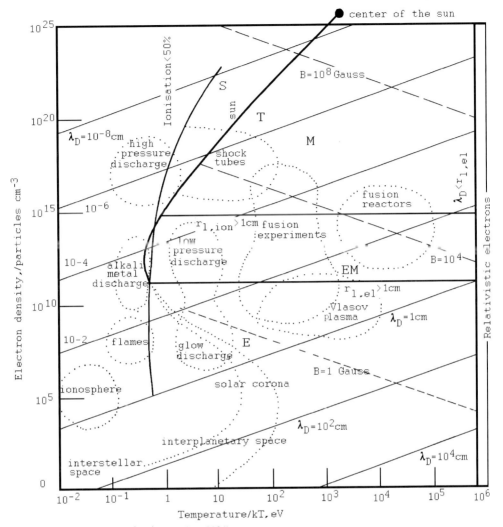

Figure 3.3 Possible states of a plasma (Cap, 1984).

where u = linear circular velocity perpendicular to B, c = velocity of light, m = particle mass, B = magnetic field strength. For M = m*, and with $m^* \times u^2 = kT$, the Larmor radius of an electron is

$$r_L^* = 3.7 \times 10^{-2} (\sqrt{T}/B) \qquad (3.14)$$

The Larmor radius of a proton is $r_{L,Prot} = 1.6\sqrt{T}/B$. Assuming local thermal equilibrium, i.e. $T_e = T_h$ it follows that $r_{L,Ion} = r_L^* (m_{Ion}/m^*)^{1/2}$. Thus it is always $r_{L,Ion} > r_L^*$. The rotation frequency (cyclotron or Larmor frequency) is $\omega_L = eB/mc$.

For a low density plasma with $\lambda_D \ll \lambda$ or $l_L \ll \lambda_D$ (plasma condition) and $l \ll \lambda$ (low density condition), and in the presence of a weak magnetic field, r_L is relatively large, but because $l \ll \lambda \times r_L$ is always small compared with λ. With increasing field strength B, r_L decreases and fulfills even more the condition $r_L \ll \lambda$. This means that a charged particle undergoes many Larmor rotations before a collision occurs that knocks it off its cyclotron path. Collisions therefore influence only marginally low density plasmas at high magnetic field strength.

A medium density plasma is characterized by $\lambda_D \ll \lambda$, $l_L \ll \lambda_D$ (plasma condition) and $l \gg \lambda$ (medium density condition). Depending on the strength of the magnetic field four plasma types can be distinguished: *electric* plasmas E ($l < r_L^*$ and $l < r_{L,Ion}$), *electromagnetic* plasmas EM ($l > r_L^*$, $l < r_{L,Ion}$), *magnetic* plasmas M ($l > r_L^*$, $l > r_{L,Ion}$; $\lambda_{El} > r_L^*$, $\lambda_{Ion} > r_{L,Ion}$) and *tensorial* plasmas T ($\lambda_{El} \gg r_L^*$, $\lambda_{Ion} < r_{L,Ion}$). The ranges of existence of these plasma types are shown in Figure 3.3. For a more detailed description, see textbooks on plasma physics and magneto-hydrodynamics.

3.1.2.3 High Density Plasmas

If the density increases so that $\lambda_{El} \ll r_L^*$, $\lambda_{Ion} \ll r_{L,Ion}$ then collisions outdo all anisotropic effects. In contrast to a tensorial plasma mentioned above any direction-dependent properties will cease to exist, and a 'scalar' (isotropic) plasma S (see Figure 3.3) evolves that can be described by the laws of fluid dynamics. Depending on the temperature, i.e. the degree of ionization, 'two-fluid' (electrons and ions) or 'three-fluid' (electrons, ions and neutral particles) plasmas exist. In plasmas of extremely high densities ($n^* > 10^{19}$ cm^{-3}) squeezing of charges can lead either to *recombination*, i.e. formation of a neutral gas or to *condensation*, i.e. formation of a solid state plasma.

The plasmas can be classified according to the characteristic plasma parameters, the electron density, temperature, magnetic field and characteristic distance l (Cap, 1984). Figure 3.3 shows the plasma states in an electron density-temperature net. Also shown are straight lines of equal Debye lengths ($10^{-8} < \lambda_D < 10^4$ cm) and magnetic field strengths ($1 < B < 10^8$ Gauss). Relativistic effects (particle velocity u > 0.9c) can be neglected for $T < 10^8$ K; quantum effects are negligible for $T_e > 6 \times 10^{-11} n^{2/3}$, $T_h > 3.2 \times 10^{-14} \times n^{2/3}$. The ranges of existence of the plasma types E, EM, M, T and S are also shown.

The plasma parameters, i.e. densities and temperatures of the charged species, can be derived, in principle, by using an electric or Langmuir probe (Smy, 1976). Such as probe consists of a single electrode whose voltage/current characteristics are measured when it is surrounded by plasma (Langmuir and Mott-Smith, 1926). In practice, the measurement of the densities and temperatures of the charged species reduce to the bulk ionization density n_0 and the electron temperature T_e.

3.1.3
Equilibrium and Nonequilibrium Plasmas

Extremely low pressure plasmas occurring in the interstellar space (10^4 e m^{-3}) and low pressure 'cold' plasmas generated under moderate vacuum conditions are *nonequilibrium or two-temperature plasmas*. They are characterized by the fact that the electron temperature T_e is much larger than the heavy particle temperature T_h (Figure 3.1). The electron temperature is measured in eV where 1 eV is the energy gained by an electron when passing through a potential difference of 1 V. The kinetic temperature can be calculated from the mean particle energy $\langle\varepsilon\rangle$: $kT_K = (2/3)\langle\varepsilon\rangle$ (1 eV corresponds to 7733 K or 1.6×10^{-19} J). Frequently the kinetic temperature of a plasma is given in terms of the most probable particle velocity, c_{Pr}: $kT_K = (1/2)\,mc_{Pr}^2$ (1 eV corresponds to 11 600 K). The electron density of a low pressure plasma is typically $<10^{20}\,\text{m}^{-3}$, the electron temperature between 10^{-1} and 10 eV.

At a soft vacuum or atmospheric pressure thermal or 'hot' plasmas can be generated whose electron temperature is on the order of the heavy particle temperature. Therefore such plasmas are called *equilibrium plasmas*.

The application of both types of plasma in materials processing is as follows:

- *Low pressure 'cold' plasmas:*
  plasma etching in semiconductor processing;

  plasma-assisted vapor deposition (PAVD);

  plasma surface modification.

- *High pressure 'hot' plasmas:*
  plasma spraying;

  plasma spheroidizing;

  plasma chemical synthesis for ultrapure ceramic powder production.

3.1.4
Maxwellian Distribution of Plasma Energies

The electrons and the heavy particles will both establish a characteristic Maxwellian energy distribution. Under steady state (stationary) conditions the 'temperature' of a plasma is defined by the mean kinetic particle energy given by the equation $(3/2)kT = (m/2)u^2$ (see Equation 3.11). Since the colliding particles change their velocities there is no universal particle velocity but only an averaged value $\sqrt{\overline{u}^2}$.

A Maxwellian distribution shows the fraction $f(\varepsilon)$ of the plasma particles with a kinetic energy or velocity within the interval ε and $\varepsilon + d\varepsilon$ (or v and $v + dv$). Despite the continuous collisions this fraction is time invariant even though different particles will contribute to the overall distribution at any one time. Figure 3.4 shows the dependence of $f(\varepsilon)$ on ε for various mean energies ε. The relationship between the velocity v and the kinetic energy ε is $\varepsilon = (m/2)v^2$.

This velocity distribution (Maxwell distribution) is

$$f(v)dv = \sqrt{2/\pi}\,(m/kT)^{3/2}v^2\exp(-mv^2/2\,kT)dv, \qquad (3.15)$$

or in terms of the kinetic energy ε,

$$f(\varepsilon)d\varepsilon = 2/\sqrt{\pi}\,(kT)^{-3/2}\sqrt{\varepsilon}\exp(-\varepsilon/kT)d\varepsilon. \qquad (3.16)$$

The Maxwell distribution of plasma particles determines how many electrons have sufficient velocity and energy, respectively to extract energy from the electric

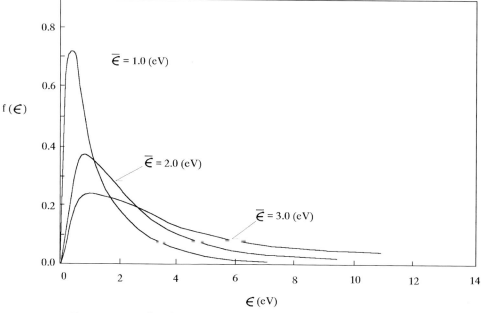

Figure 3.4 Maxwellian distribution of plasma particle energy and velocity, respectively.

field that is being transferred to enthalpy. It also determines the probability of collisions.

3.1.5
Equilibrium Composition of Plasma Gases (Phase Diagrams)

Ionization of plasma gases takes place by a variety of processes chiefly high energy radiation such as neutron beams, X-rays and ultraviolet radiation, collision processes in electric discharges, and also collision processes in intensively heated gases.

The ionization products assume a specific distribution in a particle density (number density)–temperature plot thus resulting in a phase diagram of ionization. Figure 3.5 shows such equilibrium distributions at 1 atm pressure of argon and nitrogen plasmas according to Drellishak 1963.

Figure 3.6 shows the energy content of typical plasma gases as a function of the temperature. This relationship is essentially linear but deviations from linearity occur due to ionization and dissociation. Monatomic gases such as argon and helium need only to be ionized to enter the plasma state. Bimolecular hydrogen and nitrogen must first be dissociated, and therefore need larger energy input to enter the plasma state. This enhanced energy will create increased enthalpy of the plasma. Small quantities of hydrogen or helium added as an auxiliary gas to argon lead to increased plasma enthalpy and thus increased heat transfer rates from the plasma to the powder

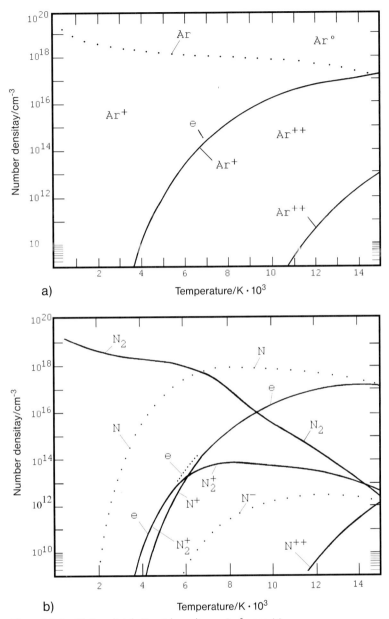

Figure 3.5 Equilibrium distribution (phase diagram) of argon (a) and nitrogen (b) plasmas (Drellishak, 1963).

particles. As a result, the plasma is hotter, and particles melt more easily and completely. Figure 3.7 shows the specific heat of various plasma gases at constant pressure as a function of gas temperature indicating that the high specific heat of helium and, in particular hydrogen increases the heat transfer rate.

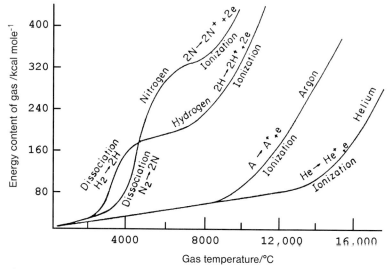

Figure 3.6 Energy content of typical plasma gases as a function of temperature (Drellishak, 1963).

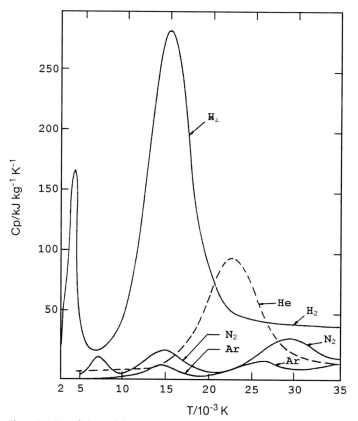

Figure 3.7 Specific heat of plasma gases at a constant pressure (1 atm) as a function of temperature (Boulos et al., 1989; Drellishak, 1963).

3.2 Plasma Generation

The different methods of plasma generation are always applications of gas ionization. In principle, energy will be transferred to atoms or molecules in an elementary process that is sufficient to initiate ionization by two basic mechanisms:

- Increase of the energy content of all internal degrees of freedom of the gas by application of heat. This application can be accomplished directly through the container walls, or indirectly through chemical processes, compression or an electrical current. Plasma is generated by collision ionization of the particles and photoionization by the electromagnetic radiation in the hot gas. Such plasmas are generally close to their thermodynamic equilibrium state (isothermal plasmas).

- Transfer of energy for effective ionization without substantial temperature increase of the gas through particle or electromagnetic radiation and electrical current, respectively. Such plasmas are nonequilibrium plasmas with a high electron temperature $T_e \gg T_{gas}$ (nonisothermal plasma).

Coupled mechanisms occur frequently during technical generation of plasmas.

3.2.1 Plasma Generation by Application of Heat

In a so-called *plasma furnace* (Figure 3.8a) a gas is heated by temperature increase of the confining walls. A true thermal equilibrium plasma forms whose degree of ionization x is determined by

$$x^2/(1-x^2) = 2(2\pi m^*/h^3)^{3/2}(g_1 E_0^{5/2}/g_0 p)(kT/E_0)^{5/2}\exp(-E_0/kT) \quad (3.17)$$

(Saha-Eggert equation), where m^* = mass of the electron, E_0 = ionization energy, p = gas pressure, and g_1 and g_0 = statistical weights of the influence of ionization and ground states, respectively.

Since the highest temperatures achievable by this method are only around 3500 K, the degree of ionization is small, about 1%. Thus this method is confined to generation of low-energy plasmas for fundamental laboratory investigations.

An increase of the degree of ionization is possible by use of the contact ionization that occurs when the ionization energy of the atoms colliding with the hot container walls is smaller than the work function, i.e. the thermionic electron emission energy of the wall material. For example, the ionization of cesium vapor atoms at heated tungsten plates leads to a noise-less ('quiet') plasma in which the reduction of the radial charge carrier losses of the plasma column is achieved through a longitudinal magnetic field (magnetic plasma confinement). The degree of ionization is up to 50%. Such a plasma generation device is called a 'Q-engine' (q = quiet) (Figure 3.8b).

Indirect heat transfer to a gas by exothermic chemical processes (flames, explosions) can yield stationary temperatures of up to 5000 K. However, dramatically

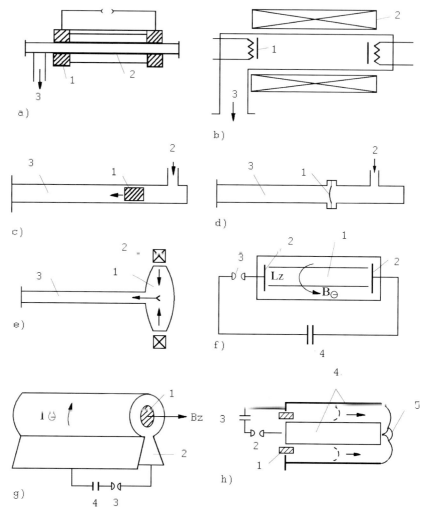

Figure 3.8 Principal ways of generating plasmas by heating (a, b) and compression of gases (c–h) (Rutscher, 1982). (a) King's plasma furnace (1 electrodes, 2 graphite tube, 3 vacuum connection); (b) 'Q engine' (1 indirectly heated tungsten plates, 2 magnetic coil, 3 vacuum connection); (c) Ballistic compressor (1 sliding piston, 2 propellant gas inlet, 3 compression and observation section)' (d) Mechanical shock tube (1 diaphragm, 2 high pressure gas inlet, 3 compression and observation section); (e) Inductive-hydrodynamic shock tube (1 discharge chamber, 2 shock coils for magnetic field, 3 compression and observation section); (f) z=pinch apparatus (1 plasma column, 2 electrodes, 3 arc gap, 4 capacitor battery); (g) Θ-pinch apparatus (1 plasma column, 2 one-turn coil, 3 arc gap, 4 capacitor battery); (h) Plasma focus (1 insulator, 2 arc gap, 3 capacitor battery, 4 electrodes, 5 focused plasma).

higher temperatures can be achieved during short times, for example, when a strategic nuclear fusion device (H-bomb) is triggered by the explosion of a nuclear fission device (Pu-bomb). Chemical reactions heating plasmas are used in the MHD generator.

3.2.2
Plasma Generation by Compression

A ballistic compressor (Figure 3.8c) consists of a tube of a few meters lengths through which a piston of up to a few kilograms mass is accelerated by a burst of a propellant gas. The kinetic energy of the piston will be transformed to heat by adiabatic compression. Instead of the moving piston the compression can be achieved by a shock wave generated by the burst of a membrane separating the high pressure from the low pressure region of the tube (Figure 3.8d). For a short time temperatures of up to 5×10^4 K can occur.

In an electromagnetic shock tube (Figure 3.8e) an oscillating magnetic field is used to generate ionization waves in a shallow chamber that propagate as a shock wave into the tube. This effect is used in the novel method of electromagnetically accelerated plasma spraying (EMAPS, Kitamura *et al.*, 2003).

The so-called *pinch effect* refers to magnetic self-compression of a current-carrying plasma. In an impulse mode, temperatures up to 10^7 K can be generated in rapidly increasing magnetic fields. There are two principal configurations:

3.2.2.1 z-Pinch (Figure 3.8f)
Two planar electrodes (2) conduct an axial current j_z. The compression of the plasma occurs in the resulting magnetic field strength B_Θ by the inwardly directed Lorentz force, $F = (j \times B_\Theta)$. In the regions of decreasing cross-section of the plasma column coupling of the Lorentz force and the pressure drop results in an axial force and consequently leads to the formation of a high velocity plasma jet. This mechanism is essential for the design of plasmatrons for plasma spraying machines.

3.2.2.2 Θ-Pinch (Figure 3.8g)
A current j_Θ flowing through a coil generates an axial magnetic field of strength B_z that compresses a plasma formed by pre-ionization.

3.2.2.3 Plasma Focus (Figure 3.8h)
This is a special variant of the z-pinch configuration in which the discharge of a capacitor battery in a coaxial electrode system leads to an electrodynamic acceleration of the plasma along the electrodes and a subsequent compression by radially directed forces. A variant of this configuration uses particle acceleration by a fast travelling unsteady gas flow caused by a pulsed high current arc discharge (electromagnetically accelerated plasma, EMAP (Kitamura *et al.*, 2003, Usuba and Heimann, 2006).

3.2.3
Plasma Generation by Radiation

Gases can be ionized by interaction with energy provided by either particle radiation such as electrons or protons, or by electromagnetic waves. This kind of ionization is found in astrophysical plasmas.

In an electron beam-plasma generator an energetic electron beam (keV range) will be focused on a diluted gas and provides, by collisional ionization, the plasma state. Interaction of the beam with the plasma can generate turbulent Langmuir oscillations. In the field of these oscillations the electrons of the plasma can acquire energy to form a nonisothermal plasma with a high electron temperature.

Gases can also be ionized in the optical range by application of powerful laser radiation. A focused laser beam produces at a threshold intensity of about $10^5\,\mathrm{MW\,cm^{-2}}$ a plasma in the focal plane in an explosive fashion by extension of the high frequency discharge beyond the microwave region (optical discharges).

3.2.4
Plasma Generation by Electric Currents (Gas Discharges)

The oldest and most important method of generating plasmas is the formation of an electric current in a gas. The conversion of electrical energy in the various energy forms of plasmas is facilitated by the acceleration of charge carriers in an electric field with subsequent energy transfer by collision. According to the temporal behavior of the current there are several options: DC discharge mode, AC discharge mode, or impulse mode.

In the AC discharge mode the frequencies used range from technical AC current to the high frequency range to microwaves and optical frequencies to X-ray and γ-ray frequencies. In the DC discharge mode the current is supplied to the gas discharge region directly through metallic electrodes whereas in the AC mode the electrodes can be separated from the gas region by solid insulating material.

In general gases are good insulators and to start a discharge free charge carriers must be generated or injected into the potential plasma. In many cases the gas creates those carriers by self discharge. In other cases a plasma is ignited by a high frequency impulse, i.e. a spark that ionizes the gas and creates enough charge carriers to sustain the electric discharge. This principle is generally be applied to ignite the plasma of a DC plasmatron.

The static discharge characteristic of a gas column is shown in Figure 3.9 in an I–U plot (Rutscher, 1982). At small voltages a very small current flows based on external sources of ionization such as space ionization. In this range Ohm's law is valid in good approximation, i.e. I is linearly proportional to U. At larger currents deviation from this linearity occur if the loss of charge carriers, due to their movement towards the electrodes, cannot be neglected anymore compared with the generation of carriers by external sources of ionization. Beyond the saturation current $I = I_s =$ constant, the characteristic enters a region of increasing current in which additional carriers will be generated in an avalanche fashion, caused by collisional ionization of electrons. The number of electrons formed by ionization increases to $\exp(\alpha x)$ along a distance x. In particular, the multiplication along the electrode distance d is $M = \exp(\alpha d)$, where α is the 1st Townsend coefficient that is determined by the ionization frequency ν_i and the drift velocity ν_e of the electrons ($\alpha = \nu_i/\nu_e$).

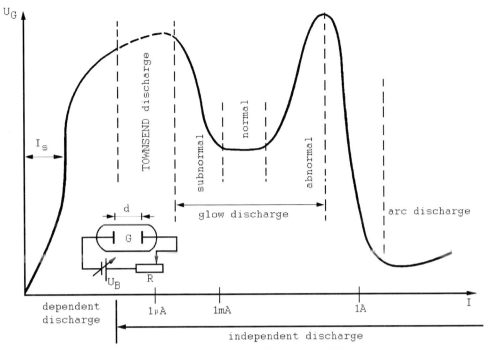

Figure 3.9 Complete static discharge characteristic of a gas (Boulos et al., 1989).

As long as the current still depends on the existence of an external ionization source to sustain the discharge despite the multiplication effect of charge carriers, the gas remains in the state of a *dependent discharge*. The ignition of an *independent discharge* is achieved by fulfilling the Townsend ignition condition, $\gamma[\exp(\alpha d) - 1] = 1$. The 2nd Townsend coefficient γ considers all processes leading to the generation of secondary electrons, for example electron emission at the cathode by ion impact, photo effect, as well as the volume ionization by fast ions or photons.

In the range of small currents in the transitional range between the dependent and independent discharge there is the dark or Townsend discharge (Figure 3.9). It is characterized by lack of luminescence, small field displacement by space charges, and an almost horizontal I, U-characteristic (dashed line in Figure 3.9). With increasing current electrical space charge domains occur with a tendency to decrease the voltage. Dependent on the value of the current, two cathode mechanisms can be distinguished, the glow cathode with a small discharge current, and the arc cathode with a large discharge current.

3.2.4.1 Glow Discharges
Glow discharges occur in diluted gases ($p \approx 1\ldots 10^4$ Pa) at small currents ($I \approx$ 0.1...100 mA). They lead to characteristic luminescent features between the electro-

des that are divided into differently colored lighter (B) and darker (D) parts depending on the nature of the gas as shown in Figure 3.10 (Aston layer, D; cathode glow light, B; cathodic (Crooke's, Hittorf's) dark space, D; negative glow light, B; Faraday's dark space, D; positive column, B; anodic dark space, D; anode glow light, B; see textbooks of physics). In front of the glow cathode light there is a strong space charge field (Aston field, D) of about 10^{-4} cm width called the *cathode fall*. In this region the electrons acquire the energy for ionization. The formation of the cathode fall at the transition from the Townsend discharge to the glow discharge is associated with the decrease of the voltage (subnormal discharge, see Figure 3.9). At higher currents the cathode fall obtains initially a minimum value independent on the current (normal cathode fall). The typical range of the normal cathode fall is between 100 and 300 V. When the cathode is completely covered by the glow light, the cathode fall and the current density increase again with increasing current (abnormal cathode fall).

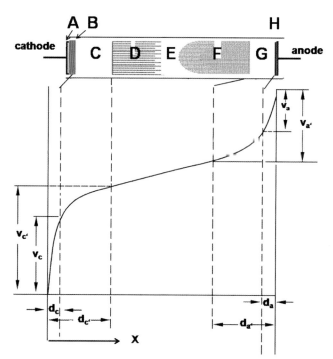

Figure 3.10 Luminescent and dark regions of an ionized gas column (top) and potential distribution in an electric arc (bottom) (after Edel, 1961). A = Aston region (D); B = cathodic glow (B); C = cathodic dark space (Crooke, Hittorf; D); D = negative glow (B); E = Faraday dark space (D); F = positive column (B); G = anodic dark space (D); H = anodic glow (B).
D: dark, B: bright.

3.2.4.2 Arc Discharges

The glow discharge is of no consequence for the plasma spray process but has an important application in the bulk elemental analysis and depth profiling analysis of thin surface layers by glow discharge optical emission spectrometry (GD-OES). At high current densities, however, the abnormal glow discharge is replaced by an *arc discharge*. The arc cathode acquires a high temperature that leads to thermal electron emission. Normally on the cathode a narrowly confined termination of the arc occurs as a bright and highly mobile spot with an extremely high current density up to $10^7\,\mathrm{A\,cm^{-2}}$ and a low cathode fall of typically 10 to 50 V. Ions impinging on the cathode are responsible for sustaining the cathode temperature and thus electron emission. Heating of the cathode (thermal arc) is complemented by a field emission of electrons (field arc) generated by the space charge field of ions flowing towards the cathode. Typical values for the current densities required for thermal emission and field emission are, respectively 10^3–$10^4\,\mathrm{A\,cm^{-2}}$ and 10^6–$10^8\,\mathrm{A\,cm^{-2}}$ with characteristic temperatures of <3500 K (hot cathode with thermal emission) and <3000 K (cold cathode with field emission). In general the production of secondary electrons by collective processes at the arc cathode is much more effective than the individual processes occurring at the glow cathode.

Recent investigations have shown that nonsteady state electron emissions from the arc cathode may also play an important role. Such a mechanism is characterized by extremely fast heating of microscopic asperities at the cathode surface by Joule heating and ion impingement. The asperities evaporate explosively as dense plasma and release a large quantity of charge carriers.

Accordingly, the electrical processes occurring in different parts of the arc discharge are complex and despite a considerable amount of research in the past (Edels, 1961; Solonenko, 2000; Becker *et al.*, 2004) still a subject of controversy.

The potential distribution in the arc is shown in Figure 3.10 with the existence of three main arc regions, namely a positive column extending over the major portion on the arc path with a constant field strength (regions D, E, F), a cathode fall region with a voltage drop $V_{c'}$ close to the cathode of a typical thickness $d_{c'}$ of about 10^{-1} cm and supporting a voltage of approximately 10 V (regions A, B), and an anode fall region of similar size with a similar voltage drop of $V_{a'}$ (regions G, H). The layers d_c and d_a immediately adjacent to the surface of the cathode and anode, respectively are the thermal boundary layers or space charge zones with a typical thickness of 10^{-4} cm (Aston region). They provide the greater part of the voltage drop. The remainders of the cathode and anode falls are also called the contraction zones (regions B, H). In general the gas pressure in arc discharges is much higher than in glow discharges (high pressure discharge, $p \geq 10^5$ Pa). The requirement for the existence of the positive column is a sufficiently large electrode distance. The high gas pressure in the positive column causes many collisions of ions that are responsible for an equalization of the electron and heavy particle temperatures (thermal plasma, see above).

3.2.5
Structure of the Arc Column

3.2.5.1 Positive Column

For a cylindrical symmetric column the field strength is constant and, according to Poisson's law the net space charge is zero. This indicates that the concentration of positive, n^+, and negative n^-, charges must be equal. The positive column (regions D to F) thus satisfies the condition for a plasma (see above). The current flowing through the column will have two components, an electron current J^- and a positive-ion current J^+. The total current density is

$$\mathbf{J} = \mathbf{J}^- + \mathbf{J}^+ = e(n^- v^- + n^+ \cdot v^+) = n \times e \times (v^- + v^+), \quad (3.18)$$

where $v-$ and $v+$ are the drift velocities produced by the electric field. The drift of the charged particles is impeded by collisions with neutral and other charge carriers. This represents a resistive phenomenon. Since $v^- \gg v^+$, the total current is carried predominantly by the electrons. Thus it follows that

$$\mathbf{J} \approx \mathbf{J}^- = n^- e \times v^- = n^- e \times \mu \times \mathbf{E} = \sigma \times \mathbf{E}, \quad (3.19)$$

where μ is the electron mobility and σ the electrical conductivity.

3.2.5.2 The Cathode Fall Region

To satisfy the continuity condition the current flowing through the positive column must also make the transit from gas to metal at both electrodes. Hence in the cathode region conditions must exist that allow this transition. For refractory cathodes the mechanism is one of thermionic emission. There is a continuous and controllable transition from a glow discharge to an arc state depending upon the cathode temperature (Figure 3.11; Wehrli, 1928). To lower the work function of the electrons and to get an abundance of electron emissions to initiate and sustain the arc discharge, the cathode of a plasmatron is usually coated with thorium.

Figure 3.12 shows the current characteristic of thermionic electron emission in a vacuum diode. It can be described by the Richardson–Dushman equation

$$\mathbf{j}_s = AT^2 \exp(-e\Phi_c/kT), \quad (3.20)$$

where A is the area of the electron emission and Φ_c is the electron work function. Because of the space charge region adjacent to the cathode the emission of electrons is impeded by the electric field on the cathode surface. Therefore the equation must be corrected. For high cathode temperature and moderate field strength the Schottky correction can be applied:

$$\mathbf{j}_s = AT^2 \exp\{-[e\Phi_c - (e^3 \mathbf{E}/4\pi\varepsilon_0)^{1/2}]/kT\}. \quad (3.21)$$

The ability to emit electrons by thermionic emission differs among different metals. Figure 3.13 shows a plot of the thermionic work function Φ_c against the boiling temperature of the metal. Metals with particularly high work functions and thus thermionic emission are molybdenum, tungsten, tantalum and rhenium (Guile, 1971).

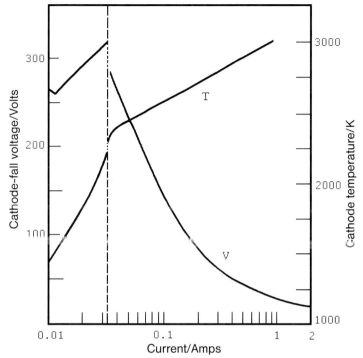

Figure 3.11 Variation of cathode fall voltage and cathode temperature with the applied current during thermionic glow discharge-to-arc transition (Wehrli, 1928).

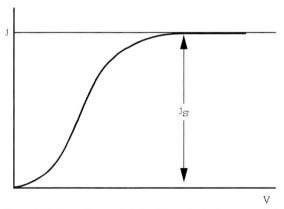

Figure 3.12 Current characteristic of thermionic electron emission in a vacuum diode.

The net energy supplied to the cathode can be described by the balance between the current carried by the impinging ions and that caused by electron emission. The energy supplied to the cathode by impinging ions is

$$j_i(V_c + V_1 - \Phi_c + (5/2)kT_i/e, \tag{3.22a}$$

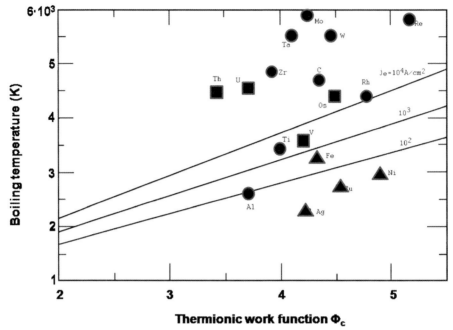

Figure 3.13 Thermionic work functions, Φ_c of various metals as a function of the boiling temperature. Circles: thermionic emitters, triangles: thermionic emission not observed, squares: no information available (Guile, 1971).

the energy removed by electron emission is

$$-j_e \Phi_c. \tag{3.22b}$$

This is, however, an idealization. The cathode acquires also energy from other sources such as neutralization energy transfer, condensation energy from the ions, heat conduction from the gas, radiation and possibly chemical reactions. The energy losses also encompass vaporization, loss of metal globules by thermal sputtering, radiation and heat conduction through both solid and surrounding gas.

The ratio j_i/j_e is between 0.15 and 0.5 depending on the cathode material and the electrical field strength.

As mentioned above, loss of material from the electrodes frequently occurs as do plasma arc jets, which are usually preferentially directed. Such plasma jets may be produced also by constricting the diameter of a small section of the arc column. In this case, the pressure gradient due to the pinch of the self-generated magnetic field increases with current density in the steady state. The constricting Lorentz force is directed inwardly and confines the plasma jet. This phenomenon will be discussed in detail later. Figure 3.14 shows the interaction of cathode and anode jets in an arc. The mode of attachment of the cathode arc jet at the surface of the cathode depends on the

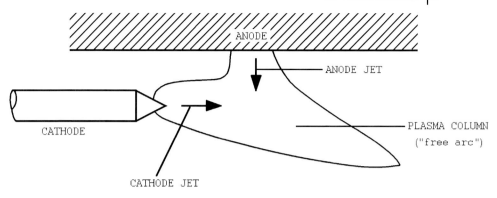

Figure 3.14 Interaction of cathode and anode jets in an electric arc (Boulos *et al.*, 1989).

current density. Diffuse attachment without localized spots occurs at 10^3–10^4 A cm^{-2} whereas distinct spots are visible at higher current densities (10^6–10^8 A cm^{-2}). In the case of the attachment of the anode arc jet, lower current densities of 10^2–10^3 A cm^{-2} produce diffuse, higher ones (10^4–10^5 A cm^{-2}) constricted, spot-like attachment. Frequently the diffuse attachment of the anode arc jet is caused by the cathode jet impinging on the anode. This is known as the cathode jet-dominated (CJD) mode.

3.2.5.3 The Anode Region

Positive-ion emission from the anode can be neglected for most arc conditions. In this region current continuity is thus achieved by the total current being fed into the anode by electrons. This necessitates the production of extra electrons in the anode region. In many ways the anode region is similar to the cathode region even though the conditions are not so extreme. For example, the contraction zone is less sharp. The calculation of the energy balance at the anode is simpler than that at the cathode since a number of the coefficients involved are more definite. Most of the power is consumed in metal vaporization so that a high anode vapor pressure exists. This is then the source for the often observed anode vapor jets (Elenbaas, 1934) (see Figure 3.14) when the cathode arc jet impinging on the anode surface causes material to evaporate.

By introducing a constricting diaphragm in front of the anode of a high-intensity argon arc the flow is interrupted thus resulting in the anode jet-dominated (AJD) mode. Temperature measurements based on line-emission coefficients and on absolute continuum-emission coefficients (Section 3.4), and modeling by simultaneously solving the conservation equations under simplifying boundary conditions, show reasonable agreement of the temperature distribution along the anode region (Figure 3.15; Sanders *et al.*, 1982).

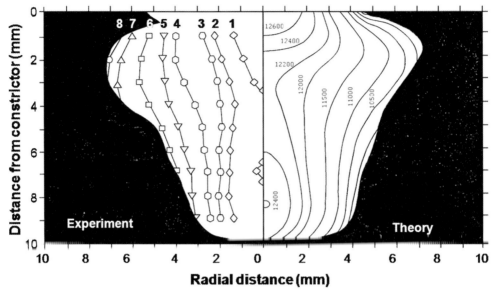

Figure 3.15 Temperature distribution in the anode region of a high intensity arc: experiment vs. theory. The isotherms are as follows: 1–12 400 K, 2–12 200 K, 3–12 000 K, 4–11 500 K, 5–11 000 K, 6–10 500 K, 7–10 000 K, 8–9500 K (after Sanders et al., 1982).

3.3
Design of Plasmatrons

Simplistically, one can visualize plasmatrons as resistance heaters in which familiar resistance elements such as NiCr, silicon carbide, molybdenum disilicide, tantalum, tungsten etc. are replaced with a consumable, conductive, and partially ionized gas. Thus the plasma column can be considered a 'consumable heating element' whose resistivity varies with operating conditions and type of gas used for ionization. The resistance range of plasmas in Ω cm is 0.01 to 0.5 for argon, 0.05 to 0.25 for nitrogen, and 0.25 to 0.0035 for hydrogen (Thorpe, 1989). These values should be compared with the resistance ranges of solid materials such as iron (0.001 to 1.00), carbon (9×10^{-4} to 3.5×10^{-3}) and copper (5×10^{-7} to 9×10^{-8}).

An important design consideration for plasmatrons is the confinement of the plasma jet along the central channel of the plasmatron by two phenomena, the thermal and the magnetic 'pinch'.

Thermal pinch occurs because at the cooled wall of the anode nozzle the conductivity of the gas is reduced thus increasing the current density at the center. The charged plasma tends to concentrate along the torch axis thus confining the jet. Simply put, this means that the cold gas surrounding the arc absorbs a significant amount of energy from the arc in proportion to the energy required to ionize it.

Magnetic pinch is related to a magneto-hydrodynamic Lorentz force created by the electrically conductive central region of the plasma jet. The moving charges induce a

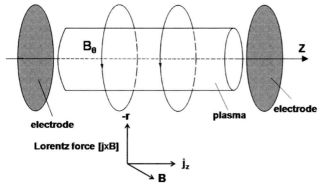

Figure 3.16 Direction of the Lorentz force (longitudinal or z-pinch).

magnetic field. According to electrodynamics, the vector cross-product of the current and the magnetic field strength defines the Lorentz force whose vector is mutually perpendicular to the former vectors. An inward moving force is created that constricts the jet even further. Its core pressure increases, and the jet is blown out of the front nozzle with supersonic speed.

Figure 3.16 shows schematically the direction of the Lorentz force $[\mathbf{j} \times \mathbf{B}]$ (longitudinal or z-pinch). The current flowing in z-direction causes circular magnetic field lines $\mathbf{B}_\theta$ that compress and thus confine the plasma.

From the momentum conservation equation Euler's flow equation can be derived:

$$\rho(d\mathbf{v}/dt) \equiv \rho(d\mathbf{v}/dt + (\mathbf{v}\nabla)\mathbf{v}) = (1/c)[\mathbf{j} \times \mathbf{B}] - \nabla p \qquad (3.23)$$

where $\rho(z,t)$ is the particle density, $\mathbf{v}(z,t)$ is the (macroscopic) particle flow velocity, p is the plasma pressure and $\rho(c;c)$ is the surface tension tensor (Cap, 1984). This equation neglects the influence of gravity and other volume forces such as centrifugal and Coriolis forces.[2] This leads to the simplified Navier–Stokes equation for momentum conservation,

$$\rho(d\mathbf{v}/dt) + \nabla p = [\mathbf{j} \times \mathbf{B}]. \qquad (3.24)$$

The first term on the lefthand side of the equation describes the time-dependent mass flux, i.e. the particle acceleration, the second term is the outward directed plasma pressure gradient. Their sum is balanced by the vector cross-product of the current and the magnetic field, i.e. the inwardly directed Lorentz force. For very high currents the z-pinch effect increases the temperature by magnetic compression considerably. Its value can be calculated by applying the Bennett equation

$$J^2 = 8\pi kTn/\mu_0, \qquad (3.25)$$

where n is the number of particles per unit length of the plasma cylinder, and μ_0 is the electron mobility. Plasma heating by magnetic compression is used in nuclear fusion reactors.

---

[2] However, it can be shown that self-confinement of a finite plasma is impossible in a gravity-free environment by its own magnetic field alone (Cap, 1984).

3.3.1
Arc Discharge Generators and their Applications

There is a wide variety of approaches towards achieving high temperatures in a plasma jet heating device, but only a few basic principles will be considered here.

We can distinguish three categories

- electrode plasmas (EPs) as either nontransferred or transferred plasmas;
- electrode-less plasmas (inductively coupled plasmas, ICPs) operating at low or high frequencies;
- hybrid devices where ICPs are superimposed on EPs (Thorpe, 1989).

3.3.1.1 Electrode-supported Plasmas

Nontransferred electrode plasmatrons (indirect plasmatrons) operate at comparatively low voltages between 20 and 150 V DC. Figure 3.17 shows the cross-section of a typical nontransferred DC plasmatron. It consists of a stick- or bullet-shaped cathode made from tungsten, and a water-cooled copper anode which forms the front nozzle of the torch. The tungsten cathode is thoriated to lower the electron work function, and to get an abundance of electron emissions that initiate and sustain an electric arc between the tip of the cathode and the internal diameter of the positive anode nozzle. The plasma rapidly diminishes in charged particle density near the nozzle exit. This results in an exponential decrease in electrical conductivity as shown in the lower part of Figure 3.17.

The actual discharge is initiated by a high frequency pulse or a high voltage field emission discharge. The arc stretches down the cathode with a constant voltage

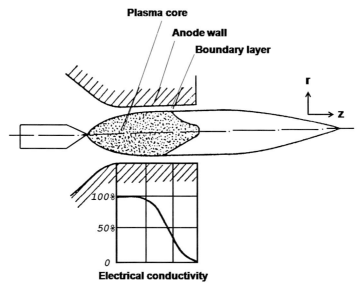

Figure 3.17 Cross-section of a typical nontransferred DC plasmatron.

3.3 Design of Plasmatrons

gradient and is attached to the anode by a voltage drop (anode fall). The flow through the torch is mildly turbulent with minimum swirl content for devices operated at low voltages. High voltages between 150 and 1000 V, however show a significant amount of highly turbulent swirl gas that produces a strong vortex with a low-pressure center core that must be traveled through by the arc.

The heat transfer rates are close to the heat flux failure limit of the anode, i.e. copper material (Thorpe, 1989). Therefore, effective cooling by a constant stream of water or cold gas is essential to keep the anode from evaporating. In this configuration the maximum cathode amperage is about 3000 A. This sets a maximum achievable power. For example, at a voltage of 200 V the maximum electrical power input would be 600 kW. The workpiece to be coated remains relatively cool and its temperature rarely exceeds 200 °C.

Transferred electrode plasmatrons (direct plasmatrons) can be operated in two modes: transfer of energy to the workpiece that is electrically connected to the power source, and transfer of energy to an intermediate electrode.

In the first type (Figure 3.18), the cathode is some distance (usually less than 15 cm) from the 'working' anode. An arc is struck between the two electrodes by either a capacitance or a high frequency discharge with the direct current following the ionized path thus generated. The workpiece/substrate is electrically connected to the power source, and the whole assembly actually constitutes an arc welding configuration. This means that the substrate is heated to high temperatures.

The heat losses by radiation or convection are high (20–40%). If the system is used for plasma spraying, care must be taken to inject the powder downstream of the arc root at the anode to avoid contamination of the electrodes. Also, the substrate may have to be cooled in order to avoid melting, warping or undesirable changes of the microstructure. Applications of a transferred arc include welding, cutting and

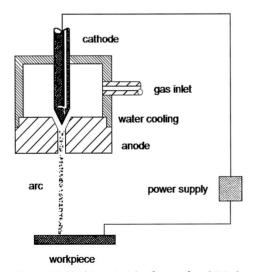

Figure 3.18 Working principle of a transferred DC plasmatron.

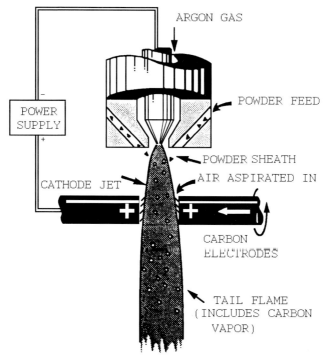

Figure 3.19 Working principle of a transferred arc terminated by an intermediate electrode (Thorpe, 1989).

melting operations of conductive materials. With severe constriction of the arc by an intermediate nozzle of small diameter, the heat flux and gas velocity are very high, and this device is therefore used for cutting metal. Less constricted arcs with larger intermediate nozzle size have lower gas velocities. Thus the arc is 'softer' and can be used for welding and melting operations.

In the second type (Figure 3.19), the transferred arc is terminated by an intermediate electrode that produces a nontransferred free plasma jet. As shown in the figure, the secondary carbon anode is self-cleaning and slowly consumed. In the Ionarc device shown there are three rotating carbon electrodes that are fed at a rate that replenishes the consumed material. Powder injected close to the cathode spot can be heated with high thermal efficiency. For example, the powder feed rate is as high as $0.9\,\mathrm{kg\,h^{-1}\,kW^{-1}}$ as opposed to $0.1\,\mathrm{kg\,h^{-1}\,kW^{-1}}$ for normal tail-flame injection type plasmatrons.

Today, transferred arcs are frequently used to clean the surface of the part to be coated by sputtering off oxide scale (for example Mühlberger, 1974, 1989; Frind et al., 1990; Döring et al., 2003). This requires a polarity change, i.e. the working anode is made more negative for some time. This procedure is particularly important if highly oxygen sensitive material is to be sprayed in a 'vacuum' environment, such as NiCrAlY for coating of gas turbine vanes and blades for aerospace applications.

In a recent development, the metal workpiece is connected to the negative polarity of the powder supply and thus serves as a cathode of a low pressure plasma coating system (Kubo et al., 2005; see also Takeda and Takeuchi, 1997). Any oxide scale rapidly evaporates by the action of cathode arc spots that move randomly across the surface of the workpiece. The surface roughness also increases. Combination of conventional grit blasting and cathode spot erosion yields maximum adhesive strength of APS coatings.

Segmented anodes allow the arc to strike from the central cathode through one or more insulated and water-cooled metal rings to the terminating, also water-cooled anode rings (Figure 3.20a). This is done to increase the arc voltage that permits higher power operation. Since the amperage is reduced at given power level electrode life is increased because erosion is decreased.

Figure 3.20b shows a water-stabilized arc (see 3.3.2.2). Water vortices in a tube section between consumable cathode and anode form a hollow passage and protect

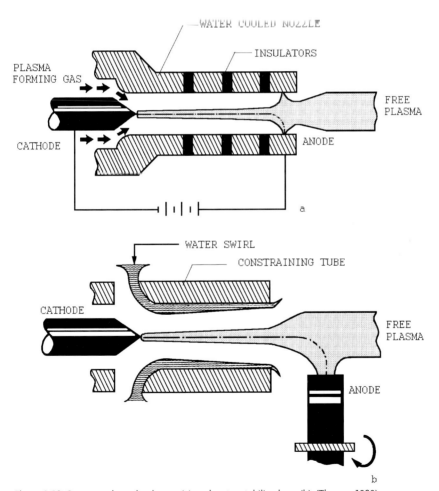

Figure 3.20 Segmented anode plasma (a) and water-stabilized arc (b) (Thorpe, 1989).

the walls from overheating. Very high power can be applied and temperatures can therefore reach 50 000 K (Gerdien arc; Gerdien and Lotz, 1922). Transferred water-stabilized arcs are used to spray alumina at powder feed rates as high as 34 kg h^{-1} compared with less than 10 kg h^{-1} for a conventional gas-stabilized nontransferred DC arc (Thorpe, 1989).

3.3.1.2 Electrode-less Plasmas

Radiofrequency (r.f.) inductively-coupled plasma (ICP) devices were conceived by Reed in 1961 (Reed, 1961), and treated extensively theoretically, for example by Eckert 1974 and Boulos 1985. They generally consist of a coil surrounding a water- or gas-cooled refractory (often silica glass) tube through which the plasma-forming gas, the sheath-gas for wall cooling, and the powder gas pass (Figures 3.21 and 3.22). The devices operate in widely varying power (0.5 kW to 1.0 MW) and frequency (9.6 kHz to 40 MHz) ranges (Boulos, 1985).

Energy coupling to the plasma is achieved through the electromagnetic field of an induction coil. The plasma gases do not come into contact with any electrode and

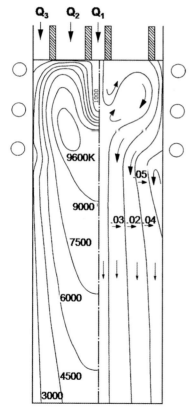

Figure 3.21 Electrode-less radiofrequency inductively-coupled plasmatron. Left: temperature distribution. Right: flow field distribution (Boulos, 1985).

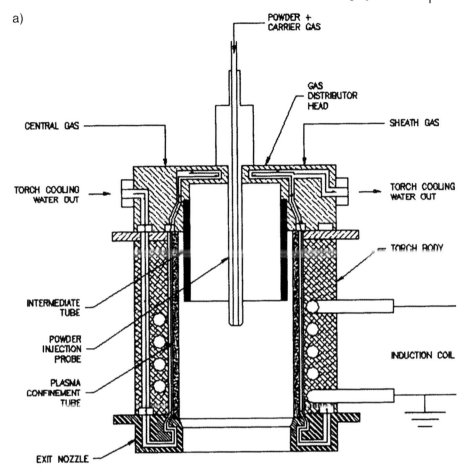

Figure 3.22 (a) Inductively-coupled plasmatron (TEKNA Plasma Systems Inc., Sherbrooke, Québec, Canada); (b) Environmental chamber to house an inductively-coupled plasmatron. The solid powder feeder A allows to operate the system in the thermal plasma chemical vapor deposition (TPCVD) mode, the liquid suspension feeder B in the suspension plasma spraying (SPS) mode (after Burlacov et al., 2006; see also Müller and Frieß, 2005).

therefore any contamination is largely excluded. Argon is commonly used because of its ease of ionization. The device, however, also works rather well with an air plasma.

The main attraction of the radio-frequency inductively coupled torch is in-flight melting of relatively large particles of refractory metals and ceramic powders at high throughputs.

Figure 3.21 shows schematically the temperature (left) and flow fields (right) in the discharge region of an inductively coupled torch (Boulos, 1985; Mostaghimi et al., 1984). The maximum plasma temperature is off-axis because the energy dissipation is limited to the outer annular region of the discharge. The argon plasma

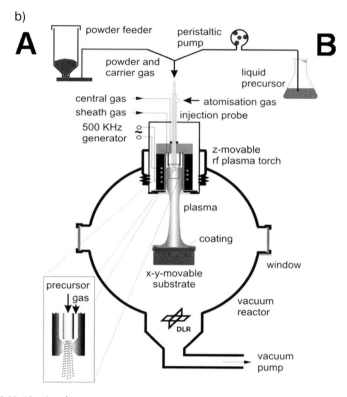

Figure 3.22 (Continued)

(power level 3 kW, oscillator frequency 3 MHz) is confined in a water-cooled vitreous silica tube with several gaseous stream introduced. Q_1 (= 3 L min^{-1}) is the axially injected powder gas that serves to introduce the material into the plasma, Q_2 is the intermediate gas that serves to stabilize the plasma and is kept constant at 3 L min^{-1}; it is often introduced with both axial and tangential velocity components. Q_3 is the sheath gas that serves to reduce the heat flux to the walls of the confinement tube, and thus protects it from damage due to overheating. In this example it had been adjusted in such a way that the total flow rate Q_0 is constant at 20 L min^{-1}.

The flow field shows low velocities (10–20 m s^{-1}) compared with a DC plasma arc torch (100–400 m s^{-1}). A recirculation eddy current exists in the coil region, caused by the electromagnetic pinch effect. In order to properly deliver the powder into the torch, it has either to be introduced at high velocity to overcome the back flow, or to be injected in the middle of the induction coil region, below the recirculation eddy current.

Advantages of this method include

- no electrodes;
- any gas possible;
- no upper temperature limit;

- low velocity plasma;
- large diameter plasmas;
- quasi-laminar flow regime, i.e. low Reynolds numbers;
- operation without gas flow because of arc stabilization by thermal pinch in the center of the tube.

Disadvantages include

- high frequencies and voltages that lead to high transmission losses;
- low power densities, and thus lower maximum temperatures (usually below 10 000 K);
- local cooling of plasma at point of powder injection;
- complex geometry of the device that makes fixation necessary.

Technical realization of the concept of induction plasma spraying has been achieved by TEKNA Plasma Systems Inc., Sherbrooke, Québec, Canada. At present the company sells four plasmatron models with different internal diameters ranging from 35 to 100 mm, and different plate powers ranging from 30 to 200 kW (Figure 3.22a). This allows deposition of coatings between 15 and 40 mm width. Figure 3.22b presents a cross-sectional view of an environmental chamber designed to operate a Tekna plasmatron at DLR Stuttgart, Germany.

3.3.1.3 Hybrid Devices

It is feasible to operate the ICP in tandem with either another IPC (Figure 3.23, bottom) or a DC discharge (Figure 3.23, top). The DC plasma forms the center core of the plasma, acting as an ignition source, and the ICP adds additional heat downstream to maintain the temperature level. It also increases the diameter of the plasma, i.e. the volume of the plasma and its energy density so that more powder per unit time can be processed by the device. There are, however, problems with plasma contamination by the material evaporated from the DC plasmatron electrodes. The lengthening of the plasma zone by several hybrid devices is schematically shown in Figure 3.24. Plasma zone lengthening provides the possibility to achieve higher particles residence time in the hot zone, and also adds design flexibility when chemical reaction times have to be increased, for example in reactive plasma processing (Smith and Mutasim, 1992). Successful approaches to hybrid devices were taken by (Yoshida *et al.* 1983) using a DC power of 4–5 kW, an r.f. plate power of 10–30 kW, and an oscillation frequency of 4 MHz, and by (Takeyama and Fukuda 1986) using an r.f.–r.f. two frequency hybrid plasma (pilot r.f. source with 2 kW and 27.12 MHz, and secondary r.f. source with 15 kW at 5 MHz).

3.3.2
Stabilization of Plasma Arcs

Since the power supply of the plasmatron normally has a negative characteristic ($R = dV/dI < 0$), it is inherently unstable. The voltage decreases with increasing

3 The First Energy Transfer Process: Electron–Gas Interactions

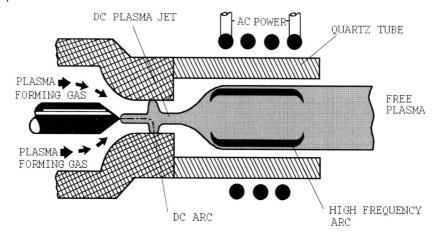

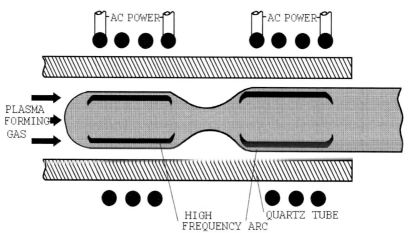

Figure 3.23 Inductively-coupled plasmatron (ICP) operated in tandem with another ICP (bottom) or a DC discharge (top). (Thorpe, 1989).

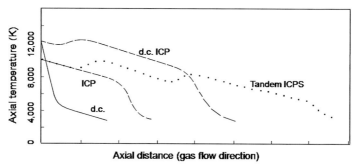

Figure 3.24 Axial temperature variations of DC, ICP, and tandem-operated DC/ICP and ICP/ICP plasmatrons (Thorpe, 1989).

current so that for steady state-conditions a resistance must be included in the circuit. The slope of the load line is given by $dv/di = -R$. The intersections of the characteristics (load line) of the current source and the plasmatron define the two stable working points A and B of the device. The Kaufman stability criterion of a plasmatron is $dv/di + R > 0$ or $R > |dv/di|$. Considering the capacitance and inductance of the arc, the criterion becomes $R < L/c\,|dv/di|$.

Electric stabilization can be achieved in several ways:

- power supply with a rectifier with a saturable core;
- thyristor-controlled silicon rectifier system + inductor to damp high frequency transient current variations leading to erratic movement of the arc root at the anode surface;
- additive stabilization resistance of the electric cables in the water cooled circuits of the torch;
- gas- or liquid-stabilized arc configurations.

3.3.2.1 Wall-stabilized Arcs

Wall-stabilized designs consist of a solid and a hollow electrode, and the plasma gas, instead of constricting the arc, becomes an integral part of it, filling the nozzle from wall to wall. Since for a cylindrical symmetric arc with fixed temperature boundary and therefore predominantly thermal conduction losses the whole discharge is determined for a given current I, the field strength is determined almost entirely by R (thus $E \propto 1/R$), and the axial temperature T_0 increases with the input power per unit length, EI. Since with forced cooling the effective wall radius decreases, for a given current EI and T_0 increase. This explains the apparent anomaly that with enhanced cooling the temperature increases.

3.3.2.2 Convection-stabilized Arcs

In the wall-stabilized arc the discharge fills the major portion of the internal nozzle tube space, and for large nozzle diameters the arc can be wall-stabilized only if very high currents and arc temperatures exist. The large tube constitutes essentially the condition of a free-burning arc that requires no external stabilization at low currents. Stabilization in this case is achieved by convective gas flow. The energy equation consists of a conductive and a convective term:

$$\sigma E^2 = \nabla(-k\nabla T) + \rho \times v_0 \times (C_p \nabla T + \nabla \Phi), \qquad (3.26)$$

where C_p is the specific heat at constant pressure, k is the thermal conductivity of the plasma gas, and v_0 is the volume of the tube space. The convective term makes it possible for arc stabilization to occur without the maintenance of thermal conduction to the wall boundaries. In the outer regions of a cylindrically symmetric arc the energy flow is constant and equals the total energy input, E, but the balance between the conductive and convective terms changes as the radius increases. Ultimately, before the wall boundaries are reached, the convective term dominates the conductive term, and the thermal conduction becomes negligible. The convective gas thus acts effectively as a boundary wall and the arc is convection stabilized. From Equation 3.26 it can be gathered that energy is transported by mass flow only when the flow has a component in the direction of the temperature or gravity gradient. In a *vertical arc* a steady-state is established between the forces of gravity and viscosity, and the gas moving upward flows into the lower arc portion thus cooling this region. When leaving the upper arc region, however, the flow heats the arc since the flow is against the temperature gradient. Thus convection tends to broaden a vertical arc at the top. In the case of a *horizontal arc* a similar process occurs where convection produces arc bowing (Weizel and Rompe, 1949).

In the *vortex-stabilized arc*, the electrodes are made from tungsten, carbon or other materials, and may be water cooled. Gas is fed tangentially into the chamber between the electrodes to produce an intense vortex within the hollow electrode, which usually is the cathode. Thus, the arc is forced to travel from the solid anode out of the nozzle and then strike back to the front face on the hollow cathode. Since the arc channel is embedded in the vortex flow intense convective cooling of the arc fringes occurs. The media producing the vortex can be liquids (water; Gerdien arc) or gases (high field-strength hydrogen arc). Modified arrangements of vortex-stabilized arcs rely on a rotating discharge vessel ('Wälzbogen') or a torus stabilized by a toroidal vortex (Boulos *et al.*, 1989). In some plasma reactors to synthesize ultrafine particles an argon gas shroud is used to convection-stabilize the plasma arc (Chang *et al.*, 1989).

3.3.2.3 Electrode-stabilized Arcs

When the electrode separation is small (1 mm) and the electrode mass is large, it is possible for the arc to be dominated and thus stabilized by the energy losses to the electrodes. In this case the electric field and current distribution may be symmetric about the mid-plane of the electrode system (Weizel and Rompe, 1949). The arc contour is elliptic with the electrodes acting as focal points.

3.3.2.4 Other Stabilization Methods

The *gas-sheath stabilized plasma jet* is between a solid tungsten cathode and a hollow water-cooled copper anode. The arc remains within the nozzle and is prevented from striking the wall by a gas sheath that is much thicker than the arc diameter. Vortex flow is not used, and the position of the arc is determined by the gas flow pattern and the turbulences.

In a *magnetically-stabilized jet*, the arc strikes radially from an inner to an outer electrode. Gas is blown axially through the annulus. The arc is rapidly rotated by a magnetic field, so that its position at various times resembles the spokes of a wheel. With this design, high pressure jets can be created without erosion of the electrodes. Since the external magnetic field interacts with the charged particles in the plasma the arc is magnetically stabilized in a cross-flow: the drag force acting on the arc is effectively compensated by the Lorentz force ($\mathbf{j} \times \mathbf{B} \propto C_D^2$), and the arc actually behaves similar to a solid body with respect of its response to drag. The arc cross-section assumes the shape of an ellipse with its major axis perpendicular to the jet axis. Within this elliptic contour, two symmetrical vortices are formed with opposite directions of rotation. Magnetically-stabilized arcs are extensively used for applications in arc gas heaters including plasmatrons, circuit breakers, arc furnaces, arc welding devices and plasma propulsion systems.

3.3.3
Temperature and Velocity Distributions in a Plasma Jet

3.3.3.1 Turbulent Jets

The complex magneto-hydrodynamic interactions within a plasma jet affect profoundly the macroscopic plasma parameters temperature and velocity. There exist strong temperature and plasma velocity gradients in a plasma arc jet. Figure 3.25 shows the temperature (a) and the axial velocity v_z (b) of a typical DC argon/hydrogen arc along the direction of propagation of the plasma jet, z (Vardelle et al., 1983).

The isotherms in Figure 3.25a show clearly that the maximum temperature exists in the core of the jet close to the nozzle exit ($z = 0$) and the center line ($r = 0$). This very hot region is followed by a transitional region in which the plasma temperature falls rapidly to less than 3000 K at $z = 80$ mm. Eventually there is the fully developed region in which the gas temperature decreases rather gradually by mixing with entrained cold ambient air. The radial temperature profile is extremely steep in the core region with temperature gradients of more than 4000 K over the distance $r = 1$ mm. It is this very steep temperature gradient that makes the problem of complete heating of the powder particles difficult (see Chapter 4) and thus necessitates close control of particle size distribution and injection conditions into the plasma jet. The extremely large radial

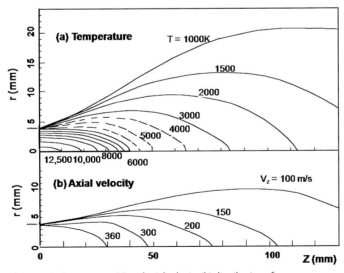

Figure 3.25 Temperature (a) and axial velocity (b) distribution of a typical DC argon/hydrogen plasma arc along the direction of propagation (Vardelle et al., 1983).

and axial temperature gradients require that in order to properly melt them, the injected particles should travel close to the center line. Also, the axial velocity isopleths indicate fast velocity decay away from the point of injection close to the nozzle exit. To measure the velocity of the plasma, small amounts of fine alumina particles (3 μm diameter) were injected as tracers under the assumption that they do not influence the flow conditions and the temperature field. The radial velocity distribution at $z = 5$ mm in an atmospheric argon jet is shown in Figure 3.26 for an arc current of 286 A. Figure 3.27 shows the associated radial temperature distributions at $z = 5$ mm for different currents ranging from 147 to 268 A. The tracer particles were injected at a rate of 0.18 g s^{-1}. Figure 3.28 finally indicates the symmetrical enthalpy distribution in kJ kg^{-1} (Boffa and Pfender, 1968). Temperatures, velocities and enthalpies can be measured with sophisticated probing techniques (for details see 3.4).

The symmetrical distributions shown in Figures 3.26 to 3.28 suggest the picture of a quiet, undisturbed and essentially laminar flow within the plasma column. This, however, is far from reality. As the gas flow increases as required in energy-efficient high temperature plasmatrons, large scale flow structures evolve that are dominated by turbulences and entrained eddies of cold surrounding air. In particular, the large difference in density of the plasma gas and the atmospheric air increases the degree of turbulence.

Figure 3.29 shows the complex structure of a plasma jet approaching transition from a laminar to a fully turbulent flow regime. Following (Pfender *et al.* 1990), the plasma jet exiting the nozzle encounters a steep laminar shear at the jet's outer edge that causes a rolling-up of the flow around the nozzle into a ring vortex. This vortex is pulled downstream by the gas flow, and thus allows the process of ring vortex formation to continue. The rings that subsequently form have the tendency to coalesce forming larger vortices. Perturbations of the latter lead to wave instabilities. Finally, the vortex interactions result in a total breakdown of the vortex structure into large scale eddies and the onset of turbulent flow.[3] Since the entrained eddies of cold air have a density and inertia much higher than that of the high temperature plasma gas, the cool eddies are left behind when the hot plasma gas is accelerated around them. Thus initially little mixing occurs. With increasing distance from the nozzle exit, however, the larger eddies break down into smaller and smaller ones, and mixing and diffusion at the eddy boundaries increases. Eventually the mixing process reaches the center line of the jet, and destroys the plasma core. At this point the jet undergoes the transition to a fully turbulent flow regime. As a result, both the temperature and the velocity of the jet decay. The consequences of this process for the melting behavior of injected powder particles will be explored in Chapter 4.

---

3) This has been humorously described, in verse, by Lewis F. Richardson: 'Big whorls have little whorls/Which feed on their velocity,/And little whorls have lesser whorls/And so on to viscosity'. This is a take-off of the well-known quip by Jonathan Swift to 'define' infinity: 'So, Nat'ralists observe, a Flea/Hath smaller Fleas that on him prey,/And these have smaller Fleas to bite 'em,/And so proceed ad infinitum' (Gleick, 1988).

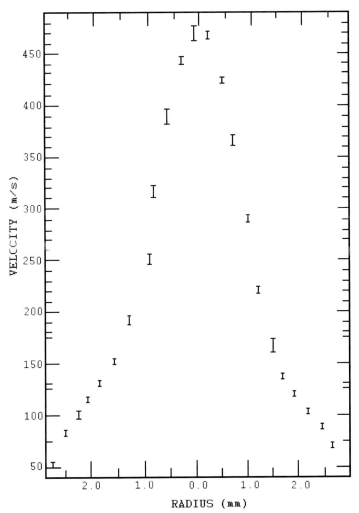

Figure 3.26 Radial velocity distribution at z = 5 mm in an atmospheric argon plasma jet at an arc current of 286 A (Boffa and Pfender, 1968).

3.3.3.2 Quasi-laminar Jets

To modify the turbulent DC plasma jet towards quasi-laminar flow condition a so-called deLaval nozzle is attached to the front of the anode (Figure 3.30). The powder is injected within this nozzle at different ports whose positions depend on the melting temperature of the material to be plasma sprayed. One port is located close to the nozzle root where dense, hot and relatively slow plasma conditions prevail that aid in effective melting of materials with high melting temperatures. The other port is close to the nozzle exit (Henne et al., 1989). Due to the special layout of the inner nozzle contour (Foelsch, 1949) a long quasi-laminar plasma jet is produced that reduces the reaction between spray powder and surrounding residual oxygen

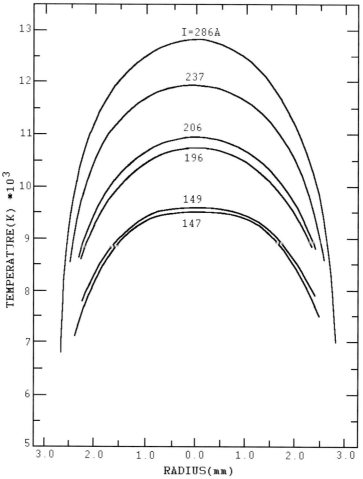

Figure 3.27 Radial temperature distribution in at z = 5 mm in an atmospheric argon plasma jet at various currents (Boffa and Pfender, 1968).

by essentially two mechanisms. First, the residence time of the particles in the jet is strongly reduced (< 1 ms for alumina particles $-22 + 5.6\,\mu m$) since the plasma jet acquires a velocity of >800 m s^{-1} (50 mbar chamber pressure, 600 A, 6 V, Ar/H$_2$ @ 60/5 SLMP) (Henne *et al.*, 1989). Secondly, because of the laminar characteristics of the plasma jet the powder particles are carried close to the central jet axis so that their divergence is low. Consequently, the particles have a reduced chance of reacting with residual oxygen surrounding the jet and hence being entrained by the remaining turbulent eddies. Thus the level of impurities within the deposited coating can be better controlled. This low divergence (>60% of the particles are deposited within a circle of 20 mm diameter at a stand-off distance of 300 mm) leads naturally to an improved deposition efficiency and thus to a more economical operation of the plasmatron.

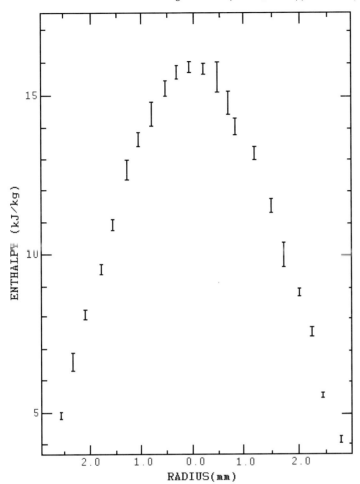

Figure 3.28 Enthalpy distribution at z = 5 mm in an argon plasma jet at 286 A (Boffa and Pfender, 1968).

The somewhat decreased gradients of temperature and velocity, and the reduced turbulence intensity of the jet fringes both for vacuum and atmospheric plasma spraying are coupled with a significant improvement of the deposition efficiency for Mach 2.5 and Mach 3 deLaval nozzles compared with standard cylindrical plasmatron nozzles (Rahmane et al., 1998).

3.4
Plasma Diagnostics: Temperature, Enthalpy and Velocity Measurements

Vardelle et al. (1988) developed a highly sophisticated system to measure, simultaneously, plasma temperature field, plasma velocity field, particle number flux, particle velocity distribution and particle temperature distribution. Figure 3.31 shows the equipment. Before discussing the details of the results of these measurements,

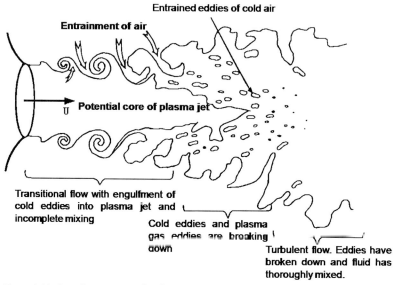

Figure 3.29 Complex structure of a plasma jet approaching transition from laminar to fully turbulent flow regime (Pfender et al., 1990).

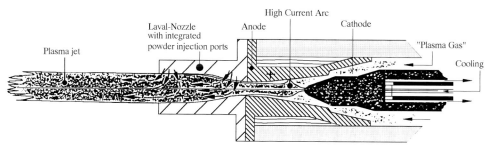

Figure 3.30 Principle of the DLR-VPS plasmatron with attached deLaval nozzle and integrated injection ports (DLR: Deutsche Forschungsgesellschaft für Luft- und Raumfahrt) (Henne et al., 1989).

the physical background of the diagnostic methods will be briefly reviewed. In this chapter, only plasma diagnostics, i.e. plasma temperature and velocity field measurements will be covered. Particle diagnostics will be treated in Chapter 4.5.

3.4.1
Temperature Measurements

3.4.1.1 Spectroscopic Methods
In the plasma jet with temperatures approaching 15 000 K conventional temperature measuring techniques are unsuccessful. Plasma temperatures can be measured

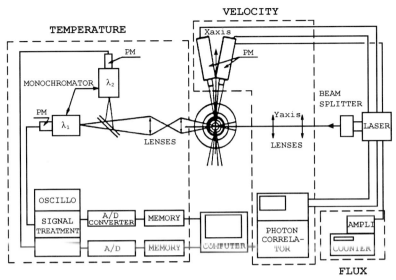

Figure 3.31 Typical layout of a system for measuring simultaneously plasma temperature field, plasma velocity field, particle number flux, particle velocity distribution, and particle temperature distribution (Vardelle et al., 1988).

assuming local thermodynamic equilibrium (LTE) using atomic or molecular spectroscopy. These methods have the advantage of high spatial resolution but can be applied only if the temperatures exceed about 6000 K. The plasma temperature field can be approximately deduced from the volumetric emission coefficient $\varepsilon_L(r)$ of suitable atomic emission lines, such as NI (746.8 nm) for a nitrogen plasma, and ArI (763.5 nm) or ArII (480.6 nm) for an argon plasma. The local emissivity ε_L is obtained from the observed side-on intensity $I_L(x)$ (Equation 3.27) by Abel's inversion (Equation 3.28) according to (Boulos et al. 1989):

$$I_L(x) = 2\int \varepsilon_L(r)dy = 2\int \varepsilon_L(r)rdr/(r^2 - x^2)^{1/2} \qquad (3.27)$$

$$\varepsilon_L(r) = -(1/\pi)\int [I_L(x)/(x^2 - r^2)^{1/2}]dx. \qquad (3.28)$$

Figure 3.32 shows the geometrical relationships of a plasma volume element of dimensions dx and $dy = rdr/(r^2 - x^2)^{1/2}$.

If oxygen is present in the plasma, the temperature can be measured using the rotational states of oxygen molecules and measuring the intensity of the rotation bands (Schumann–Runge bands) in the 336–351 nm range.

Using monochromators equipped with holographic gratings, and x–y stepping-motors to control the displacement of the plasma jet the precision of the measurement is within 5% (Vardelle et al., 1982). Using the temperature determined from the

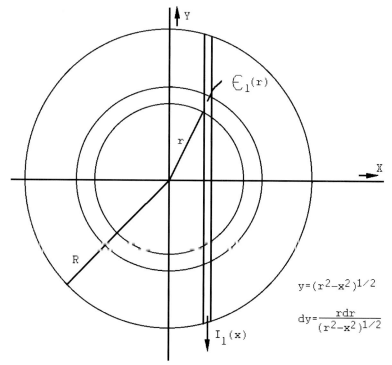

Figure 3.32 Geometric relationship of a plasma volume element used to measure local plasma emissivities ε_l by Abelian inversion (Boulos et al., 1989).

local (relative) emission coefficient, the absolute emission coefficient $\varepsilon_{L,abs}$ can be obtained as

$$\varepsilon_{L,\,abs} \propto \exp(-E_s/kT)h\nu. \tag{3.29}$$

From the ratio of the absolute emissivities of two spectral lines, $\varepsilon_{L1}/\varepsilon_{L2} \propto \exp[-(E_{s1} - E_{s2})kT]$, the relative error of the temperature measurement can be estimated as

$$\Delta T/T = \Delta(E_{s1} - E_{s2})/E_{s1} - E_{s2}. \tag{3.30}$$

With temperature measurements of this type the temperature distribution shown in Figure 3.25 was measured (Vardelle et al., 1983).

3.4.1.2 Two-wavelengths Pyrometry

The temperature of a plasma can also be measured by high-speed two-color pyrometry at two wavelengths (Vardelle et al., 1988; Mishin et al., 1987; Fincke et al., 2001). While the spectroscopic methods to measure the plasma temperature can be performed in an 'empty' plasma, i.e. without the presence of radiating particles, the two-color (wavelength) pyrometry requires seeding particles such as alumina particles of small diameter (3 μm). Figure 3.31 shows the equipment used by (Vardelle *et al.* 1988) to measure plasma temperatures. The light scattered by the moving particles was observed off-axis by two photomultipliers directed at angles of 81 and 99° to the optical axis of the laser beam. Measurements were made of the in-flight particle emission at two wavelengths, 680 and 837 nm. A monochromator placed in front of each of the photomultipliers was used to define the wavelength of the particle emission by filtering out the background plasma radiation. It is crucial to select wavelengths that are unaffected by neighboring gas emission lines and also to assume that the emissivity of the particles does not change with the wavelength, i.e. that conditions of a grey body radiation prevail. In this case, the color temperatures of the particles correspond to their true surface temperature, and the assumption is thus satisfied that the tracer particle temperature reflects faithfully the plasma temperature. By calibration using a tungsten lamp the statistical information provided by the particle emission at the two selected wavelengths can be converted into statistical information about the particle surface, and thus the plasma temperature distribution. It should be emphasized that these assumptions are only satisfied under low-loading conditions, i.e. if local cooling of the plasma due to the presence of particles can be neglected (see Section 4.4.1).

Under other simplifying assumption, for example uniform emissivity of the ideally spherical particles over their entire surface area and constant surface temperature along the entire path of observation, the particle temperature can be calculated by

$$T_P = \{C(\lambda_1 - \lambda_2)/\lambda_1\lambda_2\}/[\ln(I_{\lambda 1}/I_{\lambda 2}) + 5\ln(\lambda_1/\lambda_2)], \tag{3.31}$$

where λ_i denote the wavelengths, $I_{\lambda,i}$ the measured spectral intensities, and C a constant. While this technique shows a vast improvement over the spectroscopic methods based on line emissions of gas atoms, there are three principal limitations (Vardelle *et al.*, 1988). The first relates to the presence of background plasma radiation, in particular close to the core of the plasma jet. Therefore no measurements can be made on particles whose trajectories are close to the central axis of the jet. The second limitation results from the rapid decrease of the emission intensity with decreasing surface temperature. As a consequence the lower temperature limit for successful measurements is around 1800 K. Finally, the third limitation is due to the long depth of field that requires extremely low particle loading conditions in order to avoid saturating the photoelectric system.

3.4.1.3 Chromatic Monitoring

Measurement of particle temperatures by noncontact techniques such as emission spectroscopy (Section 3.4.1.1) are not ideal for on-line monitoring since each spectrum will need to contain typically 10^6 data points along each line of sight within each time interval. The technique of chromatic monitoring (Russell and Jones, 2001) allows cross-correlation of data acquired from different parts of the optical spectrum. The plasma conditions can thus be determined quantitatively using only three coordinates on a chromatic map. The data acquisition involves capturing the optical emission from the plasma with three broadband detectors with non-orthogonal wavelength responsivities (Russell *et al.*, 2003). The information obtained can be compressed within only three parameters R, G and B known from color science. From their raw values chromatic maps can be produced using several complex algorithms. The results are expressed in terms of the *HLS* scheme (Jones *et al.*, 2000) where *H* represents the dominant wavelength, *S* the nominal wavelength bandwidth and *L* the effective signal strength. These parameters are presented on chromatic maps in the form of *H–S* and *H–L* polar diagrams with *H* as the azimuthal angle. In this way it is possible to distinguish the emission from the air plasma jet from that of the plasma-heated particles since their signals plot at different locations on a chromatic *H–S* map. These results suggest that it may be possible to track the heating of particles within the plasma jet closer to the plasmatron nozzle. Moreover, since interaction of particles and plasma jet occurs on time scales typically between 0.1 and 1 ms it should be possible to monitor such an interaction on-line in real time with chromatic detectors of time resolution <0.1 ms.

3.4.2
Velocity Measurements

The plasma velocity can be measured with either a Pitot tube probe technique (Grey *et al.*, 1962; Cox and Weinberg, 1971) or with a time-of-flight two-point laser Doppler anemometer (LDA; Gouesbet, 1985; Drain, 1980) or a CCD detector (Fang *et al.*, 2007). The first method relies on rather stringent simplifying assumptions made about the local Reynolds numbers, plasma densities and plasma viscosities, the second one is limited by laser beam diffraction, statistical bias towards high velocity particles, inhomogeneous seeding across zones of steep velocity and temperature gradients, and slip between the particle and the plasma velocities which, however, can be reduced by reduced particle size. Despite these limitations both methods are frequently used to estimate plasma jet velocities in order to optimize plasma spray conditions and coating properties. In addition enthalpy probe measurements in thermal plasma jets can be used to obtain temperature distributions from with important aspects of deposition efficiency, and density and microstructure of coatings can be derived (Steffens and Duda, 2000).

3.4.2.1 Enthalpy Probe and Pitot Tube Techniques
A water-cooled so-called enthalpy or Grey probe is immersed in a plasma jet to extract a sample of gas. In order to minimize the disturbance of the plasma jet the dimensions of the probe should be as small as possible. Today probes with a diameter as small as 1.5 mm can be used. A tare measurement is required to account for the external heat transfer to the probe (Capetti and Pfender, 1989). This is accomplished by interrupting the gas stream (no-flow regime) through the Grey probe and measuring the increase of the temperature of the cooling water due to heat transfer from the surrounding plasma. The 'tare' measurement yields

$$\Delta Q_{nf} = (dm_c/dt)c\Delta T_{nf}, \tag{3.32}$$

with ΔQ_{nf} = difference in heat content at no-flow condition, dm_c/dt = mass flow rate of the coolant, c = specific heat, and ΔT_{nf} = temperature rise at no-flow conditions. After opening the sampling valve of the probe, the amount of heat acquired by the probe increases to

$$\Delta Q = (dm_g/dt)(h_1 - h_2) + \Delta Q_{nf} = (dm_c/dt)_c \Delta T, \tag{3.33}$$

with dm_g/dt = mass flow rate of the plasma gas, h_1 = enthalpy of the gas at the inlet temperature, and h_2 = enthalpy of the gas at standard conditions (RT). From Equation 3.33 it follows the expression for the stagnation enthalpy:

$$h_1 = h_2 + [\{dm_c/dt\,c\}/(dm_g/dt)](\Delta T - \Delta T_{nf}). \tag{3.34}$$

The temperatures are measured with thermocouples, the cooling water mass flow rate with a rotameter or a similar device, and the gas mass flow rate with either a sonic orifice or a mass flow meter. In the case of a sonic orifice the mass flow rate of the gas is, according to Capetti and Pfender (1989)

$$dm_g/dt = C_A(p_0/\sqrt{T_0})[(M/R)\gamma(2/\gamma+1)^{(\gamma+1)/(\gamma-1)}]^{1/2}, \tag{3.35}$$

where C_A = effective cross-sectional area of the orifice, p_0 and T_0 = pressure and temperature immediately upstream of the orifice, and γ = adiabatic exponent ($=c_p/c_v$).

The Grey probe cannot be operated continuously but can be used during the no-flow regime as a Pitot probe to measure the stagnation pressure, p_{st} and, according to Bernoulli's theorem, the free-stream velocity of the plasma jet:

$$v = [2(p_{st} - p_{atm})/\rho]^{1/2}, \tag{3.36}$$

where the plasma gas density ρ must be determined from thermodynamic tables using the plasma temperature determined by the methods described in Section 3.4.1. In more detail, for an argon plasma the gas density ρ can be approximated by $\rho = X_{Ar}\rho_{Ar} + X_{Air}\rho_{Air}$. X_{Air} can be determined with a lambda-type oxygen sensor probe based on stabilized zirconia, and ρ_{Ar} and ρ_{Air} from thermodynamic tables. As a check for the validity of the plasma velocity measurement the total mass flow rate of argon can be used according to Boulos et al. (1989) with

$$dm_{Ar}/dt = 2\pi \int_0^{r_0} (X_{Ar}\rho_{Ar}v)r\,dr \tag{3.37}$$

integrated between the boundaries 0 and r_0 that define the radius of the plasma jet. The energy flux through a given cross-section of such a jet can be expressed by

$$Q = 2\pi \int_0^{r_0} \rho(r)h(r)v(r)r\,dr, \tag{3.38}$$

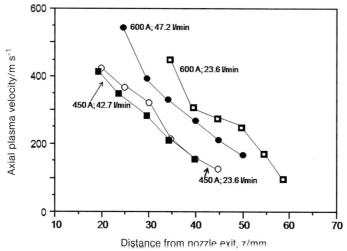

Figure 3.33 Axial velocity decay of an argon plasma jet measured by an enthalpy probe (Capetti and Pfender, 1989).

where $\rho(r)$, $h(r)$ and $v(r)$ are, respectively the radial density, enthalpy and velocity profiles of the plasma jet within the boundaries 0 to r_0. Figure 3.33 shows the decay of the axial plasma velocity for the currents of 450 and 600 A, and various argon flow rates between 23.6 and 47.2 l min^{-1}. For arc currents around 450 A the change in flow rate has only a small effect on the plasma velocity and its axial decay. An explanation for this may be found by assuming two opposing effects that affect the flow velocity with increasing argon flow rate. Increasing gas flow rates require an increase of the plasma velocity in order to satisfy the continuity equation of mass. This can be achieved only if the plasma temperature remains constant. Higher gas flow rates, however, reduce the temperature thus leading to higher plasma gas densities, and consequently to lower plasma velocities at constant cross-section of the plasma jet. These two opposing phenomena appear to compensate each other at lower arc currents. At higher currents the length of the jet varies considerably with varying gas flow rates, and compensation cannot be achieved anymore. For details on the measurement techniques and typical results the reader is referred to the paper by Capetti and Pfender (1989).

3.4.2.2 Laser Doppler Anemometry (LDA)

One of the standard plasma velocity measurement techniques, LDA has emerged as a powerful method of plasma diagnostics. Since it is an optical method it does not interfere with the flow of the plasma. However, the plasma jet must be seeded with very fine powder particles, usually alumina or zirconia (Lesinski and Boulos, 1988a), carbon or silicon (Lesinski and Boulos, 1988b) or boron nitride (Desai et al., 1968). Also, a laser pulse of 15 ns duration that produced a plasma drop of high intensity has been used as a tracer (Chen, 1966). The LDA technique can be used in two modes: *fringe and two-point measurement modes*. While the former is mostly used for velocity measurements of larger powder particles in the plasma jet (see 4.5.1) the latter is used to measure the supposedly undisturbed plasma velocity via the observation of very small tracer particles (Richards, 1977). It is essential that the particles must be small enough not to interfere with the plasma flow but large enough not to evaporate during their residence time in the hot plasma jet. Also, the method of seeding is of importance. Adding very fine powder through the powder injection port of a DC plasmatron leads to incomplete particle distribution and failure of the particles to penetrate the hot core of the jet. This can be avoided by passing a part of the plasma gas through a fluidized bed consisting of a mixture of very fine and coarse powders (Lesinski and Boulos, 1988a). The passing gas extracts selectively the fines and carries them over into the jet.

In *two-point LDA* measurements (Vardelle *et al.*, 1983) the laser beam after passing through a beam splitter is focused using a lens of a focal length of 16 mm to form two coherent light beams of 50 μm diameter and 400 μm apart. Those two beams are projected in the plasma jet by a second lens of a focal length of 50 mm (Figure 3.34). Furthermore each beam is focused onto the entrance slit of two photomultipliers equipped with a narrow-band interference filter (514.5 nm, $\Delta\lambda = 0.25$ nm). By cross-correlating the electrical output of the photomultipliers via a photon correlator, the particle (= plasma) velocity can be determined. When a particle crosses the two beams in the cylindrical measuring volume of 0.6 by 1.1 mm, it produces two successive light bursts. Their time delay can be used to calculate the velocity.

In the *fringe mode*-measurements the two laser beams produced by the beam splitter are focused and allowed to interfere thus forming an interference fringe pattern in the flow region. The inter-fringe spacing is

$$\delta = \lambda/[2\sin(\varphi/2)], \tag{3.39}$$

where λ = wavelength of the incident laser beam and φ = angle of interception of the two laser beams.

When a particle traverses the point of intersection of the two beams a light burst is produced, modulated at a frequency that is a function of its velocity and the fringe spacing δ. The signal output of the photomultiplier is used to determine the time

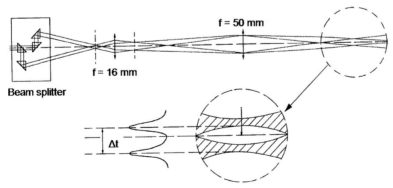

Figure 3.34 Experimental setup of a two-point laser Doppler anemometer (LDA) (after Vardelle *et al.*, 1983).

required for the particle to pass a given number of fringes. After acquisition of a sufficient number of measurements, statistical analysis is carried out resulting in the minimum velocity, maximum velocity, mean velocity and standard deviation of the particle velocities. From this the probability distribution function is obtained by dividing the velocity window into a number of segments and then counting the number of particles falling into each segment. Eventually the normalized probability distribution function can be calculated as

$$Pr(v_j) = (m_j/v_j)/\Sigma(m_j/v_j), \qquad (3.40)$$

where m_j = number of particles with a velocity between v_j and v_{j+1}.

Figure 3.35 shows the profiles of the axial plasma velocity (bottom) and the intensity of turbulence in axial and radial directions (top) along the centerline of an Ar/N_2 plasma jet (80 vol% Ar, $dm_g/dt = 23.6\,l\,min^{-1}$, I = 400 A, U = 38 V, E = 15.2 kW) (Lesinski and Boulos, 1988a). The centerline plasma velocity is 250 m s^{-1} and shows a rapid drop with increasing distance z from the nozzle. Contrary to this, the intensity of turbulence, defined as v'/v, i.e. the ratio of the standard deviation of the velocity to the mean velocity (Figure 3.35, top) first increases with increasing distance from the nozzle to about 40% at z = 40 mm, and then decreases gradually. This is consistent with other findings (Vardelle et al., 1983) of the formation of vortices immediately at the exit point of the plasma from the nozzle. This fact has let to approaches to reduce the turbulences in a plasma jet by appropriately formed nozzles such as axially symmetric deLaval nozzles that produce parallel and uniform flow, and thus prevent entrainment of cold surrounding gas (Foelsch, 1949) (see 3.3.2.2).

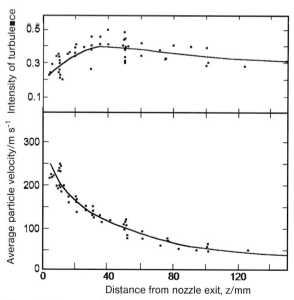

Figure 3.35 Intensity of turbulence (top) and plasma and particle velocities (bottom) along the centerline of an argon/nitrogen (80 vol% argon) plasma jet (Lesinski and Boulos, 1988a).

3.4.2.3 Other Methods

A different approach to velocity measurements requiring less sophisticated equipment is the *cross-correlation method* (Kloke, 1995). It is based on the measurement of the travel time t of powder particles between two observation points in the particle flow direction separated by the distance s. The light emitted from the two observation points will be recorded by two photodiodes. To suppress the plasma light the electrical signals will be filtered through a band pass filter. Figure 3.36 shows two typical signals 1 and 2. Shifting the signals $s_1(t)$ and $s_2(t)$ against each other by a variable time difference τ yields a maximum similarity for $\tau = \tau_{max} = t$. A quantitative expression of this similarity is the cross-correlation function CCF(τ):

$$\text{CCF}(\tau) = \varphi_{s(1),s(2)}(\tau) = \int s_1(t) \times s_2(t+\tau) dt. \tag{3.41}$$

The position of the maximum of the CCF (Figure 3.36, bottom) finally yields the velocity of the streaming plasma, $v = s/t$.

Kuroda et al. (1989) developed a technique to measure particle velocities *in situ* using a spatial filter without the need of complex optical settings. The technique

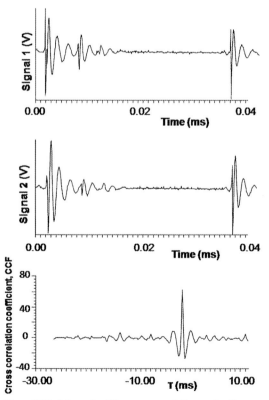

Figure 3.36 Schematic of the cross-correlation method to measure particle velocities (Kloke, 1995).

permits the average particle velocity to be measured at dense particle loading. The spatial filter consists of five parallel flat plates at equal intervals, d. A parallel beam passing through the filter is focused into an optical fiber. The filter can be moved horizontally and vertically through a sliding table. When a particle, radiating light, traverses in front of the filter assembly, four peaks of light intensity are detected by a photomultiplier placed at the end of the optical fiber. Analysis of the time interval of the peaks, t yields the particle velocity according to v = d/t, where d = known interval of the slits. Also, analysis of the power spectrum of the radiated light by an FFT analyzer allows particle velocities to be calculated by the relationship f = v/d, where f = peak frequency.

A technically simple imaging system for individual, in-flight particle temperature and velocity measurements for plasma and other thermal spray processes was introduced by Vattulainen *et al.* (2001). A custom double dichroic mirror is used to add spectral resolving capability to a single, black-and-white, fast-shutter digital charge coupled device (CCD) camera. The spectral double images produced by the individual in flight particles are processed using specialized image processing algorithms. Particle temperature determination is based on two-color pyrometry, and particle velocities are measured from the length of the particle traces during known exposure times. Dividing the imaged area into smaller sections, spatial distributions of particle temperature, velocity, and number of detected particles can be studied.

References

K.H. Becker, U. Kogelschatz, K.H. Schoembach, R.J. Barker, *Non-equilibrium Air Plasmas at Atmospheric Pressure*. CRC Press. Boca Raton, USA. ISBN 0-750-30962-8. 2004.

C.V. Boffa, E. Pfender, *Enthalpy Probe and Spectrometric Studies in Argon Plasma Jets*, HTL Tech. Rep. No.73, Univ. of Minnesota, Minneapolis, 1968.

M.I. Boulos, *Pure & Appl. Chem.*, 1985, **57** (9), 1321.

M.I. Boulos, P. Fauchais, E. Pfender, *Fundamentals of Materials Processing using Thermal Plasma Technology*. Canadian University-Industry Council on Advanced Ceramic (CUICAC) Short Course, Edmonton, Alberta, Canada, October 17–18, 1989.

I. Burlacov, J. Jirkovský, M. Müller, R.B. Heimann, *Surf. Coat. Technol.*, 2006, **201**, 255.

F. Cap, *Einführung in die Plasmaphysik I. Theoretische Grundlagen*. Vieweg + Sohn: Wiesbaden, 1984.

A. Capetti, E. Pfender, *Plasma Chem. Plasma Process.*, 1989, **9**, 329.

Y. Chang, R.M. Young, E. Pfender, *Plasma Chem. Plasma Process.*, 1989, **9**, 272.

C.J. Chen, *J. Appl. Phys.*, 1966, **37**, 3092.

J.B. Cox, F.J. Weinberg, *J. Phys. D Appl. Phys.*, 1971, **4**, 877.

S.V. Desai, E.S. Daniel, W.H. Corcoran, *Rev. Sci. Instr.*, 1968, **39**, 612.

J.-E. Döring, R. Vaßen, G. Pintsuk, D. Stöver, *Fusion Eng. Design*, 2003, **66–68**, 259.

L.E. Drain, *The Laser Doppler Technique*, Wiley: New York, 1980.

K.S. Drellishak, *Partition Functions and Thermodynamic Properties of High Temperature Gases*. Ph.D. Thesis, Northwestern University, Evanstown, Ill., 1963.

H.U. Eckert, *High Temp. Sci.*, 1974, **6**, 99.

H. Edels, in: Proc. Inst. Elec. Eng., Paper No. 3498, 1961, 108A, 55.

W. Elenbaas, *Physica*, 1934, **1**, 673.

J.C. Fang, W.J. Xu, Z.Y. Zhao, H.P. Zeng, *Surf. Coat. Technol.*, 2007, **201**, 5671.

J.R. Fincke, D.C. Haggard, W.D. Swank, *J. Thermal Spray Technol.*, 2001, **10** (2), 255.

H.K. Foelsch, *J. Aeronaut. Sci.*, 1949, March, 161.

G. Frind, P.A. Siemers, S.F. Rutkowski, US Patent 4,902,870 (Feb 1990).

H. Gerdien, A. Lotz, *Wiss. Veröff. Siemens Werke*, 1922, **2**, 489.

J. Gleick, *Chaos. Making a New Science*, Penguin Books, 1988.

R.J. Goldston, P.H. Rutherford, *Introduction to Plasma Physics*. Inst. of Physics: Bristol, 1997.

G. Gouesbet, *Plasma Chem. Plasma Process.*, 1985, **5**, 91.

H.J. Grey, P.F. Jacob, M.P. Sherman, *Rev. Sci. Instr.*, 1962, **33** (7), 738.

A.E. Guile, *IEEE Review*, 1971, **118**, 1131.

R. Henne, M.v. Bradke, G. Schiller, W. Schurnberger, W. Weber, in: Proc.12th ITSC, London, June 4–9, 1989, P112, 175.

G.R. Jones, P.C. Russell, A. Vourdas, J. Cosgrave, L. Stergioulas, R. Haber, *Measur. Sci. Technol.*, 2000, **11**, 489.

J. Kitamura, S. Usuba, Y. Kakudate, H. Yokoi, K. Yamamoto, A. Tanaka, S. Fujiwara, *J. Thermal Spray Technol.*, 2003, **12** (1), 70.

M. Kloke, in: Proc. 3rd Workshop 'Plasma Technology', TU Ilmenau, Germany, June 22–23, 1995.

Y. Kubo, S. Maezono, K. Ogura, T. Iwao, S. Tobe, T. Inaba, *Surf. Coat. Technol.* 2005, **200**, 1168.

S. Kuroda, T. Fukushima, S. Kitahara, H. Fujimori, Y. Tomita, T. Horiuchi, in: Proc.12th ITSC, London, June 4–9, 1989, P27–1.

I. Langmuir, M. Mott-Smith, *Phys. Rev.* 1926, **28**, 727.

J. Lesinski, M.I. Boulos, *Plasma Chem. Plasma Process.*, 1988a, **8**, 113.

J. Lesinski, M.I. Boulos, *Plasma Chem. Plasma Process.*, 1988b, **8**, 133.

J. Mishin, M. Vardelle, J. Lesinski, P. Fauchais, *J. Phys. E., Sci. Instr.*, 1987, 20.

J. Mostaghimi, P. Proulx, M.I. Boulos, *Plasma Chem. Plasma Process.*, 1984, **4**, 129.

E. Mühlberger, US Patent 3,839,618 (Oct 1974); US Patent 4,877,640 (Oct **1989**).

M. Müller, M. Frieß, German Patent DE0510 200422358B3 (2005).

E. Pfender, W.L.T. Chan, R. Spores, in: Proc.3rd NTSC, Long Beach, CA, May 20–25, 1990, 1.

M. Rahmane, G. Soucy, M.I. Boulos, R. Henne, *J. Thermal Spray Technol.*, 1998, **7** (3), 349.

T.B. Reed, *J. Appl. Phys.*, 1961, **32**, 821.

B. Richards (ed.), *Measurements of Unsteady Fluid Dynamic Phenomena*, McGraw Hill, 1977.

P.C. Russell, G.R. Jones, *Vacuum*, 2001, **58**, 88.

P.C. Russell, B.E. Djakov, R. Enikov, D.H. Oliver, Y. Wen, G.R. Jones, *Sensor Review*, 2003, **25** (1), 60.

A. Rutscher, *Plasmatechnik, Grundlagen und Anwendungen. Eine Einführung*. Carl Hanser-Verlag: München, Wien, 1982.

N. Sanders, K.C. Etermadi, E. Pfender, *J. Appl. Phys.* 1982, **53** (6), 4136.

R.W. Smith, Z.Z. Mutasim, *J. Thermal Spray Technology*, 1992, **1** (1), 57.

P.R. Smy, *Adv. in Physics*, 1976, **25** (5), 517.

O.P. Solonenko, *Thermal Plasma Torches and Technologies*, Vol.1 Cambridge Int. Sci. Publ.: Cambridge. 1-898-32659-2. 2000.

H.D. Steffens, T. Duda, *J. Thermal Spray Technol.*, 2000, **9** (2), 235.

K. Takeda, S. Takeuchi, *Mater. Trans. (JIM)*, 1997, **138** (7), 636.

Takeyama, T. Fukuda, K.*Development of an All-Solid State r.f. Thermal Plasma Reactor*, National Chemical Laboratory for Industry, Internal Report, 1986, 21 (4).

M.L. Thorpe, *Chem. Eng. Progress*, July 1989, 43.

S. Usuba, R.B. Heimann, *J. Thermal Spray Technol.*, 2006, **15** (3), 356.

A. Vardelle, M. Vardelle, P. Fauchais, *Plasma Chem. Plasma Process.*, 1982, **3**, 255.

M. Vardelle, A. Vardelle, P. Fauchais, M.I. Boulos, *AIChE Journal*, 1983, **29**, 236.

M. Vardelle, A. Vardelle, P. Fauchais, M.I. Boulos, *AIChE Journal*, 1988, **34**, 567.

J. Vattulainen, E. Hämäläinen, R. Hernberg, P. Vuoristo, T. Mäntylä, *J. Thermal Spray Technol.*, 2001, **10** (1), 94.

M. Wehrli, *Helvet. Phys. Acta*, 1928, **1**, 323.

W. Weizel, R. Rompe, *Theorie elektrischer Lichtbogen und Funken*, Barth, Leipzig, 1949.

T. Yoshida, T. Tani, H. Nishimura, K. Akashi, *J. Appl. Phys.*, 1983, **54**, 640.

4
The Second Energy Transfer Process: Plasma–Particle Interactions

4.1
Injection of Powders

There are several ways to inject powder particles into the plasma jet. Figure 4.1 shows injection perpendicular to the jet at the point of exit of the jet from the anode nozzle (position 1), 'upstream' or 'downstream' injection at an angle to the jet axis (position 2), or injection directly into the nozzle (position 3). Upstream injection is used when increased residence time of the particles in the jet is required, for example to melt high refractory materials with high melting points such as zirconia or hafnia. Downstream injection protects the powder material from excessive vaporization or thermal decomposition such as hydroxyapatite. Experiments with molybdenum and alumina coatings have shown that the cohesion of coatings and their tensile strengths increase with oblique injection (Hasui *et al.*, 1970).

Also, experiments are conducted with coaxial powder injection where the powder enters through a bore in the cathode, and is swept along the total length of the plasma jet for maximum residence time (position 4). This configuration, however, tends to disturb the smooth plasma flow and thus creates turbulences and eddies that may damage the nozzle walls. Problems may also occur with clogging the plasma paths with excess powder. A somewhat different approach was taken by Maruo *et al.* (1988) by feeding powder axially into the space between three thoriated tungsten cathodes set at 120° around the plasmatron axis. As three individual anode spots were formed, the thermal and erosion-damage of the anode nozzle was diminished.

Research work performed at the University of British Columbia in the early 1990s led to the successful commercialization of a three-electrode plasmatron (Axial III) with central powder injection by Northwest Mettech Corp., North Vancouver, B.C., Canada (Moreau *et al.*, 1995). In 1997, the 600 Series Axial III Plasma Torch was developed to withstand the harsh environment of nitrogen/hydrogen plasmas that have many beneficial properties, including high deposition efficiencies and feed rates that yield dense, hard coatings when ceramic materials are sprayed. The 600 series torch incorporates a new electrode set that exhibits little wear under the harshest plasma conditions at power levels up to 150 kW. While process parameters have been developed for a variety of metal, ceramic,

Plasma Spray Coating: Principles and Applications. Robert B. Heimann
Copyright © 2008 WILEY-VCH Verlag GmbH & Co. KGaA, Weinheim
ISBN: 978-3-527-32050-9

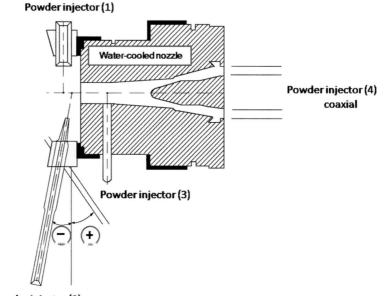

Figure 4.1 Powder injection options (for details see text).

abradable and carbide powders this torch is used in a variety of production applications (Matthäus, 2003; Fauchais *et al.*, 2006).

4.2
Characteristics of Feed Materials

Feed materials can be used in form of solid wires, rods, filled wires and powders (Table 4.1). One of the main materials requirements is that melting must occur without decomposition or sublimation.

4.2.1
Solid Wires, Rods and Filled Wires

Solid wires and rods as well as filled wires are mostly used in flame and electric arc spraying operations. Steel and nonferrous metals and alloys can be shaped into wires. A special technique (Bifilar system, OSU Oberflächentechnik KG, Bochum, Germany) allows the simultaneous spraying of two wires of different compositions to form a pseudo-alloy coating (Elvers *et al.*, 1990). Rods of metal oxide compositions are manufactured by the Rokide process by Norton Industrial Ceramics, Worcester, USA. Filled wires are composed of hollow steel or nickel tubes filled with powder of tungsten, chromium carbide or compositions of the Fe–B system (Borisov *et al.*, 1995). Also, an outer tube of polymeric material filled with metal oxide or

Table 4.1 Spray coating materials and their standards (after Elvers et al., 1990).

Parameter	Solid wires	Solid wires	Rods	Filled plastic tubes	Filled wires	Powders
German standard DIN	8566 Part 1	8566 Part 2	8566 Part 3	8566 Part 3		32529 65097
Spraying process	Flame spraying	Electric arc spraying	Flame spraying	Flame spraying	Electric arc spraying	Flame and high energy spaying
Diameter, mm	1.5–4.16	1.6–3.2	3.15, 4.75, 6.3	1.5–6.3	2 and 3.2	5–180[a]
Length, mm		450 and 600				
Main material	Steels, nonferrous metals and alloys, Mo		oxides	Metals, oxides	Fe-Cr-C, carbides	Complete range

[a]Powder particle size in µm

metal powder can be used (Sfecord, Société de fabrication d'elements catalytiques, France). For additional information on design, manufacturing, properties and applications of wire and cored wire feedstock see, for example Kaiser et al. (1994), Witherspoon et al. (1999), Wilden et al. (2000), Sacriste et al. (2001).

4.2.2
Powders

Powder particle diameters are generally between 5 and 200 µm with a preferred range of +20–100 µm, and a median diameter of 50 µm for APS applications. Many materials can be made into powders by a variety of techniques to yield spray powders with a narrow grain size distribution, good flowability and optimum deposition efficiency for optimized spray parameter ranges. Powders used for flame spraying must have melting points not exceeding 2500 K because of the relatively low temperatures in the flame. Special conditions apply for HVOF and detonation gun techniques. Powders designed for plasma spraying operations must be able to withstand the high mechanical and thermal stresses exerted on them by the hot, high speed plasma jet. Thus powders manufactured by agglomeration of finer particles must be thoroughly sintered to guarantee mechanical stability. Very fine reactive plasma spray powders with an extremely narrow grain size distribution range for vacuum spraying (LPPS) operations demand the highest quality specifications and thus very strict quality control procedures.

To produce spray powders, precursor materials must be melted or sintered with subsequent size reduction by crushing, grinding and attrition milling. Mixing of powders and classification are also important process steps. Specialized powders for a variety of industrial applications are produced by spray drying, fluidized bed

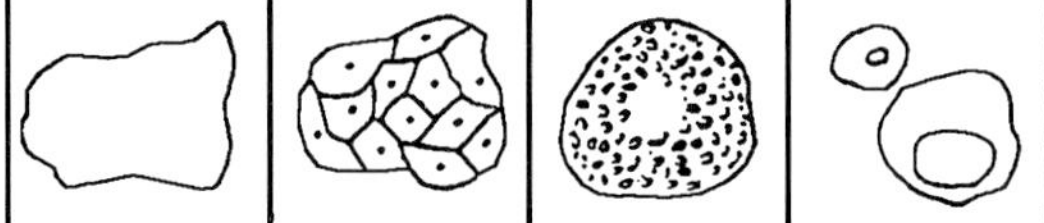

Powder type; manufacture	fused, milled	sintered, milled	agglomerated	spheroidized	atomized
shape	blocky-angular	blocky-angular	spherical	spherical	spherical-irregular
porosity	dense	dense-porous	porous	dense-hollow	porous-hollow
crystal size	coarse-fine	coarse-fine	medium-fine	medium-fine	fine
homogeneity	alloyed	alloyed	alloyed (heterogeneous)	alloyed	alloyed

Figure 4.2 Production specific powder characteristics of different spray powder types.

sintering, agglomeration, fusing/melting, plasma spheroidizing, atomizing, surface coatings and sol-gel processes. Figure 4.2 shows schematically different powder types, particle shapes and microstructures and Figure 4.3 relates the particle porosity to the structural fineness, i.e. the crystal size. Electron micrographs shown in

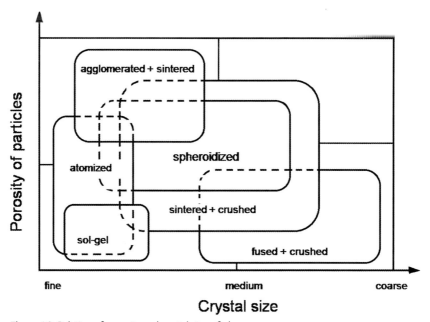

Figure 4.3 Relation of porosity and crystal size of plasma spray powder particles as a function of the production method.

4.2 Characteristics of Feed Materials

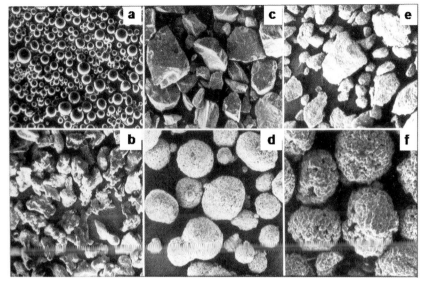

Figure 4.4 Powder morphology dependent on the production method: gas- and water-atomized (a,b) and crushed (c), clad (d), sintered (e) and agglomerated (f) powders (Kubel, 1990).

Figures 4.4 and 4.5 demonstrate the wide variability of thermal spray powders ranging from gas- and water-atomized (a, b) to fused and crushed (c) to clad (d) to sintered (e) to agglomerated (f) powders (Kubel, 1990).

For nonstandard applications special powders must be designed and thoroughly characterized in terms of chemical composition, microstructure, physical properties, morphology and spraying behavior by a wide spectrum of methods. Development of new spray powders is an extremely involved task that must be controlled at all process stages by established methods of total quality management (TQM) including statistical designs of experiments (SDE), statistical process control (SPC) and quality function deployment (QFD) (see Chapter 11). Important morphological

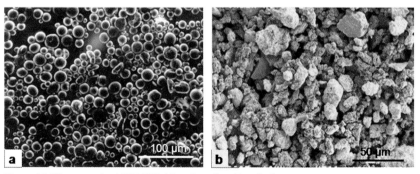

Figure 4.5 Water-atomized 85Fe15Si (a) and arc-fused/crushed 88WC12Co (b) powder particles (Heimann, 1992).

parameters are shape, surface properties, porosity, homogeneity and phase composition (Eschnauer, 1982); for more details on methods of powder production and characterization see Pawlowski (1995).

The size of the particles is of critical importance. The size distribution is usually statistical, i.e. follows a Gaussian or log-normal distribution. While the median size for an average powder used in plasma spraying is around 40 μm, much smaller and much larger particles are available. The small particles will evaporate rather quickly prior to impact at the surface, whereas the larger ones do not melt completely, and may actually rebound from the surface or fall out of the jet due to gravitational forces. In either case these small and large particles are lost from the flame or plasma jet and thus reduce the deposition efficiency. This deposition efficiency is defined as the ratio between the numbers of particles deposited at the surface to the number of particles injected. Evidently high deposition efficiency is of overriding economical importance so that its optimization constitutes a crucial step in any coating development operation.

4.2.2.1 Atomization

This technique is frequently used to produce powders based on iron, cobalt, nickel and aluminum alloys. The particles exhibit a distinct difference in shape depending on the method of quenching. Gas-atomized particles (Figure 4.4a) are highly spherical whereas water-atomized particles (Figures 4.4b; 4.5a) have a more angular shape. As a consequence, different coating characteristics are obtained when these two powder types are thermally sprayed at identical feed rates and plasma parameters even if comparable particle-size distributions are used. This also means that the parameters optimized for gas-atomized powder cannot be uncritically transferred to water-atomized powders. In this case, clogging of the spray nozzle, discoloration, and/or cracking and spalling of the coating can result. On the other hand, spraying gas-atomized powders with parameters optimized for water-atomized powder will generally result in low deposition efficiency, increased coating porosity and poor adhesion.

4.2.2.2 Fusion and Crushing

This technique of manufacturing is applied to ceramics and cermet (carbide) powders as well as to brittle metals. Adjusting the solidification rate during fusion (casting) yields differences in the crystallinity. Subsequent crushing results in a wide range of particle shapes and consequently in apparent particle density. These variations in processing parameters lead to a variability in coating density, macro-hardness and wear resistance, and also thermal-cycling fatigue of oxide thermal barrier coatings (TBCs). Special techniques to manufacture yttria-stabilized zirconia powders for TBCs include the arc-fuse/crush, hollow-spherical-powder (HOSP)/plasma-fuse and arc-fuse/spray-dry processes (Kubel, 1990). In particular, *arc-fuse/crush powders* (Figure 4.4c) are produced by fusing, crushing and screening a premixed powder of 7–8 mol% yttrium oxide and zirconia. The particles obtained are irregularly shaped and have a rather broad size distribution. Thus they do not flow very well, and because of their angular shape can be difficult to melt in a non-optimized plasmatron configuration.

HOSP powders are produced by premixing, spray drying, fusing the product in a plasma jet and screening. The hollow spherical particles thus produced show excellent flowability and their low mass guarantees easy melting. However, the thin-shelled particles can be destroyed in the plasma jet due to the large forces excerted on them. *Arc-fuse/spray-dry powders* combine micronizing and spray-drying with the arc-fuse/crush process. Even though the particles are often hollow, their walls are porous in contrast to the walls of the HOSP powders. This results in a tenfold increase in surface area. Such powders are advantageous if very high coating porosities are required as needed in effective thermal barrier and bioceramic coatings (see Chapters 8.1, 9.3).

4.2.2.3 Compositing

Combining two or more dissimilar materials is achieved by chemical cladding (Figure 4.4d), organic bonding and/or sintering. Typically Ni–Al composite powders are produced by this method. Spatial variation of the ratios of the materials in the cladding, or loss of material due to preferential vaporization/evaporation may result in nonuniform chemical composition, hardness and fracture toughness of the coating. Clad-powder coatings frequently have exceptionally high adhesion strengths surpassing those of coatings produced from gas- or water-atomized powders.

Composite powders are the first stage of composite materials. At the Institute of Materials Science, RWTH Aachen, Germany, ceramics, metals and polymers can be combined into micropellets with one or more reinforcing material (fibers, particles) by spray drying (Powder Technology, 1992). The starting powders are distributed homogeneously in the micropellets that show an excellent flowability. Also, a special technique to coat powders of any morphology and from submicron up to several hundred micrometer size was developed using the so-called HYPREPOC (HYdrogen Pressure-REducing POwder Coating) process. An inorganic or organic metal salt (sulfates, nitrates, acetates, carbonates, chlorides etc.) solution is reduced in an autoclave by hydrogen in the presence of a powder of the core material. The coatings such produced are smooth, dense and highly adhering to the core particles. Successive precipitation steps result in either very thick coatings or multilayer coatings of dissimilar materials. The advantage of the HYPREPOC process is that nearly any material can be used either as core material or as a coating. Core material can be coated regardless whether the particles are angular or spherical.

4.2.2.4 Agglomeration

The agglomeration process is carried out through a sequence of different steps including pelletizing, pressing or spray drying. Variation of the processing parameters leads to large differences in powder densities and surface areas (Figure 4.4f). Powder particles with a low density and a high surface area behave like very fine particles and thus do not flow easily compared with dense particles of the same grain-size distribution. As a consequence the deposition efficiency is decreased and porous coatings are produced.

4.3
Momentum Transfer

4.3.1
Connected Energy Transmission

As mentioned above, plasma spraying can be described as a connected energy transmission process, starting with the transfer of energy from an electric potential field to the plasma gas (plasma heating), proceeding with the transfer of energy (momentum and heat transfer) from the plasma to the powder particles, and concluding with the transfer of energy from the particles to the substrate.

In the following section, principles of momentum transfer will be discussed. Modeling of momentum transfer will be dealt with in Chapter 6. The process of particles being accelerated in the plasma jet can be described as a two-phase fluid flow, but is much more complex because of the presence of charged particles, chemical reactions and large temperature gradients. Particle velocities can be measured by time-of-flight laser Doppler anemometry (LDA) (cp. Chapter 3.4.2.2).

4.3.2
Transfer of Plasma Velocity to Particles

Figure 4.6 shows the velocities of differently sized alumina particles in an argon/hydrogen plasma jet (Vardelle *et al.*, 1982). The larger the particles the lower is the response to the gas velocity. The 3-μm particles are used as 'tracers' to measure

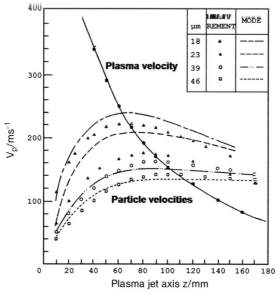

Figure 4.6 Transfer of plasma velocity ($v_{max} = 600 \, m \, s^{-1}$) to alumina particles of different sizes (18–46 μm) suspended in an Ar/H$_2$ plasma (Vardelle *et al.*, 1982).

the velocity of the gas alone in the absence of larger particles. The gas velocity decreases continuously on exiting from the nozzle. The 18-μm particles reach a maximum velocity of $220\,\text{m}\,\text{s}^{-1}$. If the target distance is 15 cm, than the particle velocity is around $170\,\text{m}\,\text{s}^{-1}$, i.e. one half of the velocity of sound at atmospheric pressure in air. Larger particles, like the 46-μm particles, reach a maximum of only $140\,\text{m}\,\text{s}^{-1}$ but this velocity does not decay quickly because of the high inertia of the large particles.

4.3.3
Surface Ablation of Particles

For liquid or plastically deformable material a global approximation of the force balance on the side walls of the deforming particle is

$$(4\pi r^2)(C_D \rho_p v^2/4) \approx m(d^2 r/dt^2), \tag{4.1}$$

where C_D = drag coefficient, ρ_p = density of plasma gas, v = particle velocity, m = particle mass and r = particle radius.

Assuming the density of the material, ρ_m remains constant, the expression

$$r(d^2 r/dt^2) = C_D \rho_p v^2 / 2\rho_m \tag{4.2}$$

can be derived. The surface of the sphere is heated by convection and shed by ablation. The resulting mass change is

$$Q_A(dm/dt) = -(1/2) h \rho_p A v^3, \tag{4.3}$$

where Q_A = heat of ablation and h = heat transfer coefficient. It can be seen that the mass change is proportional to the third power of the particle velocity and thus the materials loss becomes significant at high velocities. Also, the maximum ablation rate can be determined (Zahnle, 1992) as

$$Q_A(dm/dt) \approx A\sigma T^4, \tag{4.4}$$

and finally

$$Q_A(dm/dt) = -A\,min[(h/2)\rho_p v^3, \sigma T^4]. \tag{4.5}$$

4.4
Heat Transfer

4.4.1
Heat Transfer under Low Loading Conditions

If the mass of powder particles injected into a plasma jet is small enough so as to not disturb the flow of the plasma and to cool it down by radiative losses, the heat transfer

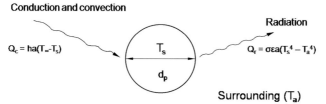

Figure 4.7 Heat transfer equilibrium of a powder particle in a thermal plasma.

can be estimated under simplifying assumptions. In particular, the Navier–Stokes equation for the flow around a sphere is highly nonlinear even under the simplifying assumptions made above, i.e. for constant physical property plasmas. Thus no completely exact solutions exist for these equations (Sayegh and Gauvin, 1979).

In many cases an analytical solution is not available, and only numerical models can be successfully used. As shown in Figure 4.7, a particle in contact with a heated plasma jet acquires energy by convective (and conductive) processes, q_c and loses energy to the surroundings by radiation, q_r. The net energy contribution to the heating and melting of particles is the difference between these two values (eqs. 4.6, 4.7).

The amount of heat gained by convective energy transfer from the plasma to the particle is

$$q_c = h \times a \times (T_\infty - T_S), \qquad (4.6)$$

the amount of heat lost by radiative energy transfer to the surrounding is

$$q_r = \sigma \times \varepsilon \times a \times (T_S^4 - T_a^4), \qquad (4.7)$$

where h = plasma/particle convective heat transfer coefficient, a = surface area of the particle, T_∞ = free-stream plasma temperature, T_S = particle surface temperature, T_a = temperature of the surrounding, σ = Stefan–Boltzmann coefficient and = particle emissivity.

The heat transfer coefficient can be expressed as

$$h = k_f \mathbf{Nu}/d_p, \qquad (4.8)$$

where $k_f = C_g \eta/\mathbf{Pr}$, or $h = [C_g \eta/d_p][\mathbf{Nu}/\mathbf{Pr}]$ (see Appendix A).

The symbols **Nu** and **Pr** denote the Nusselt and Prandtl numbers and k_f is the thermal conductivity of the particles. C_g, d_p and η are, respectively the specific heat of the plasma gas, the particle diameter and the gas viscosity.

The choice of the numerical value of the Nusselt number is crucial for a realistic estimate of the heat transfer rates from the plasma to the particles in the same way the choice of the drag coefficient is important for the momentum transfer. For the heat transfer between a gas and a spherical particle the Ranz–Marshall expression

$$\mathbf{Nu} = 2.0 + b\mathbf{Re}^m \mathbf{Pr}^n \qquad (4.9)$$

was given by Frösling (1938), Ranz and Marshall (1952) and Raithby and Eckert (1968). The exponents m and n as well as the pre-exponential factor b were given by Ranz and Marshall (1952) (b = 1.0, m = 0.5, n = 0.33), Sayegh and Gauvin (1979) (b = 0.473, m = 0.552, n = 0.78 Re$^{-0.145}$),[1] Vardelle et al. (1983) (b = 0.6, m = 0.6, n = 0.33) and Liu et al. (1993) (b = 1.08, m = 0.5, n = 0.5). Other more complex expressions for **Nu** were introduced by Fiszdon (1979), Lee et al. (1981) and Lewis and Gaudin (1973). Also, noncontinuum effects in heat transfer over the range of the Knudsen number 0.01 < **Kn** < 1.0 have been proposed by Chen and Pfender (1983) as

$$q_{noncont} = q_{cont}[1/\{1 + (2Z^*/d_p)\}], \quad (4.10)$$

with Z^* = temperature 'jump' distance. Because the particle size is of the same order of magnitude as the molecular mean free path length in the plasma gas, this 'rarefaction effect' (Pfender, 1989) may exert a strong influence on the heat transfer.

The net energy acquired by a particle in a plasma is obtained from Equations 4.6 and 4.7 as

$$q_n = q_c - q_r \gg 0, \quad (4.11)$$

since q_r is generally small compared with q_c because of the small value of the Stefan–Boltzmann coefficient $\sigma = 5.67 \times 10^{-8}$ W m^{-2} K^{-4}.

The requirement for an efficient energy economy during the plasma spray operation is that all particles melt completely during their very short residence time in the plasma jet. This means that the total energy received by the particles, q_n must be equal or larger than the energy required to melt the particles. This energy quantity consists of two components, (i) the energy needed to heat the particles from their initial temperature to their melting point and (ii) the latent heat of fusion:

$$\int_0^\tau q_n dt > \sum \{m_p c_p (T_m - T_0) + m_p H_m\}, \quad (4.12)$$

where m_p = mass of particle, c_p = specific heat of particle, T_m = melting point, T_o = initial temperature, H_m = latent heat of fusion and τ = residence time of the particle in the plasma jet.

Normally, superheating is required to achieve sufficiently high temperatures of the particles and to decrease the viscosity of the liquid droplet. The degree of superheating depends on the Biot number **Bi**, which is the ratio of the thermal conductivity of the plasma gas, k_g to the thermal conductivity of the particles, k_p:

$$\mathbf{Bi} = k_g/k_p \quad (4.13)$$

The requirement of a uniform particle temperature implies that **Bi** < 0.01, i.e. the thermal conductivity of the particle material must be much higher than that of the plasma gas. Thus the numerical value of the Biot number can be used to determine *a*

---

[1] Actually, instead of the limiting Nusselt number of 2 an expression $2f_0 = \{2(1-[T_w/T_\infty]^{1+x}\}/\{(1+x)(1-[T_w/T_\infty])(T_w/T_\infty)^x\}$ with $x = 0.8$ was used, where T_w is the surface temperature of the particle and T_∞ is the plasma temperature.

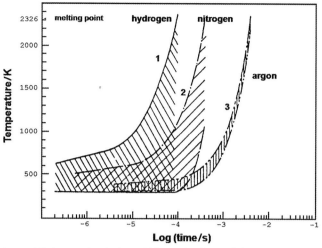

Figure 4.8 Temperature history of 50-μm alumina particles at a temperature of 10 000 K in various plasma gases (Bourdin et al., 1983).

priori whether large temperature differences may occur between the surface and the center of a (spherical) particle that may compromise the conjecture of a uniform temperature distribution and thus uniform melting behavior. According to Bourdin *et al.* (1983), a characteristic temperature $T^* = \Delta T_s/(T_\infty - T_s) = f(\mathbf{Bi})$ can be defined. For $\mathbf{Bi} < 0.02$, the temperature between the surface and the center of the particle becomes less than 5% of $(T_\infty - T_s)$.

Porous particles and particles with low thermal conductivity will develop large internal temperature gradients that can often exceed 1000 K. Figure 4.8 shows the temperature history of alumina particles of 50 μm radius at a plasma temperature of 10 000 K in hydrogen, nitrogen and argon plasmas (Bourdin *et al.*, 1983). The left lines depict the particle surface temperature, the right lines the temperature at the center of the particles. The temperature difference depends strongly on the heat flux to the particle. For example, the temperature difference in a hydrogen plasma can be so high that the thermal stresses induced can lead to shattering of the particles. As is evident from Figure 4.8, the radial temperature differences across a particle are quite large for a hydrogen plasma and decrease for nitrogen and even more for argon plasmas. This is because of the large differences in the thermal conductivities, k_g of the plasma gas, being 4.6, 1.6 and 0.2 W m^{-1} K^{-1} for hydrogen, nitrogen and argon, respectively. The value of the thermal conductivity influences the particle surface temperature because the solution of the equation

$$dT_s/dt = -12k_g(T_s - T_\infty)/\rho_s c_s d_p^2 \qquad (4.14)$$

($\mathbf{Nu} = 2$, i.e. only conductive heat transfer, $q_r = 0$, i.e. no radiative heat losses) is

$$(T_s - T_\infty)/(T_0 - T_\infty) = \exp[12k_g t/\rho_s c_s d_p^2], \qquad (4.15)$$

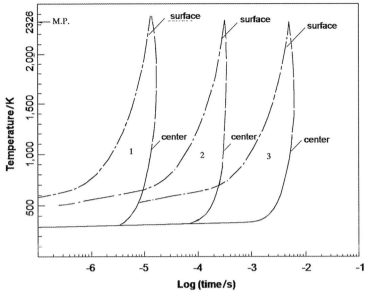

Figure 4.9 Temperature history of alumina particles of different radii (1 = 10 μm; 2 = 50 μm; 3 = 200 μm) suddenly immersed in a nitrogen plasma at 10 000 K (Bourdin et al., 1983).

where ρ_s and c_s are the density and the specific heat, respectively of the particle, d_p the particle diameter and T_0 and T_∞ the temperature of the surrounding and the plasma, respectively (Bourdin et al., 1983). Figure 4.8 indicates that the time required to melt an alumina particle of 50 μm radius ranges from 80 μs (hydrogen plasma) to 0.3 ms (nitrogen plasma) to 4 ms (argon plasma).

Figure 4.9 shows the temperature history of alumina particles of different radii (10, 50 and 200 μm) suddenly immersed in a nitrogen plasma at 10 000 K (Bourdin et al.,1983). Again, the leading curves refer to the surface temperature of the particle, the trailing curves to the center of the particles. The radius of the particles has little influence on T_s.

A somewhat easier approach to making a qualitative estimate of the materials performance in terms of ease of plasma melting of particles is to consider the maximum particle diameter for complete melting, d_{max}. This is a useful parameter for comparing materials with widely differing densities and thermophysical properties.

Based on the equation of transient heat conduction in solid spheres (assumption: surface of particle is brought instantaneously to its melting point), the following expression can be evaluated (Marsh et al., 1961):

$$(T_m - T_R)/(T_m - T_1) = f(\mathbf{Bi}, \mathbf{Fo}) = f[hr/k, \alpha\Theta/r^2], \tag{4.16}$$

where T_m = melting point of material, T_R = T at points along the radius of a sphere ($r = 0$ at center, $r = r_0$ at surface), T_1 = initial temperature of particle, α = thermal diffusivity = $k_p/\rho c_p$, Θ = time, $\mathbf{Bi}$ = Biot number and $\mathbf{Fo}$ = Fourier number.

Table 4.2 Relative ease of plasma spraying for various oxide, carbide and metallic coatings (Marsh et al., 1961).

Material	Thermal diffusivity, a (cm^2 s^{-1})	C_p ($T_m - T_s$), (cal·g^{-1})	$d_{max}°$ (μm)	Relative ease of spraying
ZrO$_2$	0.005	430	26	5
UO$_2$	0.025	214	58	4
TiC	0.04	645	72	3
TaC	0.09	115	110	1
ZrC	0.05	525	82	3
TiN	0.07	556	96	2
B$_4$C	0.06	109	90	2
AISI304	0.05	226	82	3
W	0.63	111	280	1

At the particle center ($r = 0$), the equation becomes approximately

$$1 - (T_0/T_m) = \Phi\{h\, r_0/k, \alpha\Theta/r_0^2\}. \tag{4.17}$$

A comparison of different materials in terms of the relative ease of plasma spraying is given in Table 4.2. Assuming constant time (100 μs), a center temperature of the particle $T_0 = 0.9\, T_m$, a Biot number of 10 and a Fourier number of 0.3, an expression can be obtained for the maximum diameter d_{max} of a particle that melts completely in 100 μs:

$$d_{max} = 2r_0 = 2[\alpha \times \Theta/0.3]^{1/2}. \tag{4.18}$$

For a qualitative comparison of the melting behavior of different materials Marynowski et al. (1965) introduced an 'ease of melting' parameter $\Phi = 10^{-3}\, T_m/\sqrt{\rho}$. Easy-to-melt materials have $\Phi < 1.0$, difficult-to-melt materials $\Phi > 1.4$ (see Table 4.3). The parameter Φ is based on considerations by Engelke (1962) involving equating the time required for the particle to traverse the plasma of length X to the time required

Table 4.3 'Ease of melting' parameter Φ (Engelke, 1962; Marynowski et al., 1965).

	Material	Φ	T_m (K)	ρ (g·cm^{-3})
Difficult (carbides, nitrides)	TiC	1.53	3,410	4.93
	TiN	1.40	3,200	5.21
	ZrC	1.58	3,840	6.73
Easier (oxides)	ZrO$_2$	1.38	2,680	4.40
	Al$_2$O$_3$	1.26	2,318	3.97
	SiO$_2$	1.21	1,975	2.65
	(TaC)	1.09	4,150	14.53
Easy (metals)	Ni	0.58	1,728	8.90
	Ta	0.80	3,270	16.60
	W	0.83	3,650	19.30
	Mo	0.90	2,883	10.20

to melt this particle completely. An expression can be evaluated that leads to

$$[X(k_p \Delta T)^2 / V\mu] \geq [L^2 D^2 / 16\rho], \quad (4.19)$$

where k_p = mean boundary layer thermal conductivity, V = plasma velocity, μ = plasma viscosity, L = particle heat content, D = particle diameter and ρ = particle density. Since the left hand side of Equation 4.19 increases proportional to the square of the plasma enthalpy it follows that the enthalpy should be proportional to $(LD/\sqrt{\rho})$ (Engelke, 1962).

4.4.2
Exact Solution of Heat Transfer Equations

According to Chen and Pfender (1982) exact solutions for the heat transfer from a plasma to a particle under low-loading conditions can be obtained by considering the following simplifying assumptions.

- particles are exposed to a uniform atmospheric pressure argon plasma;
- heat transfer process is steady;
- particles are spherical;
- free convection can be neglected (**Re** = 0, **Gr** <<1);
- radiation from and to particle can be neglected (**St** = 0);
- influence of vapor from evaporating particles can be neglected.

Two cases can be distinguished: heating of a particle with and without evaporation. However, realistically a certain degree of vaporization of material below the boiling point always must be taken into account. As the temperature of a particle in the plasma jet increases its vapor pressure also increases. Then the mass loss by vaporization (Pfender, 1989) becomes

$$dm/dt = \rho \ h_m M \ \ln(p/p - p_v), \quad (4.20)$$

where h_m = mass transfer coefficient, M = molecular weight of the material, p = partial vapor pressure with respect to saturation and p_v = partial vapor pressure with respect to the particle surface temperature. The mass transfer coefficient can be expressed similarly to the Nusselt number (eq. 4.9) by related dimensionless groups as

$$\mathbf{Sh} = h_m d_p / D_{ij} = 2.0 + 0.6 \ \mathbf{Re}^{0.5} \mathbf{Sc}^{0.33}, \quad (4.21)$$

with **Sh** = Sherwood number, **Re** = Reynolds number, **Sc** = Schmidt number, d_p = particle diameter and D_{ij} = interdiffusion coefficient.

4.4.2.1 Particle Heating without Evaporation
The simplified heat transfer equation is

$$4\pi r^2 (k \ dh/c_p dr) = Q_0 \quad (4.22)$$

where k = thermal conductivity, c_p = specific heat, h = specific enthalpy of plasma and r = radial coordinate.

Introducing the 'heat conduction potential', S as

$$S = \int_{T_0}^{T} k\, dT = \int_{h_0}^{h} (k/c_p)\, dh, \qquad (4.23)$$

eq. 4.22 can be rewritten as

$$Q_0 = 4\pi r_S^2 \, dS/dr \qquad (4.23a)$$

and on integration

$$Q_0 = 4\pi r_S (S_\infty - S_S), \qquad (4.24)$$

where S_∞ and S_S are the values of S at the surface and far away from the surface and r_S is the particle radius.

The specific heat flux is $q_0 = Q_0/4\pi r^2_S$ and thus

$$q_0 = (S_\infty - S_S)/r_S. \qquad (4.25)$$

It can be seen that the specific heat flux is inversely proportional to the particle radius.

The Nusselt number can be redefined as

$$Nu = q_0 2 r_0/(S_\infty - S_0) = (h\, d_p)/k_f \qquad (4.26)$$

Under pure heat conduction condition (**Re** = 0) is **Nu** = 2. For **Re** ≠ 0 follows

$$\mathbf{Nu} = q_0 2 r_S / k(T_\infty - T_S) = q_0 2 r_S / (kc_p)(h_\infty - hS). \qquad (4.26a)$$

From Equation 4.26 it follows that S is only a function of the temperature for a given plasma gas and a given solid material. Figure 4.10 relates the thermal conductivities of the pure plasma gases argon, nitrogen and hydrogen and the heat conduction potentials as functions of the plasma temperature (Bourdin et al., 1983).

Figure 4.11 shows in addition the plot of the heat conduction potential, S of an argon/hydrogen mixture as a function of the plasma temperature. The high thermal conductivity of the added hydrogen leads to a substantially increased heat conduction potential of the mixture compared with pure argon. For this reason argon/hydrogen mixtures are frequently used as a plasma gas to improve the melting characteristics of powders with high melting points such as zirconia.

4.4.2.2 Particle Heating with Evaporation

The simplified convection–diffusion equation can be written as

$$-4\pi r^2 D\rho\, df/dr + G f = G \qquad (4.27)$$

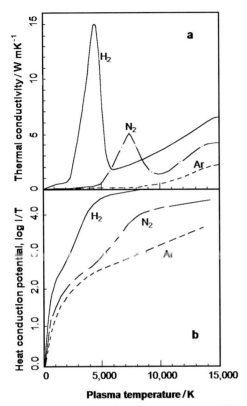

Figure 4.10 (a) Thermal conductivities k(T) and (b) heat conduction potentials S = I(T) of some pure gases (Bourdin et al., 1983).

or

$$4\pi r^2 D\rho\, df/dr = G(f-1) \tag{4.27a}$$

where G = total mass flux due to evaporation, f = mass fraction of evaporated species, D = diffusion coefficient and ρ = gas density. Boundary conditions are f = 0 as r → ∞.

The related energy equation is

$$4\pi r^2 (k/c_p)(dh/dr) = G(h - h_S + L), \tag{4.28}$$

where h = enthalpy corresponding to the surface of particle and L = latent heat of evaporation. Boundary conditions for solving Equation 4.28 are

$$r = r_S\, h = h_S$$
$$r \to \infty\, h = h_\infty$$

The solution of Equation 4.28 is

$$G = 4\pi r_S \int (k/c_p)(dh/h - h_S + L) = 4\pi r_S \int (kdT)/h - h_S + L. \tag{4.29}$$

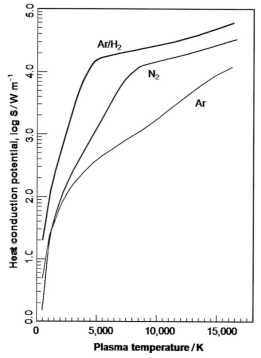

Figure 4.11 Heat conduction potentials of pure (argon, nitrogen) and mixed (argon/hydrogen) plasmas at atmospheric pressure (Chen and Pfender, 1982).

The total mass flow rate is related to the total heat flux as

$$Q_1 = G\,L. \tag{4.30}$$

Therefore:

$$Q_1 = 4\pi r_S L \int (k dT)/h - h_S + L, \tag{4.31}$$

and, because $q = Q/4\pi r^2$, the specific heat flux becomes

$$q_1 = (L/r_S)\int (k dT)/h - h_S + L. \tag{4.31a}$$

Since the integral is only a function of T for a given plasma gas and a given material it can be replaced by the function

$$\int (k dT)/h - h_S + L = M(T_\infty, T_S, L) \tag{4.32}$$

From Equations 4.25 and 4.31a, the ratio of the heat flux with and without evaporation becomes

$$q_1/q_0 = L/(S_\infty - S_S) M(T_\infty, T_S, L). \tag{4.33}$$

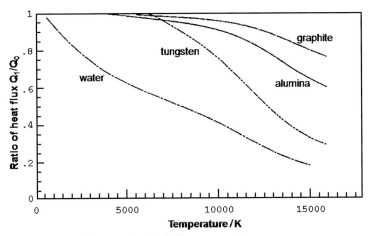

Figure 4.12 Heat flux ratios Q_1/Q of different materials as a function of the argon plasma temperature (Chen and Pfender, 1982).

Figure 4.12 shows a plot of Q_1/Q_0 as a function of the plasma temperature for different materials. It can be seen that evaporation has a strong effect if the plasma temperature is high and if the material has a low latent heat of evaporation.

4.4.2.3 Evaporation Time of a Particle

For quasi-steady state evaporation it follows

$$Q_1 = 4\pi r_S^2(-dr_S/dt)\rho_c L \tag{4.34}$$

or

$$q_1 = -\rho_c L(dr_S/dt). \tag{4.34a}$$

With Equations 4.34a, 4.31a and 4.32 one obtains

$$q_1 = (L/r_S)M(T_\infty, T_S, L) = -\rho_c L(dr_S/dt) \tag{4.35}$$

or

$$-(dr_S^2/dt) = K \tag{4.36}$$

where $K = 2M(T_\infty, T_S, L)/\rho_c$ is the so called *evaporation constant*. This constant is only a function of temperature, not a function of the radius of the particle. Figure 4.13 shows the evaporation constant K for several materials as a function of the plasma temperature.

Integration of Equation 4.36 leads to

$$r_{S0}^2 - r_S^2 = K \times t \tag{4.37}$$

where r_{S0} is the initial radius of the droplet at $t = 0$.

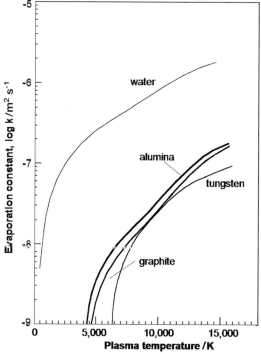

Figure 4.13 Evaporation constants for various materials suspended in an argon plasma as a function of the plasma temperature (Chen and Pfender, 1982)

The time for complete evaporation of a droplet ($r_S = 0$) then becomes

$$t_V = r_{s0}^2 / K. \tag{4.38}$$

This time does not include the time needed for initial heating and melting of the particle. From Figure 4.13 we obtain for an alumina particle of a radius $r = 50\,\mu m$, heated in an argon plasma of $T_\infty = 10\,000\,K$ the time for complete evaporation $t_V = (5.0 \times 10^{-5})^2/(3.0 \times 10^{-8}) = 80\,ms$. This result shows the need to design the plasmatron nozzle and to set the spray parameters in such a way that the residence time of a 50 µm-alumina particle in the 10 000 K zone of the plasma jet is much shorter than 80 ms, to avoid substantial evaporation. On the other hand the dwell time must be substantially larger than the 4 ms required to completely melt an alumina particle of that size as shown in Figures 4.8 and 4.9.

The evaporation of particles in the plasma gas can drastically reduce the specific heat flux in the plasma (eq. 4.33, Figure 4.12). On the other hand, the heat transfer calculations by Chen and Pfender (1982) assumed that there is no influence onf the vapor from evaporating or sublimating particles on the thermophysical properties of the plasma.

There is, however, a noticeable effect on the temperature of the plasma under dense particle loading conditions (see for example Proulx *et al.*, 1985). Local cooling of the plasma takes place due to the presence of particles. This will be described in the following section.

4.4.3
Heat Transfer under Dense Loading Conditions

In dilute systems, i.e. systems at low-loading conditions, it has been tacitly assumed that particle movement, including momentum and heat transfer can be considered in a ballistic manner. This means that the stochastic single-particle trajectories calculations, as well as those of the continuum flow, temperature and concentration fields can be performed under the simplifying assumptions used above. However, under realistic conditions the mass of particles injected into a plasma jet is far from those idealized dilute systems. Indeed, in-flight processing of powders to produce plasma-sprayed coatings must be performed under high loading conditions in order to make efficient use of the thermal energy stored in the plasma. On exceeding a certain critical loading it is no longer warranted to treat the individual impulse and thermal histories of the particles in a stochastic way. As a result particles interact with each other, and the momentum and temperature of the plasma jet decrease with increasing mass of powder. The temperature decrease has three sources: (i) the heat extracted from the plasma to melt a larger mass of powder could cause substantial local cooling of the plasma; (ii) the evaporated fraction of the powder (see. 4.4.2.2) can drastically alter the thermophysical, thermodynamic and transport properties of the plasma gas; (iii) small powder particles with high optical emissivities evaporate easily and radiate away plasma energy.

In order to arrive at an analytical solution for plasmas under dense loading conditions, the continuity, momentum, energy and species conservation equations, as well as the Maxwell electromagnetic field equations must be solved simultaneously. This is beyond the scope of this text. Detailed descriptions of the numerical techniques employed to solve the plasma conservation equations and the electromagnetic field equations can be found by Mostaghimi *et al.* (1985), Proulx *et al.* (1985) and Crowe *et al.* (1977) as well as in Chapter 6.

4.4.4
Heat Transfer Catastrophy

It is a puzzling but well-known fact that the properties of plasma-sprayed coatings vary widely even though the spray parameters have supposedly been fixed within narrow ranges using sophisticated microprocessor-controlled metering devices and stringent quality control measures have been applied using, for example, principles of statistical process control (SPC) and Taguchi analysis. A study on the reproducibility of the APS process via control of the particle state was reported by Srinivasan *et al.* (2006). The particle state of 8%Y-stablized zirconia powder was controlled via feedback from a plasma plume sensor (DPV 2000, Tecnar Automation Ltd., Québec, Canada) and a torch diagnostic system-spray plume trajectory sensor (TDS-SPT, Inflight Ltd., Idaho Falls, ID, USA). Despite the fact that the results obtained from averaged individual particle measurements showed exceptionally high stability of the temperature (<1%) and velocity (<4%), variability of coating properties such as thickness and mass deposited were surprisingly large. In other words, infinitesimally small changes to the input parameters cause large, and generally nondeterministic changes in the output parameters, i.e. those parameters influencing the properties and hence the performance of the coatings. This is the hallmark of nonlinearity.

To search for the cause of nonlinear behavior of a collection of particles under dense loading conditions in a plasma jet, several approaches may be taken that all relate to very complex interactions within the plasma. It may suffice to outline in the context of this book only one possible chain of events leading to nonlinearity.

In a thermal plasma cooperative processes occur through numerous types of waves (transverse electromagnetic, longitudinal plasma-acoustic Langmuir, magneto-hydrodynamic and drift waves) that interact and cause stochastic fluctuations in which the phases of the waves exhibit probabilistic distributions. Such nonequilibrium distributions can be roughly described, respectively as *electromagnetic and magneto-hydrodynamic turbulences* (see 3.1.1). Since both of these turbulences affect the local magnetic field strength, **B** and the electrical current density, **j**, their cross-product, the Lorentz force (**j** × **B**) fluctuates rapidly and with it the plasma compression ('z-pinch'). The Lorentz force is compensated for by the gas kinetic pressure gradient, grad p; consideration of the plasma velocity and viscosity yields the momentum (Navier–Stokes) conservation equation that is known to be nonlinear:

$$(\mathbf{j} \times \mathbf{B}) = \rho_m u(\text{grad } u) + \text{grad } p - \text{div}(\eta \text{ grad } u). \tag{4.39}$$

Hence changes in the magnitude of the Lorentz force influence the plasma velocity u, the pressure gradient grad p, as well as the dynamic plasma viscosity, η. In particular, the fluctuations of the plasma-confining Lorentz force generate a plasma jet pulsating perpendicular to the jet axis with frequencies that are on the order of the residence time of the particles in the jet, i.e. tenths of microseconds. The rate with which turbulent eddies of cool air surrounding the plasma jet are entrained by the pumping action of the plasma changes accordingly. In this way a rapidly fluctuating temperature profile is obtained, probably with rapidly changing heat transfer coefficients, h, of the plasma-particle system. Through arc root fluctuation time-dependent periodic variations of the mean particle temperature and velocity up to $\Delta T = 600\,^\circ\text{C}$ and $\Delta v = 200\,\text{m s}^{-1}$ were experimentally observed in the center of the plasma jet during a voltage cycle (Bisson *et al.*, 2003).

As a consequence the degree of melting of the particles is affected because the relative proportion of heat transferred by convective and radiative processes is severely and instantaneously disturbed. Finally, the local thermal equilibrium breaks down on a scale that is small compared with the overall volume of the plasma jet. Then the system enters the realm of a *heat transfer catastrophy*. This phenomenon can be tentatively described in terms of stability theory by an elementary cusp catastrophy of co-dimension two (Riemann–Hugoniot catastrophy; Thom, 1975). The order parameters X and Y on the control surface, and the parameter Z separating the control from the behavior surface can be selected in such a way that only dimen-

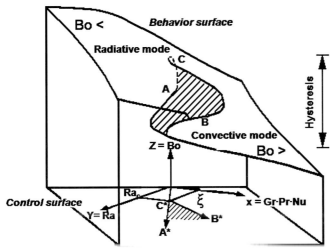

Figure 4.14 Control and behavior surfaces of a plasma undergoing a heat transfer catastrophy (see text). The coordinates X and Y of the control surface and Z of the behavior surface are expressed by dimensionless numbers (see Chapter 6.1 and Appendix A).

sionless groups (see Appendix A) as real variables occur (Figure 4.14). The first ordering parameter X is the heat flux that can be expressed as the product of the Grashof, Prandtl and Nusselt numbers, **Gr·Pr·Nu**. The second ordering parameter Y is the temperature difference, expressed through the thermal Rayleigh number, **Ra**. The two ordering parameters define a bifurcation set of local competition between two stable states of heat transfer: a convective heat transfer mode and a radiative heat transfer mode. On the control surface, lines A* (radiative mode) and B* (convective mode) meet at a temperature-dependent point C* with the critical Rayleigh number, **Ra$_c$** and a critical heat flux, ξ beyond which the bimodality ceases to exist. As the parameter defining the Z-coordinate the Boltzmann number **Bo** can be chosen, given by **Bo** = [**Re Pr/St**]/e* = [**Pe/St**]/e*, where **Re** = Reynolds number, **Pr** = Prandtl number, **Pe** = Peclet number and **St** = Stefan number.

The behavior surface shows the characteristic shape of the fold of a cusp catastrophy that outlines the region of inaccessibility: heat transfer from plasma to particles occurs either in the convective mode at lower temperature, or in the radiative mode at higher temperature. Consequently, the lower sheet of the behavior surface corresponds to the convective regime B, characterized by a large Boltzmann number **Bo**. The upper sheet corresponds to a radiative regime A with a small Boltzmann number.[2] The Boltzmann number defines the ratio of the bulk heat transport to the heat transported by radiation. For dense loading conditions

---

2) The expression for the Boltzmann number is **Bo** = $\rho_m c_p v_s/e^* \sigma T^3$, with ρ_m = mass density, c_p = specific heat, v_s = plasma velocity, e^* = surface emissivity of the particles, σ = Stefan–Boltzmann coefficient of radiative heat transfer, and T = local plasma temperature.

the Bouguet number ($\mathbf{B} = 3C_B\lambda_r/4\rho_D r$, with C_B = mass ratio of particles to unit bed volume, λ_r = mean path of radiation, ρ_D = particle density, r = mean particle radius) must be used. This dimensionless group describes the radiant heat transfer to a particle-loaded plasma gas stream. The approach taken here to relate the nonlinear characteristics of a plasma to plasma turbulences with high frequencies, and the ensuing heat transfer catastrophy is similar to that of the thermo-diffusive mass transport in systems with rapidly changing chemical potential and temperature gradients (Heimann, 1987).

4.4.5
Energy Economy

At this point it should be mentioned that only a small fraction of the energy supplied to the plasma gas is stored in the particles. A macroscopic energy balance (Houben, 1988) shows:

Energy supplied to the torch	42.0 kW
Cooling water losses (66%)	− 27.7 kW
Losses due to convection and radiation to surroundings	− 13.3 kW
Net energy stored in particles	1.0 kW

The energy efficiency is thus only 2.4%. This low number should be compared with a typical energy balance in welding using heat flow density data. Table 4.4 shows process variables for welding and plasma spraying operations. For a *welding process*, the heat flow density is q_w = (efficiency × current × voltage × time)/(transverse speed × weld width × time). For the *plasma spraying process*, it is q_p = (deposit mass flow rate × heat content of particles)/(traverse speed × spray width × time). Using the data of Table 4.4 the following values are calculated:

$$q_w = (0.8 \times 200[A] \times 25[V] \times 1[s])/(3.6[mm\,s^{-1}] \times 10[mm] \times 1[s]) = 111\,J\,mm^{-2}$$
$$q_p = (40[g\,min^{-1}]/60[s] \times 1500[J\,g^{-1}])/(40[mm\,s^{-1}] \times 25[mm] \times 1[s]) = 1\,J\,mm^{-2}$$

Table 4.4 Comparison of process variables for welding and plasma spraying operations.

Variable	Unit	CO_2 welding	Hand welding	Plasma spraying
Deposit mass flow rate	g min^{-1}	42	40	40
Current	A	200	200	400
Voltage	V	25	25	40
Traverse speed	mm s^{-1}	5.5	3.6	40
Weld/spray width	mm	11.5	10	25
Heat content of spray particles	kJ kg^{-1}			1,500
Energy efficiency	%	80	80	2.4

Consequently, the heat flow density during plasma spraying is only about 1% of that of the standard welding process. In conclusion this means that (i) plasma spraying has a low energy efficiency and (ii) plasma spraying has a low heat flow density compared with plasma welding.

Similarly low efficiencies have been calculated for plasma-arc wire-spraying operations (Rykalin and Kudinov, 1976). Approximately 40% of the input energy is spent to heat the plasmatron and hence is removed by water-cooling whereas the majority of the energy is used to heat the plasma gas. Only 2–5% of this energy is available for melting the wire and accelerating the liquid droplets and another 6 to 9% is used to heat the substrate.

The results of the modeling of the momentum and heat transfer could be used on-line for estimating the powder velocities and temperatures at the time of impact. Then a computer-assisted feedback loop could be installed to directly control the input parameters such as

- plasma power;
- particle feed rate;
- plasma gas composition;
- stand-off distance and so on.

To verify the feedback parameters, time-of-flight laser-Doppler anemometry (LDA) and multi-wavelength pyrometry could be employed. This advanced concept of plasma spray process control has been treated theoretically. Its implementation on the shop floor, however, is still in the future. Areas of improvement and developments are (Mostaghimi, 1991)

- internal particle heat transfer;
- 3D models for DC plasma spraying (see Ghafouri-Azar *et al.*, 2003);
- verification and control of drag and Nusselt correlations;
- modeling of particle impact on substrate (see Salimijazi *et al.*, 2007);
- low pressure, supersonic models for DC plasmas.

4.5
Particle Diagnostics: Velocity, Temperature and Number Densities

The methods of measurement of plasma velocities and plasma temperature, and the equipment used for this purpose have been described in Chapter 3. In this section some typical results related to particle velocities and temperatures and number densities will be described.

4.5.1
Particle Velocity Determination

As described in detail above, the highly accelerated plasma jet transfers momentum to the powder particles injected near the arc root. The resulting particle velocities must be chosen in such a way that most particles travel in trajectories that assure optimum residence time in the hot zone of the jet. To obtain axial trajectories an optimum injection velocity is likewise required and thus the powder gas pressure must be adjusted to the remaining selected plasma parameters. As the flow of the carrier gas controls the particle trajectory through the plasma jet, a balance must be maintained between plasma gas and powder carrier gas velocities to correctly position the particles in the plasma for proper melting and acceleration. For a given powder there is an optimum value of powder feed rate that will produce maximum deposition efficiency. In-situ particles property (velocity, temperature, size) measurements also showed that axial injection of powder into the plasma jet provides more uniform and axisymmetric particle property distributions compared with the customary transverse injection (Cetegen, 1999).

Figure 4.15 shows that for experiments conducted in an argon/hydrogen plasma at a power of 29 kW alumina particles (mean diameter of $18 \pm 3\,\mu m$) injected with a carrier gas (argon) flow rate of $5.5\,L\,min^{-1}$ move in a trajectory close to and grazing the plasma jet axis. On increasing the powder carrier gas flow rate to $10\,L\,min^{-1}$, the particles completely cross the axis of the jet at a distance of approximately 8 mm from the point of injection (Vardelle et al., 1988). Thus they enter the hotter region of the jet but are blown quickly out of it again so that their residence time in the jet actually decreases. Figure 4.16 shows the particle velocity profiles at an axial distance of $z = 75\,mm$ for carrier gas flow rates between 5.5 and $8.5\,L\,min^{-1}$. Maximum velocities were obtained with a low carrier gas flow rate ($5.5\,L\,min^{-1}$). Under these conditions the majority of the powder particles did occupy trajectories close to the jet axis (see Figure 4.15) where they acquired the highest momentum and temperature.

It should be noted, however, that the real trajectory of a particle may deviate considerably from the calculated one based on the momentum conservation equation. This has been explained by Vardelle et al. (1982) by the influence of *thermophoretic forces*. Thermophoresis, however, is strongly dependent on the (free stream) temperature and the particle size (Pfender, 1989). Also, thermophoresis becomes comparable to viscous drag forces only for small relative velocities (around $1\,m\,s^{-1}$).

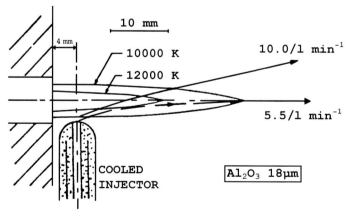

Figure 4.15 Dependence of the particle trajectories on the flow rate of the argon carrier gas (after Vardelle et al., 1988).

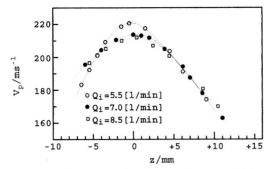

Figure 4.16 Alumina particle velocity profiles at an axial distance of z = 75 mm for different argon carrier gas flow rates (Vardelle et al., 1988).

Therefore this effect is only significant for very small particles (< 10 μm) that follow very quickly the flow field.

Particle injection can be described in two stages. In the first stage (penetration stage) the particles still maintain their velocities acquired by momentum transfer from the carrier gas. In addition they pick up momentum from the highly accelerated plasma gas. In this stage thermophoretic effects will be negligible since the particle velocities are relatively high even though the temperature gradients are on the order of 10^7 K m^{-1}. But during the following stage of deceleration (relaxation) thermophoretic effects may also be rather insignificant because the temperature gradient decreases rapidly; the displacement of particles will also be small during their short residence time of around 1 ms in the plasma jet (Pfender, 1989).

The particle velocity distribution assumes frequently a quasi-Gaussian distribution. Figure 4.17a shows a histogram of the velocity of alumina particles of a mean diameter of 79 ± 18 μm injected with a feed rate of 1.2 g min^{-1} into an inductively coupled r.f. argon plasma operated under soft-vacuum conditions at a plate power of 15 kW (Sakuta and Boulos, 1987). The powder carrier gas was helium with a flow rate of 8 L min^{-1}. The velocities were measured in-flight simultaneously with particle temperatures (Figure 4.17b) using the analysis of the change in the time- and amplitude domains of the waveform of a He–Ne laser light pulse. A powder particle with a diameter d_p traverses the observation window of height H and width W. The half-width, t_w of the laser light pulse corresponds to the time-of-flight of the particle through the slit height. Thus the particle velocity, V_p is:

$$V_p = H/(t_w R), \qquad (4.40)$$

where R = image ratio determined by the optical setup.

The rise time t_r and the fall time t_f of the trapezoidal profile of the light pulse correspond to the time elapsed between which the particle is entering or leaving the observation window:

$$t_r = t_f = d_p/V_p. \qquad (4.41)$$

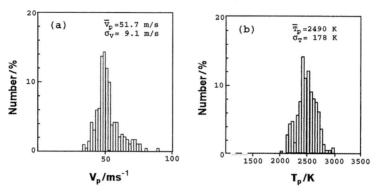

Figure 4.17 Histograms of the velocity (a) and temperature (b) of alumina particles (79 ± 18 μm diameter) injected with a feed rate of 1.2 g min^{-1} into an inductively-coupled r.f. argon plasma (Sakuta and Boulos, 1987).

The pulse intensity, i.e. the amplitude can give information on the surface temperature of the particle, T_p as shown in Equation 3.31.

The particle velocities in the plasma jet are also dependent on the residual chamber pressure. Work by Steffens *et al.* (1986) showed that momentum and heat transfer between plasma and particles are reduced at low pressure compared with atmospheric pressure. Modeling of the particle velocity under low pressure ('vacuum') plasma spray conditions require noncontinuum assumptions about the flow conditions (Wei *et al.*, 1984; see also Pfender, 1989). As shown in Figures 4.18 and 4.19 the particle velocities are strongly dependent on density and grain size of the powder materials used. The particle materials, residual chamber pressures, average grain sizes, and nozzle types used shown in Figure 4.18 are as follows: 1 (W, 1013 mbar, 23 μm, nozzle I), 2 (W, 50 mbar, 23 μm, nozzle II), 3 (Al_2O_3, 1013 mbar, 54 μm, nozzle I), 4 (Al_2O_3, 50 mbar, 54 μm, nozzle II), 5 (Al_2O_3, 50 mbar, 54 μm, nozzle III) and 6

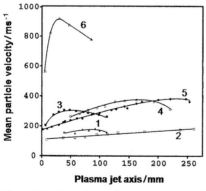

Figure 4.18 Dependence of particle velocity on density and grain size of plasma spray powders (Wei *et al.*, 1984). The designation of the different curves 1 to 6 are described in the text.

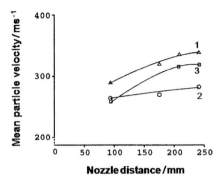

Figure 4.19 Dependence of particle velocity on density of powders used in spraying of thermal barrier coatings (Wei et al., 1984). (1: $ZrO_2 \cdot 8Y_2O_3$, <45 µm; 2: CoCrAlY, <45 µm; 3: $ZrO_2 \cdot 8Y_2O_3$ – 50 vol% CoCrAlY).

(kaolin, 50 mbar, 0.55 µm, nozzle III). Also, the acceleration of the particles in a low pressure-plasma jet is considerably less than under atmospheric pressure conditions. Figure 4.19 in particular illustrates the problems associated with the production of graded ceramics/metal coatings as suggested for TBCs on gas turbine blades for aerospace applications (see also Chapter 8.1).

4.5.2
Particle Temperature Determination

In principle, the analytical techniques applied to measure in-flight particle temperatures have already been reviewed in the preceding chapter (see 3.4.1). In order to measure the plasma temperature by two-color pyrometry, very small particles have to be introduced into the plasma jet as seeds that do not disturb the flow of the plasma (3.4.1.2).

4.5.3
Particle Number Density Determination

It is of great interest to study simultaneously the velocity, temperature and number density of powder particles in a plasma jet. Any successful modeling requires input parameters that give reliable information on the numerous forces acting on the powder particles. Equipment has been developed to accomplish the difficult task to measure simultaneously the three parameters mentioned above (see Figure 3.31). Typical particle trajectory data and particle flux number density distributions obtained with this equipment are shown in Figure 4.20.

Alumina powder particles (18 ± 3 µm diameter) were injected with a powder gas flow rate of 5.5 L min^{-1} into an Ar/H$_2$ plasma (29-kW arc power) with a feed rate of 4 g min^{-1}. The tip of the noncooled powder injection probe was 4 mm from the edge of the nozzle, i.e. 8 mm from the axis of the jet. The particles spread over a large region of the plasma (Figure 4.20c) as shown by the isopleths of the particle flux in the

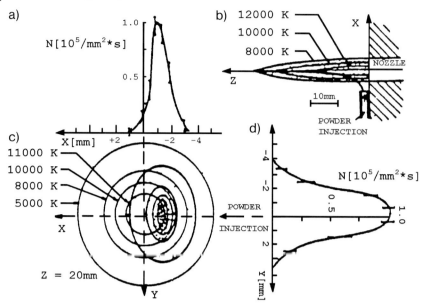

Figure 4.20 Determination of particle flux number densities N in axial and radial directions (alumina, 18 ± 3 μm) (Vardelle et al., 1988).

X–Y plane. The profiles of the particle flux number densities are shown in the orthogonal X–Z (Figure 4.20a) and Y–Z (Figure 4.20d) planes. There is a noticeable off-center asymmetry in the X–Z plane of injection, indicating the disturbance of the smooth plasma flow by the injected powder. On the other hand, the number densities N display Gaussian distributions both in axial and radial directions.

Comparison of these results with those shown in Figure 4.15 shows that by using a water–cooled injection probe and advancing the point of injection of the powder by only 4 mm towards the plasma jet axis, a substantial modification of the particle trajectories can be observed. As a consequence, in spite of identical carrier gas flow rates in this case the particles penetrate deeper into the jet, and that results in turn in the acquisition by the particles of larger momentum and higher temperatures.

In the preceding paragraphs the turbulent nature of the plasma jet was not explicitly considered. As the smooth and steady flow field of the plasma is disturbed by turbulences, the particle trajectories also become dispersed, i.e. the plasma velocity field is locally modified by randomly oriented turbulent eddies (Lee and Pfender, 1985).

For typical eddy sizes of 1 mm, Figure 4.21 illustrates the dispersed particle trajectories of alumina particles with mean diameters of 10, 30 and 50 μm injected with a velocity of $10\,\mathrm{m\,s^{-1}}$ into a turbulent free plasma argon jet (Pfender, 1989). Obviously, smaller particles will be more dispersed than larger ones since the smaller particles are following more easily the turbulent motions within the eddies. Also, the dispersion becomes more prominent when moving downstream from the point of injection due to the accumulated 'random walk' influence. For sufficiently large

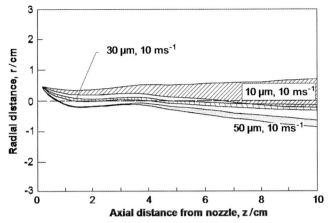

Figure 4.21 Dispersed alumina particle trajectories in a turbulent free argon plasma jet (Pfender, 1989).

particles the variable property corrections in the equation of motion (see Equation 6.11) may still exceed the turbulent dispersion term. Therefore, a correction for turbulent motion of the plasma can be neglected for larger particles.

References

J.F. Bisson, B. Gauthier, C. Moreau, *J. Thermal Spray Technol.*, 2003, **12** (1), 38.

Y. Borisov, N. Voropaj, I. Netesa, V. Korzhyk, I. Kozyakov, Proc. 14th ITSC'95, Kobe, May 22–26, 1995, 1115.

E. Bourdin, P. Fauchais, M.I. Boulos, *Int. J. Heat Mass Transfer*, 1983, **26**, 567.

B.M. Cetegen, *J. Thermal Spray Technol.*, 1999, **8** (1), 57.

X. Chen, E. Pfender, *Plasma Chem. Plasma Process.*, 1982, **2**, 185.

X. Chen, E. Pfender, *Plasma Chem. Plasma Phys.*, 1983, **3**, 351.

C.T. Crowe, M.P. Sharma, D.E. Stock, *J. Fluid Eng.*, 1977, **99**, 325.

B. Elvers, S. Hawkins, G. Schulz (eds.), Ullmann's Encyclopedia of Industrial Chemistry, 5th edition, *Metals, Surface Treatment*, 1990, Vol. A16, p. 433, VCH Weinheim.

J.L. Engelke, Proc. AIChE. Meeting, Los Angles, CA, Feb 5, 1962;

H. Eschnauer, *Die morphologische Struktur von pulverförmigen Spritzwerkstoffen*. Habilitationsschrift, RWTH Aachen, 1982.

P. Fauchais, G. Montavon, M. Vardelle, J. Cedelle, *Surf. Coat. Technol.*, 2006, **201**, 1908.

J.K. Fiszdon, *Int. J. Heat Mass Transfer*, 1979, **22**, 749.

N. Frösling, *Gött. Beitr. Geophys.*, 1938, **52**, 170.

R. Ghafouri-Azar, J. Mostaghimi, S. Chandra, M. Charmchi, *J. Thermal Spray Technol.*, 2003, **12** (1), 53.

A. Hasui, S. Kitahara, T. Fukushima, *Trans. NRIM*, 1970, **12** (1), 9.

R.B. Heimann, *Appl. Geochemistry*, 1987, **2**, 639.

R.B. Heimann, *Development of Plasma-Sprayed Silicon Nitride-Based Coatings on Steel*. Research Report to NSERC and EAITC Canada, December 15, 1992.

J.M. Houben, *Relations of the Adhesion of Plasma Sprayed Coatings to the Process Parameters Size, Velocity and Heat Content of* the Spray Particles. Ph.D.Thesis, Technical University Eindhoven, The Netherlands, 1988.

J.J. Kaiser, Z. Zurecki, K.R. Berger, R.B. Swan, E.A. Hayduck, US Patent 5,294,462 (1994).

E.J. Kubel, *Adv. Mater. Process.*, 1990, **12**, 24.

Y.C. Lee, K.C. Hsu, E. Pfender, 5th Int. Symp. on Plasma Chemistry, 1981, 2, 795.

Y.C. Lee, E. Pfender, *Plasma Chem. Plasma Process.*, 1985, **5**, 3.

J.A. Lewis, W.H. Gaudin, *AIChE J.*, 1973, **19**, 982.

X. Liu, L.A. Gabour, J.H. v.Lienhard, *J. Heat Transfer*, 1993, **115**, 99.

H. Maruo, Y. Hirata, J. Kato, Y. Matsumoto, Proc. Intern. Symp. Adv. Thermal Spraying Technol. and Allied Coatings (ATTAC'88) 1988, 153.

C.W. Marynowski, F.A. Halden, E.P. Farley, *Electrochem.Technol.*, 1965, **3**, 109.

D.R. Marsh, N.E. Weare, D.L. Walker, *J. of Metals*, 1961, July, 473.

G. Matthäus, Proc. Wiss. Koll., TU Ilmenau, Germany, Sept 24–26 2003, session 7.7.

C. Moreau, P. Gougeon, A. Burgess, D. Ross, in: S. Sampath, C.C. Berndt,(eds.), *Advances in Thermal Spray: Science and Technology*, ASM Intern., Materials Park, OH, USA, 1995, 41.

J. Mostaghimi, P. Proulx, M.I. Boulos, *Int. J. Heat Mass Transfer*, 1985, **28**, 187.

J. Mostaghimi, Trans. 17th Workshop CUICAC, October 1–2, 1991, Quebec.

L. Pawlowski, *The Science and Engineering of Thermal Spray Coatings*, Chapter 1, pp. 1–20, J. Wiley: Chichester, New York, Brisbane, Toronto, Singapore, 1995.

E. Pfender, *Plasma Chem. Plasma Process.*, 1989, **9**, 167S.

Powder Technology, Product brochure, Institute of Materials Science, Rheinisch-Westfälische Technische Hochschule Aachen, 1992.

P. Proulx, J. Mostaghimi, M.I. Boulos, *Int. J. Heat Mass Transfer*, 1985, **28**, 1327.

G.D. Raithby, E.R. Eckert, *Int. J. Heat Mass Transfer*, 1968, **11**, 1233.

W.E. Ranz, W.R. Marshall, *Chem. Eng. Prog.*, 1952, 48.

N.N. Rykalin, V.V. Kudinov, *Pure & Appl.Chem.*, 1976, **48**, 229.

D. Sacriste, N. Goubot, M. Ducos, A. Vardelle, *J. Thermal Spray Technol.*, 2001, **10** (2), 352.

T. Sakuta, M.I. Boulos, Proc. 8th Int. Symp. Plasma Chem., ISPC-8, Tokyo, 1987, paper BVII-02, p. 371.

H.R. Salimijazi, M. Raessi, J. Mostaghimi, T.W. Coyle, *Surf. Coat. Technol.*, 2007, **201** (18) 7924.

N.N. Sayegh, W.H. Gauvin, *AIChE Journal*, 1979, **25**, 522.

V. Srinivasan, A. Vaidya, T. Streibl, M. Friis, S. Sampath, *J. Thermal Spray Technol.*, 2006, **15** (4), 739.

H.-D. Steffens, K.-H. Busse, M. Schneider, *Adv. Therm. Spraying*, Pergamon Press: New York, 49, 1986.

R. Thom, *Structural Stability and Morphogenesis*, W.A. Benjamin, 1975.

A. Vardelle, M. Vardelle, P. Fauchais, *Plasma Chem. Plasma Process.*, 1982, **2**, 255.

M. Vardelle, A. Vardelle, P. Fauchais, M.I. Boulos, *AIChE Journal*, 1983, **29**, 236.

M. Vardelle, A. Vardelle, P. Fauchais, M.I. Boulos, *AIChE Journal*, 1988, **34** (4), 567.

D. Wei, D. Apelian, M. Paliwal, S.M. Correa, Proc. Symp.Mat. Res. Soc., *Plasma Processing and Synthesis of Materials*, 1984, **30**, 197.

J. Wilden, A. Wank, F. Schreiber, in: Proc. ITSC2000, Montreál, Canada, May 8–11, 2000.

F.D. Witherspoon, D.W. Massey, R.W. Kincaid, US Patent 6,001,426 (1999).

K.J. Zahnle, *J. Geophys. Res.*, 1992, **97**, 10, 243.

5
The Third Energy Transfer Process: Particle–Substrate Interactions

5.1
Basic Considerations

Coatings are built up particle by particle. The solidification time, t_{sol} of a hot molten particle arriving at the cold substrate surface is orders of magnitude shorter than the intermission time, t_i, i.e. the time lag between two arriving particles flying in the same trajectory. Therefore an arriving particle does not encounter a permanent liquid melt pool such as exists in welding processes. The cooling time from the freezing point down to ambient temperature is two to three orders of magnitude longer than the solidification time. This means that a limited degree of particle interaction by surface diffusion as well as relief of some stress is still possible.

The properties of the layered deposited material are determined by the

- velocity and temperature of the particles on impact;
- relative movement of the plasmatron and the substrate;
- substrate and coating cooling during spraying.

The wetting and flow properties of the liquid droplets are of great importance. They influence the

- coating porosity;
- morphology of the coating/substrate interface;
- cohesive bonding among splats and successive layers;
- adhesive bonding to the substrate.

Flow and solidification of molten droplets on impact are difficult to treat theoretically because of interaction between heat transfer and crystal growth kinetics (Apelian, 1984; Madejski, 1976; McPherson, 1980; Houben, 1984; Pasandideh-Fard et al., 1998; Hadjiconstantinou, 1999). It is also necessary to take into consideration the propagation of shock waves into the flattened particles and the substrate.

Plasma Spray Coating: Principles and Applications. Robert B. Heimann
Copyright © 2008 WILEY-VCH Verlag GmbH & Co. KGaA, Weinheim
ISBN: 978-3-527-32050-9

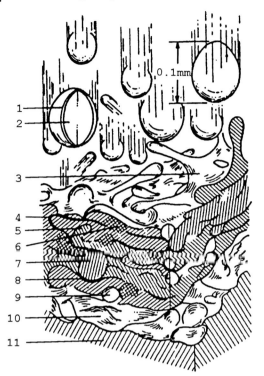

Figure 5.1 Schematic rendering of the chaotic structure of a plasma-sprayed coating layer. 1: Thin molten shell; 2: unmelted core; 3: liquid splash; 4: 'pancake' splat; 5: interlocked splat; 6: oxidized particle; 7: unmelted particle; 8: pore; 9: void; 10: roughened substrate surface; 11: substrate (Herman, 1988).

The study of particle splats on glass or ceramic slides ('wipe test') (Gruner, 1984) is a useful method to determine quickly the degree of superheating and thus the viscosity of the liquid particle droplets and the morphology and contiguity of the solidified splats.

Figure 5.1 shows an artistic rendering of the chaotic plasma spay process (Herman, 1988). The molten droplets are thought to arrive one at a time, propagate along ballistic trajectories, but do not interfere with each other during flight ('ballistic model', see Section 10.2.2). A spherical droplet with a diameter d_p impacts the solid substrate surface, and deforms to a so-called 'pancake' or 'Mexican hat' shape with a splat diameter D_s (Figure 5.2). Entrained gases, partially melted and oxidized particles, and pores and voids are mixed with the particle splats, and tend to degrade the coating properties because they provide points of stress concentration during in-service loading that will act as crack initiation centers. Particles with sizes substantially larger than the mean do not melt completely to their core and thus do not spread on impact. In their wake porosity can build up as well as trains of other only partially

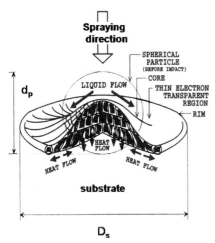

Figure 5.2 Deformation of a plasma-sprayed particle (Herman 1988)

melted particles (Figure 5.3). Thus it is extremely important to use spray powders with a narrow particle size distribution with a sufficiently small standard deviation. The extremely complex relationships between the flattening behavior of particles and the dynamics of the coating formation and their properties have been recently reviewed by Fauchais *et al.* (2004).

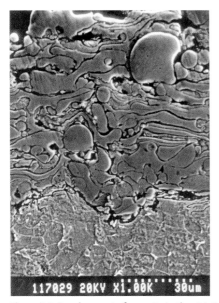

Figure 5.3 Development of porosity in the vicinity of large unmelted particles (85Fe15Si on low carbon steel, etched for 20 s in Nital).

5.2
Estimation of Particle Number Density

A quantitative estimation of the number of particles arriving at a defined surface area of the substrate in unit time, and the build-up of lamellae of the coating has been given by Houben (1988) by considering spraying molybdenum powder (specific mass $\rho_{Mo} = 10.2$ Mg m^{-3}) of a mean particle size d_p of 50×10^{-6} m with a powder feed rate of 40 g min^{-1} (6.7×10^{-4} kg s^{-1}) onto a steel substrate. Thus, the number of particles injected into the plasma jet, is $N_1 = 10^6$ particles s^{-1}.[1] The traverse speed of the plasmatron moving relative to the substrate is assumed to be $v = 4 \times 10^{-2}$ m s^{-1}, and the spray width be $w = 2.5 \times 10^{-2}$ m. The lamella diameter D_s of a single splat is 125×10^{-6} m ($\xi = D_s/d_p = 2.5$). With these parameters, the average number of lamellae on top of each other, N_2 after one pass of the plasmatron can be calculated under the assumption that all particles N_1 injected will arrive at and stick to the substrate surface (deposition efficiency 100%) as follows.

The N_1 particles produce a totally coated spray surface/unit time of

$$A_p = N_1 \times (\pi/4) D_s^2 \\ = 10^6 [p/s] \times (\pi/4)(125 \times 10^{-6})^2 [m^2] = 1.23 \times 10^{-2} m^2 s^{-1}. \quad (5.1)$$

Because the plasma jet moves, the covered substrate area will be

$$A_s = v \times w = 1.0 \times 10^{-3} \text{ m}^2 \text{ s}^{-1}. \quad (5.2)$$

Then the average number of lamellae on top of each other is

$$N_2 = A_p/A_s = 12. \quad (5.3)$$

The deposition time, i.e. the time required to lay down the N_2 lamellae is $t_{dep} = w/v = 0.625$ s. The intensity of the bombardment of the surface is $n = N_1/w^2 = 1.6 \times 10^9$ particles s^{-1} m^{-2}. The surface receives per unit area $N_2 = t_{dep} \times n = N_1/v \times w = N_1/A_s = 1.0 \times 10^9$ particles m^{-2}.

Furthermore, the time elapsed between the arrival of two particles belonging to the i th and (i + 1)th lamella plane is defined as t_i or 'intermission' time. It follows:

$$t_i = w/v \times N_2 = 0.052 \text{ s}. \quad (5.4)$$

The solidification time for a 50 μm Mo droplet can be calculated using an approximate solution of the heat diffusion equation at the interface solid substrate/liquid droplet:

$$t_{sol} = x^2/4p^2 \times a, \quad (5.5)$$

where x = lamella thickness, p = constant,[2] and a = thermal diffusivity.

---

[1] Mass of single particle $m_p = V_p \times r_{Mo} = (4/3)\rho (d_p[m]/2)^3 r_{Mo}[kg \text{ m}^{-3}] = (\rho/6)(50 \times 10^{-6}[m])^3 \times 10.2 \times 10^3 [kg \text{ m}^{-3}] = 6.7 \times 10^{-10}$ kg. Powder feed rate $m = 6.7 \times 10^{-4}$ kg s^{-1}. Number of particles injected $N_1 = m/m_p = 10^6$ s^{-1}.

[2] According to Houben (1988), p is the Neumann–Schwartz parameter that can be estimated as the fitting parameter from the solidification time vs. layer thickness relationship (Figure. 5.13).

Numerically, for a lamella thickness of 7 μm,[3] the solidification time becomes $t_{sol} = (7 \times 10^{-6}\,[m^2])^2/4(0.582)^2 \times (5.61 \times 10^{-5}\,[m^2/s]) = 6.5 \times 10^{-7}$ s.

The ratio between intermission time and solidification time is $t_i/t_{sol} = 5.2 \times 10^{-2}$ [s]/6.5×10^{-7}[s] = 80 000! This means that the time between two consecutive collisions is orders of magnitude longer than the time needed to solidify the droplets. Thus the liquid droplets will probably not encounter a liquid surface, i.e. a weld pool does not exist. This is consistent with the view of plasma spray technology as a rapid solidification technology with cooling rates exceeding 10^6 to 10^7 K s^{-1}. It should be emphasized that Equation 5.5 is only a very rough zero-order solution of the heat transfer (Fourier) equation. An exact solution is presented in Section 5.4.1, and a numerical example will be given in Section 5.4.2 and Appendix B.

With similar simple assumptions further insight can be gained into the particle 'economy' and distribution at the substrate surface. Questions to be answered include how far two particles, on average, are apart, i.e. what their flight distance along a common trajectory might be. This distance, z_s can be calculated simply by multiplying the intermission time, t_i with the flight velocity u_p of the particles. If the flight velocity is 50 m s^{-1}, than with the time of $t_i = 0.052$ s calculated above, the flight distance between two particles is

$$z_s = u_p \times t_i = 50[m/s] \times 0.052[s] = 2.60\,m$$

On the other hand, the required flight distance between two particles moving in the same trajectory in order to meet a liquid preceding particle cannot be longer than $z_l < u_p \times t_{sol} = 50[m/s] \times 6.5 \times 10^{-7}$ [s] $= 3.25 \times 10^{-5}$ m. Again, it is not likely that a particle will meet another liquid one.

While one particle solidifies at the surface, the number of particles arriving will be

$$\begin{aligned}N_3 &= t_{sol} \times N_1 \\ &= 6.5 \times 10^{-7}[s] \times 10^6 \times [p/s] = 0.65\,\text{particles}.\end{aligned} \quad (5.6)$$

These particles are distributed over the surface area $A_0 = (\pi/4)w^2 = 4.9 \times 10^{-4}\,m^2$. Thus the number of particles arriving per unit area during solidification of one particle are

$$\begin{aligned}N_4 &= N_3/A_0 \\ &= 0.65[p]/4.9 \times 10^{-4}[m^2] = 1326\,\text{particles m}^{-2}.\end{aligned} \quad (5.7)$$

These N_4 particles will start to solidify simultaneously. Since they are very small but spread out over a large area, they will not interact with each other, i.e. the initial assumption of a ballistic deposition process is warranted. Experimental determination of the particle number density for model systems has been shown in Section 4.5.3.

---

3) Assuming that the originally spherical droplet with a diameter $d_p = 50 \times 10^{-6}$ m and a volume V_p of 65.4×10^{-15} m^3 deforms on impact to a cylindrical disc of diameter $D_s = 125 \times 10^{-6}$ m of a volume $V_c = h(\pi/4)D_s^2 = 65.4 \times 10^{-15}$ m^3, the height h of the disc, i.e. the lamella thickness is 5.3×10^{-6} m. Accounting for the 'Mexican hat' shape of the splat the geometrical factor of 1.3 (Houben, 1984) transforms this value to an actual thickness of 7×10^{-6} m.

5.3
Momentum Transfer from Particles to Substrate

Molten particles arrive with high velocities at the substrate surface, will be deformed and, as shown below, partially solidify due to the increase of the melting temperature with shock pressure. The flattening ratio $\xi = D_s/d_p$ (D_s = splat diameter, d_p = original particle diameter; see Figure 5.2) depends not only on intrinsic materials properties such as viscosity μ and density ρ of the liquid phase, but also on the impact velocity v_i and the impact angle. The classic semi-quantitative expression developed by Madejski (1976) for a particle impact at an angle of 90° on a flat substrate reads as follows

$$\xi = D_s/d_p = 1.2941 \times (\rho \times v_i/\mu)^{0.2}. \tag{5.8}$$

Since the Reynolds number is given by $\mathbf{Re} = \rho v d/\mu = vd/\nu$, an approximate relationship exists for the flattening ratio as

$$\xi = D_s/d_p = A \times \mathbf{Re}^z \tag{5.8a}$$

assuming an isothermal particle (Trapaga and Szekely, 1991; McPherson, 1989). However, because there are pronounced thermal gradients across the particles, this relationship cannot be taken for granted in reality (Fantassi et al., 1993). Also, the flattening ratio increases with increasing initial particle radius. Hence many values for the pre-exponential coefficient A and the exponent z have been proposed in the literature, for example A = 1.2941, z = 0.2 (Madejski, 1976; Equation 5.8); A = 0.82, z = 0.2 (Watanabe et al., 1992); A = 0.5, z = 0.25 (Pasandideh-Fard and Mostaghimi, 1996); A = 0.5, z = 0.25 (Hadjiconstantinou, 1999); A = 1.21, z = 0.125 (Li et al., 2005) or more recently, A = 0.925, z = 0.2 (Dyshlovenko et al., 2006) and A = 1.00, z = 0.22 (Kang and Ng, 2006).

To account for the suspected dependence of the flattening (spreading) ratio ξ on droplet mass loss and roughness of the substrate surface, Sobolev et al. (1996) developed an expression

$$\xi = 0.8546\sqrt{\chi}\mathbf{Re}^{1/4}\{1-0.06\sqrt{\gamma}\sqrt{\mathbf{Re}}\}, \tag{5.8b}$$

where χ is the (dimensionless) parameter of droplet mass loss and γ is the dimensionless roughness parameter defined as the ratio of the roughness height ε to the particle radius r_p. An important consequence of this relationship is that the spreading ratio decreases with increasing surface roughness, i.e. a rough surface promotes friction between a spreading droplet and the surface thus retarding the spreading process (see Section 6.4.2).

With deviation of the spray angle from 90° the properties of the coatings change depending on the nature of the sprayed material. For example, in case of molybdenum the cohesion within the coating increases by a factor of two for an angle of 45° whereas for alumina coating the cohesion increases slightly but the adhesion is not affected. The deposition efficiency decreases for both materials with increasing deviation of the spray angle from 90° (Hasui et al., 1970).

Kang and Ng (2006) presented a study on morphology and spreading behavior of Y-stabilized zirconia splats impacting obliquely at the substrate surface. In particular, the effect of the angle of incidence on the spreading ratio ξ and the aspect ratio ψ (defined as the ratio of the major to the minor diameter of an elliptical splat) were considered, indicating a parabolic relationship between spreading ratio and angle of incidence θ to yield

$$\xi = -9 \times 10^{-4}\theta^2 + 5.57 \times 10^{-2}\theta + 4.6024 (r^2 = 0.945). \tag{5.9a}$$

The spreading ratio ξ was found to be 4.60 at $\theta = 0°$ and reaches a maximum of 5.50 at $\theta = 30°$ before dropping to 4.80 at $\theta = 60°$. The aspect ratio ψ increased from 1.00 (circular splat) at $\theta = 0°$ to about 1.82 at $\theta = 60°$ and could be fitted to the following parabolic equation:

$$\psi = 1.0252 + 3 \times 10^{-4}\theta^2 - 6.2 \times 10^{-3}\theta (r^2 = 0.982) \tag{5.9b}$$

The relationships obtained are in qualitative agreement with those reported by Montavon et al. (1997).

Many attempts have been made to study theoretically the time-dependent deformation and solidification of molten droplets according to the Madejski model (Maruo et al., 1995; Fukumoto et al., 1995; Solonenko et al., 1995; Pasandideh-Fard et al., 1998; Hadjiconstantinou, 1999; Pasandideh-Fard et al., 2002; Ghafouri-Azar et al., 2004). For example, Maruo et al. (1995) solved the conservation law of mechanical energy using the Marker and Cell method (Harlow and Welch, (1965)). As heat from the molten droplet is transported by the high speed flow of the spreading, cooling and solidification occurs at a much higher rate compared with the Madejski model that operates with the assumption of pure heat conduction. Also, a substrate with higher thermal conductivity results in a lower deformation ratio of the droplets. The work by Solonenko et al. (1995) showed that for particle sizes of $d_p = 0.31$ mm the theoretical solution of the heat transfer equations compares satisfactorily with experimental results without taking into account any empirical constants under the conditions of a main regime parameters (MRP) control prior to impact. Those parameters were the velocity, temperature, size, density, thermal conductivity and latent heat of melting of the particles, as well as the substrate temperature. One of the important results of the study were that the parameter

$$F_{p,s} = \rho_{pm} L_p / \rho_{sm} L_s, \tag{5.10}$$

where the subscripts 'p' and 's' refer to particle and substrate, respectively, and 'm' refer to the melting point, affects the dynamics of simultaneous solidification and particle flattening in a profound way. The parameter c_ξ (dimension: length) characterizes the thickness h_s of the molten layer of the substrate according to $h_s = c_\xi \sqrt{Fo}$, and can be computed from the equation:

$$c_\xi = F_{p,s}[c_\zeta + 2(q_2 - q_1)/\mathbf{K}u_p], \tag{5.10a}$$

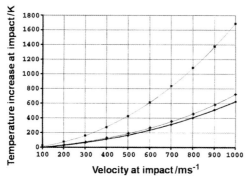

Figure 5.4 Dependence of the temperature increase due to particle impact on the particle velocity (Pawlowski, 1995).

where q is the heat flux and **Ku** = $L_p/c_{pm} T_{pm}$ is the Stefan–Kutateladze dimensional group (L_p = latent heat of phase transition, c_p = molar heat at constant pressure, T_p = temperature of phase transition). If the heat flux into the melt, q_1 and into the substrate, q_2 are comparable than $F_{p,s} = c_\xi/c_\zeta$. The parameter c_ζ is a function of the Fourier number **Fo** and the Peclet number, **Pe**, and according to Equation 5.10 equals ρ_{sm}/L_s. It should be emphasized that the parameter $F_{p,s}$ is a criterion of the relative ease of melting, Φ (see Section 4.4.1). A second important finding of the study was that the theoretical and experimental results differ appreciably from those reported by Madejski (1976) and Jones (1971).

Particles traveling along a high-velocity plasma jet as produced in a D-Gun™ or under HVOF conditions acquire a high kinetic energy $m \times v^2/2$ that on impact can be partially transformed adiabatically into thermal energy $m \times c_p \times \Delta T$ to yield

$$m \times v^2/2 = m \times c_p \times \Delta T, \quad (5.11)$$

where c_p = specific heat at constant pressure and m = mass of particle. The temperature increase reaches appreciable values only for particle velocities exceeding about 400 m s^{-1} as shown in Figure 5.4 (Pawlowski, 1995). The collision of particles with the substrate surface and with already deposited solidified particles, respectively causes a deformation-induced melting within a thin surface layer thus generating a quasi-metallurgical bond at very high particle velocities (Kreye et al., 1986). Particle velocities of around 100 m s^{-1} lead to negligible temperature increase of only 5 °C on deceleration to zero speed assuming c_p = 1000 J kg^{-1} K^{-1}.

Apart from the transformation of kinetic impact energy to heat by an adiabatic process there is, however, another process that can lead to substantial particle and substrate heating by isentropic energy changes facilitated by planar shock waves.[4] These shock waves obey the conservation equations of mass, momentum, and energy (Rankine–Hugoniot equation of state).

---

4) According to Trapaga and Szekely (1991) the shock approach is not fully legitimate since the Mach numbers in normal plasma spray jets appear to be rather low (< 0.05). However, with supersonic particle velocities obtained in D-Gun™ or HVOF processes the assumption of the existence of shock waves is valid.

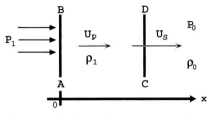

Figure 5.5 Shock wave front CD moving through a solid material (see text).

The shock wave concept can be illustrated by the idealized situation of Figure 5.5.

A uniform pressure P_1 is suddenly applied to the thick slab of compressible solid material. This pressure is transmitted to the interior of the slab through a high-amplitude stress wave. If the material behaves normally in compression, i.e. if the compressibility decreases with increasing pressure, the wave front will essentially be a discontinuity in the stress and materials velocity fields, and in density and internal energy, and travels at supersonic speed with respect to the material ahead of the shock. This discontinuity is called a *shock wave*, characterized mathematically by a set of nonlinear hyperbolic differential equations. If P_1 remains time-invariant the state variables are constant between the boundary AB and the shock front CD. As shown in Figure 5.5 the particle velocity is U_p and the shock wave front velocity is U_s. The mass flux density in [g cm^{-2} s^{-1}] into and out of the shock front is the product of the material density and the particle velocity relative to the shock front, $\rho_0 (U_s - U_{p0})$ ahead of the shock, and $\rho (U_s - U_p)$ behind the shock. Conservation of mass requires that

$$V/V_0 = \rho_0/\rho = (U_s - U_p)/(U_s - U_{p0}). \tag{5.12}$$

Momentum conservation is expressed through Newton's second law,

$$F = m(du/dt). \tag{5.13}$$

The force per unit area across the shock front is the pressure difference $(P - P_0)$, the mass flux per unit area is again $\rho_0 (U_s - U_{p0})$, and the materials velocity change is $(U_p - U_{p0})$. Substituting into Equation 5.13 the momentum conservation equation

$$P - P_0 = \rho_0 (U_s - U_{p0})(U_p - U_{p0}) \tag{5.13a}$$

results.

The conservation of energy across the shock front can be expressed by equating the work done per unit area and time by the pressure forces, $(PU_p - P_0 U_{p0})$ to the sum of the kinetic energy change, $(1/2)\rho_0 (U_s - U_{p0})(U_p^2 - U_{p0}^2)$ and the internal energy change $\rho_0 (U_s - U_{p0})(E - E_0)$, i.e.

$$PU_p - P_0 U_{p0} = (1/2)\rho_0 (U_s - U_{p0})(U_p^2 - U_{p0}^2) + \rho_0 (U_s - U_{p0})(E - E_0). \tag{5.13b}$$

It can be shown (Rice et al., 1958) that combining Equation 5.13b with Equations 5.12 and 5.13a yields the common form of the energy conservation equation

$$\Delta E = (E - E_0) = (1/2)(P + P_0)(V_0 - V), \tag{5.14}$$

where P and P_0 are the pressures behind and ahead of the shock front, and E and E_0 are the internal energies of the substrate material behind and ahead of the shock. The conservation equation in the form of Equation 5.14 is called the Rankine–Hugoniot equation.

From the conditions of conservation of mass, momentum and energy, assuming $U_{p0} = 0$, it follows that

$$U_s/U_s - U_p = V_0/V, \quad (5.15a)$$

$$U_s U_p/V_0 = P - P_0. \quad (5.15b)$$

The conservation equations have intuitive geometric interpretations. Assuming $P_0 = 0$ in Equations 5.15a and 5.15b, two definitions of the shock front velocity, U_s and the particle velocity, U_p can be derived as

$$U_s^2 = V_0^2/(P/V_0 - V) \quad (5.16a)$$

$$U_p^2 = P(V_0 - V). \quad (5.16b)$$

Any set of values for U_s and U_p correspond in the P–V diagram (Figure 5.6) to a straight line $U_s = $ constant and a hyperbola $U_p = $ constant. The intersection of the curves fixes the state of the shock compression with parameters P_1, V_1 in A through which the Hugoniot adiabatic P_H passes. The shock wave with a pressure P_1 results in a decrease of the specific volume $V_1 < V_0$. The pressure P_1 is composed of two contributions P_c and P_t. P_c is the 'cold' pressure resulting from the strong repulsive force of the interatomic potential, and P_t is the 'thermal' pressure associated with the thermal motion of atoms and electrons due to the shear compression. Therefore, as follows from Equation 5.14, the total increase in internal energy (ΔE) is equal to the area of the triangle OAB in Figure 5.6. This energy increment consists of an elastic component, ΔE_c (curvilinear triangle OCB)

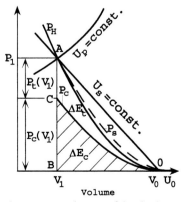

Figure 5.6 P–V diagram of the shock compression of a solid material. P_H is the Rankine–Hugoniot shock adiabat (see text) (Heimann and Kleiman, 1988).

which is a result of the elastic (cold) pressure developing in the material, and a thermal energy component, ΔE_t, represented by the likewise curvilinear triangle OAC.

The shock transition provides both kinetic and internal energy to the material through which the shock wave propagates. Moreover, irreversible work is done on the material as it passes through the shock front. It can be shown that the entropy increases monotonically with pressure along the Rankine–Hugoniot shock adiabatic that lies above the adiabatic passing though the initial state.

Inasmuch as $\Delta E > \Delta E_c$, the shock–compression process is accompanied by heating of the substance and by an increase in its entropy which in turn leads to the appearance of the thermal pressure component P_t. It can be seen in Figure 5.6 that the thermal energy and the thermal pressure increase progressively with increasing shock pressure. The shock temperature can be determined by the equation

$$T_A = T_C \exp[S_A/c_v], \qquad (5.17)$$

where T_A = shock state temperature at state A (P_c, V_1) and T_C = adiabatic compression temperature at state C (P_1, V_1). On the other hand during relaxation the heated material cools down, and the volume increases along the expansion isentrope P_s as shown by the dashed line in Figure 5.6. The temperature after relaxation can be determined by

$$T_E = T_0 \exp[S_A/c_p], \qquad (5.18)$$

where T_0 = temperature before the shock, and S_A = entropy in the shocked state.

The mechanics of a spherical molten particle with radius r colliding with the flat surface of a substrate at a velocity U_p can be approximated by Figure 5.7. The inward moving shock front defines that portion of the material initially within the volume BOB that has been compressed into the volume BCB at time t. The contact face perimeter, a_t moves outward with a velocity $å_t$.

It can be simply calculated by

$$a_t^2 = 2rU_p t - [U_p t]^2 \qquad (5.19)$$

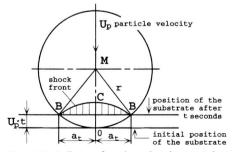

Figure 5.7 Collision of a spherical molten particle with a flat surface (Houben, 1988).

and the time derivative, i.e. the velocity at which the perimeter of the contact area moves outwards can be expressed as

$$da_t/dt = å_t = U_p[r-U_pt]/[2rU_pt-(U_pt)^2] \qquad (5.20)$$

The following conclusions can be drawn (Houben, 1988).

- The perimeter velocity $å_t$ increases with decreasing contact area radius a_t. For larger particles $å_t$ may even exceed the shock velocity, U_s.
- The initial collision phase, arbitrarily defined by a contact area radius of 1 μm, lasts only 2×10^{-9} to 1.3×10^{-10} s for small (5 μm), and 5×10^{-10} to 3×10^{-11} s for larger (25 μm) particles.
- The spreading velocity of the particles increases with size. The spreading process is about two orders of magnitude faster than the solidification time, t_{sol}.
- During the process, adiabatic conditions prevail. Superheated particles will spread as a liquid from the start of the collision. This will provide good conditions for adherence of the coating to the substrate.
- Due to adiabatic shock heating the particle splats can spread more easily over the substrate surface. This improves adhesion by establishment of a thin diffusion layer, i.e. a metallurgical bond.

An important conclusion from the increase of the melting temperature of most substances with pressure (Clausius–Clapeyron's equation) is that a completely molten particle arriving at the substrate surface can partially solidify during collision, not by conductive heat loss to the cold substrate, but through adiabatic processes. This was used by Houben (1988) to develop a tractable model to explain the occurrence of different types of splats (pancake, flower and exploded type).[5] Figure 5.8 shows how the generation of a shock wave on particle impact modifies the spreading pattern of this particle. In the (subcritical) phase 1 ($t<t_c$) compression but no flow occurs. This situation corresponds to that shown in Figure 5.7.

As the shock front moves inward the pressure increases and partial solidification sets in because of the pressure dependence of the melting temperature expressed by the Clausius–Clapeyron equation. In the post-critical phase 2 ($t>t_c$) the shock wave front squeezes out solid material laterally forming the typical flower pattern often observed in the 'wipe' test (see Section 10.2.2.1). In phase 3 the rarefaction wave accompanying the shock pressure release moves in the opposite direction towards the shock wave front. The compressed solid material of the droplet can relax and thus reliquefy. This liquid material now flows out sideways over the splat surface and forms a corona of spray material surrounding the splat. Depending on the critical radius of the outflow and the angle α material tends to escape from the pressurized zone. The larger the critical radius the larger the potential energy stored in the pressurized zone. Instantaneous release of this pressure may lead to an explosive splat pattern that is detrimental to the coating cohesion as well as adhesion to the substrate.

---

5) More modern approaches to determine the splat morphology using FE models are discussed in Chapter 6.

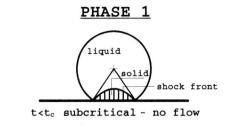
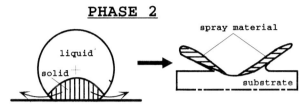
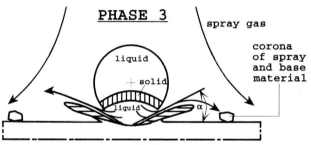

Figure 5.8 Spreading pattern of highly accelerated molten particles (Houben, 1988).

Another type of pattern occurs during impact of highly superheated material presumably without the involvement of a partially solidified portion. Figure 5.9 shows frozen-in-time traces of superheated alumina splats with material ejected on impact through the shock process (Heimann, 1991). This ejected matter leaves behind voids that build up secondary microporosity in the plasma-sprayed layer. Very complex sequences of shock compression and rarefaction waves are set up which generate nonequilibrium relaxation temperatures in 'hot spots'. These high local temperatures may delay solidification of the particle, and help to spread a liquid film across the surface of the substrate. In this way the adhesion of the coating to the substrate can be improved because the surface of the latter can partially melt to form a reaction boundary layer of intermediate composition (Kitahara and Hasui, 1974).

The complex spreading and solidification pattern of molten particles has a profound impact on the chemical homogeneity of a plasma-sprayed coating. This has been shown recently by Heimann (2006) for the phase composition of calcium phosphate coatings for biomedical application as bioconductive coating for implants

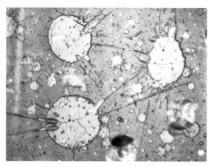

Figure 5.9 Frozen-in-time traces of superheated alumina splats with materials ejected in impact (Heimann, 1991).

(Heimann, 2007). Figure 5.10 shows a schematic model of the thermal decomposition of a spherical hydroxyapatite (HAp) particle at a defined water partial pressure according to the phase diagram of CaO–P_2O_5 (Riboud, 1973).

This model has recently been modified by Dyshlovenko *et al.*, (2004) to include the product of the solid state transformation of HAp into tricalcium phosphate (TCP) and tetracalcium phosphate (TetrCP) between 1633 and 1843 K.

During impact of the particle at the solid surface of Ti alloy the phase separation still present in the molten particle will be lost, and an extremely inhomogeneous calcium phosphate layer will be deposited in which HAp, oxy/oxyhydroxyapatite (OAp/OHAp), TCP, TetrCP, CaO and amorphous calcium phosphate (ACP) of various composition are interspaced on a nano- to microcrystalline scale (Götze *et al.*, 2001, Li *et al.*, 2001). On impact, the morphology of the deposited particle splat follows the laws of fluid dynamics under supersonic conditions (Fauchais *et al.*, 2004; Heimann, 2006). Figure 5.11 shows the hypothetical sequence of events,

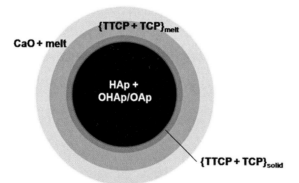

Figure 5.10 Schematic model of the thermal decomposition of a spherical hydroxyapatite particle subjected to high temperature in an argon plasma jet at a water partial pressure of 1.3 kPa (Dyshlovenko *et al.*, 2004).

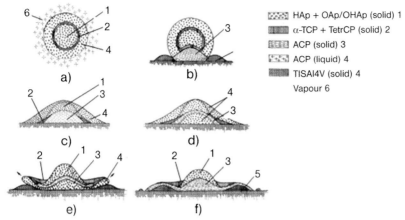

Figure 5.11 Sequence of events (a to f) occurring during deposition of a supersonically accelerated semi-molten hydroxyapatite particle impinging on a roughened Ti6Al4V substrate (Heimann, 2006).

starting with stage 1 (free flight and evaporation phase, a) and progressing through stage 2 (contact and compression phase, b), stage 3 (compression and flow phase, c), stage 4 (remelting phase, d), stage 5 (deformation and splashing phase, e) to stage 6 (relaxation and cooling phase, f). When the spherical semi-molten particle (Figure 5.11a) impacts the solid planar substrate with supersonic velocity an inward moving shock front compresses molten material within the contact area radius (McPherson, 1980) (Figure 5.11b) that solidifies by rapid conductive heat transfer into the substrate as well as by an increase of the melting temperature with pressure (Clausius–Clapeyron's law) through an adiabatic process. The boundary of the hemispherical solidified material is defined by the locus of the shock front. During stage 3 (Figure 5.11c) the molten material of the outer shell not in contact with the substrate starts to flow down and over the solidified ACP forming the characteristic sombrero-like shape of a plasma-sprayed particle. In stage 4 (Figure 5.11d) the shock wave has mostly dissipated and spontaneous partial remelting of the ACP by isentropic relaxation occurs (Heimann and Kleiman, 1988). This liquid material will be squeezed out laterally (stage 5, Figure 5.11e) and the splat takes on a pancake-like shape. The spray of liquid superheated material frequently forms a corona of tiny spherical beads surrounding the splat (Houben, 1988). In the final stage 6 (Figure 5.11f) the splat will relax, expand laterally and cool down. Much of the molten material will solidify as amorphous calcium phosphate (ACP) with minor crystalline contributions of OAp, TCP and TetrCP.

The next particle arriving at the substrate surface will splash over the previously deposited, already cooled particles compressing them further. The time lag between arrival of subsequent particles at the substrate interface has been calculated by Houben (1988). Since the arrival time and the very short solidification time of an individual particle differ by several orders of magnitude a molten particle will never meet a molten pool at the surface. As a consequence any particle stresses will be retained and accumulate as the coating grows in thickness.

Complications arise since the solidified part of an impacting particle on collision may undergo shattering by catastrophic fragmentation because of differential pressures across the particle. The leading face of the arriving particle is subjected to an average dynamic pressure

$$p_{dyn} \approx C_D \rho_p v^2/2, \tag{5.21}$$

where C_D = drag coefficient, ρ_p = particle density and v = particle velocity. The pressure on the trailing face is much smaller. Integration of Equation 5.21 over the surface of the spherical particle, $4\pi r^2$ yield the drag force

$$m(dv/dt) = -(1/2)C_D \rho_p A v^2 + (g/m) \sin\varphi, \tag{5.22}$$

where m = particle mass, $A = \pi r^2$ = cross-sectional area, φ = angle between particle trajectory and plasma jet axis (Chyba et al., 1993). The second term in Equation 5.22 can be neglected because of the short distance of travel of the particle. Fragmentation of the particle occurs when p_{dyn} in Equation 5.21 exceeds the characteristic yield strength of the solid material. This mechanism has been confirmed by Reisel and Heimann (2004) by a fractal approach (see Section 5.6.6).

5.4
Heat Transfer from Particles to Substrate

Equation 5.5 above is used to calculate the solidification time but is only approximate. It is the solution of the second-order differential equation of the thermal diffusivity (Fourier's law); in exactly the same way the expression $D = x^2/t$ is the approximate solution of the second-order differential equation of the chemical diffusivity (Fick's law).

Exact solutions require some mathematical 'inconveniences'. For the sake of completeness, and to show the elegance of the solution steps, the complete treatment will be given in Appendix B adopted from the work by Houben (1988).

5.4.1
Generalized Heat Transfer Equation

The temperature at the substrate surface can be investigated by solving the heat transfer equations in a coordinate system [x,y,z]. The temperature at the position x,y,z at time t is $\Theta(x,y,z,t)$. The heat flux j per unit time and unit area is

$$j = -a \, \text{grad} \, \Theta, \tag{5.23}$$

and for the change of temperature with time follows

$$\partial\Theta/\partial t = -\text{div} \, j, \tag{5.24}$$

where a is the thermal diffusivity. Eliminating the flux term j from Equations 5.23 and 5.24 by forming the divergence on both sides of Equation 5.23, div j = −a div grad Θ, and inserting this expression into Equation 5.24, yields

$$\partial\Theta/\partial t = a\,\Delta\Theta \text{ or}$$
$$\Delta\Theta - (1/a)[\partial\Theta/\partial t] = 0, \qquad (5.25)$$

where $\Delta a = \text{div grad } a = \partial^2 a/\partial x^2 + \partial^2 a/\partial y^2 + \partial^2 a/\partial z^2$ (Laplace operator).

Equation 5.25 is a partial differential equation of second order for the temperature Θ as a function of space and time. The starting condition is $\Theta(x,0) = f(x) = \Theta_0 = T_{s0}$. The solution of Equation 5.25 follows the Bernoulli method by anticipating $\Theta(x,t) = X(x) \times T(t)$ (Zachmann, 1981). Introducing this assumption in Equation 5.25 and dividing by XT, it follows

$$X''/X - (1/a)[T'/T] = 0. \qquad (5.26)$$

Equation 5.26 leads to two ordinary differential equations:

$$X''/X = k^2 \qquad (5.27)$$

$$(1/a)[T'/T] = k^2, \qquad (5.28)$$

where k^2 is a constant. The general solutions of those differential equations are

$$X = A_1 \exp(kx) + A_2 \exp(-kx) \qquad (5.29a)$$

and

$$T = C_k \exp(ak^2 t). \qquad (5.29b)$$

While there are no boundary conditions required for a one-dimensionally unlimited space, physically meaningful solutions must assure that the temperature Θ will not be infinite at $x = +\infty$ and $x = -\infty$. This is fulfilled if Equation 5.29a does not contain an exponential function but only trigonometric functions. The term k^2 must therefore be negative, and consequently k must be imaginary. With $k^2 = -\kappa^2$, one obtains for the partial integral

$$(a_\kappa \exp(i\kappa x) + a_{-\kappa} \exp(-i\kappa x)) \exp(-\kappa^2 a t). \qquad (5.30)$$

From this expression a coefficient a_κ can be evaluated so that the starting condition $\Theta(x,0) = \Theta_0$ is fulfilled:

$$a_\kappa = (1/2\pi) \int f(\xi) \exp(-i\kappa\xi) d\xi. \qquad (5.31)$$

Integration over κ, pulling out (at), and quadratic addition in the exponent of Equation 5.30 eventually leads to

$$\begin{aligned}\Theta(x,t) &= (1/2(\pi a t)^{1/2}) \int \delta(\xi - x_0) \exp(-[x-\xi]^2/4at) d\xi \\ &= (1/2(\pi a t)^{1/2}) \exp[-x^2/4at],\end{aligned} \qquad (5.32)$$

with $\delta(\xi - x_0) = \Theta(x,0)$ (Dirac delta function). Full details of the derivation of Equation 5.32 can be found in the literature (Zachmann, 1981).

From Equation 5.32 the solution for the temperature profile of the substrate and the coating, respectively can be derived as

$$\Theta(x,t) = \Theta(x,0) + \beta\{1 + \text{erf}[x/(4at)^{1/2}]\}, \quad (5.33)$$

where β = 'contact' thermal conductivity at the interfaces and $\text{erf}[x/(4at)^{1/2}] = (1/(\pi at)^{1/2}) \int \exp[-x^2/4at]$ (error function). These expressions will be used to calculate the temperature profile across a plasma-sprayed Mo coating on a steel substrate as shown in Section 5.4.2 and Appendix B.

5.4.2
Heat Transfer from Coating to Substrate

Figure 5.12 shows the coordinate system for the thermal diffusion from the particles into the substrate. At $x = 0$ is the substrate surface, at $x = X(t)$ the (moving) interface between solid and liquid deposit. It will be assumed that for $x < 0$ the temperature approaches ambient conditions, i.e. $\Theta_0 = T_{s0}$ for $x \to -\infty$.

The initial temperature condition of the particles in the plasma jet is $\Theta = T_3 > T_m$ for $x \gg 0$, where T_m is the melting temperature of the particle. At the interface between liquid and solid deposit $\Theta = T_m$ for $x \geq 0$.

The heat transfer equations can be solved numerically under the following simplifying assumptions:

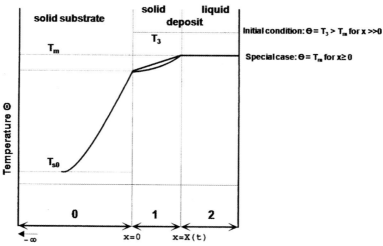

Figure 5.12 Coordinate system for solving the equations of heat transfer from the liquid/solid deposit into the substrate (Houben, 1988).

Table 5.1 Numerical values of thermophysical quantities for iron, molybdenum and AISI–316 stainless steel used to calculate temperature profiles across the coating/substrate interface as well as contact temperatures (Houben, 1988).

Quantity	Fe	Mo	AISI316	Unit
λ	75	146	18	$J\,s^{-1}\,m^{-1}\,K^{-1}$
ρ	7,870	10,200	7,670	$kg\,m^{-3}$
$C = C_p$	460	255	489	$J\,kg^{-1}\,K^{-1}$
$a\,(\times 10^5)$	2.07	5.61	0.48	$m^2\,s^{-1}$
T_m	1,536	2,610	1,375–1,400	°C
$L\,(\times 10^{-3})$	272	288	297	$J\,kg^{-1}$
$(\lambda \rho C_p)^{1/2}$	16,478	19,478	8,217	$J\,m^{-2}\,K^{-1}\,s^{-1/2}$
$C_p\,(T_m - T_{s0})/L(\pi)^{1/2}$	1.4464	1.2938	1.2797	–
		Mo on Fe	AISI316 on Fe	
B		1.1828	0.4987	
$(B + \mathrm{erf}\,p)\,pe^{p^2}$		1.2938	1.2703	
p		0.5487	0.6835	
pe^{p^2}		0.7415	1.0905	
$\mathrm{erf}\,p$		0.5622	0.6662	
$(a_1/a_0)^{1/2}$		1.6463	0.4815	

See also Appendix B.

- heat transfer takes place only by conduction;
- thermophysical properties of materials are not temperature dependent;
- particle disks have uniform temperature (=melting temperature, T_m);
- supercooling or pressure dependent effects are absent in the contact area;
- melting of substrate does not take place.[6]

Initial conditions are that the region $x > 0$ is liquid at the uniform temperature T_3, and the region $x < 0$ is solid at the uniform temperature T_{s0}. The treatment of the thermal diffusion (heat conduction) equations for the solid substrate (0), the interface substrate/solid deposit (1) and the interface solid deposit/liquid deposit (2) are given in Appendix B.

Table 5.1 summarizes the thermophysical quantities for Mo, low carbon steel and AISI–316 (Houben, 1988).

The simplified solution of the thermal diffusion equations for the solidification time, $t_{sol} = x^2/4p^2 a$ (Equation 5.5), is plotted in Figure 5.13 against the thickness of the deposit of Mo and stainless steel AISI–316 on low carbon steel. Because of the much lower thermal conductivity of the stainless steel, the solidification time of a

---

6) As we will see later this requirement must be relaxed for the deposition of a coating of a material with an extremely high melting point (molybdenum) on a steel substrate.

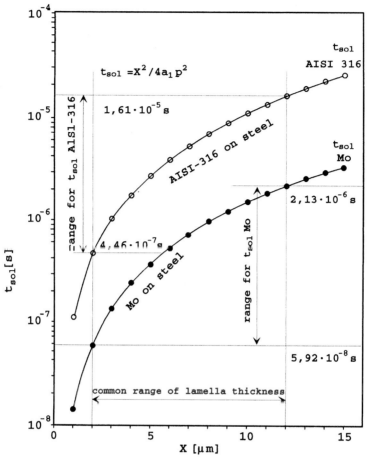

Figure 5.13 Solidification time of molybdenum particles and AISI–316 stainless steel particles impacting on mild steel (Houben, 1988).

Mo droplet is considerably shorter than that of AISI–316 by t_{sol} (AISI–316)/t_{sol}(Mo) = 7.55.

Figure 5.14 shows the temperature profiles across the interface solid Mo deposit/steel substrate (top) and solid AISI–316 deposit/steel substrate (bottom). The data points were calculated using the procedure shown in Appendix B for a thin AISI–316 layer (profiles 5 and 7) and a thick Mo layer (profiles 2 and 4). From Figure 5.14 it can be seen that the contact temperature between Mo and steel of 1775.5 °C is sufficiently high to melt a thin (1–2 μm) layer of substrate material. Consequently adhesion will be improved because existence of a solid solution of Mo in α–Fe can be assumed with a diffusion depth of approximately 20 nm (Houben, 1988).

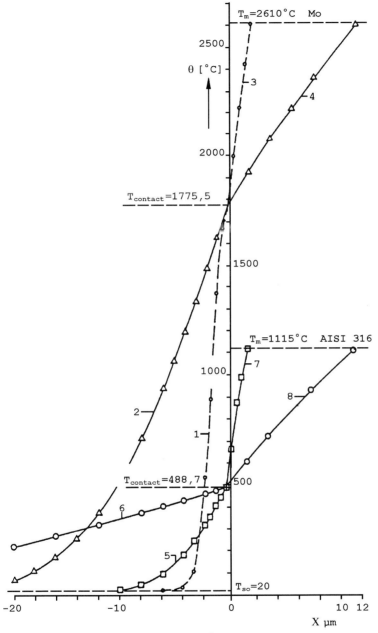

Figure 5.14 Calculated contact temperatures and temperature gradients at the interface Mo/mild steel and AISI–316/mild steel, respectively (Houben, 1988).

5.5
Crystallinity of Coatings

Determination of the crystallinity and phase composition of plasma-sprayed coatings is straightforward and requires well-known techniques such as X-ray diffraction (XRD), or electron or neutron diffraction. For thin coatings with smooth surfaces glancing-incidence X-ray diffraction (GIXRD) is a suitable tool to detect minor phases in the coating as well as preferred orientation of crystalline constituents. However, in many cases the rapid quenching of molten spray droplets results in formation of substantial amounts of an amorphous ('glassy') phase whose characterization is no longer straightforward and thus requires more advanced analytical techniques such as Rietveld refinement with addition of a standard (for example Gualtieri *et al.*, 2006) or wide-angle X-ray scattering (WAXS) to quantify the amount and composition of glass within the coatings.

The WAXS technique involves the characterization of the short range-ordered (SRO) structures as present in plasma-sprayed coatings but also in quasi-amorphous glasses and gels. The scattering factors $S(Q)$ for structural ordering of intermediate range, i.e. $Q = (4\pi\sin\Theta)/\lambda \approx 0.5\ldots 3\,\text{Å}^{-1}$ of glasses are characterized by the so-called first sharp diffraction peak (FSDP) that can be deconvoluted into up to three partial profiles whose maximum positions and intensities are dependent on the chemical composition (Figure 5.15).

The method should be introduced using a simple sodium silicate glass acting as a placeholder for more complex plasma-sprayed coating compositions. It is thought that the individual partial profiles obtained by deconvolution of the spectral envelope represent connecting polyhedra consisting of several basic structural units such as $SiO_{4/2}$ tetrahedra of differing degrees of condensation and topology. After radial symmetric summing up of the 2D scatter images obtained with an image plate

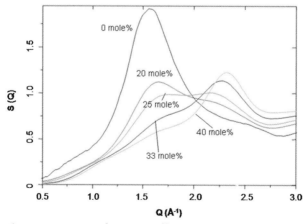

Figure 5.15 Scattering factor S(Q) of binary sodium silicate glasses for small scattering vectors Q as a function of the Na_2O content (Pentinghaus *et al.*, 2004; Heimann *et al.*, 2007).

system and corrections applied for background, distance and polarization the structure factors are calculated by

$$S(Q) - 1 = i(Q) = \frac{I_{cor}(Q) - N \sum_{i}^{uk} f_i(Q)^2}{N f_e(Q)^2}. \tag{5.34}$$

where $f_e(Q) = [\sum_{i}^{uk} f_i(Q)]/[\sum_{i}^{uk} Z_i]$, $I_{cor}(Q)$ is the corrected scatter intensity, $f_i(Q)$ is the shape factor of particle i, $\sum_{i}^{uk} f_i(Q)^2$ is the independent scatter intensity, N is the number of basic structural units per scattering volume, and Z_i is the atomic number of particle i. Separating from the structure factor $S(Q^*)$ the base line $S_0(Q^*)$ by linear interpolation between adjacent local minima, the function $\hat{S}(Q^*)$ can be divided by a nonlinear least-squares fit into a minimum number ($j \leq 3$) of partial profiles $\hat{S}_j(Q^*)$ (see Figure 5.16):

$$S(Q^*) = S_0(Q^*) + \widehat{S}(Q^*), \qquad \widehat{S}(Q^*) \approx \sum_j \widehat{S}_j(Q^*), \tag{5.35}$$

with $S_j(Q^*) = f(Q_j, \Delta Q_j, A_j)$ and $a_j = A_j / \sum A_j$.

Figure 5.16 provides an example of the fitting procedure applied to a sodium silicate glass, Na$_2$O×4SiO$_2$ (Pentinghaus et al., 2004). Since a glass can be described quantitatively by the peak positions Q$_j$, peak areas A$_j$ and half widths ΔQ$_j$ of these partial profiles, its chemical composition can be determined by appropriate calibration with glasses or gels of known composition. The numerical values of the temperature-dependent parameters Q$_j$ (peak position), A$_j$ (peak area) and ΔQ$_j$ (peak half width) are Q$_1$ = 1.013, A$_1$ = 0.027, ΔQ$_1$ = 0.454; Q$_2$ = 1.600, A$_2$ = 0.379, ΔQ$_2$ = 0.554; and Q$_3$ = 2.158, A$_3$ = 0.221, ΔQ$_3$ = 0.616. The scale of these values is given in Å^{-1}. Their transformation into real space leads to pseudo d-values, d$_{ps}$ (Pentinghaus et al., 2004). The chemical composition of the glassy phase of a coating

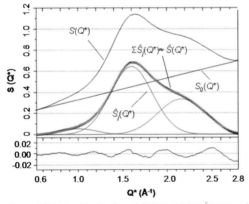

Figure 5.16 Graphic display of $S(Q^*)$, $S_0(Q^*)$, $\hat{S}_j(Q^*)$ and $\hat{S}(Q^*)$ from Equation 5.35 for Na$_2$O·4SiO$_2$ at room temperature. In the panel below the residual values $\hat{S}(Q^*) - \sum \hat{S}_j(Q^*)$ of the fitting procedure are given (Pentinghaus et al., 2004; Heimann et al., 2007).

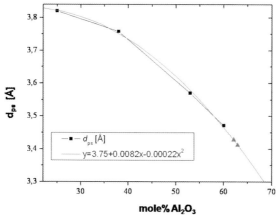

Figure 5.17 Position of d_{ps} values for aluminum silicate glasses as a function of the alumina content. Squares refer to data by Okuno et al. (2005) (Heimann et al., 2007).

can then be determined quantitatively from the peak positions alone after generating a calibrated scale with glasses of known composition. With decreasing silica content the number of pseudo-Bragg reflexes in the medium range order (MRO) is reduced to only one.

This treatment has been applied to plasma-sprayed mullite coatings (Seifert, 2004; Seifert et al., 2006a; Seifert et al., 2006b). As in the example of sodium silicate glass the glassy aluminum silicate phase yields a simple nonlinear d_{ps} vs. mole% Al_2O_3 relation from which the composition of the glass can be determined (Heimann et al., 2007).

Figure 5.17 shows the position of the d_{ps} values obtained by fitting the scattering curves with an asymmetric Couchy function using data by Okuno et al. (2005) (squares). This calibration curve can be used to determine the composition of glass found ubiquitously in atmospheric plasma sprayed (APS) mullite coatings (triangles). Figure 5.18a shows the synchrotron radiation diffraction pattern of an APS mullite coating on an aluminum substrate. The sharp lines correspond to the

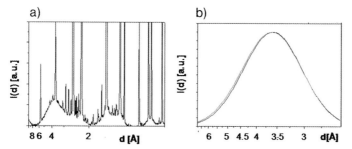

Figure 5.18 Synchrotron radiation diffraction pattern of an APS mullite coating after background correction (a) and plot of the pre-peak fitted with a symmetric Gauss function (b) (Heimann et al., 2007).

crystalline 2:1-mullite content of the coating whereas the elevated background can be assigned to the aluminum silicate glass produced during quenching of the molten droplet at impact with the substrate surface. This prominent pre-peak is centered at a d_{ps} value of about 3.4 Å.

The profile fitting of the pre-peak performed with a symmetric Gauss function led to the two d_{ps} values of 3.428 Å and 3.412 Å, respectively (Figure 5.18b). Plotting these pseudo-Bragg reflections on the calibration diagram (Figure 5.17) results in glass compositions between 62 and 63 mole% Al_2O_3 (triangles). While this composition is close to that of the starting 2:1-mullite material its deviation from the theoretical value (66.7 mole% Al_2O_3) signals a shift of the overall composition towards that of 3:2-mullite (60 mole% Al_2O_3).

To estimate the amount of mullite glass in the coating the total intensity of the diffraction pattern shown in Figure 5.18a was calculated after background correction and subtraction of the accumulated intensities of the four peaks in the scattering curve of the glass component. Since the crystalline and amorphous components have almost the same composition this procedure is warranted. The substantial amount of $32 \pm 2\%$ of glass in the coating is to be expected for a rapidly quenched plasma-sprayed aluminum silicate material.

5.6
Fractal Properties of Surfaces

Fractal geometry is a natural description for disordered objects ranging from macromolecules to the earth's surface (Mandelbrot, 1982). These objects often display 'dilatation symmetry', that means that they look geometrically self-similar under transformation of scale such as changing the magnification of a microscope. Many structures can be simply characterized by a single parameter D, the fractal dimension that is defined as the exponent that relates the mass M of an object to its size R (Schaefer, 1989):

$$M \propto R^D. \tag{5.36}$$

Equation 5.36 applies to Euclidian objects such as rods, discs and spheres, for which the exponent D equals 1, 2 and 3, respectively consistent with the common notion of dimensionality (topological dimensions). However, for fractal objects the exponents need not be integral. While the objects described by Equation 5.36 are called 'mass fractals' or polymers, 'surface fractals' are uniformly dense, i.e. colloidal, $D = 3$ as opposed to polymeric but have a rough surface. Such surface fractals share the self-similarity property[7] in the sense that if the surface is magnified, its geometric features do not change. Mathematically, surface self-similarity can be expressed analogously to Equation 5.36 by

$$S \propto R^{D(S)}, \tag{5.37}$$

---

7) Note, however, that fractal sets are infinitely detailed but not necessarily self-similar.

where S is the surface area and D(S) is the surface fractal dimension. For an ideally smooth object, D(S) = 2. For fractally rough surfaces, however, D(S) varies between 2 and 3, so that D(S) is a measure of roughness (Avnir *et al.*, 1985; Underwood and Banerji, 1986; Siegmann and Brown, 1998; Reisel and Heimann, 2004). It should be emphasized here that fractals describe only the principle of ordering but do not give any information on the mechanism that leads to this ordering. A comprehensive treatise on fractals, in particular methods to determine fractal dimensions was provided by Stoyan and Stoyan (1995).

A popular way to describe the deposition of coatings is a ballistic model. In this model particles are added, one at a time, to the growing deposit through randomly selected linear or ballistic trajectories (Meakin, 1989). In the most simple of these models the particles stick to the surface at the position at which contact is first made, *i.e.* surface diffusion is excluded. Also, since the conditions in the direction perpendicular to the surface are very different from those parallel to the surface, we will deal essentially with *self-affine geometries* (Mandelbrot, 1985). Self-affine fractals are structures that can be rescaled by a transformation that involves a different change in length of scales in different directions. For example, whereas the scaling factor perpendicular to a plasma-coated surface may be on the order of micrometers, the scaling factor along the coated surface will be measured in millimeters or even centimeters. For structures of rough surfaces such as plasma-sprayed ones, the approaches developed for self-similar structures are not strictly valid anymore and must be modified to account for this self-affinity. Since a rough surface can be represented by a single-valued function h(x) of the position x in the lateral directions parallel to the surface, it is convenient to use the height difference correlation function $C_h(x)$ defined by

$$C_h(x) = <h(x) - h(x_0 + x)>_{|x|=x}. \tag{5.38}$$

to characterize the surface. It is reasonable to expect that the surface roughness $\xi_\perp$ will grow with some power of time according to

$$\xi_\perp \sim t^\beta, \tag{5.39}$$

assuming that the surface advances with a constant rate. The variance ξ^2, i.e. the amplitude of the waviness is defined by the expression $\xi^2 = <h_i - h>^2$ and is often used as a quantitative measure of the surface 'thickness' or roughness. For a growing fractal plasma-sprayed surface layer there will also be a characteristic correlation length $\xi_\parallel$ describing the lateral distance over which surface height fluctuations occur. This length is related to the former by

$$\xi_\perp \sim \xi_\parallel^\alpha, \tag{5.40}$$

and from Equations 5.39 and 5.40 it follows that

$$\xi_\parallel \sim t^{\beta/\alpha} \sim t^{1/z}. \tag{5.41}$$

Using these self-affinity scaling laws the height difference correlation function (Equation 5.38) can be expressed as

$$C_h(x,t) \sim t^\beta f(x/t^{\beta/\alpha}). \tag{5.42}$$

The theoretical values of α and β are 1/2 and 1/3, respectively (Kardar et al., 1986). However, the true value of β may be obtained from the time dependence of the surface thickness (Equation 5.39). Numerical simulations of the dependence of the surface roughness (ξ) on the deposited layer thickness, h may be obtained from Meakin (1989).

It is quite tempting to apply the fractal approach to other plasma-sprayed coating properties. Fractal dimension should provide information that may be used to describe not only surface roughness (Amada et al., 1995; Guessasma et al., 2003; Reisel and Heimann, 2004) but also surface area (Avnir et al., 1985), fracture toughness (Fahmy et al., 1991; Mecholsky et al., 1989), adhesion strength (Amada et al., 1995), hardness, porosity (Llorca-Isern et al., 2001), thermal conductivity and frictional properties of wear-resistant coatings. There are to date only a few papers available in the literature that attempt to apply the concept of fractal geometry to describe coating properties, even though the fractal approach will provide very fundamental answers to questions about the microstructure of thin surface coatings. For example, Yehoda and Messier (1985) discussed the nature of very thin CVD layers in terms of a fractal structure. Their criterion was that fractal films should be self-similar at least over three orders of magnitude. This was not observed, however, but the development of pores in the film showed a scaling similar to the percolative scaling found by Voss et al. (1985) in thin evaporated gold films.

Fractal dimensions can be determined from a variety of relationships that are unique to systems conforming to fractal geometries. Several of these relationships will be described in more detail below.

5.6.1
Box Counting Method

This method is the simplest way to evaluate fractal dimensions. The interface line of a surface roughness profile of a coating is shown in Figure 5.19a. It is covered with boxes of side length d. If the interface line is completely covered with N squares than according to the rules of fractal geometry it holds

$$N(d) = \mu \, d^{-D} \tag{5.43}$$

or $\log N(d) = -D \log d + \log \mu$. (5.44)

By continuously changing the size of the boxes d, i.e. the magnification scale the number of squares N(d) covering the interface line is counted. On plotting log N versus log d the slope −D of the straight line obtained is the fractal dimension sought (Figure 5.19b). Fractal geometry was used to investigate the dependence of the fractal dimension on the angle of grit blasting (Amada et al., 1995) and the coating adhesion as a function of the surface roughness (Amada and Yamada, 1996; Amada and Hirose, 1998). Even though the average roughness R_a did not change with blasting angle the fractal dimension was maximized at a blasting angle of 75° (Figure 5.19c). This points to a more detailed rough surface with more undercuts and hook-like

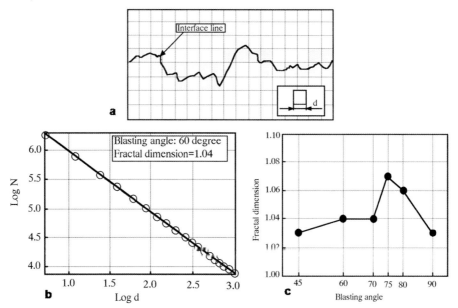

Figure 5.19 Box counting method to determine the fractal dimension of a plasma-sprayed coating (Amada et al., 1995).

protrusions that will anchor a plasma-sprayed coating more strongly and thus improve coating adhesion by mechanical interlocking. The measurement and prediction of the fractal dimensions of grit-blasted surfaces will therefore be a very valuable tool to maximize coating adhesion. In particular, the R_a values assessed normally will not give a true measure of the effective surface roughness but should be replaced by the fractal approach (see also Section 10.2.4.2).

5.6.2
Density Correlation Function

The principle of this method is to find experimentally a density correlation function C(r)

$$\langle C(r) \rangle \propto r^{-\alpha}, \tag{5.45}$$

$$C(r) = (1/N) \sum \rho(r_i)\rho(r_i + r), \tag{5.46}$$

where N is the number of splats, i is the observation point, $\rho(r_i)$ is the splat density at the observation point = 1, and $\rho(r_i + r)$ is the splat density at a point located a distance r from the observation point, and

$$-\alpha = D(\alpha) - d, \tag{5.47}$$

where $D(\alpha)$ is the fractal dimension, and d is the Euclidian dimension of the system (LaRosa and Cawley, 1992). In a 2D section the density at the distance r is determined by the number of splats N divided by the area of a ring containing the N splats. For an inner ring of radius r and an outer ring of radius $(r + \Delta r)$ the density correlation function can be approximated by

$$\langle C(r) \rangle = \{N(r)\}/2\pi r \Delta r. \tag{5.48}$$

Plotting $\ln \langle C(r) \rangle$ versus $\ln(r)$ results in a straight line of slope $-\alpha$ from which the fractal dimension $D(\alpha)$ can be calculated according to Equation 5.47.

5.6.3
Mass Correlation Function

This method is very similar to the preceding one. Fractal dimensions can be obtained from the relation

$$\langle M(r) \rangle \propto r^{D(\beta)}, \tag{5.49}$$

that and may be obtained by integration of Equation 5.49. M is the mass enclosed by some distance r. Assuming that each molten particle arriving at the substrate surface has equal mass, M can be replaced by the particle number $\langle N(r) \rangle$ contained within a sphere of radius r. Again, plotting $\ln \langle N(r) \rangle$ against $\ln(r)$ results in the slope $D(\beta)$. If $D(\alpha) \equiv D(\beta)$ than proof exists that the coating microstructure has a fractal nature.

5.6.4
Slit Island Analysis (SIA)

This method is based on measuring the ratio perimeter/area of 'islands' that appear during successive removal of thin surface layers of a metal-coated fracture surface (Mandelbrot et al., 1984). The surfaces obtained after each step can be investigated with modern image analysis software, and the perimeters P and areas F of newly appearing surface features recorded. Plotting of log P against log F of individual islands results in the slope $(D-1)/2$ (Mecholsky et al., 1989).

5.6.5
Fracture Profile Analysis (FPA)

Fracture profile analysis uses a Fourier analysis approach applied to a fracture surface to determine fractal dimensions (Passoja and Amborski, 1978). Since the fracture profile can be considered a spectrum of microstructural information all signal-processing analytical tools developed for electric signals can be applied to fracture profiles. Many spectral details observed during analysis reflect the 'fundamental length', i.e. the correlation length (see Section 5.6) of the microstructure and their related high-order harmonics. According to fractal hypothesis the integrated spectrum, i.e. the squared sum of the amplitudes takes the form $k^{-B'}$ where k is the wave number (inverse length) and $B' = B - 1 = 6 - 2D$ (Mecholsky et al., 1989). The fractal character of fractured coating surfaces can then be tested by plotting log (amplitude)^2 vs. log k to obtain the slope B′ from which the fractal dimension D can be calculated.

5.6.6
Scale-sensitive Fractal Analysis

To determine the fractality of plasma-sprayed coating surfaces the area-scaled fractal complexity (ASFC) method can be applied using the so-called 'patchwork' method (Brown *et al.*, 1993; Siegmann and Brown, 1998; Reisel and Heimann, 2004) that is analogous to the better known Richardson method (Mandelbrot, 1984). It is based on the general principle of fractal geometry that the numerical value of the area of a rough surface is not unambiguously determined but depends on the measuring scale, i.e. the magnification. Calculation of the measured and projected area is performed using 3D topographical data obtained by recording the surface roughness profile along a multitude of linear scans. These scans can be produced either by a diamond stylus-type surface roughness tester or by a scanning laser profilometer.

Figure 5.20 shows two 3D topographical maps of the surfaces of a plasma-sprayed chromium oxide coating with different roughness values $R_z = 9.0\,\mu m$ (a) and $R_z = 17.8\,\mu m$ (b).

The ASFC calculation algorithm uses repeated virtual tiling of the geometrical surface by successively smaller triangular patches to determine the apparent surface

a

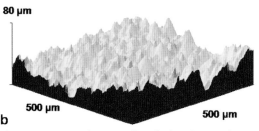

b

Figure 5.20 3D roughness profiles of relatively smooth (a) and rough (b) plasma-sprayed chromium oxide coatings (Reisel and Heimann, 2004).

5.6 Fractal Properties of Surfaces

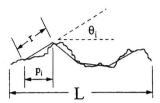

$$\text{Relative Length } (r) = \sum_i^N \frac{1}{\cos\theta_i} \frac{p_i}{L}$$

Figure 5.21 Definition of the relative length r.

for a certain triangular size. The surface fractal dimension (Equation 5.37) can then be determined by the relation

$$D(S) = \log N / \log(1/r) \tag{5.50}$$

where N = number of triangles and r = linear scaling ratio.

The total measured area for a given scale equals the number of triangles N multiplied by the area of an individual triangle, i.e. the patch size. According to Delesse's principle the relative area is proportional to the relative length r, and it follows that

$$r = \sum (\cos\Theta_i)^{-1}(p_i/L), \tag{5.51}$$

where Θ_i = angle between the apparent (geometric) surface and the surface of an individual triangle, p_i = projected area of a triangle, and L = total length of the measured profile (Figure 5.21).

Plotting the logarithm of r against the logarithm of p_i results ideally in a straight line whose slope m is connected to the fractal dimension D(S) by $D(S) = (2-2m)$, and $ASFC = -1000m$. For $r=1$ the so-called 'smooth-rough crossover' (SRC) is obtained that can be used, among others, to estimate the maximum size of the surface sampling area for topographic data acquisition (Siegmann and Brown, 1998).

Figure 5.22 shows a plot of the relative area for $2\,\mu m^2$ patch size (RAPS) vs. patch area in μm^2 for a series of plasma-sprayed chromium oxide coating deposited by a statistical design of second order (Draper Lin-Cube-Star design) with four parameters at five levels ($-2, -1, 0, +1, +2$) with 8 cube points ($-1, 0, +1$; 2^3 full factorial), 8 star point ($-2, 0, +2$), and 2 center points (0,0,0) (Zimmermann, 2000). The parameters varied were the plasma power (33–45 kW), the spray distance (90–130 mm), the argon/hydrogen ratio of the plasma gas (2.6–3.4), and the powder carrier gas flow rate (2.4–3.6 slpm).

For large scale factors (patch area >10 000 μm^2) the RAPS tends towards 1.00 and hence the surface appears microscopically smooth. For finer scale factors the RAPS deviates significantly from 1. The smooth-rough crossover (SRC) occurs at a patch area size of 2000 μm^2. Below the SRC the RAPS increases monotonically with decreasing logarithm of the scale factor. This is the region of self-similarity, and the slope m of the line is a measure of the area-scaled fractal complexity, i.e. $ASFC = -1000m$ and $D(S) = (2-2m)$. Different spray powder grain size distributions produce different

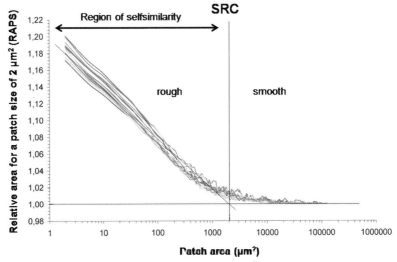

Figure 5.22 Relative area for a patch size (RAPS) of $2\,\mu m^2$ vs. patch area (in μm^2) of the surface roughness of plasma-sprayed chromium oxide coatings (powder grain size $-45 + 22.5\,\mu m$) (Reisel and Heimann, 2004: Zimmermann, 2000).

surface roughness and hence different surface fractal dimensions D(S), i.e. with increasing R_a D(S) increases (Table 5.2).

Figure 5.23a shows the RAPS in (%) as a function of the surface roughness R_a for the three spray powder grain sizes used. It can be concluded that the relation between the surface roughness and the fractal quantity RAPS is close to linear. Figure 5.23b reveals that the SRC for small R_a values is nearly constant at approximately $200\,\mu m^2$ and hence correlates well with the median grain size of $14\,\mu m$ of the grain size range $-22.5 + 5.6\,\mu m$. However, higher R_a values result in increased SRC values between 400 and $1400\,\mu m^2$, corresponding to grain size diameters between 20 and $37\,\mu m$. Surprisingly, these SRCs are much lower than would be expected for maximum grain sizes of approximately $75\,\mu m$. This suggests that coarser particles may suffer from large thermal stresses in the hot plasma jet and hence may disintegrate in-flight before melting.

Table 5.2 Correlation between powder grain size, surface roughness R_a and fractal dimension D(S) (Reisel and Heimann, 2004).

Grain size (μm)	R_a(μm)	D(S)
−22.5 +5.6	−4.5 +3.4	2.05
−45 +22.5	−5.8 +4.7	2.06
−75 +25	−7.3 +6.4	2.08

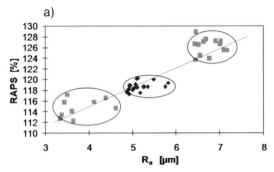 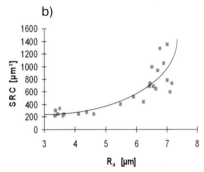

Figure 5.23 Relative area of patch size (RAPS) as a function of the arithmetic mean surface roughness R_a of plasma-sprayed chromium oxide coatings (a: clusters relate, from left to right, to grain sizes of $-22.5 + 5.6$, $-45 + 22.5$ and $-75 + 25$ μm, respectively) and smooth–rough crossover (SRC) (b: grain size range $-22.5 + 5.6$ μm: $R_a < 5$ μm; grain size range $-75 + 25$ μm: $R_a > 5.5$ μm) (Reisel and Heimann, 2004).

5.7
Residual Stresses

The particle–substrate interactions characterized by momentum and heat transfer from the solidifying particle after impacting the surface of the substrate lead eventually to the development of residual stresses in the coating. These stresses can be divided into microscopic, mesoscopic and macroscopic stresses.

Microscopic stresses are found inside individual splats, and are generated due to the gradient of the coefficient of thermal expansion between the hot particle and the cooler substrate. Figure 5.24 shows the etched cross-section of an r.f. inductively-sprayed chromium coating with stress-induced cracks in the individual particle splats perpendicular to the splat boundaries (Müller *et al.*, 1995).

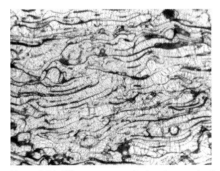

Figure 5.24 Microstructure of splats in an etched cross-section of a chromium coating on copper (Müller *et al.*, 1995).

Mesoscopic stresses occur between particle splats inside a lamella, and are responsible for reduced coating cohesion. These stresses result principally from frozen-in contraction of the rapidly quenched molten particles at the substrate interface (see below).

Macroscopic stresses occur between the coating as a whole and the substrate. They depend on the temperature gradients within the coating between passes of the torch which decrease when both the thickness of the lamella and the spray width are decreased. Macroscopic stresses can be controlled by

- cooling with gas jets: compressed air or carbon dioxide (Steffens *et al.*, 1981);
- cooling with liquid gases such as argon or carbon dioxide;
- adjustment of the relative velocities of torch/substrate;
- grooving of substrate.

Since residual coating stresses influence the quality and the service life of coatings, in particular their adhesion and wear performance, it is the goal of any coating development to minimize such stresses. The origin of residual stresses is twofold. First, rapid quenching of the molten particles at the substrate interface results in frozen-in particle contraction according to

$$\sigma_{qu} = \alpha_c \times E_c \times (T_m - T_s) \text{ or} \tag{5.52a}$$

$$\sigma_{qu} = E_c(T) \int_{T_s}^{T_m} \alpha_c(T) dT \tag{5.52b}$$

where T_m is the melting temperature of the coating materials and T_s the temperature of the substrate (Stokes and Looney, 2004). For $T_c > T_s$ tensile forces develop between individual particles and add up to residual stress states of the first order ('macro' residual stresses) throughout the coating (Knight and Smith, 1993).

Secondly, the differing coefficients of thermal expansion of coating and substrate contribute to the total stress state of the system. If this stress state exceeds the adhesive or cohesive bonding forces of the coating crack formation or delamination occur (Drozak, 1992). Thermal stresses at the coating/substrate interface can be determined approximately by the Dietzel equation (Salmang and Scholze, 1982; Levine *et al.*, 1981) that uses the differences in the coefficients of thermal expansion of coating and substrate, the temperature gradient, and the thickness ratio to calculate the coating stress:

$$\sigma_c = \{E_c(\alpha_c - \alpha_s)\Delta T/(1-v_c)\} + \{[(1-v_s)/E_s]d_c/d_s\}, \tag{5.52c}$$

where α is the linear coefficient of thermal expansion, v is the Poisson ratio, E is the modulus of elasticity, and d is the thickness. The subscripts 'c' and 's' refer to coating and substrate, respectively. The sign of the stresses depends on the sign of $(\alpha_c - \alpha_s)$: for $\alpha_c > \alpha_s$ tensile stresses develop in the coating and compressive ones in the substrate (Godoy *et al.*, 2002) that can be minimized by maximizing the ratio d_s/d_c. Thus for given values of v and E the coating stresses increase with increasing coating thickness (cp. Figure 5.26b below).

Residual tensile stresses are particularly severe in ceramic–metal composite coatings because of the generally large difference in the coefficients of thermal expansion of ceramic and metal. Their states can be determined, in principle, by borehole (blind hole), X-ray or neutron diffraction measurements as well as specimen curvature measurements.

5.7.1
Blind Hole Test

A hole is drilled into the coating and the measurement of the relaxation occurring allows to estimate the original stress state (Mathar, 1932; Gadow et al., 2005; Santana et al., 2006). A strain gauge based either on a semiconductor circuit or a modified Wheatstone bridge and glued to the coating surface is employed to record the relaxation. The measuring principle is based on the expansion-resistance effect of metallic conductors according to Wheatstone and Thomson. The change of the resistance of a conductor subjected to tensile or compressive forces can be attributed to the deformation of the conductor as well as to the change of its specific resistance due to changes in the microstructure. The relaxation of the coating material on drilling of a borehole results in an expansion or contraction of a measuring grid of strain gauges that in turn is recorded as a change in resistance.[8] Since this method is affected in various ways by external parameters it can only be used as a qualitative or at best semi-quantitative indicator of the stress state. This method is recommended for use concurrently with other techniques such as X-ray diffraction measurement (Noutomi, 1989). A more quantitative treatment of the method can be found by Bialucki et al. (1986). The technique has also been used to estimate the stress level of HAp coatings deposited on Ti alloy implant surfaces (Han et al., 2001).

5.7.2
X-ray Diffraction Measurements (sin²ψ-technique)

Determination of the stress state is based on the measurement of the lattice deformation of a polycrystalline material subjected to stresses. This is accomplished by measuring the change of the D-values of the interplanar spacings of selected lattice planes {hkl} relative to the stress-free state, D_0:

$$dD = D - D_0. \tag{5.53}$$

Since the penetration of the radiation into the coating is rather limited (1–10 μm) only the stress state of the coating surface can be measured with accuracy. To obtain a stress distribution profile the surfaces must be consecutively removed by polishing, sputtering or etching, and the measurement then repeated.

Differentiation of the well-known Bragg equation $n\lambda = 2D\sin\theta$ yields the (relative) lattice deformation

$$\varepsilon^L = dD/D_0 = -\cot\theta_0 d\theta. \tag{5.54}$$

---

[8] Actually the released surface strain measured with the strain gage rosette is converted to stress using a calibration curve and Hooke's law (Cox, 1988).

From Equation 5.54 it follows that the change of the Bragg angle, $d\theta$ is maximized for a given ε^L when θ_0 is large. Therefore, interplanar spacings with the largest possible Bragg angles must be chosen. Also, $d\theta$ increases with increasing stress σ and decreasing Young's modulus E. However, ceramic materials have generally large E values. Therefore, very small shifts of the interplanar spacings must be recorded with high accuracy. This requires highly sophisticated X-ray diffraction hardware and appropriate software (Eigenmann et al., 1989).

Figure 5.25a shows the sample-based coordinate system used to deduce the basic equations for stress measurements. The lattice deformations $\varepsilon^L_{\varphi,\psi}$ measured close to the surface obtained according to Equation 5.54 in the directions φ,ψ are taken as the deformations $\varepsilon_{\varphi,\psi}$ expected from the theory of elasticity. The angle φ is the azimuthal angle to the x-axis, and the angle ψ is the distance angle to the z-axis, i.e. the surface normal N of the sample. For a triaxial stress state $\{\sigma_1, \sigma_2, \sigma_3\}$ with σ_1 and σ_2 parallel to the surface the distribution of the deformation is given by

$$\varepsilon_{\varphi,\psi} = [(1+\nu)/E][\sigma_1\cos^2\varphi + \sigma_2\sin^2\varphi - \sigma_3]\sin^2\psi - (\nu/E)[\sigma_1+\sigma_2] + \sigma_3/E, \tag{5.55}$$

where ν = Poisson's ratio and E = Young's modulus.

In the plane stress state of ceramic coatings, $\sigma_3 = 0$, so the general equation to determine the stress state is obtained as

$$\varepsilon_{\varphi,\psi} = (s_2/2)\sigma_\varphi\sin^2\psi + s_1(\sigma_1+\sigma_2), \tag{5.56}$$

with $\sigma_1\cos^2\varphi + \sigma_2\sin^2\varphi = \sigma_\varphi$ (see Figure 5.25a) and the elastic compliances $s_1 = -\nu/E$, $s_2/2 = (1+\nu)/E$. Figure 5.25b shows an example of a plot of $\varepsilon_{\varphi,\psi}$ versus $\sin^2\psi$ for the interplanar spacing (416) of alumina. From the slope of the lines the surface stress component σ_φ, and from the intersection with the ordinate axis ($\varepsilon_{\varphi,\psi=0}$) the sum of the main stresses ($\sigma_1 = -100$ N mm^{-2}, $\sigma_2 = -200$ N mm^{-2}) can be obtained (Eigenmann et al., 1989).

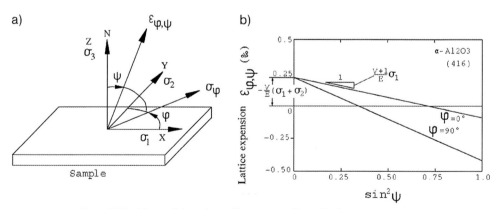

Figure 5.25 (a) sample-based coordinate system for residual stress measurements. (b) lattice expansion against $\sin^2\psi$ for (416) of alumina (Eigenmann et al., 1989).

Figure 5.26a displays the often observed dependency of the residual stresses on the stand-off distance for an APS Al_2O_3 coating on St 38 steel using an NiAl bond coat (Tietz et al., 1993). As the stand-off distance increases, the increasing coating porosity may cause the material to acquire quasi-elastic behavior that can compensate for differences in the coefficients of thermal expansion of the coating and the steel substrate. This has been confirmed by Berndt (1986) who suggested that plasma-sprayed coatings have a high compliance since the bonding force between lamellae is low enough to permit some relative sliding. In Figure 5.26b the linear increase of the residual coating stresses with the coating thickness is shown (see also Equation 5.48). It should be emphasized that the results shown in Figure 5.26 do not consider the different contributions that the alumina and titania phases may make to the total stress pattern. Such different stress responses have been demonstrated for t–ZrO_2/α-Al_2O_3 (Eigenmann et al., 1989), (Ti,Mo)C/NiCo (Thiele, 1994) and ZrO_2/8%Y_2O_3 (Matejicek et al., 1998) coatings. In particular, the former example has shown that stress equilibrium exists between individual ceramic phases that can be described by a simple additive mixing rule.

A major problem exists in unequivocally describing the surface stress states by X-ray diffraction measurements. Since the deformation ε is dependent on material parameters such as the Poisson ratio, ν and the modulus of elasticity, E (see Equation 5.55), those quantities must be known with rather high accuracy (Tsui et al., 1998). However, the bulk modulus of elasticity is certainly different from that of a coating owing in particular to the increased porosity of the latter (Noutomi, 1989). Furthermore, free-standing coatings must be used to determine the value of E experimentally. This means that for every coating system the dependency of E on the porosity must be determined in parallel to the determination of the stress

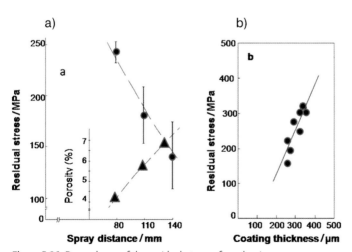

Figure 5.26 Dependence of the residual stress of an alumina coating on St 38 steel on the spray distance (a) and of an alumina/2.5% titania coating on St 37 steel on the coating thickness (b) (Tietz et al., 1993).

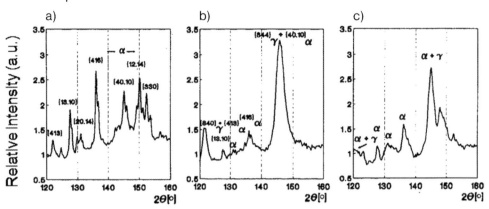

Figure 5.27 Interplanar XRD spacings of the high angle range of as-received alumina powder (a), as-sprayed coating (b) and annealed coating (c) (Kraus et al., 1997).

state for each plasma spray parameter set. As a consequence the method becomes extremely time consuming and cannot be applied routinely in industrial practice. Also, the continuous removal of surface layers changes the stress state, and the surface roughness established by cutting or lapping may lead to scattering of the X-rays and thus to line broadening that influences the accuracy of the measurement. Furthermore, since the modulus of elasticity, E, is a function sensitive to the stresses present in the coating[9] whereas E is used as a determining parameter in the calculation of the coating stress, conditions of nonlinearity exist.

Figure 5.27 shows XRD pattern of the high diffraction angle range of as-received alumina spray powder (a), as-sprayed 500 μm thick alumina coating consisting mainly of γ-Al_2O_3 (b) as well as the powdered coating after annealing for 30 min at 1000 °C (c). The interplanar spacings $(40.10)_\alpha$ and $(844)_\gamma$ were used for determination of residual stresses by the $\sin^2\psi$-technique (Kraus et al., 1997).

Figure 5.28 shows the lattice strain of HAp coatings deposited on Ti6Al4V substrates measured by the $\sin^2\psi$ method, $\varepsilon = (d - d_0)/d_0 \times 10^{-3}$ as a function of $\sin^2\psi$, where ψ is the tilt angle towards the X-ray beam. The as-sprayed coatings exhibit rather strong compressive surface stresses owing to thermal mismatch developed during cooling of the deposited layer to room temperature. This compressive stress will relax during incubation in simulated body fluid when high levels of amorphous calcium phosphate (ACP) thought to be a main contributor to the residual stress will undergo transition to crystalline calcium phosphate phases, most notably TetrCP and HAp. Hence the layer of bone-like secondary apatite deposited at the outermost rim of the samples will be decoupled from the declining stress field and show close to zero stress (Heimann et al., 2000; Härting et al., 2001; Figure 5.28a).

---

9) The modulus at low stress levels or under conditions of the existence of a complex tension/compression stress pattern (Siemers and Hillig, 1981) is greater than that existing at pure tension and plastic flow of the material (Berndt, 1986).

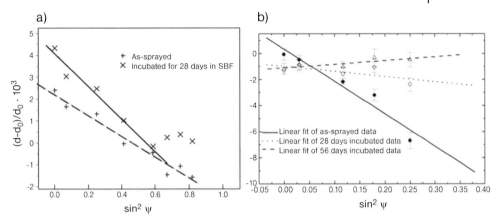

Figure 5.28 Development of residual stresses in as-sprayed HAp coatings (a: dotted line, b: solid line) (Heimann et al., 2000; Härting et al., 2000) and coatings incubated in SBF for 28 days (a) and up to 56 days (b).

This is confirmed by the results shown in Figure 5.28a with a stress level of as-sprayed coatings at -221 ± 32 MPa that decreases after incubation for 28 days to -34 ± 20 MPa (dotted line). Continuing incubation up to 56 days results in a slightly tensile stress at 24 ± 10 MPa (dashed line in Figure 5.28b).

According to Equation 5.56 the calculation of the stress state requires the knowledge of the lattice spacing D_0 of the material in the stress free state. This value is generally not accessible. In particular, for alloys the lattice spacing depends sensitively on the concentration of the alloying elements. Since only one Bragg reflection is usually measured to determine 2D stress states, the assumed (apparent) lattice spacing of the unstressed material is strongly correlated with the stress tensor components. This means that it is not possible to distinguish whether the lattice spacing selected is the true D_0 value for the unstressed material or merely a composite value that includes contributions from the existing residual stresses. This has been accounted for by Härting (1998) who combined the conventional experimental $\sin^2\psi$-procedure with a new semi-numerical method to determine nondestructively the full 3D stress state with depth resolution. Applying a differential equation of equilibrium and the surface boundary conditions for every stress tensor component σ_{ij}, a model function $\sigma_{ij}(z)$ can be extracted that describes the depth dependence of all six stress tensor components.

Another factor affecting the evaluation of the $\sin^2\psi$-plots is the often observed fact that severe internal residual stresses can result in deviation from the true crystal symmetry. In titanium nitride (Rickerby, 1986) or hafnium nitride (Sproul, 1984) coatings stresses in the coating plane cause densification and account for the deviation from cubic symmetry as evidenced in Nelson–Riley extrapolation function plots (Rickerby, 1986; Barrett and Massalski, 1966).

The frequently strong residual stress gradients within the coatings as well as problems with penetration of the X-ray beam through a very thin coating have prompted Peng et al. (2006) to apply a pseudo-grazing incidence X-ray diffraction (GIXRD) method to measure near-surface stresses and stress gradients by a modified 'sin$^2\psi^*$' technique. Classical GIXRD combines the Bragg condition of the classical sin$^2\psi$ method with the condition of the total external reflection of X-rays by the crystal surface.[10] The sample is investigated on a 4-circle goniometer set up in a ψ-geometry. The penetration depth of the X-rays can be expressed as

$$\tau = \frac{\cos\psi[\sin^2\theta - \sin^2(\theta-\Omega)]}{2\mu \sin\theta \cos(\theta-\Omega)}, \tag{5.57}$$

where ψ is the tilt angle towards the X-ray beam, θ is the Bragg diffraction angle, Ω is the incidence angle of the X-ray and μ is the absorption coefficient of the materials at the X-ray wavelength selected. When using GIXRD the geometry of the reflection conditions changes (see footnote 10). Hence the new tilt angle ψ^* differs from the angle ψ of the conventional diffraction geometry, and can be defined by the Nernst–Descartes law as

$$\cos\psi^* = \cos\psi_0^* \cos\psi \tag{5.58}$$

with $\psi_0^* = 2\theta/2 - \Omega$ for $\psi = 0°$. Also, the expressions for the generalized X-ray elastic compliances s_1, s_2 and elastic moduli E_{ij} must be redefined by replacing the values of ψ by the equivalent values ψ^*. Supposing biaxial planar stress state with $\sigma_1 = \sigma_2$ (homogeneous stress) and $\sigma_3 = 0$ the XRD elastic modulus in only one direction yields

$$E_{11} = (s_2/2)(\sin^2\psi^* \sin^2\beta) + s_1, \tag{5.59}$$

with $\beta = (\pi/2) - \theta + \Omega$, and s_1 and s_2 elastic compliances. E_{ij} is plotted against the measured strain $\varepsilon_{\varphi\psi}$ and the residual stress can be calculated from the slope in the form $\varepsilon_{\varphi\psi} = E_{ij} \times \sigma_{ij}$ (see Equation 5.56). Numerically, $\sigma(\psi^*)$ can be obtained by a linear regression of the form

$$b = \frac{n\sum xy - (\sum x)(\sum y)}{n\sum x^2 - (\sum x)^2}, \tag{5.60}$$

where $b = \sigma(\psi^*)$, $x = \sin^2\psi^*$ and $y = 1 - \sin\theta/\sin\theta_0$. The relationship between the stress value calculated by the classical sin$^2\psi$ method and the new sin$^2\psi^*$ method using grazing incidence is

$$\sigma(\psi^*) = (\sin^2\beta)^{-1}\sigma(\psi). \tag{5.61}$$

10) For X-rays incident on solid materials from the air total external reflection occurs at incident angles smaller than a critical angle α_c given by cos $\alpha_c = n$, where n is the index of refraction, $n = 1 - \delta - i\beta$. For angles larger than the critical angle α_c the measured Bragg diffraction angle $2\theta^*$ for GIXRD is different from 2θ measured for conventional XRD and yields $2\theta^* = 2\theta + (\alpha - \alpha')$, where α' is the angle between the surface and the refracted X-ray beam. The angle α' is defined by the Nernst–Descartes law as cos $\alpha = n \times \cos \alpha'$ (Dudognon et al., 2006).

Verification of this novel approach has been obtained by Peng et al. (2006) on thin biphasic films containing both cubic α-Ta and tetragonal β-Ta, varying film thickness between 0.55 and 1.16 μm, measuring direction, interplanar spacings (α-Ta: {211}, {321}, {310}; β-Ta: {513}), type of goniometer, and X-ray wavelength (CrK_α, CuK_α). In general, good correlation was noted even though the authors caution that for further use of this method the influences of surface roughness, optical effects of incident beam and radiation from the sample surface, and mechanical effects must be considered.

5.7.3
Curvature Monitoring Technique (Almen-type test)

To circumvent the drawbacks of the $\sin^2\psi$-technique described under 5.7.2 including the problem of a nonlinear modulus of elasticity attempts have been made to apply other tests to obtain estimates of the residual stresses. The method of specimen curvature measurement is arguably the most widely used method for determining residual stresses. It involves measuring the bending of the coated sample in response to both quenching and thermal stresses. From the measured radii of curvature the stress can be calculated according to the modified Stoney equation (Stoney, 1909; Pina et al., 2003),

$$\sigma = \frac{E_s}{6(1-\nu_s)} \frac{t_s^2}{t_c} \left[\frac{1}{R} - \frac{1}{R_0}\right], \tag{5.62}$$

where σ is the average stress induced in the coating, E_S is Young's modulus of the substrate material, ν_S denotes Poisson's ratio of the substrate, t_S is the thickness of the substrate, t_c is the thickness of the coating, R and R_0 are the radii of curvatures of the substrate and coated sample, respectively. The two main assumptions underlying the Stoney equation, relating the substrate curvature R to mismatch strain, are (i) that the coating is very thin compared with the substrate and (ii) that the deformations are infinitesimally small. If either assumption is relaxed a corrected equation has been introduced by Freund et al. (1999) in which the term shown in brackets in Equation 5.62 is replaced by $(R \times K)^{-1}$ where the reduced curvature $K = (1/16)(L^2/R \times t_s)$ with L = width of the sample lamella investigated. In a recent study geometry-based criteria were developed that constrain the Stoney formula even further (Mézin, 2006). This work yielded a modified Stoney equation as

$$\sigma_0 = -\frac{1}{6} \frac{E_s}{(1-\nu_s)} \frac{t_s^2}{t_c} \frac{1}{R} \left[1 + \left(\frac{E_c(1-\nu_s)}{E_s(1-\nu_c)} - 1\right)\frac{t_c}{t_s}\right], \tag{5.62a}$$

in which the corrective term in brackets accounts for the fact that the condition $t_c \ll t_s$ is not always fulfilled. In fact for a coating/substrate system consisting of the same materials with $t_c/t_s = 0.1$ there the error made by not using the corrective term may be as large as 30%.

The experimental verification of these relations is found in the so-called Almen test, for example to determine the efficacy of shot peening applied to induce compressive stresses at the surface of metallic workpieces. This technique has been adapted to thermally sprayed thin coatings (Knight and Smith, 1993). A thin test strip is mounted on a holder (SAE Handbook, 1977) and blasted with shot. This treatment leads to curving after removal from the fixture. The peened side of the curved strip will be convex. The degree of curvature, measured with an appropriate gauge or a laser detection system is a function of the residual compressive stresses developed in the strip. For estimation of residual stresses in thermally sprayed coatings a thin foil of the substrate material will be clamped to the Almen fixture (Figure 5.29) and treated exactly in the same way as the bulk substrate to be sprayed, i.e. using grit blasting, ultrasonically cleaning, and spraying. After each working step the strip is removed from the fixture and its stress state as displayed by the degree and direction of its curvature is determined.

Figure 5.29a shows the mounting block for the Almen test (SAE Handbook, 1977), Figure 5.29b shows a thin foil before (1) and after grit blasting (2). The induction of compressive stresses led to a convex curvature of the grit-blasted side of the foil that was reduced by the amount of the tensile stresses induced during coating (3). The reduction of the arc height, in this example 0.20 mm, can be directly related to the tensile residual stress state in the coating (Brandt, 1995; Knight and Smith, 1993).

As described in more detail in Chapter 9, bioceramic HAp coatings tend to develop large residual stresses when deposited onto Ti alloy implant surfaces. According to Equation 5.52c these stresses ought to be tensile since $\alpha_c > \alpha_s$ where c refers to coating and s to substrate. Indeed stress levels measured by the specimen curvature technique as reported by Tsui et al. (1998) revealed tensile stresses in the range of 20–40 MPa. However, recent work done on APS HAp coatings deposited on Ti6Al4V substrates showed deviating results (Topić et al., 2006).

Figure 5.30 shows that the curvature increased for all as-sprayed and grit-blasted samples as well as for samples incubated up to 28 days whereas incubation for 56 days caused a significant decrease in the radius of curvature, R. Accordingly, a compressive residual stress was induced in the uncoated, grit-blasted Ti6Al4V samples, the as-sprayed coatings, and all incubated coatings with the exception of the samples incubated for 28 days.

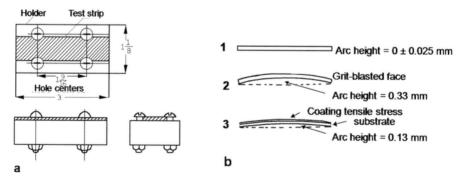

Figure 5.29 Almen test configuration to measure the specimen curvature. (a) mounting block, (b) a thin foil before (1) and after grit blasting (2). The convex curvature will be reduced by tensile stresses induced during coating (3). After SAE standard, 1977.

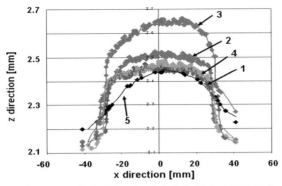

Figure 5.30 Determination of curvatures of a Ti6Al4V sample as-sprayed with hydroxyapatite (1) and a grit-blasted Ti6Al4V sample (5) compared with hydroxyapatite coatings incubated in simulated body fluid for various times (2–4) (Topić et al., 2006).

Furthermore, as shown in Figure 5.31 the level of the compressive stress varied from −48.5 MPa in grit-blasted to −120.8 MPa in as-sprayed to −117.7 MPa in samples incubated for 7 days. It should also be noted that grit-blasting itself did not induce a significant stress gradient. After immersion in simulated body fluid for 28 days, the stress changed sign, becoming slightly tensile at + 17.8 MPa. The dramatic changes in the magnitude of the compressive stress during incubation between 7 and 28 days could be explained by stress relaxation (Yang et al., 2003; Lu et al., 2004). This relaxation is thought to originate from preferential dissolution along microcracks already introduced during cooling of the as-sprayed coating to room temperature. It is conceivable that infiltration of narrow microcracks by simulated body fluid led to

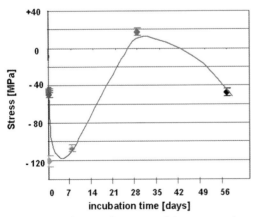

Figure 5.31 Relative surface stress of plasma-sprayed hydroxyapatite coatings as a function of incubation time in simulated body fluid (Topić et al., 2006).

localized dissolution and thus widening of the cracks (Sun et al., 2002). This could be caused by the fact that owing to the Gibbs–Thompson relation of the stability of a solid–liquid interface the highly convex curvature of the edges of a crack generally shows a higher dissolution rate than the adjacent flat coating region (Heimann, 1975). Furthermore, it is suggested that the thin APS layer formed originally at the interface substrate-coating by quenching of the impacting molten HAp particles (Heimann and Wirth, 2006) controls the residual stress state to a large extent. When on incubation this ACP layer undergoes structural changes towards transformation into crystalline calcium phosphate such as TetrCP and eventually HAp the compressive stress relaxes to zero as observed previously on samples prepared and incubated under close to identical conditions (Heimann et al., 2000; Härting et al., 2001). However, extending the incubation time to 56 days, the coating experienced another rise in compressive stress to $-50\,\text{MPa}$, presumably owing to continuous precipitation of a secondary HAp layer. Hence in coatings incubated for only up to 28 days the newly formed layer of secondary HAp is decoupled from the stress state of the original layer, showing essentially zero or even slightly tensile ($+18\,\text{MPa}$) stress. Increasing the incubation time may also increase the compressive stress state again caused by reorganization, consolidation, and densification of the biomimetically precipitated Ca-deficient 'bone-like' apatite layer.

Some authors have directly compared the results of specimen curvature measurements and those of the $\sin^2\psi$ technique (Kesler et al., 1998; Matejicek et al., 1998; Totemeier and Wright, 2006). In general, qualitative agreement was found but the magnitude of the residual stresses determined by the different methods varied widely. For example, Totemeier and Wright (2006) deposited AISI 316L and Fe$_3$Al-based alloy coatings by HVOF onto AISI 1020 low-carbon steel and type 316 stainless steel substrates, and measured the residual stresses by the specimen curvature method and by the classical $\sin^2\psi$ technique. For the former, four curvature-stress models were employed: the two-beam elastic bending model (Clyne and Gill, 1996), a model based on Stoney's equation (Equation 5.62), the Tsui–Clyne progressive deposition model (TCPDM) for planar geometry (Tsui and Clyne, 1997), and an extension of the TCPDM modified to include substrate plasticity by implementing an FEM-based analytical code. The study clearly indicated that the two-beam model is unsuitable to reproduce the experimentally obtained stress levels. The calculated coating surface stresses vary from compressive in thin coatings to tensile for thicker coatings, in marked contrast to the near constant surface stresses obtained from XRD measurements. While the use of the Stoney equation correctly reflected the trends measured by XRD it failed to predict the magnitude of surface stresses. Roughly the same can be said about the two remaining models used. Here the velocity of the spray particles appears to play a decisive role: at low particle velocity ($520\,\text{m}\,\text{s}^{-1}$) the compressive stresses predicted by the Stoney approximation, and the elastic progressive and elastic-plastic progressive Tsui–Clyne models agree well with the XRD results whereas at higher particle velocities ($640\,\text{m}\,\text{s}^{-1}$) the Stoney approximation underestimates the measured stresses and the Tsui–Clyne models overestimate them. An explanation of these discrepancies may be found by considering a 'deposition stress' uniquely determined by the coating parameters but not by the surface and/or coating temperatures. For example, impact of semi-molten or oversized solid rigid particles may induce compressive near-surface stresses by a mechanism akin to shot peening that superpose with thermal and quenching stresses.

5.7.4
Photoluminescence Piezospectroscopy

This method is based on measuring the shift that the optical energy levels of fluorescent elements such as Cr^{3+} undergo as the result of altering the distance of ions within the strained crystal structure of the host lattice in response to a change of the stressed state (Yu and Clarke, 2002).

Equipment used to record photoluminescence spectra include confocal laser-Raman spectrometers equipped with a liquid nitrogen cooled CCD detector and a motorized X–Y microscope table to allow point-by-point mapping.

Piezospectroscopy is a powerful tool to detect, characterize and quantify local strain fields in crystals and hence provides a convenient means to record the in-plane components of residual stress of plasma-sprayed coatings. Cr^{3+} ions present as impurity in the spray powder will enter Al^{3+} sites in corundum and mullite crystals, and the resulting stressed state will cause a linear shift of the optical energy levels R1 and R2 of Cr^{3+} towards lower wave numbers (higher wavelengths) for compressive and higher wave numbers (lower wavelengths) for tensile stresses, respectively (Figure 5.32).

This shift can be expressed by

$$\Delta \nu = \prod_{ij} a_{ik}\, a_{jl}\, \sigma_{kl} \tag{5.63}$$

Assuming that for in-plane biaxial stress $\sigma_{xx} = \sigma_{yy} = \sigma_b$ and $\sigma_{zz} = 0$ (Ma and Clarke, 1993) the residual stress can be estimated following Margueron and Lepoutre (2002) to be $\sigma_b(GPa) \approx 0.2\Delta\nu\ (cm^{-1})$ for the corundum structure.

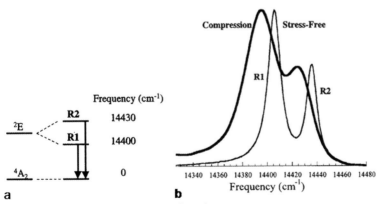

Figure 5.32 Chromium optical energy levels in aluminum sites in the corundum structure (a) and R1 and R2 spectra in the stress-free state and under compression (b) (Margueron, 2002).

Figure 5.33 shows surface scans of $30 \times 20\,\mu m^2$ areas of two APS mullite coatings deposited on aluminum substrates with coarse (< 136 μm; average grain size 51 μm; median grain size 47 μm) (Figure 5.33a) and fine (< 89 μm; average grain size 23 μm; median grain size 17 μm) (Figure 5.33b) starting powders (Seifert, 2004; Heimann et al., 2007). The shift of the Cr^{3+} photoluminescence line R1 is shown in the strained samples relative to the position of the stress-free R1 line in the corundum structure at 694.2 nm. Green and light blue colors in the line shift scans reveal essentially stress-free zones with no or small shifts whereas dark blue colors (negative shift) point to areas under tensile stress. On the other hand, yellow and red colors (positive shift) relate to compressive stresses shown by a shift of the luminescence line of Cr^{3+} to higher wavelengths.

The integrated intensities of the R1 line as a function of the wavelength are depicted in Figure 5.34. The bars were fitted to a distribution curve using a pseudo-Voigt algorithm. The wavelengths belonging to the maxima of these curves were taken to calculate the Raman shifts Δν and from these values the in-plane stress

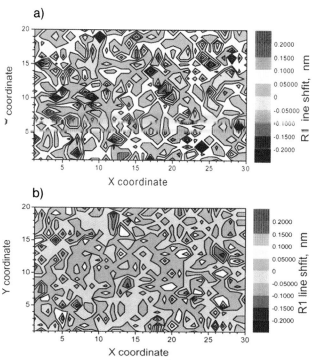

Figure 5.33 Surface scans ($30 \times 20\,\mu m^2$ area) of APS mullite coatings deposited with (a) coarse powder (<136 μm; average grain size 51 μm; median grain size 47 μm) and (b) fine powder (<89 μm; average grain size 23 μm; median grain size 17 μm) showing the Cr^{3+} photoluminescence R1 line shift relative to the stress-free position at 694.2 nm (Heimann et al., 2007).

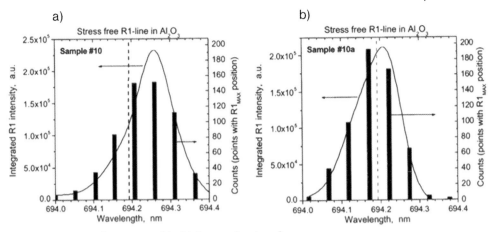

Figure 5.34 Integrated intensities of the R1 line as a function of the wavelength. (a). coarse powder. (b). fine powder.

components σ_b were determined to yield values between 100 and 300 MPa. Comparison of samples deposited with coarse and fine powders shows that the use of coarse spray powders result in higher stresses compared with the fine ones, and that a higher powder feed rate produces higher stresses. However, the grain size effect can be overridden by the effect of lower spray distance, higher plasma power and/or lower traverse speed that would lead to more pronounced substrate heating and thus higher surface stresses (Heimann et al., 2007). In conclusion, the development of residual compressive stresses in APS mullite coatings on aluminum substrates appears to be strongly influenced by the powder grain size, the plasma power and the travelling speed of the plasmatron but only marginally by the powder feed rate.

In contrast to the statements made above, Selcuk and Atkinson (2003) proposed that the wavelength of R1 is nonlinearly dependent on the stress level whereas R2 appears to have a fairly good linearity under compressive load. Consequently, the shift of R2 was evaluated to assess stress levels in TBCs deposited by EB-PVD (electron beam physical vapor deposition) (Zhao and Xiao, 2006; Lee et al., 2006).

5.7.5
Neutron Diffraction

The presence of significant porosity in plasma-sprayed coatings affects the measured residual stresses. In particular, flat planar imperfections that do not significantly contribute to the overall porosity (see Section 10.2.3.4) are believed to have a major influence on the elastic constants of the coating material. Indeed, elastic constants measured in coatings are much lower than those of the corresponding bulk material. The residual stresses measured by the conventional $\sin^2\psi$ technique (see Section 5.7.2) using X-rays are frequently ambiguous and difficult to interpret precisely because of the existence of flat, crack-like pores that open up during 4-point bending tests applied to measure elastic properties of coatings including their Young's moduli. Owing to the limited penetration of X-rays they will be scattered at surface cracks thus preventing measurement of stresses in compression. In this case use of neutrons with their higher penetration will yield much more realistic results unaffected by the dynamics of surface crack opening or closing in response to the mechanical deformation during bending tests. Experimental results by Dubský et al. (2002) confirm that the results of measurement by neutron diffraction of stresses in plasma-sprayed chromium oxide coatings correspond to those obtained by X-ray diffraction. However, differences in magnitude of the stresses are thought to be due to differences in the measured volumes with respect to stress gradients across the coating thickness. Figure 5.35 shows the deformation ε in % as a function of the loading force F in N for coatings in tension (a) and in compression (b), respectively. In the rather thick chromium oxide coating (1.53 mm) many cracks exist that in tension will open up. Hence no increase in lattice spacing $\Delta d/d_0$ is observed (Figure 5.35a). On the other hand, in compression up to a force of approximately 200 N the imposed deformation is taken up by the closing cracks

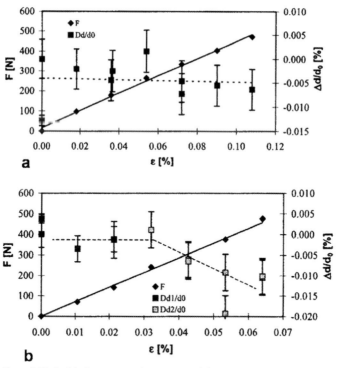

Figure 5.35 Residual stresses in plasma-sprayed chromium oxide coatings subjected to a 4-point bending test in tension (a) and compression (b) measured by a monochromatized neutron beam of 0.313 nm wavelength. The lattice deformation $\Delta d/d_0$ was evaluated from the shift of the (113) interplanar spacing of Cr_2O_3 (Dubský et al., 2002).

and hence the lattice does not deform significantly. With increasing force the compressive stresses in the lattice increase beyond a strain of about 0.03%.

Theoretical analyses of residual stresses and modeling of the complex interaction of the elastic and thermophysical coating parameters can be found in Chapter 6.4.3.

References

D. Apelian, *Mat. Res. Soc. Symp. Proc.*, 1984, 30.

S. Amada, T. Hirose, K. Tomoyasu, in: Proc.14th ITSC, Kobe, May 22–26, 1995, 885.

S. Amada, H. Yamada, *Surf. Coat. Technol.*, 1996, **78**, 50.

S. Amada, T. Hirose, *Surf. Coat. Technol.*, 1998, **102**, 132.

D. Aviir, D. Farin, P. Pfeifer, *J. Coll. Interface Sci.*, 1985, **103** (1), 112.

C.S. Barrett, T.B. Massalski, *Structure of Metals*, 3rd edn, McGraw-Hill New York, 1966, 465.

C.C. Berndt, in: *Adv. in Thermal Spraying* (ed. N.F. Eaton,), Welding Institute of Canada, Pergamon Press, 1986, 149.

P. Bialucki, W. Kaczmar, J. Gladysz, in: *Adv. in Thermal Spraying* (ed. N.F. Eaton,), Welding Institute of Canada, Pergamon Press, 1986, 837.

O. Brandt, in: Proc.14th ITSC, Kobe, May 22–26, 1995, 639.

C.A. Brown, P.D. Charles, W.A. Johnson, S. Chesters, *Wear*, 1993, **161**, 61.

C.F. Chyba, P.J. Thomas, K.J. Zahnle, *Nature*, 1993, **361**, 40.

T.W. Clyne, S.C. Gill, *J. Thermal Spray Technol.*, 1996, **5**, 401.

L.C. Cox, *Surface Coat. Technol.*, 1988, **36**, 807.

J. Drozak, *Haftung und Schichtaufbau von Spritzschichten*. Moderne Beschichtungsverfahren. DGM-Verlag 1992.

J. Dubský, H.J. Prask, J. Matějíček, T. Gnäupel-Herold, *Appl. Phys. A*, 2002, **74** [Suppl.], S1115.

J. Dudognon, M. Vayer, A. Pineau, R. Erre, *Surf. Coat. Technol.*, 2006, **200**, 5058.

S. Dyshlovenko, B. Pateyron, L. Pawlowski, D. Murano, *Surf. Coat. Technol.*, 2004, **187**, 408.

S. Dyshlovenko, L. Pawlowski, B. Pateyron, I. Smurov, J.H. Harding, *Surf. Coat. Technol.*, 2006, **200** (12/13), 3757.

B. Eigenmann, B. Scholtes, E. Macherauch, *Mat.-wiss.u. Werkstofftech.*, 1989, **20**, 314.

Y. Fahmy, J.C. Russ, C.C. Koch, *J. Mat. Res.*, 1991, **6** (9), 1856.

S. Fantassi, M. Vardelle, A. Vardelle, M.F. Eichinger, P. Fauchais, in: Proc.TS'93, Aachen 1993, DVS 152, 387.

P. Fauchais, M. Fukumoto, A. Vardelle, M. Vardelle, *J. Thermal Spray Technol.*, 2004, **13** (3), 337.

L.B. Freund, J.A. Floro, E. Chason, *Appl. Phys. Lett.*, 1999, **74** (14), 1987.

M. Fukumoto, S. Kato, I. Okane, in: *Proc.14th ITSC*, Kobe, May 22–26, 1995, 353.

R. Gadow, M.J. Siegert-Escribano, M. Buchmann, *J. Thermal Spray Technol.*, 2005, **14** (1), 100.

R. Ghafouri-Azar, J. Mostaghimi, S. Chandra, *Int. J. Comput. Fluid Dynamics*, 2004, **18** (2), 133.

C. Godoy, E.A. Souza, M.M. Lima, J.C.A. Batista, *Thin Solid Films*, 2002, **420/421**, 438.

J. Götze, R.B. Heimann, H. Hildebrandt, U. Gburek, *Mat.-wiss.u. Werkstofftechn.*, 2001, **32**, 130.

H. Gruner, *Thin Solid Films*, 1984, **118**, 409.

M.L. Gualtieri, M. Prudenziati, A.F. Gualtieri, *Surf. Coat. Technol.*, 2006, **201** (6), 2984.

S. Guessasma, G. Montavon, C. Coddet, *Surf. Coat. Technol.*, 2003, **173**, 24.

N.G. Hadjiconstantinou, in: *Proc. IMECE'99 (Intern. Mech. Eng. Congress and Exposition)*, Nov 14–19, 1999, Nashville, USA.

M. Härting, *Acta Mater.*, 1998, **46** (4), 1427.

M. Härting, A. Hempel, M. Hempel, R. Bucher, D.T. Britton, *Mater. Sci. Forum*, 2001, **363–365**, 502.

Y. Han, K. Xu, J. Lu, *J. Biomed. Mater. Res.*, 2001, **55** (4), 596.

H. Harlow, J.E. Welch, *The Physics of Fluids*, 1965, **8**, 12.

A. Hasui, S. Kitahara, T. Fukushima, *Trans. Nat. Res. Inst. for Metals*, 1970, **12** (1), 9.

R.B. Heimann, *Auflösung von Kristallen (Dissolution of Crystals)*, Appl. Miner. Vol. 8 Springer: Wien, New York, 1975, 270 pp.

R.B. Heimann, J. Kleiman, in: *Crystals. Growth, Properties, and Application*, H.C. Freyhardt, (ed.)., Springer: Berlin, Heidelberg, New York, London, Paris, Tokyo, 1988, **11**, 1–73.

R.B. Heimann, *Process. Adv. Mat.*, 1991, **1**, 181.

R.B. Heimann, O. Graßmann, M. Hempel, R. Bucher, M. Härting, in: Rammlmair, *et al.* (eds.), Proc. 6th Intern. Congress on Applied Mineral., ICAM 2000, Göttingen, Germany, Applied Mineralogy, Vol. 1, Balkema: Rotterdam p. 1555.

R.B. Heimann, *Surf. Coat. Technol.*, 2006, **201**, 2012.

R.B. Heimann, R. Wirth, *Biomaterials*, 2006, **27**, 823.

R.B. Heimann, in: *Trends in Biomaterials Research*, P.J. Pannone, (ed.), Nova Science Publishers, Inc., Hauppauge, N.Y., USA, 2007, ISBN 1-60021-361-8, Chapter 1, 1–81.

R.B. Heimann, H.J. Pentinghaus, R. Wirth, *Eur. J. Mineral.*, 2007, **19**, 281.

H. Herman, *Sci. Am.*, 1988, 112.

J.M. Houben, in: *Proc.2nd NTSC*, Oct 31–Nov 2, 1984. Long Beach, CA.

J.M. Houben, *Relation of the Adhesion of Plasma Sprayed Coatings to the Process Parameters Size, Velocity and Heat Content of the Spray Particles*. Ph.D.Thesis, Technical University Eindhoven, The Netherlands, 1988.

H. Jones, *J. Phys. D: Appl. Phys.*, 1971, **4**, 1657.

C.W. Kang, H.W. Ng, *Surf. Coat. Technol.*, 2006, **200**, 5462.

M. Kardar, G. Parisi, Y.-C. Zhang, *Phys. Rev. Lett.*, 1986, **56**, 889.

O. Kesler, J. Matejicek, S. Sampath, S. Suresh, T. Gnaeupel-Herold, P.C. Brandt, H.J. Prask, *Mater. Sci. Eng. A*, 1998, **257**, 215.

S. Kitahara, A. Hasui, *J. Vac. Sci. Technol.*, 1974, **11**, 747.

R. Knight, R.W. Smith, in: *Proc. 3rd NTSC* 1993, Anaheim, CA, 607.

I. Kraus, N. Ganev, G. Gosmanová, H.-D. Tietz, L. Pfeiffer, S. Böhm, *Adv. Perf. Mater.*, 1997, **4**, 63.

H. Kreye, D. Fandrich, H.H. Müller, G. Reiners, in: *Advances in Thermal Spraying* (ed. N.F. Eaton), Proc.11th ITSC, Montreal, Sept. 8–12, 1986, Welding Inst. of Canada, 1986, p. 121.

J.L. LaRosa, J.D. Cawley, *J. Amer. Ceram. Soc.*, 1992, **75** (7), 1981.

G. Lee, A. Atkinson, A. Selçuk, *Surf. Coat. Technol.*, 2006, **201**, 3931.

S.R. Levine, R.A. Miller, M.A. Gedwill, in: Proc. 2nd Conf. on Advanced Mater. for Alternative Fuel-Capable Heat Engines, Monterey, CA, 1981.

C.J. Li, H.L. Liao, P. Gougeon, G. Montavon, C. Coddet, *Surf. Coat. Technol.*, 2005, **191**, 375.

H. Li, K.A. Khor, P. Chang, *Biomaterials*, 2004, **25** (17), 3463.

N. Llorca-Isern, G.B. Vidal, J. Jorba, L. Bianchi, D. Sánchez, *J. Thermal Spray Technol.*, 2001, **10** (2), 287.

Y.P. Lu, M.S. Li, S.T. Li, Z.G. Weng, R.F. Zhu, *Biomaterials*, 2004, **25** (18), 4393.

Q. Ma, D.R. Clarke, *J. Am. Ceram. Soc.*, 1993, 1433.

J. Madejski, *Int. J. Heat Mass Transfer*, 1976, **19**, 1009.

B.B. Mandelbrot, *The Fractal Geometry of Nature*, Freeman San Francisco, 1982.

B.B. Mandelbrot, *Les Objects Fractals*, Paris: Flammarion, 1984.

B.B. Mandelbrot, *Physica Scripta*, 1985, **32**, 257.

B.B. Mandelbrot, D.E. Passoja, A.J. Paullay, *Nature*, 1984, **308**, 721.

S. Margueron, J. Scient. Barriéres Therm., ONERA, 23 mai 2002, 22.

S. Margueron, F. Lepoutre, in: Proc. 45èmes Colloque de Métallurgie 'Surfaces, interfaces et rupture', Saclay, France, 2002. juin 25.27; Report ONERA TP 2002–2007.

H. Maruo, Y. Hirata, Y. Matsumoto, in: Proc. 14th ITSC'95 Kobe, May 22–26, 1995, 341.

J. Matejicek, S. Sampath, J. Dubsky, *J. Thermal Spray Technol.*, 1998, **7** (4), 489.

J. Mathar, *Arch. Eisenhüttenwesen*, 1932, **6**, 277.

R. McPherson, *J. Mat. Sci.*, 1980, **15**, 3141.

R. McPherson, *Surf. Coat. Technol.*, 1989, **39/40**, 173.

P. Meakin, *J. Mat. Educ.*, 1989, **11**, 105.

J.J. Mecholsky, D.E. Passoja, K.S. Feinberg-Ringel, *J. Am. Ceram. Soc.*, 1989, **72** (1), 60.

A. Mézin, *Surf. Coat. Technol.*, 2006, **200** (18/19), 5259.

G. Montavon, S. Sampath, C.C. Berndt, H. Herman, C. Coddet, *Surf. Coat. Technol.*, 1997, **91**, 107.

M. Müller, F. Gitzhofer, R.B. Heimann, M.I. Boulos, in: Proc. NTSC'95, Houston, TX, Sept 11–15, 1995.

A. Noutomi, *Welding Intern.*, 1989, **11**, 947.

M. Okuno, N. Zotov, M. Schmücker, H. Schneider, *J. Non-Cryst. Solids*, 2005, **351**, 1032.

M. Pasandideh-Fard, J. Mostaghimi, *Plasma Chem. Plasma Proc.*, 1996, **16**, 83S.

M. Pasandideh-Fard, R. Bhola, S. Chandra, J. Mostaghimi, *Int. J. Heat Mass Transfer*, 1998, **41**, 2929.

M. Pasandideh-Fard, V. Pershin, S. Chandra, J. Mostaghimi, *J. Thermal Spray Technol.*, 2002, **11** (2), 206.

D.E. Passoja, D.J. Amborski, *Microstruct. Sci.*, 1978, **6**, 143.

L. Pawlowski, *The Science and Engineering of Thermal Spray Coatings*, J. Wiley: New York, 1995.

J. Peng, V. Ji, W. Seiler, A. Tomescu, A. Levesque, A. Bouteville, *Surf. Coat. Technol.*, 2006, **200**, 2738.

H.J. Pentinghaus, U. Precht, J. Göttlicher, in: Proc. ICG Kyoto, ISBN 4-931298-43-5, 2004, C3858.

J.A. Pina, J. Dias, L. Lebrun, *Mater. Sci. Eng. A*, 2003, **547** (1/2), 21.

G. Reisel, R.B. Heimann, *Surf. Coat. Technol.*, 2004, **185**, 215.

P.V. Riboud, *Ann. Chim.*, 1973, **8**, 381.

M.H. Rice, R.G. McQueen, J.M. Walsh, in: *Solid State Physics, VI* (eds. F. Seitz and D. Turnbull), Academic Press, New York, 1958.

D.S. Rickerby, *J. Vac. Sci. Technol.*, 1986, **4**, 2809.

SAE Handbook, *Test Strip, Holder and Gage for Shot Peening*, SAE Standard 442 SAE Handbook, Part I, SAE Inc., Warrendale, PA, 1977, 9.05.9.06.

H. Salmang, H. Scholze, *Keramik. Teil I*. Springer Berlin, Heidelberg, New York, 1982, 237.

Y.Y. Santana, J.G. La Barbera-Sosa, M.H. Staia, J. Lesage, E.S. Puchi-Cabrera, D. Chicot, E. Bemporad, *Surf. Coat. Technol.*, 2006, **201**, 2092.

D.W. Schaefer, *Science*, 1989, **243**, 1023.

S. Seifert, *Development of Ceramic Lightweight Thermal Protection Coatings for Reusable Space Launch Vehicles*. Unpublished Master Thesis, TU Bergakademie Freiberg, June 2004.

S. Seifert, E. Litovski, J.I. Kleiman, R.B. Heimann, *Surf. Coat. Technol.*, 2006a, **200** (11), 3404.

S. Seifert, J.I. Kleiman, R.B. Heimann, *J. Spacecraft Rockets*, 2006b **43** (29), 439.

A. Selcuk, A. Atkinson, *Acta Mater.*, 2003, **51**, 535.

S.D. Siegmann, C.A. Brown, in: Proc. 15th ITSC, 1998, May 25.29, Nice, France, p. 831.

P.A. Siemers, W.B. Hillig, *Thermal Barrier-Coated Turbine Blade Study*, Final report, NASA CR-165351, 1981. pp. 123.

V.V. Sobolev, J.M. Guilemany, A.J. Martin, *J. Thermal Spray Technol.*, 1996, **5** (2), 207.

O.P. Solonenko, A. Ohmori, S. Matsuno, A.V. Smirnov, in: Proc.14th ITSC'95, Kobe, May 22–26, 1995, 359.

W.D. Sproul, *Thin Solid Films*, 1984, **118**, 279.

H.-D. Steffens, H.-M. Höhle, E. Ertürk, *Schweissen & Schneiden*, 1981, **33**, 159.

J. Stokes, L. Looney, *Surf. Coat. Technol.*, 2004, **177–178**, 18.

G.G. Stoney, *Proc. Royal Soc. Lond.*, 1909, **A82**, 172.

D. Stoyan, H. Stoyan, *Fractals, Random Shapes and Point Fields*. Methods of Geometrical Statistics. J. Wiley & Sons New York, Toronto, Chichester, 1995.

L. Sun, C.C. Berndt, K.A. Khor, H.N. Chang, K.A. Gross, *J. Biomed. Mater. Res.*, 2002, **62** (2), 228.

S. Thiele, *Mikrohärte, Mikrostruktur und Haftung vakuumplasmagespritzter TiC/Mo$_2$C/Ni, Co- Verbundschichten*. Unpublished Master thesis. Technische Universität Bergakademie Freiberg, June 1994.

H.-D. Tietz, B. Mack, L. Pfeiffer, in: Proc. TS'93, Aachen 1993, DVS 152, 205.

M. Topić, T. Ntsoane, R.B. Heimann, *Surf. Coat. Technol.*, 2006, **201**, 3633.

T.C. Totemeier, J.K. Wright, *Surf. Coat. Technol.*, 2006, **200**, 3955.

G. Trapaga, J. Szekely, *Metall. Transactions B*, 1991, **22**, 901.

Y.C. Tsui, T.W. Clyne, *Thin Solid Films*, 1997, **306**, 23.

Y.C. Tsui, C. Doyle, T.W. Clyne, *Biomaterials*, 1998, **19**, 2015.

E.E. Underwood, K. Banerji, *Mat.Sci.Eng.*, 1986, **80**, 1.

R.F. Voss, R.B. Laibowitz, E.I. Alessandrini, in: *Scaling Phenomena in Disordered Solids* (eds. R. Pynn, A. Skjoltorp), Plenum Press, 1985, 279.

T. Watanabe, I. Kuribayashi, T. Honda, A. Kanazawa, *Chem. Eng. Sci.*, 1992, **47**, 3059.

Y.C. Yang, E. Chang, S.Y. Lee, *J. Biomed. Mater. Res.*, 2003, **67** (7), 886.

J.E. Yehoda, R. Messier, *Appl. Surf. Sci.*, 1985, **22/23**, 590.

H. Yu, D.R. Clarke, *J. Am. Ceram. Soc.*, 2002, **85** (8), 1966.

H.G. Zachmann, *Mathematik für Chemiker*, 4.Auflage, VCH: Weinheim, New York, Basel, Cambridge, 1981.

X. Zhao, P. Xiao, *Surf. Coat. Technol.*, 2006, **201**, 1124.

J. Zimmermann, *O Untersuchungen zur Haftfestigkeit und Rauheit plasmagespritzter Cr_2O_3-Schichten*, 4th year thesis, Technische Universität Bergakademie Freiberg, 2000.

6
Modeling and Numerical Simulation

6.1
Principal Aspects of Modeling

Much information has been collected experimentally that connects plasma properties and parameter variations with their influence on the behavior of particles carried by the plasma stream. Of prime importance for the design of coatings that perform successfully was the study of the efficient transfer of thermal energy and momentum from the highly accelerated plasma jet to the injected particles. Likewise, the effects of particle impact at the substrate surface to be coated were of interest, including splat development in terms of size and morphology, and the extent and distribution of porosity and residual stresses. All these complex processes are now available for modeling, and the interactions of numerous intrinsic and extrinsic plasma spraying parameters are being investigated by numerical simulation.

Dimensional analysis is a key step in any modeling process (Bridgeman, 1922; Isaacson and Isaacson, 1975). Since there are usually fewer dimensionless groups (see Appendix A) than physical quantities (Buckingham's rule; Ashby, 1992) the required algorithms are characterized by mathematical economy. When a model is evaluated or displayed, the dimensionless groups are the correct axes to choose. It should be emphasized that dimensionless groups should always be used for exponential, logarithmic or power law arguments, since failure to do so may lead to pseudo-constants with no physical significance.

Because of their ability to simplify complex modeling problems dimensionless groups are frequently used in chemical engineering to solve complex equations of heat and mass transfer. In particular, equations of heat, mass and impulse transfer that occur throughout the description of plasma spray processes and numerical solutions of modeling approaches can be deduced from existing solutions of geometrically similar systems (Rosenberger, 1979). The term 'geometric similarity' refers to the interchangeability of streamlines and boundaries of systems of widely varying dimensional extension by linear scaling laws. Motion of a fluid such as a plasma is dynamically similar when the solutions of the related dimensionless transfer equations are identical. For example, the simplified

Plasma Spray Coating: Principles and Applications. Robert B. Heimann
Copyright © 2008 WILEY-VCH Verlag GmbH & Co. KGaA, Weinheim
ISBN: 978-3-527-32050-9

Navier–Stokes equation

$$\rho(\delta_v/\delta_t) = \rho(v \times \nabla)v - \nabla p + \eta \nabla^2 v + f, \tag{6.1}$$

where v = mass average velocity (barycentric velocity), ρ = plasma density, p = plasma gas pressure, and η = plasma gas viscosity, can be transformed to its dimensionless version by using the free-stream plasma velocity V, a characteristic length L and a typical pressure p_0 to normalize v(x,y,z) and p by setting the velocity to W = v/V, the lengths to X = x/L, Y = y/L, Z = z/L, and the pressure to P = p/p_0.

Substitution of these dimensionless ratios into Equation 6.1 yields

$$\rho(V^2/L)(W \times \mathrm{grad})W = -(p_0/L)\mathrm{grad}(P) + (\eta V/L^2)\nabla^2 W \tag{6.2}$$

or

$$(W \times \mathrm{grad})W = -(p_0/\rho V^2)\mathrm{grad}(P) + \mathbf{Re}^{-1}(\nabla^2 W), \tag{6.2a}$$

with $\mathbf{Re} = VL/\eta$.

It should be noted that the second term on the right side of Equation 6.2a contains all parameters that govern the fluid dynamics of a plasma jet. Dynamic similarity is assured if, for two geometrically similar systems A and B, $\mathbf{Re}(A) = \mathbf{Re}(B)$. The similarity principle was discovered by Reynolds and has overriding importance for any modeling approach in fluid dynamics (Boucher and Alves, 1959).

6.2
Modeling of Plasma Properties

In the case of a thermal equilibrium, modeling by numerical simulation of the characteristic plasma parameters can be performed assuming thermodynamic equilibrium conditions. In particular, rather simple modeling of the arc column has already been performed in the mid-1930s by Elenbaas (1934) and Heller (1935) by equating the temperature gradient in the arc column with the generating electrical field:

$$\mathrm{div}(-\lambda \mathrm{grad}\, T) - \sigma E^2 = 0, \tag{6.3}$$

where λ is the thermal conductivity, σ is the electrical conductivity, and E is the electrical field. It should be particularly emphasized that the actual situation in an arc column is too complex to be solved through the conservation equations of mass, momentum and energy. Hence in the Elenbaas–Heller approach thermal diffusion and the effects of radiation have been neglected. Assuming rotational symmetry and thus introducing cylindrical coordinates the equation then reads

$$(1/r)d/dr\{r\lambda(dT/dr)\} + \sigma E_z^2 = 0. \tag{6.4}$$

With the definition of the heat flux potential, $S = \int \lambda dT$ it follows

$$(1/r)d/dr\{r(dS/dr)\} + \sigma E_z^2 = 0; \quad S = S(\sigma). \tag{6.5}$$

With Ohm's law, $I = 2\pi E_z \int \sigma r dr$ we arrive at the closed-form solutions (Maecker, 1959)

$$I = Rf_1(S); \quad E = (1/R)f_2(S), \quad (6.6a)$$

and also

$$IE = f_3(S) = f_4(T), \quad (6.6b)$$

from which follows

$$T_{max} = f(IE), \quad (6.6c)$$

i.e. the maximum temperature in an arc depends only on the power input per unit length. Thus in contrast to a flame torch, whose maximum achievable temperature is limited by the internal enthalpy of the combustion gases, the maximum temperature of a plasmatron is basically unlimited. It only depends on the power input that by itself is limited by the cross-section of the power leads. The modeling approach was taken from Boulos *et al.* (1989). An experimental confirmation of the statement made above is shown in Figure 6.1 that displays the almost linear dependence of the axial temperature of a hydrogen arc on the electric power input in kW m^{-1}.

More involved modeling using the simultaneous solution of the conservation equations as well as species diffusion and the Maxwell equations can only be performed by simplifying assumptions including:

- 2D rotational symmetry of the arc column;
- turbulent flow for DC plasmas;
- laminar flow, i.e. low Reynolds number **Re** for r.f. plasmas;
- optically thin arc;
- local thermal equilibrium, i.e. $T_e = T_h$;
- no viscous dissipation of the arc power;
- negligible diffusion of species;
- wall-stabilized arc conditions.

Under these assumptions the temperature and velocity profiles in a plasma arc column can be modeled, and particle trajectories, and heat and momentum transfer from the plasma to the particles be predicted (see Chapter 4). For details see, for example, Mostaghimi (1992) and Proulx *et al.* (1987), and the paragraphs below.

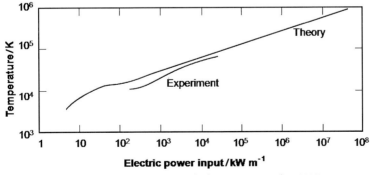

Figure 6.1 Maximum axial temperature in a hydrogen arc (Maecker, 1959).

Diagnostics and modeling of an argon/helium plasma spray process has been performed by Duan et al. (2000) to account for the natural instability of the arc in a DC plasmatron as this is one of the most important causes for nonlinear variations in heat transfer to the powder particles, leading to inconsistencies in the final coating quality. A diagnostic system was used to monitor the plasma jet instability, as well as some other important process characteristics. The effects of the operating parameters and the anode condition on the properties of plasma jets, particle properties and coatings have been measured, showing that plasma jet instability influences the plasma spray process in several critical ways. A series of high speed images revealed that coating porosity and deposition efficiency can be correlated to an average jet length, and selected frequency peaks in the power spectrum of the acoustic signal were found to correlate with the average jet length. These experimental results were used to derive a simple control scheme adopting a fuzzy look-up model indicating the condition of the anode. In a more practical application, it was concluded that an important way to counteract the negative effects of anode erosion is to increase the arc current.

6.3
Modeling of the Plasma–Particle Interaction

6.3.1
Modeling of Heat Transfer

6.3.1.1 Conservation Equations

To account for the particle interaction in a densely loaded plasma jet, the PSI (particle-source-in-cell) model by Crowe et al. (1977) has been adopted by Proulx et al. (1985). By coupling the stochastic single-particle trajectory calculations with those describing the continuum flow, temperature and concentration fields were obtained by introducing source-sink terms for mass, S_p^c, momentum, S_p^m (in the r- and z-directions), and enthalpy, S_p^h.

The four plasma equations used to model the plasma–particle interactions under dense loading conditions (Mostaghimi et al., 1985) are the

1. continuity (mass conservation) equation

$$\text{div}(\rho u) = S_p^c, \tag{6.7}$$

with $S_p^c = \Sigma c(\Delta m_p/\tau)$ (c = particle concentration, Δm_p = amount of mass evaporated, τ = residence time of a particle in the plasma jet), or, in cylindrical coordinates,

$$(1/r)[\delta(r\rho v)/\delta r] + (\delta(\rho u)/\delta z) = S_p^c. \tag{6.7a}$$

2. momentum conservation (Navier–Stokes) equation

$$\rho u(\text{grad } u) = -\text{grad } p + \text{div}(\eta \text{grad } u) + (j \times B) + S_p^m, \tag{6.8}$$

with $S_p^m = \Sigma c[\Delta(m p v_{zp})/\tau]$ (axial direction, z) and $\Sigma c[\Delta(m_p v_{rp})/\tau]$ (radial direction, r).

3. energy conservation equation

$$\rho u(\text{grad } h) = -\text{grad}(k/c_p)\text{grad } h + \sigma E_\theta^2 - q_r + S_p^h, \quad (6.9)$$

with $S_p^h = \Sigma c(q_p + q_v)$, where the expression in parentheses is the energy balance between convective and radiative energy transfer (see above). The term σE_θ^2 describes the energy gain of the plasma by Ohmic heating (see Equation 6.5), the term q_r the energy loss by radiation. The source-sink term S_p^h can be evaluated for the local cooling generated under dense loading conditions.

4. species conservation equation

$$\rho u(\text{grad } Y_i) = \text{div}(D \text{ grad } Y_i) + S_p^c. \quad (6.10)$$

The general term Y_i refers to the concentration in different reference systems. In the barycentric (mass-centered) system, Y_i equals W_i, i.e. the mass fraction of species i; in the mole-centered system, Y_i is the mole fraction, X_i, and in the volume-centered system, Y_i becomes $c_i v_i$, i.e. the product of the concentration of species i and its partial molar volume.

6.3.1.2 Experimental Validation of Modeling under Dense Loading Conditions

Proulx et al. (1985) injected copper particles of 70 μm diameter downwards through the central tube of an inductively coupled plasmatron. Copper evaporates easily, and the vapor changes the electrical and radiative properties of the plasma gas drastically.

Figure 6.2 shows the isotherms (A) and stream lines (B), and the isopleths of the concentration of copper vapor (C) for a powder feed rate of 5 g min^{-1} copper powder.

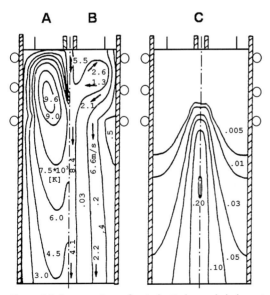

Figure 6.2 Cross-sections of an inductively-coupled plasmatron after injection of copper powder (5 g/min). Left: isotherms (A) and stream lines (B). Right: isopleths of concentration of copper vapor (C) (Proulx et al., 1985).

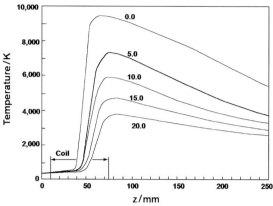

Figure 6.3 Cooling of a plasma under dense loading conditions of copper particles ($m_p = 0$–$20.0\,\text{g min}^{-1}$) measured along the centerline of a plasma jet (Proulx et al., 1985).

Since the trajectories of the particles are very close to the axis of the jet, the plasma gas is significantly cooled down in this region as shown in Figure 6.3 for powder feed rates between 0 and $20.0\,\text{g min}^{-1}$. Even a feed rate of copper as low as $5\,\text{g min}^{-1}$ leads to a temperature drop of about 2000 K. Higher feed rates lead to even lower plasma temperatures. Note that in the upper part of the confining cylindrical plasmatron an eddy develops (Figure 6.2B) that leads to impaired powder transport into the plasma column proper. Hence it is essential that powder injection occurs below the position of the eddy.

Table 6.1 gives numerical data of the feed rate m_p in g min^{-1}, the mass of copper evaporated, m_v, the heat absorbed by the solid particles, Q_p, and the heat absorbed by the copper vapor, Q_v. Q_v is generally much smaller than Q_p, and the ratio Q_v/Q_p changes from 30% at $1\,\text{g min}^{-1}$ feed rate to 5% at $20\,\text{g min}^{-1}$ feed rate. The total energy absorbed by the powder is between 3.1 and 17.8% of the plasma power input.

Plasma–particle interaction in APS has been studied by a Lagrangian two-phase fluid model by Gawne et al. (2005) assuming that under certain special conditions powder particles are heated and accelerated so much that downstream in a plasma jet they will transfer heat and momentum to the cooler and slower surrounding gas flow.

Table. 6.1 Numerical data of plasma power absorbed by copper particles, Q_p and copper vapor, Q_v under dense loading conditions (Proulx et al., 1985).

m_p (g min^{-1})	m_v (g min^{-1})	Q_p (W)	Q_v (W)	% of total energy absorbed
1.0	0.50	71.0	21.0	3.1
5.0	1.40	255.0	42.0	9.9
10.0	1.40	386.0	35.0	14.0
15.0	1.16	460.0	28.0	16.3
20.0	0.94	511.0	23.0	17.8

Then the direction of energy transfer has been reversed. In particular, the dominating factor for plasma–particle interaction is the powder feed rate. For feed rates of Ni powder below 40 g min^{-1} the interaction is negligible, i.e. the substantial decrease in temperature and velocity observed above (Figure 6.3) for an ICP situation at already much lower feed rates does not occur.

6.3.1.3 Modeling of Heat Transfer in Two-fluid Interfacial Flow

Modeling of heat transfer and fluid flow becomes a formidable task when the position of the moving interface between fluids has to be taken into account simultaneously. Medhi-Nejad *et al.* (2004) developed a method to calculate heat transfer across the boundary of two immiscible liquids, e.g. molten tin droplets falling through an oil bath using a volume tracking method to model the simultaneous movement of mass, momentum and energy across cell boundaries. A second-order van Leer approach (van Leer, 1979) was applied to calculate the advected temperature based on the cell-averaged temperatures of a series of donor and acceptor cells in the presence of a steep temperature gradient between the two fluids. The model has some bearings on modeling of pressure and temperature discontinuities in space such as those encountered during the propagation of shock waves (van Leer, 1999) as well as boundaries within two-fluid plasmas (Helenbrook *et al.*, 1999).

6.3.2
Modeling of Momentum Transfer

Momentum transfer from the plasma to the particles can be modeled with the following simplifying assumptions:

- the molten droplets are spherical;
- the molten droplets are carried in a gas stream with constant temperatures along its symmetry axis;
- the gas jet is not influenced by particles, i.e. the operation is carried out in a diluted system as opposed to the dense loading conditions described above;
- continuum gas flow.

The governing equation applied to describe the impulse (momentum) transfer is the Basset–Boussinesq–Oseen equation of motion[1] (Schoeneborn, 1975). This equation determines the time dependency of the particle velocity, dV_p/dt,

---

1) The equation given here is a simplification. The complete equation of motion of a particle injected into a plasma jet can be described in a summary form by $F_P = F_D + F_{PG} + F_{AM} + F_H + F_E$, where F_P = mass-acceleration product, i.e. particle inertia, F_D = Stokesian drag force, F_{PG} = drag due to the plasma pressure gradient, F_{AM} = drag due to the added mass, F_H = Basset history term due to the non-steady motion of the particle, and F_E = external potential forces such as gravity, electric or magnetic forces (Lewis and Gauvin, 1973). Since in a thermal plasma the density of the gas is rather low compared with that of the particle, all terms but the viscous drag force F_D and the history term F_H can be neglected. Thus the equation reduces to $F_P = F_D + F_H$. This applies essentially to pure Stokesian motion with low Reynolds numbers ($C_D = 24/\mathbf{Re}$). According to Pfender (1989) additional turbulence and thermophoretic effects must be considered when **Re** exceeds a critical threshold value.

considering only the viscous drag force, F_D and the velocity gradient, $V_g - V_p$:

$$dV_p/dt = V_p(\partial V_p/\partial x) = [3C_D\rho_g/4d_p\rho_p]|V_g-V_p|(V_g-V_p), \quad (6.11)$$

where

$$C_D = [F_D/A_p]/[1/2\rho_g u_R^2]. \quad (6.12)$$

The velocity term $u_R = V_g - V_p$, i.e. the relative velocity of a particle with respect to the plasma gas, can be more accurately described by considering the velocity components of the particle in axial (z) and radial (r) directions: $u_R = [(V_g^{(z)} - V_p^{(z)})^2 + (V_g^{(r)} - V_p^{(r)})^2]^{1/2}$.

The decay of the plasma velocity V_g with distance from the nozzle exit is inversely proportional to the distance x, and can hence be described by the simple relation

$$V_g = V_0/(c_1 + c_2 x), \quad (6.13)$$

where V_0 is the exit velocity, and c_1, c_2 are constants (see Figure 4.6).

A semi-empirical approach to the particle velocity in a plasma jet was developed by Nikolaev (1973) for a Reynolds number $\mathbf{Re} < 1$

$$V_p = V_g\{1 - \exp[18\nu\rho_g t/(d_p^2 \rho_m)]\}, \quad (6.14a)$$

with ν = kinematic viscosity of the plasma, ρ_g = plasma gas density, d_p = particle diameter, ρ_m = particle density, and t = dwell time of the particle, and for $\mathbf{Re} > 2$ yielding

$$V_p = (V_g t)/t^* + t; \quad \text{with } t^* = 4d_p\nu/3C_D V_g \rho_m. \quad (6.14b)$$

6.3.2.1 Modeling of the Drag Coefficient

Since knowledge of the prevailing viscous drag coefficient C_D is essential for accurate modeling of the momentum transfer, many attempts have been made to evaluate C_D for different Reynolds numbers ranging from $C_D = 24/\mathbf{Re}$ for $0 < \mathbf{Re} < 0.2$ (Stokesian motion) to $C_D = (24/\mathbf{Re})(1 + 0.189 \, \mathbf{Re}^{0.63})$ for $21 \leq \mathbf{Re} \leq 200$ (Beard and Pruppacher, 1969). A nonlinear expression for even higher turbulent plasmas ($0.15 \leq \mathbf{Re} \leq 500$) was given as $C_D = (23.7/\mathbf{Re})[1 + 0.165 \, \mathbf{Re}^{2/3} - 0.05 \, \mathbf{Re}^{-0.1}]$ (Fiszdon, 1979). It should be noted that there are several problems with modeling the forces acting on a particle in a plasma jet that require modification of the Basset–Boussinesq–Oseen approximation. These problems relate to (1) the generally large temperature gradient present, (2) the change of the particles diameter d_p by surface ablation and thus the uncertainty of the shape of the projected surface area (see Section 4.3.3), thermal expansion, decomposition and vaporization, and (3) non-continuum effects for small particles with $d_p < 10 \, \mu m$.

1. Corrections for temperature effects were introduced by Lewis and Gauvin (1973) who modified the drag coefficient assuming a local Reynolds number

$$C_D = C_{Df}[(\rho_\infty/\rho_f)(\nu_f/\nu_\infty)^{0.15}], \quad (6.15)$$

with C_{Df} = drag coefficient evaluated at the 'mean film temperature' ($T_f = (T_\infty T_s)/2$), ρ_f = gas density at T_f, ρ_∞ = gas density at T_∞, ν_f = kinematic viscosity at T_f, and ν_∞ = kinematic viscosity at T_∞. Another correction approach was reported by Lee *et al.* (1981) with

$$C_D = C_{Df}[\rho_\infty \mu_\infty / \rho_s \mu_s]^{-0.45} = C_{Df}(\nu_\infty / \nu_s)^{-0.45} = C_{Df} f_1, \quad (6.16)$$

where the subscripts ∞ and s refer to the plasma temperature, T_∞ and the surface temperature of the particle, T_s, respectively.

2. The drag coefficient depends on the projected surface area of the particle perpendicular to the flow, A_p (Equation 6.12). When the particle diameter decreases by ablation or evaporation, the surface area also decreases, and the drag coefficient and hence the drag force change accordingly.

3. Noncontinuum effects for small particles ($d_p < 10\,\mu m$) can be accounted for through the Knudsen number, **Kn** = (λ/d_p), where λ = molecular mean free path. In the slip flow regime 0.01 < **Kn** < 1, a correction of the drag coefficient has been proposed (Chen and Pfender, 1983) as

$$C_{D,slip} = C_{D,cont}[1/1 + A \times B(4\,\mathbf{Kn}/\mathbf{Pr})]^{0.45} = C_{D,cont} f_2 \quad (6.17)$$

where A = $(2 - a)/a$ with a = thermal accommodation coefficient, B = $(\gamma/1 + \gamma)$ with $\gamma = c_p/c_v$, and **Pr** = Prandtl number. According to Pfender (1989), the corrections expressed in Equations 6.16 and 6.17 can be combined to arrive at a general expression for the drag coefficient:

$$CD = C_{Df} f_1 f_2. \quad (6.18)$$

The acceleration of the particles caused by the viscous drag forces is proportional to the relative velocity of the particles, and inversely proportional to the density and the square of the diameter of the particle. The correction for noncontinuous effects is particularly important for low-pressure plasma spraying where the mean free path of the gas molecules, λ is close to the diameter of the particles, d_p.

Figure 6.4 shows the dependency of the viscous drag coefficient, C_D expressed in Equation 6.12 as a function of the Reynolds number, **Re**. It can be seen that the

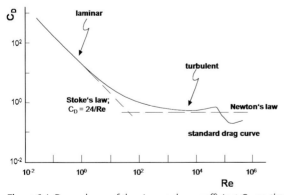

Figure 6.4 Dependence of the viscous drag coefficient C_D on the Reynolds number **Re**.

standard drag curve is limited by the Stokes' equation for low Reynolds numbers, i.e. laminar flow regime, whereas it is limited by Newton's law at high Reynolds numbers, i.e. turbulent flow regime as present in most plasma arc jets. The turbulent flow causes a nonsteady flow field around the immersed particles and thus a rapid change of the particle Reynolds number **Re** with time. According to Lewis and Gauvin (1973) this results in an excess drag that was predicted to range from 20% for 30 μm particles to 100% for 150 μm particles. Together with the effect of the Basset history term (see footnote 1) the effect of excess drag must be considered in momentum transfer modeling.

6.3.3
Modeling of Particle Dispersion

Particle behavior during plasma spraying was numerically modeled to account for the relative importance of effects introduced by plasma turbulence, particle size, particle density, and injection velocity and direction (Williamson et al., 2002). A steady-state plasma jet conforming to a commercial plasmatron was first modeled using the LAVA computer code (Idaho National Engineering and Environmental Laboratory, Idaho Falls, ID), and single particle composition (ZrO_2) and injection location were assumed. To these standard conditions, real world complexity in terms of turbulent dispersion, particle size, and injection velocity and direction, were added 'one phenomenon at a time' to account for the different contributions of individual parameters. A final calculation considered all phenomena simultaneously. The modeling thus provided insight into particle behavior as a function of multiparameter interaction. It was found that turbulent particle dispersion is most significantly affected by (in order of decreasing importance) particle size distribution, injection velocity distribution, degree of turbulence, and injection direction distribution or particle density distribution.

Similar work was performed by Zhang et al. (2006) using a Monte Carlo approach to model particle motion and heat transfer in a plasma jet. The effects of several process parameters such as arc power, gas flow rate, particle size and powder injection velocity on particle properties such as average temperature, velocity and impact position and their standard deviations were determined. The modeled data were plotted on process–property maps to assist optimization of the thermal spray process. It was found that particle size distribution and powder injection position play decisive roles in melting of the feedstock particles. Plasma volume flow rate and powder injection velocity were found to be critical parameters and their combined effect to have a major influence on in-flight properties and hence the expected coating performance.

6.4
Modeling of Particle–Substrate Interaction

Much work has been done to model the complex processes that accompany the impact of molten and/or semi-molten particles with the solid substrate surface.

6.4.1
3D Simulation of Coating Microstructure

A 3D stochastic simulation based on the Monte Carlo method (for example Shreider, 1966; Robert and Casella, 2004) was presented by Ghafouri-Azar *et al.* (2003) to predict coating porosity, thickness, surface roughness, and their variation with varying spray parameters (particle velocity, particle size, traverse speed of the plasmatron). Since the plasma spraying process is inherently three-dimensional, previous 2D simulations (Knotek and Elsing, 1987; Cirolini *et al.*, 1991, Kanouff *et al.*, 1998) could reasonably predict only the structure of a single cross-section through the deposited layer and hence provided only limited information on the true nature of a plasma-sprayed coating and its complexity.

The approach taken by Ghafouri-Azar *et al.* (2003) started from the assumption that particle velocity V, particle diameter D, temperature T, and the point of impact (defined by the dispersion angle ω and the azimuthal angle θ) have random, continuously varying values. In particular, the distributions of V and T can be represented by a Gaussian probability density function (PDF) whereas D is better described by a log-normal distribution. After assigning to each particle propagating along the plasma jet axis values for V, T and D the splat size can be calculated after impacting with the (flat) substrate surface using a simple analytical model developed by Aziz and Chandra (2000). This model considers deformation of a spherical molten droplet to form a cylindrical disk of diameter d_{max} and height h (see also Section 5.2, footnote 3; Figure 6.8), and neglects any break-up and splashing[2] but takes into account particle overlap, fusion, and other shape-conserving effects (see Ghafouri-Azar *et al.*, 2003b). The maximum spreading (flattening) ratio $\xi_{max} = d_{max}/D$ was shown to be a complex function of several dimensionless numbers (**We**, **Re**, **St** and **Pe**) as well as the liquid–solid contact angle θ.[3]

$$\xi_{max} = \{[\mathbf{We} + 12]/[3(1-\cos\theta) + 4(\mathbf{We}/\sqrt{\mathbf{Re}}) + \mathbf{We}(3\mathbf{St}/4\mathbf{Pe})^{1/2}]\}^{1/2}, \quad (6.19)$$

where $\mathbf{We} = \rho V^2/\gamma$ (Weber number[4]), $\mathbf{Re} = VD/\nu$ (Reynolds number), $\mathbf{St} = C_p(T_m - T_{sub})/H_f$ (Stefan number), and $\mathbf{Pe} = VD/\lambda$ (thermal Péclet number); (γ: surface tension, ν: kinematic viscosity, C_p: molar heat at constant pressure, T_m: melting point, T_{sub}: substrate temperature, H_f: latent heat of melting, λ: thermal conductivity) (see also Appendix A).

It should be mentioned that a somewhat improved approach was presented by Hadjiconstantinou (1999) considering the effect of solidification, i.e. the formation of

---

[2] The effect of splashing on the splat shapes has been simulated and experimentally verified by Pasandideh-Fard *et al.*,2002a and Mehdizadeh *et al.*, 2004a, 2004b.

[3] Since ω is very small under thermal spray conditions, the term (1 − cosω) can be safely neglected (Aziz and Chandra, 2000).

[4] The Weber number, $\mathbf{We} = rV^2/\gamma$ expresses the relative importance of the inertia of a fluid, rV^2 compared with its surface tension γ. Numerical measurements of surface tension and viscosity of melt drops heated in an r.f. plasma have been reported by Moradian and Mostaghimi (2005).

a thin solidified layer of thickness s. In the nonisothermal case the ratio of s and the viscous dissipation boundary layer $\delta = 2D/\sqrt{Re}$ is important, and a modified Equation 6.19 was proposed by including the non-dimensional solidified layer thickness $s^* = s/D$:

$$\xi_{max} = \{[We + 12]/[(3/2)We \times s^* + 3(1 - \cos\theta) + 4(We/\sqrt{Re})(2/(2 + s^*\sqrt{Re}))]\}^{1/2} \quad (6.20)$$

Since the Stefan problem approach is inadequate to describe the solidification kinetics and their effect on spreading of the molten droplets (Madejski, 1976; Bennett and Poulikakos, 1993, 1994; Pasandideh-Fard et al., 1998) it was assumed by Hadjiconstantinou (1999) that $St \ll 1$. The Nusselt number based on the total heat transfer coefficient h_t is $Nu_t = (h_t \times D)/k = Re \times St$. Hence a more accurate non-dimensional expression for the solidified layer thickness s^* can be derived to yield

$$s^* = [4Nu_t \times St/3Re \times Pr(1 + St_T)], \quad (6.21)$$

where $Pr = Pe/Re$, and $St_T = C_p(T - T_{sub})/H_f$. Table 6.2 shows a comparison of the modeled values of the spreading ratio ξ_{calc} (Hadjiconstantinou, 1999) with experimentally obtained values ξ_{exp} for tin droplets impinging on a steel plate with varying temperature T_{sub} (Pasandideh-Fard et al., 1998). The excellent agreement confirms the validity of the approach.

Coating porosity was simulated by Ghafouri-Azar et al. (2003a) assuming that splat curvature was the sole source of porosity. Owing to thermal stresses splats tend to detach from the underlying surface, and pores form under the elevated edges of these splats at a distance 0.6R from the center (Fukanuma, 1994) where $R = d_{max}/2$ is the splat radius. The angle of detachment α was defined as $\alpha = \tan^{-1}(h_{g,R})/(0.4R)$ where $h_{g,n}$ is the thickness of the gap at the edge of the curled-up splat (Figure 6.5).

To model subsequent coating build-up a 3D Cartesian grid was used to define the computational domain and to track its evolution during spraying. Coatings build-up occurs by agglomeration of splats within this 3D grid together with pores trapped between impacting splats and the pre-existing coating layer (Figure 6.5).

Table. 6.2 Maximum spreading ratios ξ_{max} of tin droplets impinging on a steel plate having different temperature T_{sub}. Experimental values (Pasandideh-Fard et al., 1998) vs. simulated values (Hadjiconstantinou, 1999).

T_{sub} (°C)	s^*	ξ_{calc}	ξ_{exp}
Case A*: 240	0	3.24	3.3
150	0.014	3.06	3.1
25	0.035	2.8	2.9
Case B: 240	0	3.24	3.3
150	0.014	3.17	3.1
25	0.035	2.89	2.9

*Case A: viscous dissipation uniformly distributed to solidified and liquid parts of the droplet;
Case B: heat generated by viscous dissipation retained within the viscous boundary layer (equation 6.20)

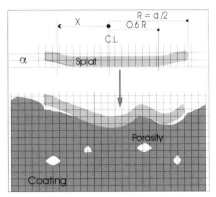

Figure 6.5 Definition of the curl-up of a splat. A splat deposited at a pre-existing coating with an irregular surface detaches thus forming a pore. Detachment starts at a distance 0.6R from the splat center, with an angle α to produce a gap of height $h_{g,R}$ (Ghafouri-Azar et al., 2003a).

The result of the simulation of deposition of Ni particles from a stationary plasmatron on a stainless steel substrate (Figure 6.6) conforms closely to the cone-shaped deposit formed experimentally.

In addition, it was found that more uniform coatings can be deposited by moving the plasmatron in a sinusoidal fashion, i.e. with lowest speeds near the ends of the

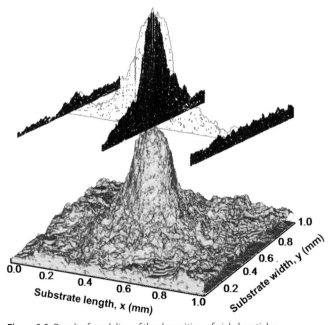

Figure 6.6 Result of modeling of the deposition of nickel particles by a stationary plasmatron showing a Gaussian distribution of the deposit (Ghafouri-Azar et al., 2003a).

substrate and maximum speed at its mid-point. Increasing the average particle velocity led to slightly decreased values of coating porosity, thickness, and roughness while increasing the average particle size strongly decreased both coating porosity and thickness, but increased the average roughness. Variation of the traverse speed of the plasmatron did not appreciably affect coating porosity and thickness but showed a general increase of roughness with increasing speed.

In conclusion the modeling approach taken by Ghafouri-Azar *et al.* (2003a) was able to simulate rather well the microstructure of plasma-sprayed coatings as a function of the spray process parameters. The predicted trends were in reasonable agreement both with theoretically expected effects and their experimental verification.

Since deposition of coatings usually requires a critical substrate surface roughness to ensure sufficient adhesion, such rough surfaces influence the spreading and solidification behavior of molten droplets impacting the surface. Consequently, modeling approaches were developed that consider the effect of surface roughness on splat shapes (Feng *et al.*, 2002; Raessi *et al.*, 2005; see Section 6.4.2) and on solidification of droplets over already solidified splats (Ghafouri-Azar *et al.*, 2004). For example, a study of the solidification behavior and splat morphology of LPPS Ti6Al4V by 3D numerical modeling showed good agreement between calculated values and results obtained by experimental observation of single splat and as-sprayed coating microstructures (Salimijazi *et al.*, 2007). The average splat cooling rate obtained from numerical simulation was as high as $10^8\,°C\,s^{-1}$, the solidification front velocity was $0.63\,m\,s^{-1}$, the splat thickness around $3\,\mu m$, and the deposition efficiency was calculated to be 70%.

Perturbation of the plasma jet by the carrier gas flow and movement of particles under dense loading conditions are common causes of nondeterministic behavior of the coating process. Hence a 3D computational model was developed to describe the plasma jet coupled with the orthogonal injection of carrier gas and particles (Xiong *et al.*, 2004). This model considered in-flight particle physical phenomena such as acceleration, heating, melting and evaporation. The effects of carrier gas flow rates on the characteristics of plasma jet and particle spray pattern were simulated. The simulation results compare well with experimental data for two common particle materials, NiCrAlY and ZrO_2.

Recently Wilden *et al.* (2006) simulated the process of low pressure plasma spraying (LPPS) and the development of coating microstructure following the complete succession of process steps, starting with plasma generation, heating and acceleration of particles, splat flattening at the substrate surface, nucleation and solidification, coating formation and development of residual stresses during cooling. The results of this modeling approach demonstrated that experimental investigation of coating development, for example by statistical experimental design methodology (see Chapter 11) can partly be replaced or at least augmented by simulation. Hence new coating systems with defined properties can be developed more easily and economically by adjusting critical process parameters. In particular, optimization of such process parameters by modeling will simultaneously lead to efficient control of coating microstructure, increase in deposition rate to satisfy economical constraints and reduction of residual coating stresses to improve system reliability.

6.4.2
Modeling of Splat Shapes

The relationships between viscous dissipation, surface tension, contact angle, contact heat resistance and solidification kinetics, and the maximum spreading diameter ξ_{max} have been elaborated on in the preceding paragraph. In this paragraph the actual shape of the solidified droplets will be considered.

Splat morphology has been shown to have an important effect on coating quality in terms of cohesion and adhesion strengths, modulus, porosity, fracture toughness, and compliance. Experiments showed that the temperature of the substrate determines to a large degree the splat shape. A transition temperature has been defined for several material combinations above which the splat morphology changes from irregularly splashed or 'exploded' morphology to perfectly circular, disk-like shapes (Bianchi et al., 1994; Fukomoto et al., 1998; Sakakibara et al., 2000; see also Figures 5.9, 7.19, 10.3). This transition temperature appears not only to be a materials property but also to depend on surface contamination (Li et al., 1998; Li et al., 2006) and oxidation (Pech et al., 2000; Pasandideh-Fard et al., 2002a; McDonald et al., 2007).

Since there is no general consensus yet on the mechanism(s) leading to transition from irregular to disk-shaped splats modeling approaches of splat shapes by a fully 3D free surface flow model have been developed (Bussmann et al., 1999) that was later extended to include heat transfer and solidification (Pasandideh-Fard et al., 2002a, 200b).

Fluid flow in an impacting droplet was modeled by a finite difference solution of the Navier–Stokes' equation in a 3D Cartesian coordinate system, assuming a flow Reynolds number $\mathbf{Re} = 10^4$ that is below the onset of turbulence motion. Furthermore liquid density and surface tension were assumed to be constant whereas liquid viscosity and thermal properties of the substrate were allowed to vary with temperature.

Heat transfer in the droplet was simulated via solving the energy equation by converting it to an equation in which enthalpy appears as the sole dependent variable (Cao et al., 1989). Viscous dissipation of energy was excluded and the densities of liquid and solid were assumed to be constant and of equal values.[5] At the free surface of the impinging droplet an adiabatic boundary condition was imposed.

Solidification was modeled by tracking the moving, irregularly shaped solidification front, and the continuity and momentum equations solved by applying the method developed by Pasandideh-Fard (1998).[6]

Validation of the numerical simulations was done by plasma spraying Ni powder (+53–63 µm) onto a stainless steel substrate using an SG-100 plasmatron (Miller

---

5) Later this condition was relaxed by including the effect of density variation during phase change liquid–solid, and by validation against the Stefan and plane solidification problems (Raessi and Mostaghimi, 2005). See also Hadjiconstantinou (1999) who in his modeling approach considered the effect of surface tension, viscous dissipation, and contact angle.

6) A somewhat simpler approach was developed by Dhiman et al. (2007) who introduced a dimensionless solidification parameter Θ defined by a set of other dimensionless parameters such as **St**, **Pe**, **Bi**, **We** and **Re** as well as the splat contact angle u (see equations 6.22, 6.23, 6.23a).

Thermal, Appleton, WI) operated at 20.5 kW. In-flight particle properties were monitored with a DPV-2000 system (Tecnar Automation Ltd., Montreal, Canada), yielding an average particle temperature of $1600 \pm 220\,°C$ and a velocity of $73 \pm 9\,m\,s^{-1}$ (Pasandideh-Fard *et al.*, 2002a, 2002b). The substrate was preheated to $400\,°C$ before spraying the Ni powder through a 4 mm diameter hole of a steel shielding plate. The substrate was moved rapidly across the narrow spray plume as to create a 'wipe' test condition (see 5.1). The experiments revealed that Ni particles splash when sprayed onto a substrate preheated to below $300\,°C$ but form circular disks on a substrate surface heated to $400\,°C$.

Modeling showed that the solidification of the spreading droplets on a substrate heated to only $290\,°C$ produces finger-like protrusions (Figure 6.7) similar to those experimentally observed by Heimann (1991) for solidified alumina splats (see Figure 5.9). Small fluctuations of the velocity of the liquid around the periphery of the droplet created a fluid instability that after 1.4 μs led to the onset of the finger-like protrusions. These fingers grew in size with increasing contact time and eventually solidified after about 10 μs. Separation of material from the fingers formed a corona of spherical tiny droplets akin to those observed and phenomenologically explained by

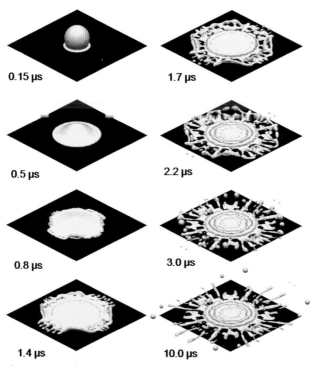

Figure 6.7 Simulations showing a molten Ni particle (60 μm diameter, 1600 °C) impacting a preheated stainless steel plate (290 °C) with a velocity of 73 m s^{-1}. The thermal contact resistance at the interface was assumed to be $10^{-7}\,m^2\,K\,W^{-1}$ (Pasandideh-Fard *et al.*, 2002a).

Houben (1988). However, increasing the preheating temperature to 400 °C eliminated these protrusions and nearly perfectly circular disks were formed instead. The reason for this behavior was seen in the tenfold increased value of the thermal contact resistance that delayed the onset of solidification so that spreading had abated before solidification could set in. Such large thermal contact resistance and thus reduced heat transfer could in reality be caused by formation of an iron oxide scale during plasma spraying. As a consequence the onset of splashing was delayed, and disk-shaped splats developed.

The requirement of a change of the thermal contact resistance to explain changes in the splat morphology has been relaxed in a model developed by Dhiman *et al.* (2007) that assumed the contact resistance R_c at the melt-substrate interface to be constant so that the splat surface temperature can be expressed by $T_s = T_m - q_0 R_c$ where q_0 is the heat flux through the bottom surface of the solidifying splat. In the model a dimensionless solidification parameter $\Theta = s/h$ was introduced where s describes the thickness of the solidified layer and h the total splat thickness (Figure 6.8).

The layer thickness h and the solidified layer thickness s can be further non-dimensionalized by setting $h^* = h/D_0$ and $s^* = s/D_0$, respectively. The reduced layer thickness h^* can be expressed after several simplifications[7] by

$$h^* = [(s^*/4) + (4/\mathbf{We}) + (8/3\sqrt{\mathbf{Re}})]. \quad (6.22)$$

The dimensionless solidified layer thickness s^* can be obtained as a function of time from the equation (Poirier and Poirier, 1994)

$$s^* = \frac{2}{\sqrt{\pi}} \mathbf{St} \sqrt{\frac{\gamma_s t^*}{\gamma_D \mathbf{Pe}}} \left\{ 1 - \frac{1}{\mathbf{Bi}} \sqrt{\frac{\gamma_s \mathbf{Pe}}{\gamma_d \pi \times t^*}} \ln \left[1 + \mathbf{Bi} \sqrt{\frac{\gamma_d \pi \times t^*}{\gamma_s \mathbf{Pe}}} \right] \right\}. \quad (6.23)$$

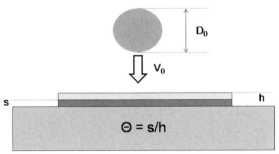

Figure 6.8 Schematic rendering of the deformation of a spherical droplet with diameter D_0 impacting a solid substrate with the velocity V_0 to yield a cylindrical splat of thickness h that on freezing forms a solid deposit layer of thickness s (after Dhiman et al., 2007).

7) Setting $h^* = (2/3 \times \xi^2_{max})$ (see 6.19) and considering the solid quasi-isothermal ($\mathbf{St} \ll 1$) and further assuming that the melt does not wet the substrate ($\theta \approx 180°$; $1 - \cos\theta = 2$) with $\mathbf{We} \gg 12$ Equation 6.22 is obtained.

Setting $t^* = 8/3$ (Dhiman and Chandra, 2005) Equation 6.23 reduces to

$$s^* = \frac{16}{3}\left(\frac{\mathbf{St}}{\mathbf{Pe} \times A}\right)\left\{1 - \frac{\ln(1 + \mathbf{Bi} \times A)}{\mathbf{Bi} \cdot A}\right\}, \quad (6.23a)$$

with $A = (8\pi\gamma_d/3\mathbf{Pe}\gamma_s)^{1/2}$ where γ_d and γ_s are the surface energies of droplet and substrate, respectively.

From Equations 6.22 and 6.23a the values of Θ can be estimated and interpreted as follows:

- If the solidified layer is very thin ($s \ll h$ or $\Theta \ll 1$) the splat will spread into a thin circular liquid sheet that subsequently on freezing ruptures. Hence a small central splat will be produced surrounded by a ring of fragmented debris (see Figure 9.5A).
- If the thickness of the solidified layer increases during spreading ($s \sim 0.1\,h$ to $0.3\,h$ or $\Theta \sim 0.1$ to 0.3) a disk-shaped splat shape will be attained since the solid layer will prevent further spreading (see Figure 9.5B,C).
- Very rapid solidification ($s \sim h$ or $\Theta \sim 1$) will obstruct the outward moving liquid producing extensive splashing with finger-like eruptions extending from the splat into the surrounding (see Figures 5.9, 9.5D).

This model was validated by examining splats produced from plasma-sprayed powders of Ni, Mo and ZrO_2 on glass, steel and Inconel substrates.

Further complications will be introduced if the effect of surface roughness on splat shape and size is being considered (Raessi *et al.*, 2005; Syed *et al.*, 2005). Droplet solidification rate was found to be the predominant mechanism responsible for changing splat shapes, whereas increasing surface roughness up to a certain value increases the splat diameter. Moreover, splat morphology can be altered independent of substrate temperature by exploiting the combined effect of spray distance and plasma powder as found experimentally by plasma spraying hydroxyapatite powder (see Figure 9.5). Here the degree of overheating, given by the numerical value of the Biot number, **Bi** appears to play a dominant role.

6.4.3
Modeling of Residual Coating Stresses

Since most coating/substrate combinations show a pronounced mismatch in their thermophysical and mechanical properties, the dynamic heat and mass transfer processes encountered during plasma spraying lead to formation of residual stresses at the interface but also in the bulk of the coating (see Chapter 5.7). The magnitude, distribution, and sign (tensile or compressive) of the stresses influence coating reliability and in-service performance. Consequently analysis of residual stresses and understanding their origin are essential for optimization of coating properties and the plasma spraying process leading to these improved properties. Many attempts have been made to model residual coating stresses.

Because of the complexity in terms of error possibilities, time and cost requirements of the XRD-based measurements, attempts have been made to solve the stress state determination in coatings by mathematical means. The mathematical models applied are subsequently validated by experiments. Thus the influence of individual parameters on the residual thermal stress state of the coatings can be studied, and the in-service behavior of the coating/substrate system predicted (Elsing et al., 1990).

A model developed by Knotek et al. (1988) is based on compartmentalization of the coating deposition process into time increments of 10^{-6} s. The coating thickness will be calculated for each time element and also the temperature distribution for the entire coating/substrate tandem from the difference of the amounts of heat introduced and dissipated. The thermal constraints are the convective heat losses at the surfaces, ideal contact between substrate and coating, and negligible radiation losses. Then the internal thermal strain state can be described by a two-stage model as follows.

In the first stage the total lateral expansion of the plate-shaped coated workpiece is found by numerical integration of the equation

$$\partial^2 x_i / \partial t^2 \Sigma m_i = x_i \Sigma E_i A_i / L_{0i} - \Sigma E_i A_i \alpha_i (\Theta_i - \Theta_{0i}), \tag{6.24}$$

where x = path coordinate along the substrate surface, m = mass of individual elements i, E = Young's modulus, A = cross-sectional area of individual elements i, α = temperature-dependent coefficient of thermal expansion, Θ_i = instantaneous temperature of individual elements i, L_{0i} = length of element in the unstressed state, and Θ_{0i} = maximum temperature of the (i–1)st coating lamella. Equation 6.24 is a special form of the fundamental Newton momentum equation. Each particle arriving at the surface is incorporated at the precise moment when the supporting (i–1)st lamella has reached its maximum temperature Θ_{0i}. Therefore the total lateral expansion can be calculated from the instantaneous thermal expansion and the size of the substrate prior to deposition.

In the second stage, after termination of the deposition process the length L_i of the individual elements i can be calculated by

$$L_i = L_{0i}[1 + \alpha_i(\Theta_i)(\Theta_i - \Theta_{0i})], \tag{6.25}$$

where L_i is the length that an element would attain at the temperature Θ if it could adjust freely, i.e. unconstrained by the surrounding elements. The internal strain among individual elements can be calculated considering the bending moments and applying equations developed for bimetallic strips (Knotek et al., 1988).

Verification of the model was done using experimental data obtained by depositing alumina (100 μm thick coatings) and stabilized zirconia (40 μm thick coatings) on ferritic and austenitic steel substrates. Internal strains were measured by the blind hole method (see Section 5.7.1). Even though the measured internal strain values were found to be lower than the calculated ones due to variations introduced by the drilling process, there is rather good agreement between the measured and the calculated internal strains.

Additional theoretical work has been performed at UKAEA's Harwell laboratory by Eckold et al. (1987) on plasma-sprayed stabilized zirconia coatings deposited onto

AISI 304 stainless steel. The model considers transfer of heat to the substrate from the molten impinging particles as well as radiative and/or convective redistribution of heat to the surrounding atmosphere. The heat transfer model applied uses a number of constraints such as

- continuity of the coating process;
- coating area is large compared with its thickness;
- heat loss can be described by standard convection and radiation heat loss equations;
- start of spraying approximates to a coating of infinitesimal thickness;
- thermophysical and elastic properties are temperature-invariant.

The heat transfer model generates data that predict the thermal history of a coating during deposition. From this data the stress history of the coating is calculated, that aids in the prediction of the final residual stress state in the coating. Sensitivity analyses consider the influence of the following parameters on the stress state:

- deposition rate;
- plasma temperature (enthalpy);
- substrate temperature;
- final coating thickness;
- coating area;
- substrate area;
- thermal conductivity, specific heat and latent heat of coating;
- density, emissivity, Poisson ratio, Young's modulus and expansivity of the coating;
- heat capacity and expansivity of the substrate.

The model produces a graphical output of the stress distribution within the coating dependent on a path coordinate x. It indicates that (i) the stress is everywhere compressive, (ii) the stress maximum occurs at the coating's free surface, and (iii) the magnitude of the stress (maximum values 25 MPa) is much lower than the operational stress in a diesel engine where such thermal barrier coatings (TBCs) may be applied. As a first approximation it appears that the residual stresses occurring in such zirconia coatings are not at all life limiting. However, because of the rather low Weibull modulus of ceramic coatings in general, failure probability is high. In particular, coating fracture stresses in ceramic coatings can be of the same order of magnitude as the residual stresses predicted by this model.

As pointed out by Steffens *et al.* (1987) rather large errors will be introduced into thermal and residual stress calculations by neglecting the temperature dependence of the modulus of elasticity. Furthermore since TBCs are designed to be highly porous to minimize their thermal diffusivity, the strong dependence of the modulus on porosity must be considered. In the stress calculations performed to determine the influence of residual and thermal stresses on the thermal shock resistance of yttria-stabilized zirconia TBCs, the following equation was used:

$$\sigma = [E_c(\alpha_s - \alpha_c)\Delta T]/(1 - v_c), \tag{6.26}$$

where the subscripts 'c' and 's' refer to coating and substrate, respectively. This simplified equation does not consider the substrate rigidity E_s, Poisson ratio of the substrate and the coating thickness as done in the more comprehensive Dietzel equation discussed above (Equation 5.52c). Since the residual stresses can be somewhat controlled by substrate preheating, the stress relaxation that normally occurs by micro- or even macrocracking of TBCs can be suppressed. Furthermore, the steep stress gradients at the coating/substrate interface where the stress state changes from tensile to compressive can be smoothed by inserting a soft compliant intermediate bond coat layer or a graded metal–ceramic coating (Kaczmarek et al., 1983).

The generation and dynamic development of thermally induced stresses can be followed by coupling heat transfer with an elastic–plastic finite element method (FEM). Work by Gan et al. (2004) on NiCoCrAlY coatings deposited on an aluminum substrate yielded close agreement between the numerically predicted deflected shapes of the specimen and the experimental results. This indicates that the stress–strain distribution from the elastic–plastic FEM model can be applied with reasonable accuracy to predict residual stresses in the plasma-sprayed model.

The need to reduce the weight of passenger car engine blocks to increase fuel economy has led to efforts to replace the customary steel by lightweight materials including aluminum alloys and, in the future, magnesium and its alloys. However, the limited tribological performance and the risk of corrosion by fuel gasses of these novel materials require hard, environmentally stable coatings such as alumina, molybdenum or alumina nano-composites (Dearnley et al., 2003) to ensure low friction properties of the piston ring/cylinder liner contact area. To estimate residual stresses in these crucial parts of an engine Wenzelburger et al. (2004) have performed finite element modeling (FEM) of the layer deposition, heat transfer, and process cooling with the aim of minimizing residual coating stresses. Commercial software (ABAQUS™, version 5.8) was used to model an aluminum tube (external diameter and height of 100 mm, wall thickness of 10 mm) with its inner cylindrical surface coated by a 200 µm thick layer of APS alumina. Since residual stresses are strongly influenced by the efficiency of substrate cooling, air jet cooling and CO_2 cooling with different geometries were studied. The temperature distribution was calculated taking into account the surface heat flux as well as convection and radiation, and other thermophysical properties (density, specific heat capacity, thermal conductivity) evaluated. The stress analysis yielded the temporary, multiaxial strains and stresses for each node of the FE model, based on linear isotropic elasticity theory.[8] The calculated data were validated by experimentally obtained residual stress measurements by a microhole drilling and milling method (Buchmann et al., 2002). It could be shown that air cooling is the least efficient cooling technique compared with CO_2 cooling of which line cooling and spot cooling were evaluated. Of those, spot cooling performs slightly better than line-shaped cooling in which the cooling

---

8) Thermally sprayed coatings may be described more accurately as nonlinear elastic. To account for this property a suitable stress–strain model has been developed leading to a nonlinear bi-material beam solution. An inverse analysis technique was used to process measured curvature–temperature profiles (Nakamura and Liu, 2007).

nozzle is placed in an angle offset by 180° to the plasmatron. However, the highest cooling efficiency and hence to lowest residual coating stresses are expected for 30°-line-shaped cooling.

The development of residual stresses in stainless steel (Metco-Diamalloy™ 1003) and WC12Co coatings deposited by HVOF onto AISI316 stainless steel substrates was investigated by Ghafouri-Azar et al. (2006). Comparison of experimental coating microstructure (thickness, porosity and surface roughness) and modeling of these properties by a 3D stochastic Monte Carlo model (see 6.4.1; Ghafouri-Azar et al., 2003a) showed that measured coating thickness and porosity values agreed well with predicted ones whereas the calculated roughness was higher than that obtained from experiments. The residual coating stresses were calculated using an object oriented finite element model (OOF; Langer et al., 2001) based on an adaptive meshing technique to discretize the coating microstructure. SEM micrographs of coating cross-sections were digitized graphically into triangular elements. Plane stress analyses with complete isotropy were carried out to estimate stress generated by coefficient of thermal expansion (CTE) mismatch for different temperatures and substrate preheating. As expected the highest stresses were found to develop at the interface between coating and substrate owing to the large difference in the CTE of AISI316 stainless steel ($1.62 \times 10^{-5}\,\mathrm{mm\,m^{-1}\,°C^{-1}}$) and WC12Co ($7.3 \times 10^{-6}\,\mathrm{mm\,m^{-1}\,°C^{-1}}$).

Figure 6.9 shows the calculated contours of equivalent (von Mises) stresses[9] of WC12Co coatings on AISI316 substrates at initial coating temperatures T_c ranging from 250 to 550 °C. For all coating temperatures the maximum stresses were measured close to the interface substrate/coating. It is evident that increasing coating temperatures increase the magnitude of the von Mises stresses and hence increase the risk of coating failure. This is also valid for increasing coating thickness. Consequently thinner coatings and preheated substrates will reduce the residual stresses and thus increase the durability and in-service life performance of such coatings.

Finite element coupled heat transfer and elastic–plastic thermal stress analyses were applied to predict as-sprayed residual stresses of plasma-sprayed duplex TBCs (Ng and Gan, 2005) as well as NiCoCrAlY coatings on aluminum substrates (Gan et al., 2004) with the FEM software ANSYS (Ansys Inc., Southpoint, PA, USA). In the simulation approach the coating geometry was approximated by a multiple layer-on-layer structure, and this geometry was iteratively updated using the results of the previous layer analysis. Both the predicted coating temperature and curvature agreed well with measured values. In addition, post-depositional treatment was proposed to reduce the residual stress level.

Recently a hybrid nonlinear explicit–implicit FE methodology was developed by Bansai et al. (2007) to predict residual stresses occurring during HVOF deposition of a 316 stainless steel coating on a substrate of the same composition.

---

[9] If the equivalent stress σ_c exceeds the yield stress σ_y (von Mises criterion; $\sigma_1^2 - \sigma_1\sigma_2 + \sigma_2^2 \leq \sigma_y^2$) coating failure will occur by ductile tearing (σ_1, σ_2 are the principal stresses or eigenvalues under plane stress conditions, where $\sigma_3 = 0$).

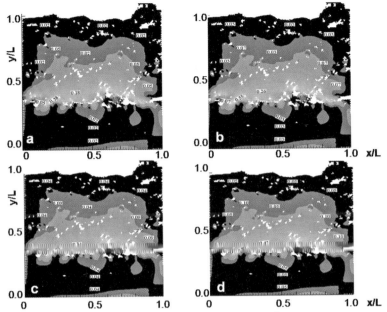

Figure 6.9 Contours of von Mises (equivalent) stress in GPa of WC12Co coatings deposited on AISI316 stainless steel substrates at different coating temperatures $T_c = 250\,°C$ (a), $350\,°C$ (b), $450\,°C$ (c) and $550\,°C$ (d). The substrate temperature was kept constant at $25\,°C$ (Ghafouri-Azar et al., 2006).

6.4.4
Modeling of Thermal Conductivity

Porous thermal barrier coatings (TBCs, see Section 8.1) reduce the operating temperature of last generation hot path components of stationary and aerospace gas turbines by up to $300\,°C$. Since the thickness, microstructure and materials selection of the coatings positively affect the efficiency of the gas turbine by either reducing the required coolant flow or allowing an increase of the turbine inlet temperature (TIT) it is mandatory to design *a priori* the thermal properties of the coating material as well as the deposition conditions. Hence modeling of thermal conductivity of two-phase (pore-free solid and pores) systems is an important step in optimizing the efficiency of TBCs. Such a two-phase composite structure can be described either by asymmetric dispersion of isolated grains embedded in a continuous matrix or by a symmetrical penetration of grains of the two phases occupying the entire volume (Cernuschi et al., 2004). The dependence of the thermal conductivity of porous ceramic coatings on the number and orientation of differently shaped pores has been determined earlier by Hasselman 1978 (see Section 8.1.3.1, Equations 8.2a to c), and Litovsky and Shapiro (1992).

Expansion in a Taylor series of the classical asymmetric Maxwell model (Maxwell, 1881) yields for the thermal conductivity of a composite material the expression

$$\frac{k}{k_m} = 1 - \left[\frac{3f(1 - k_d/k_m)}{\left(2 + \frac{k_d}{k_m}\right)} \right], \qquad (6.27a)$$

where k_m, k_d and f are the thermal conductivities of the matrix, the dispersed phase (pores), and the dilute volumetric fraction, respectively. However, this Maxwell approach is valid only for very dilute situation, i.e. $f < 0.1$. This constraint has been removed by Bruggeman (1935) who showed that the radius of the (spherical) pores varied within an infinite range and that each sphere is embedded within the continuous matrix. With these assumptions the dilute dispersion can be varied within the full range $0 \leq f < 1$, and the equation then becomes

$$\frac{\left(k/k_m - \frac{k_d}{k_m}\right)}{\left[\left(\frac{k}{k_m}\right)^{1/3}\left(1 - \frac{k_d}{k_m}\right)\right]} = 1 - f \qquad (6.27b)$$

Pores of volumetric fraction f_1 and thermal conductivity k_d are thought to be symmetrically dispersed in a matrix of volumetric fraction $f_2 = 1 - f_1$ and thermal conductivity k_m.

For porous TBC coatings k_d is assumed to be negligible ($k_d \approx 0$) if the temperature is low enough to exclude a contribution of radiation to heat transfer (Cernuschi et al., 2004). With this assumption the Voigt–Reuss limits (Maxwell, 1881) are $0 \leq k \leq (1-f)k_m$ for an in-parallel mode, and consequently the Maxwell and Bruggeman (Bruggeman, 1935) models of thermal conductivity expressed by Equations 6.27a and 6.27b are reduced to the simple expressions

$$k/k_m = \{1 - (3/2)f\} \quad \text{(Maxwell)} \qquad (6.27c)$$

$$k/k_m = (1-f)^{3/2} \quad \text{(Bruggeman)}. \qquad (6.27d)$$

In particular, pores can be modeled as spheroids with different aspect ratios, and for dispersion of such spheroids (rotational ellipsoids with either oblate, $c > a$ or prolate, $c < a$ characteristic) the Bruggeman model reduces further to

$$k/k_m = (1-f)^X, \qquad (6.27e)$$

where $X = [1 - \cos^2\alpha/(1-F)] + \cos^2\alpha/2F$ is a geometrical factor describing the shape and the orientation of a spheroid relative to the uniform heat flux field. The angle α is the angle between the rotational axis of the spheroid and the vector of the unperturbed heat flux, the shape factor F is $1/3$ for a sphere ($a = c$), $0 < F < 1/3$ for an oblate spheroid ($c > a$), and $1/3 < F < 1/2$; for a prolate spheroid ($c < a$). Note that in case of a sphere ($F = 1/3$) for both $\alpha = 0°$ and $\alpha = 90°$ X is $3/2$, corresponding to Equation 6.27d.

It was found that Equation 6.27e is sufficient to describe a wide range of porosity configurations by taking into account different pore space geometries and orientations (Schulz, 1981). However, complications arise when in more realistic cases more than one pore type in more than one orientation must be considered. This has recently been achieved by Cernuschi *et al.* (2004) who assumed manifold porosity systems with three- and four-phase mixtures that can be iteratively calculated. For example, if f_0 is the total volumetric fraction of porosity, and f_1 and f_2 are the volumetric fractions of pores of type 1 and 2 ($f_0 = f_1 + f_2$), the thermal conductivity of the three-phase mixture can be expressed by

$$\frac{k}{k_m} = \frac{1}{2}\left\{\Phi\left[\frac{f_2}{(1-f_1)}\right]\Psi(f_1) + \Psi\left[\frac{f_1}{(1-f_2)}\right]\Phi(f_2)\right\}, \tag{6.28}$$

where $\Psi(f_1)$ and $\Phi(f_2)$ are functions accounting for the effect of porosity of type Ψ and Φ, described by Equation 6.27e, on the thermal conductivity of the matrix. For the special case $\Phi(f) = \Psi(f) = (1-f)^x$ Equation 6.28 reduces to Equation 6.27c.

The expression for a four-phase mixture (three types of pores Ψ, Φ and Θ) is much more complex. Such a three pore-type model has been successfully applied to different TBC coatings consisting of zirconia stabilized with yttrium oxide (8Y-PSZ), magnesium oxide (22MSZ), and a mixture of cerium and yttrium oxides (25C2.5YSZ), with the functions Ψ, Φ and Θ relating to inter-, intra- and trans-lamellar microcracks (disk-shaped oblate spheroids, i.e. 'penny-shaped' pores with c/a = 15; X > 5); open, randomly oriented pores (X = 1.66); and spheroids that are not flat, with c/a = 3 (X = 1.7) oriented parallel to the direction of heat flux, respectively. In particular, in case of the pore population Ψ, 75% were assumed to have their rotational axes parallel to the coating thickness, i.e. the direction of heat flux and 25% perpendicular to the heat flux. These two families of pores have also been found to be of decisive importance for the pronounced anisotropic behavior of the elastic properties of plasma-spayed coatings (Sevostianov and Kachanov, 2000).

Experimental values of the thermal conductivities of the three types of TBC chosen were obtained under the constraint that other than some densification during heating of the coatings no other reaction will occur. The measured values before heat treatment were between 1.1 and 1.2 W m^{-1} °C^{-1} for all three types of TBC but increased during thermal cycling to 1250 °C to 1.62 (8Y-PSZ), 4.37 (22MSZ) and 2.05 (25C2.5YSZ), respectively. The modeled values were in excellent agreement with the measured ones before heat treatment. This was also true after three heating cycles up to 1250 °C for 8Y-PSZ (1.68 W m^{-1} °C^{-1}) and 25C2.5YSZ (1.68 W m^{-1} °C^{-1}) but not for 22MSZ whose large increase in thermal conductivity to 4.37 W m^{-1} °C^{-1} could not be accounted for by modeling that yielded a value of only 1.37 W m^{-1} °C^{-1}. This can be explained by a phase change from c/t-ZrO_2 to m-ZrO_2 due to destabilization and precipitation of MgO.

An image-based extended finite element model (XFEM) was developed by Michlik and Berndt (2006) to predict the effective in-plane Young's modulus, the global thermal conductivities, and the fracture behavior of YSZ and multilayer TBCs. The XFEM served to enhance the OOF used by incorporating real microstructural features including crack networks. As a result, heat transfer analysis permitted to predict the effective thermal conductivities of as-sprayed and thermally cycled TBCs. In particular, the significance of inter-lamellar microcracks for a decrease of the thermal conductivity (see Hasselman and Singh, 1979; McPherson, 1984; Litovsky and Shapiro, 1992; Litovsky et al., 1999; Seifert et al., 2006) of conventional YSZ TBCs was modeled and the thermal conductivity of multilayer TBCs consisting of a succession of SiC substrate-BSAS (barium strontium aluminum silicate)/mullite bond coat-$ZrO_2/8\%Y_2O_3$ top coat (see also Zhu et al., 2003) calculated. The predicted range of thermal conductivities k of multilayer coatings consisting of an YSZ top coat and a 50% mullite-50% BSAS bond coat was k = 2.93 W m^{-1} K^{-1} for a conventional YSZ top coat without cracks, 2.64 W m^{-1} K^{-1} for a conventional YSZ top coat with 0.5 µm wide cracks, 2.50 W m^{-1} K^{-1} for a conventional YSZ top coat with 1.5 µm wide cracks, and 2.17 W m^{-1} K^{-1} for a nanostructured YSZ top coat.

References

M.F. Ashby, Mat. Sci. Technol., 1992, **8**, 102.

S. Aziz, S. Chandra, Int. J. Heat and Mass Transfer, 2000, **43** (16), 2841.

P. Bansai, P.H. Shipway, S.B. Leen, Acta Mater., 2007, **55** (15), 5089.

K.V. Beard, H.R. Pruppacher, J. Atmos. Sci., 1969, **26** 1066.

T. Bennett, D. Poulikakos, J. Mater. Sci., 1993, **28**, 963.

T. Bennett, D. Poulikakos, J. Mater. Sci., 1994, **29**, 2039.

L. Bianchi, F. Blein, P. Lucchese, M. Vardelle, A. Vardelle, P. Fauchais, in Thermal Spray Industrial Applications C.C. Berndt, S. Sampath (eds.), ASM International: Materials Park, OH, 1994.

D.F. Boucher, G.E. Alves, Chem. Eng. Progr., 1959, **55**, 55.

M.I. Boulos, P. Fauchais, E. Pfender, Fundamentals of Materials Processing Using Thermal Plasma Technology. Canadian University-Industry Council on Advanced Ceramic (CUICAC) Short Course, Edmonton, Alberta, Canada, October 17–18, 1989.

P.W. Bridgeman, Dimensional Analysis, 2nd ed., Yale University Press: New Haven, CT, 1922.

D.A.G. Bruggeman, Ann. Phys., 1935, **24**, 636.

M. Buchmann, R. Gadow, D. López, Abstr. 8th Europ. Interreg. Conf. on Ceramics (CIEC8), Sept 3–5, 2002, Lyon, France.

M. Bussmann, J. Mostaghimi, S. Chandra, Phys. Fluids, 1999, **11**, 1406.

Y. Cao, A. Faghri, W.S. Chang, Int. J. Heat Mass Transfer, 1989, **32**, 1289.

F. Cernuschi, S. Ahmaniemi, P. Vuoristo, T. Mäntylä, J. Eur. Ceram. Soc., 2004, **24**, 2657.

X. Chen, E. Pfender, Plasma Chem. Plasma Phys., 1983, **3**, 351.

S. Cirolini, H. Harding, G. Jacucci, Surf. Coat. Technol., 1991, **48**, 137.

C.T. Crowe, M.P. Sharma, D.E. Stock, J. Fluid Eng., 1977, **99**, 325.

P.A. Dearnley, K. Panagopoulos, E. Kern, H. Weiss, Surf. Eng., 2003, **19** (5), 373.

R. Dhiman, S. Chandra, Int. J. Heat Mass Transfer, 2005, **48**, 5625.

R. Dhiman, A.G. McDonald, S. Chandra, Surf. Coat. Technol., 2007, **201**, 7789.

Z. Duan, L. Beall, J. Schein, J. Heberlein, M. Stachowicz, J. Thermal Spray Technol., 2000, **9** (2), 225.

G. Eckold, I.M. Buckley-Golder, K.T. Scott, in: Proc. 2nd Conf. Surf. Eng., Stratford-on-Avon, June 16.18, 1987 433.

W. Elenbaas, Physica, 1934, **1**, 673.

R. Elsing, O. Knotek, U. Balting, Surf. Coat. Technol., 1990, **43/44**, 416.

Z.G. Feng, M. Domaszewski, G. Montavon, C. Coddet, J. Thermal Spray Technol., 2002, **11** (1), 62.

J.K. Fiszdon, Int. J. Heat Mass Trans., 1979, **22**, 749.

H. Fukanuma, J. Thermal Spray Technol., 1994, **3** (1), 33.

M. Fukomoto, Y. Huang, M. Ohwatari, in: *Thermal Spray: Meeting the Challenges of the 21st Century* C. Coddet (ed.) ASM International: Materials Park, OH, 1998.

Z. Gan, H.W. Ng, A. Devasenapathi, *Surf. Coat. Technol.*, 2004, **187**, 307.

D.T. Gawne, B. Liu, Y. Bao, T. Zhang, *Surf. Coat. Technol.*, 2005, **191**, 242.

R. Ghafouri-Azar, J. Mostaghimi, S. Chandra, M. Charmchi, *J. Thermal Spray Technol.*, 2003a, **12** (1), 53.

R. Ghafouri-Azar, S. Shakeri, S. Chandra, J. Mostaghimi, *Int. J. Heat Mass Transfer*, 2003b, **46**, 1395.

R. Ghafouri-Azar, J. Mostaghimi, S. Chandra, *Int. J. Comput. Fluid Dynamics*, 2004, **18** (2) 133.

R. Ghafouri-Azar, J. Mostaghimi, S. Chandra, *Comput. Mater. Sci.*, 2006, **35**, 13.

N.G. Hadjiconstantinou, in: Proc. IMECE'99 (Intern. Mech. Eng. Congress and Expos., Nov 14–19, 1999, Nashville, USA.

D.P.H. Hasselman, *J. Comp. Mater.*, 1978, **12** (19), 403.

D.P.H. Hasselman, J.P. Singh, *Am. Ceram. Soc. Bull.*, 1979, **58** (9), 856.

R.B. Heimann, *Process. Adv. Mat.*, 1991, **1**, 181.

B.T. Helenbrook, L. Martinelli, C.K. Law, *J. Comput. Phys.*, 1999, **148**, 366.

G. Heller, *Physics*, 1935, **6**, 389.

J.M. Houben, *Relation of the Adhesion of Plasma Sprayed Coatings to the Process Parameters Size, Velocity and Heat Content of the Spray Particles*. Ph.D. Thesis, Technical University Eindhoven, The Netherlands, 1988.

E. Isaacson, M. Isaacson, *Dimensional Models in Engineering and Physics*, Wiley: New York, 1975.

R. Kaczmarek, W. Robert, J. Jurewicz, M.I. Boulos, S. Dallaire, in: Proc. Symp. Mater. Res. Soc., Boston, 1983.

M.P. Kanouff, R.A. Nieser, T.J. Roemer, *J. Thermal Spray Technol.*, 1998, **7** (2), 219.

O. Knotek, R. Elsing, *Surf. Coat. Technol.*, 1987, **32**, 261.

O. Knotek, R. Elsing, U. Balting, *Surf. Coat. Technol.*, 1988, **36**, 99.

S.A. Langer, E.R. Fuller, Jr., W.C. Carter, *Comput. Sci. Eng.*, 2001, **3** (3), 15.

Y.C. Lee, K.C. Hsu, E. Pfender, Proc. 5th Int. Symp. Plasma Chem., 1981, 2, 795.

B. van Leer, *J. Comput. Phys.*, 1979, **32**, 101.

B. van Leer, *J. Comput. Phys.*, 1999, **153**, 1.

J.A. Lewis, W.H. Gauvin, *AIChE Journal*, 1973, **19**, 982.

C.J. Li, J.L. Li, W.B. Wang, in: *Thermal Spray: Meeting the Challenges of the 21st Century*, C. Coddet (ed.), ASM International: Materials Park, OH, 1998.

H. Li, S. Costil, H.L. Liao, C.J. Li, M. Planche, C. Coddet, *Surf. Coat. Technol.*, 2006, **200**, 5435.

E.Yu. Litovsky, M. Shapiro, *J. Am. Ceram. Soc.*, 1992, **75**, 3425.

E. Litovsky, T. Gambaryan-Roisman, M. Shapiro, A. Shavit, *J. Am. Ceram. Soc.*, 1999, **82** (4), 994.

J. Madejski, *Int. J. Heat Mass Transfer*, 1976, **19**, 1009.

H. Maecker, *Z. Phys.*, 1959, **157**, 1.

J.C. Maxwell, *A Treatise on Electricity and Magnetism*, 2nd ed., **Vol. 1**, Clarendon Press: Oxford, 1881.

A. McDonald, C. Moreau, S. Chandra, *Surf. Coat. Technol.*, 2007, **202** (1), 23.

V. Medhi-Nejad, J. Mostaghimi, S. Chandra, *Int. J. Numer. Methods in Eng.*, 2004, **61**, 1028.

N.Z. Mehdizadeh, M. Raessi, S. Chandra, J. Mostaghimi, *J. Heat Transfer*, 2004a, **126**, 445.

N.Z. Mehdizadeh, S. Chandra, J. Mostaghimi, *J. Fluid Mech.*, 2004b, **510**, 353.

R. McPherson, *Thin Solid Films*, 1984, **112**, 89.

P. Michlik, C. Berndt, *Surf. Coat. Technol.*, 2006, **201**, 2369.

A. Moradian, J. Mostaghimi, *IEEE Trans. Plasma Sci.*, 2005, **33** (2), 410.

J. Mostaghimi, P. Proulx, M.I. Boulos, *Int. J. Heat Mass Transfer*, 1985, **28**, 187.

J. Mostaghimi, in: Trans. 17th Workshop CUICAC (R.B. Heimann), Quebec, Canada, October 2, 1992.

T. Nakamura, Y. Liu, *Inter. J. Solids Struct.*, 2007, **44**, 1990.

H.W. Ng, Z. Gan, *Finite Elements in Analysis and Design*, 2005, **41**, 1235.

A.V. Nikolaev, in: *Plasma Processing in Metallurgy and in Process Engineering of Non-Organic Materials*, Nauka, Moscow, 1973.

M. Pasandideh-Fard, *Droplet Impact and Solidification in a Thermal Spray Process*, Unpublished Ph.D. thesis, University of Toronto, 1998.

M. Pasandideh-Fard, R. Bhola, S. Chandra, J. Mostaghimi, *Int. J. Heat Mass Transfer*, 1998, **41**, 2929.

M. Pasandideh-Fard, V. Pershin, S. Chandra, J. Mostaghimi, *J. Thermal Spray Technol.*, 2002a, **11** (2), 206.

M. Pasandideh-Fard, S. Chandra, J. Mostaghimi, *Int. J. Heat Mass Transfer*, 2002b, **45**, 2229.

J. Pech, B. Hannoyer, A. Denoirjean, P. Fauchais, in: *Thermal Spray: Surface Engineering via Applied Research*, C.C. Berndt (ed.), ASM International: Materials Park, OH, 2000.

E. Pfender, *Plasma Chem. Plasma Process*, 1989, **9**, 167S.

D.R. Poirier, E.J. Poirier, *Heat Transfer Fundamentals for Metal Casting*, 2nd edn, Minerals, Metals and Materials Society: Warrendale, PA, 1994.

P. Proulx, J. Mostaghimi, M.I. Boulos, Proc. ISPC-8, Tokyo, Japan, 1987, 1, 13.

P. Proulx, J. Mostaghimi, M.I. Boulos, *Int. J. Heat Mass Transfer*, 1985, **28**, 1327.

M. Raessi, J. Mostaghimi, M. Bussmann, *Thin Solid Films*, 2005, **506.507**, 133.

M. Raessi, J. Mostaghimi, *Numerical Heat Transfer, Part B*, 2005, **47**, 1.

C.P. Robert, G. Casella, *Monte Carlo Statistical Methods*, 2nd edn, Springer: New York, 2004.

F. Rosenberger, *Fundamentals of Crystal Growth I*, Springer: Berlin, Heidelberg, New York, 1979, p. 267ff.

N. Sakakibara, H. Tsukuda, A. Notomi, in: *Thermal Spray: Surface Engineering via Applied Research*, C.C. Berndt (ed.), ASM International: Materials Park, OH, 2000.

H.R. Salimijazi, M. Raessi, J. Mostaghimi, T.W. Coyle, *Surf. Coat. Technol.*, 2007, **201** (18), 7924.

P.R. Schoeneborn, *Int. J. Multiphase Flows*, 1975, **2**, 307.

B. Schulz, *High Temp. High Press.*, 1981, **13**, 649.

S. Seifert, E. Litovsky, J.I. Kleiman, R.B. Heimann, *Surf. Coat. Technol.*, 2006, **200** (11), 3404.

I. Sevostianov, M. Kachanov, *Acta Mater.*, 2000, **48**, 1361.

Yu.A. Shreider, *The Monte Carlo Method*, Oxford University Press: Oxford, 1966.

H.-D. Steffens, Z. Babiak, U. Fischer, in: Proc. 2nd Conf. Surf. Eng., Stratford-on-Avon, June 16–18, 1987, 471.

A.A. Syed, A. Denoirjean, B. Hannover, P. Fauchais, P. Denoirjean, A.A. Khan, J.C. Labbe, *Surf. Coat. Technol.*, 2005, **200**, 2317.

M. Wenzelburger, M. Escribano, R. Gadow, *Surf. Coat. Technol.*, 2004, **180/181**, 429.

J. Wilden, H. Frank, J.P. Bergmann, *Surf. Coat. Technol.*, 2006, **201** (5), 1962.

R.L. Williamson, J.R. Fincke, C.H. Chang, *J. Thermal Spray Technol.*, 2002, **11** (1), 107.

H.B. Xiong, L.L. Zheng, S. Sampath, R.L. Williamson, J.R. Fincke, *Int. J. Heat Mass Transfer*, 2004, **47** (24), 5189.

T. Zhang, Y. Bao, D.T. Gawne, B. Liu, J. Karwattzki, *Surf. Coat. Technol.*, 2006, **201**, 3552.

D. Zhu, S.R. Choi, J.I. Eldridge, K.N. Lee, R.A. Miller, *Eng. Sci. Proc.*, 2003, **24** (3), 469.

7
Solutions to Industrial Problems I: Structural Coatings

This chapter deals with basic information on the structural coatings that are widely applied to solve mechanical and frictional performance and maintenance problems in industry and which impart new functional properties to materials surfaces. The material covered is by no means exhaustive but is intended to elucidate some fundamental trends in coating design and to illustrate various physical processes that occur during plasma spraying of wear- and corrosion-resistant coatings. The literature available is vast and the reader is referred to numerous databases available on the Internet. Very recent information on technology and potential of wear- and corrosion-resistant coatings has been provided by Bach *et al.* (2005). An evaluation of recent patents can be found in Heimann and Lehmann (2008).

Structural coatings for wear and corrosion protection are most frequently based on group 4 to 6 transition metal carbides (WC, TiC, VC, TaC, NbC, Mo_2C, Cr_3C_2) and also some hard oxides (Al_2O_3, TiO_2, Cr_2O_3), metals (W, Mo, Ti, Ta) and alloys (NiCoCrAlY) and diamond. Since the melting temperatures of the carbides are extremely high, and oxidation/decarburization generally occurs at such high temperatures, pure carbide powders cannot be properly melted and deposited even in high enthalpy plasma jets. Instead, carbide particles are embedded into easily melted binder metals such as Ni, Co, Cr and their mixtures and alloys, respectively. There are, however, restrictions on the use of binder metals. For example, VPS-B_4C coatings on graphite and CFC or steel tiles were tested for walls of nuclear fusion reactors without the use of binder metals that would be activated by the high neutron flux (Malléner and Stöver, 1990).

A broad range of competing technologies exists to apply hard ($HV \leq 40\,GPa$), superhard ($40 < HV < 80\,GPa$) and ultrahard ($HV \geq 80\,GPa$) coatings including pulsed and reactive magnetron sputtering, arc and E-beam coating, CVD coating, laser assisted and ion beam coating, molecular beam epitaxy, laser cladding and many more. Most of these coating materials consist of transition metal carbides, nitrides or borides that may deteriorate by thermal decomposition at high working temperatures. Hence their thermal stability must be determined and possibly improved (Raveh *et al.*, 2007). Despite this plethora of competing deposition techniques, wear- and corrosion-resistant coatings applied by plasma spraying still maintain their importance as a versatile, economic and widely adaptable tool to protect valuable industrial and public investments.

Plasma Spray Coating: Principles and Applications. Robert B. Heimann
Copyright © 2008 WILEY-VCH Verlag GmbH & Co. KGaA, Weinheim
ISBN: 978-3-527-32050-9

7.1
Carbide Coatings

Carbides with simple structures are considered interstitial compounds characterized by a strong interaction of metal to carbon: the small carbon atom is located either in an octahedral interstitial site or at the center of a trigonal prism within the close-packed transition metal atoms.

7.1.1
Pure Carbides

The crystal structures of the carbides under discussion are determined by the radius ratio $r = r_X/r_T$, where X = carbon and T = transition metal (Hägg, 1929, 1931). If $r < 0.59$, the metal atoms form very simple structures with close-packed cubic or hexagonal arrangement. The carbon atoms are situated at interstitial sites. Those interstitial sites must be somewhat smaller than the carbon atom, otherwise there will be insufficient bonding resulting in an essentially unstable structure (Baumgart, 1984). The following monocarbides crystallize in the cubic B1 (NaCl) structure: TiC, ZrC, HfC, VC, NbC, TaC. Although molybdenum carbide has an r-value of 0.556 its cubic phase is nonstoichiometric MoC_{1-x} with a low temperature α-form and a high temperature β-form. The more stable structure with higher hardness and wear resistance is Mo_2C that crystallizes in the hexagonal T_2C structure in which only half of the interstitial sites are occupied by carbon. The same structure type is found for β-W_2C.

Although *titanium carbide* crystallizes only in the cubic B1 structure it has an extraordinarily wide compositional range, and is thus stable between $TiC_{0.97}$ and $TiC_{0.50}$. The melting point is 3067 °C near a composition $TiC_{0.80}$. *Tungsten carbide* is stable at room temperature as hexagonal α-WC that melts at 2867 °C. The hexagonal β-W_2C-phase melts at a slightly lower temperature of 2750 °C and transforms to the cubic B1 structure just below its melting point. As mentioned above, WC loses carbon at appreciable rate above 2200 °C and will form a surface layer of W_2C.

If $r > 0.59$, more complicated structures arise in which the transition metal atoms no longer form a close-packed arrangement. An example is chromium carbide whose complicated phase relationships include a peritectically melting cubic $Cr_{23}C_6$ high temperature phase, a hexagonal Cr_7C_3 phase and a peritectically melting orthorhombic low temperature Cr_3C_2 phase ($T_P = 1810$ °C). All three phases show a narrow range of homogeneity. Precipitation-strengthening of Cr_3C_2 by formation of Cr_7C_3 due to decarburization may be the reason for the exceptional solid-particle erosion resistance of those coatings (see Section 7.1.2.3).

Hard and dense crystalline boron carbide (B_4C) coatings with high adhesion strength were deposited on mirror-polished stainless (SUS304) steel substrates by electromagnetically accelerated plasma spraying (EMAPS) (Kitamura *et al.*, 2003). The system consisted of a pulsed high-current arc-plasma gun and a large flow-rate pulsed powder injector. The plasma gun with a co-axial cylindrical electrode configuration generated an electromagnetically accelerated arc plasma with a typical velocity and maximum pressure of 1.5–3.0 km s^{-1} and 1 MPa, respectively, by discharging a pulsed peak current of about 100 kA for about 300 μs duration.

Water-stabilized plasma spraying (WSP) has been used to deposit B$_4$C coatings as a low-Z material for potential application as neutron absorber in nuclear fusion reactors (Sasaki and Takahashi, 2006) and biological shields for microwave heating systems (Nanobashvili *et al.*, 2002).

Boron carbide was deposited on Ti6Al4V alloy by low pressure plasma spraying (Salimijazi *et al.*, 2005). The microstructure consisted of equiaxed boron carbide grains, microcrystalline particles and amorphous carbon regions. The amount of boron oxide and amorphous carbon in the coating was increased compared with the initial powder. The measured microhardness was 1,033 ± 200 HV. There was significant variation in measured nanohardness (−100 + 39 GPa) from point to point caused by multiple phases, splat boundaries and porosity in the deposited structure. Carbon segregation to grain boundaries and/or splat boundaries in boron carbide and the presence of nonstoichiometric B$_{4-x}$C was observed directly using spatially resolved electron energy loss spectroscopy (EELS).

7.1.2
Cemented Carbides

Cemented carbides are composite materials of pure carbides with a binder metal of low melting point and high ductility. The principal use of these materials is to produce cutting tools but plasma-sprayed coatings of cemented carbides enjoy wide applications as surface layers to protect an extraordinarily wide range of machinery and tools from wear, erosion and corrosion. The term refers to carbides of group 4b–6b elements of the Periodic Table together in a metal matrix such as cobalt or nickel. Mixtures of these metals, also together with chromium are frequently used. The selection of the binder metal depends to a large extent on its ability to wet the surface of the carbide particles to ensure secure coating cohesion. Addition of TiC, TaC and NbC as well as VC (Lyckx and Machio, 2007) and Cr (López Cantera and Mellor, 1998) to WC/Co-cemented carbide causes important property changes in terms of surface reactivity and/or melting behavior. Such multicarbide materials may contain TiC-TaC-NbC[1] solid solutions that improve the oxidation resistance at high temperatures as well as hardness and hot strength. Often such a modification of the hard phase content of the composite is augmented by incorporating chromium into the carbide mix or by modifying the binder phase, e.g. cobalt-chromium or chromium-nickel-cobalt alloys. Like the hardness the transverse rupture strength of a coating can be strongly influenced by the binder metal content that can be varied by the degree of dispersion of the carbide and binder metal phase. It is evident that the starting spray powder must be highly homogeneous in the first place to avoid layering of the coating and thus introduction of potential weak zones. The corrosion resistance of cemented carbide coatings is determined by the corrosion resistance of both

---

[1] In the American engineering literature on cemented carbide coatings the name columbium (symbol: Cb) instead of niobium is still being used.

the carbide(s) and the binder metals. The latter are generally soluble in acids so that their corrosion performance limits the application of such coatings in operations where highly acidic solutions or gases occur. In such cases the design engineer often resorts to oxide ceramic coatings such as alumina. But although alumina is highly corrosion-resistant its low fracture toughness and compliance pose problems (see Section 7.3.1).

In addition to mixed carbides and carbide solid solutions, ternary complex carbides have been developed (Lugscheider *et al.*,1987). These hexagonal carbides T_2MC (T = group 3b–6b elements of the Periodic Table; M = group 2b, 3a–6a elements) are called H-phases. Their crystal structure was determined for Cr_2AlC (Lugscheider and Kuelkens, 1986) and Ti_2AlC (Jeitschko *et al.*, 1964). The latter was used as hard phase in a pseudo-alloy matrix together with a hard Ni-Cr-10B-Si alloy (Knotek *et al.*, 1975) to flame- and plasma-spray wear-resistant coatings onto steel. Such tribological coatings showed improved wear resistance.

7.1.2.1 Tungsten Carbide based Coatings

The performance profile of WC-based thermally sprayed coatings includes high indentation hardness in excess of 15 GPa, excellent resistance to abrasion, solid particle erosion (SPE), cavitation and chemical corrosion, low porosity, low friction coefficient as well as the option of powder processing by a broad spectrum of thermal spray techniques (EN 657, 1994). Increasingly WC-based coatings are used to answer the call for replacement of hard chrome coatings applied by environmentally hazardous electrochemical processes (for example Bolelli *et al.*, 2006).

These materials are widely applied in industry as wear-resistant coatings because of their favorable properties, that is, high hardness and excellent abrasion resistance. The spray powder exhibits complex interactions with the plasma jet, the environment and the substrate material, so the coating process requires careful control of the powder characteristics and the plasma parameters (Rangaswami and Herman, 1986; Chandler and Nicoll, 1987). The coating techniques which are applied include conventional flame spraying, HVOF techniques (D-Gun™, Gator Gard™, Jet Cote™), arc spraying and plasma spraying including conventional APS, LPPS, SPS and VPS, respectively, reactive plasma spraying, underwater plasma spraying (UPS) and more. Recent developments indicate that both DJH (diamond jet hybrid)-HVOF and CDS (continuous detonation spraying)-HVOF flame spraying yield coatings that are highly optimized in terms of density, hardness, adhesion strength and phase purity compared to APS or VPS (for example Smith and Novak, 1991a; Kreye *et al.*, 1986; Henke *et al.*, 2004; Marx *et al.*, 2006). However, for the development of 'designed coatings' under exclusion of oxygen that may lead to coating degradation, VPS is also being considered (Tu *et al.*, 1985; Shimizu and Nagai, 1991).

The coatings deposited must be thin enough so as not to compromise given tolerances of the machined parts and tools. A high coating density is desired for wear applications but sometimes this requirement must be relaxed. The surfaces of plungers and barrels of reciprocating and centrifugal pumps, for example, must be somewhat porous so that the lubricants can adhere (Heimann *et al.*,1990a). Cobalt-based tungsten carbide coatings are most useful for achieving high sliding-wear resistance and better friction properties (D'Angelo and El Joundi, 1988; Henke et al., 2004, Marx et al., 2006). They can be used to prevent general wear, fretting wear (Koiprasert et al., 2004), cavitation erosion (Lima et al., 2004) and also chemical corrosion (Perry et al., 2002) of a variety of oil-field equipment including protection against severe particle erosion of choke nozzles of tungsten carbide alloy that are used to control high-velocity flow of heavy oil contaminated with sand (Wozniewski, 1990). This requires secure bonding of a thin layer (<10 µm) of compatible tungsten carbide-based material to the nozzle substrate (Heimann et al., 1990a). As the nozzle is supposed to rotate within a steel sleeve, problems of stringent clearance control must also be addressed. The present technology has become sophisticated to protect these choke nozzles from failure due to severe erosion, corrosion and scaling caused by hot, corrosive fluids and sand and this is illustrated by a study performed by ESSO Resources Canada Limited (Wozniewski, 1990) that concluded that the massive tungsten carbide/cobalt substrate is best protected by a system of a CVD titanium nitride coating of 10 µm thickness followed by a 7 µm alumina and then a boron diffusion coating. It is obvious that a single conventional 'off the shelf' coating is no longer sufficient to protect machinery from nonlinear effects of corrabrasion, i.e. the synergistic interaction of corrosive and abrasive destruction of the material. Also, while it is still common practice to develop coatings by establishing a set of spray parameters adapted to different substrates by making only minor modifications to information about parameter sets obtained from the widespread literature on this subject, there is an increasing need to design coating/substrate systems as a single entity (Sivakumar and Mordike, 1989). Since this requires optimization of coating properties, statistical multifactorial experimental designs should be applied (see Chapter 11).

WC-based coatings, for example Sulzer Metco's Sume®Turb are being applied by HVOF to the large blades of Kaplan turbines to combat severe solid particle erosion (SPE) associated with the high sediment loads of rivers in China and in particular in the Himalaya region of India (Nestler et al., 2007).

Since proper adhesion of the WC/Co coatings requires an optimized D/d ratio (see Equation 5.8) and therefore optimized particle viscosity and velocity, the plasma enthalpy and thus the plasma temperature become the determining factors. The plasma jet temperature must be substantially higher than the melting point of cobalt (1495 °C) but not so high that the tungsten carbide undergoes decomposition (Vinayo et al., 1985) or η-carbide (Co_nW_mC) formation (Hajmrle and Dorfman, 1985). The changes that a stoichiometric WC/Co powder is subjected to during thermal exposure in the plasma spray process is illustrated in Figure 7.1 that shows the low carbon portion of the ternary phase diagram W-C-Co. The WC phase reacts with the Co melt (see Figure 10.35) and loses carbon by reaction with oxygen in the flame. The composition of the coating matrix is close to the η-carbide phase Co_3W_3C. This phase presumably leads to a deterioration of the mechanical strength due to its brittleness. Furthermore, W_2C formed by decarburization of WC tends to embrittle the coating. Even though under VPS conditions due to the absence of oxygen the formation of W_2C can be completely suppressed (Rangaswamy and Herman, 1986; Vinayo et al., 1985), carbon can evaporate and form mixed tungsten carbides (Tu et al., 1985).

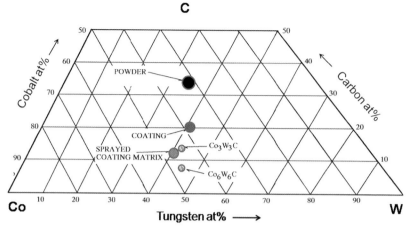

Figure 7.1 Low carbon portion of the ternary phase diagram W-C-Co showing the shift of the chemical composition of the starting powder towards the η-carbide phases Co_3W_3C and Co_6W_6C during spraying (Hajmrle and Dorfman, 1985).

Figure 7.2 shows how the carbon loss and the oxygen pick-up increased with increasing spray distance and the associated increase of the residence time of the particles in the flame and the Rockwell hardness, HRC, substantially decreased (Hajmrle and Dorfman, 1985). Suppression of the decomposition of tungsten carbide has been attempted by selecting high plasma gas flow rates at reduced hydrogen content (Kirner, 1989). However, even under those conditions trace amounts of oxygen contained in the plasma gas (up to 50 ppm) or in voids of the spray powder can lead to oxidative processes under formation of $CoWO_4$ (Vinayo et al.,1985). Decrease of decomposition of tungsten carbide with increasing gas volumes at the same energy level because of reduced dwell time has also been confirmed by Chandler and Nicoll (1987). They also found that the use of an argon/hydrogen plasma dramatically reduced the amount of remaining WC and a concurrent increase in M_6C (Co_3W_3C) and $M_{12}C_4$ ($Co_3W_9C_4$) phases. Such an increase has been also observed with negative, i.e. upstream injection that has been found to disappear using positive, i.e. downstream injection. Again, the reduction of dwell time reduces decomposition.

Another way to suppress degradation of coating performance due to matrix alloying is to inject the 88WC12Co phase and a Ni-based alloy (73Ni15Cr4Si4-Fe3B1C) separately into a DC plasmatron with two injection ports (Lenling et al., 1990). The resulting composite coating deposited onto AISI 5150 steel is very dense and the carbide phase composition is close to that of the initial spray powder. In particular, decarburization products such as brittle W_2C and η-carbides are absent. The reason for this may be that with dual injection, the heat input to the carbide and the matrix metal can be independently tailored for optimum results by adjustment of the injection angles and locations as well as the plasmatron operating parameters.

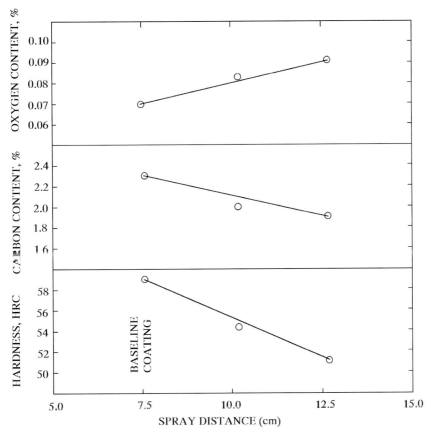

Figure 7.2 Oxygen pick-up and carbon loss and associated decrease of the Rockwell hardness of WC-Co(Ni) coatings as a function of spray distance (Hajmrle and Dorfman, 1985).

The results of conventional co-spraying of a mechanical blend of 50WC50'A' (A = 73Ni15Cr4Si4Fe3B1C) onto very fine-grained powder metallurgically densified tungsten carbide (WC12NiCo) are shown in Figure 7.3. Figure 7.3a shows a cross-section of the interface between the WC/NiCo substrate (left) and the 50WC50'A' coating (right) with an X-ray line scan of cobalt. Figure 7.3b shows a back-scattered image of the polished coating subjected to a Charpy impact test, illustrating the coarse-grained WC grains within the NiCo matrix (Kamachi et al., 1990). As shown in Figure 7.4 the main crack is situated within the fine-grained substrate with a second crack running close to the interface but always within the substrate material. Such tungsten carbide composite coatings were developed to combat severe solid-particle erosion (SPE) problems of choke nozzles in heavy crude oil production in the oil patch in Northern Alberta, Canada (Wozniewski, 1990) as well for use in offshore gate valves (Wheeler and Wood, 2005) and other petrochemical applications (Scrivani et al., 2001).

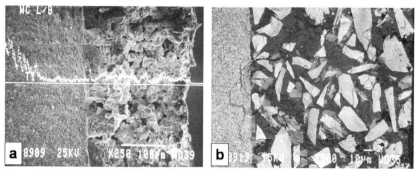

Figure 7.3 Tungsten carbide-based coating (right) on a fine-grained WC12(NiCo) substrate (left). Panel a shows an X-ray line scan of Co, panel b shows a backscattered electron image of a polished coating illustrating the very coarse WC grains within the fine-grained 'A' matrix.

It has been mentioned above (see Section 7.1.2) that the modification of tungsten carbide by addition of highly refractory and extremely hard carbides such as TiC and TaC yielded coatings with clearly improved wear resistance, frictional properties and fracture toughness (Smith and Novak, 1991b). At low frictional velocities the wear resistance is rather low but increases with increasing velocity and therefore also temperature (Shimizu and Nagai, 1991). This can be attributed to surface oxidation that decreases friction. In TiC-reinforced WC/Co coatings the TiC particles tend to collect near the surface of the coating due to their low specific gravity thus creating a top layer that has an extremely good abrasion resistance.

The addition of TaC leads to three-phase alloys with improved high temperature properties in terms of oxidation and diffusion resistance against ferrous substrate materials (Kolaska and Dreyer, 1989).

Despite an enormous amount of research invested into thermally sprayed coatings based on tungsten carbide, many aspects are not yet fully understood, particularly

Figure 7.4 Crack behavior of a WC-based coating (right)/fine-grained WC12(NiCo) substrate (left) system subjected to a Charpy impact test.

with regard to the chemical, microstructural and phase changes that occur during spraying and their influence on properties such as wear resistance. This is due to widely varying starting powders, spray system types, spray parameters and other variables that influence the coating structures and cause difficulties when comparing results from different workers. A review of WC-12Co and WC-17Co coatings has been provided that aimed to compare the properties of coatings obtained from cast and crushed, sintered and crushed and agglomerated and densified powder types (Lovelock, 1998).

Chemical degradation of the WC phase by oxidation, decarburization and matrix-alloying as discussed above yields coatings with noticeably decreased performance in terms of abrasion resistance. Coating microhardness may be, under simplified assumptions, indicative of abrasion resistance even though proportionality between microhardness and sliding and abrasive wear, respectively is rarely observed (Sandt, 1986; Factor and Roman, 2002a, 2002b). Much more important is the interaction of wear mode and microstructure of the coating. Spraying of WC/Co powder at high velocities in an inert atmosphere should considerably minimize the degradation of the WC/Co system. Indeed, experiments by Mutasim et al. (1990) using a high velocity plasma spray process, modified with a plasmatron extension operating in a low pressure chamber (Figure 7.5) pointed to a substantial decrease of the amounts of W_2C and Co_3W_3C and thus to an increase in microhardness approaching that of the powder-metallurgically densified WC12Co material (Figure 7.6). It was concluded that the higher amounts of WC retained contributed to the observed increase in microhardness. Figure 7.7 shows a cross-section of a dense, well-adhering WC12Co wear coating (right) on a mild steel substrate (left) produced by APS with careful parameter optimization (Heimann et al., 1990a).

The hardness of the WC/Co composite depends strongly on the ratio WC to Co as well as, for monolithic materials, its grain size (Figure 7.8). Therefore, different areas of application emerge.

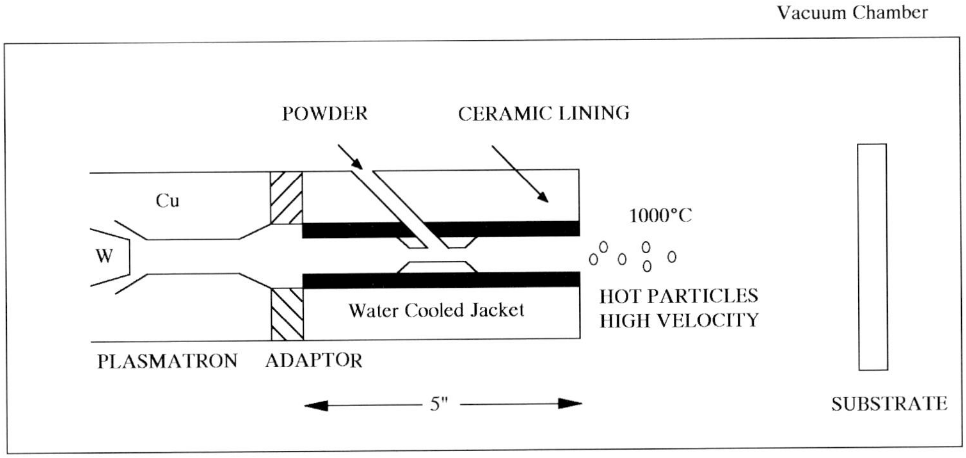

Figure 7.5 Modified plasmatron to reduce decarburization of WC-Co coatings (see text) (Mutasim et al., 1990).

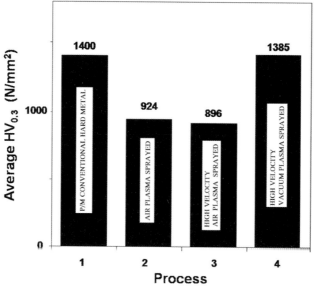

Figure 7.6 Comparison of the microhardness (HV$_{0.3}$) of differently processed WC12Co powders (2–4) and powder-metallurgically densified materials (1) (Mutasim et al., 1990).

In the metal casting, construction materials and ceramics industries machines are used to form, shape and densify hard granular precursor materials (Stehr et al., 2000). Hence to extend the service life of forming tools hard coatings are being applied to protect them from abrasive sliding wear. In a recent study (Henke et al., 2004) several WC-based coatings (WC-12Co, WC-17Co, WC-10Co4Cr, WC-20Cr$_3$C$_2$7Ni) differing in their compositions, grain size distribution and production routes were applied by HVOF-DJH to the surface of hardened X210Cr12 tool steel, and their performance tested under three different regimes of wear that were simulated by a 2-body Taber® Abraser abrasive-rolling integral wear test (ASTM F1978), a 2-body dry sand/rubber wheel abrasion test (ASTM G65-04, 2004) and a 3-body sliding wear test (block-on-

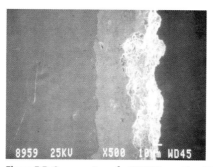

Figure 7.7 Cross-section of a WC12Co coating (right) on mild steel (left) (Heimann et al., 1990a).

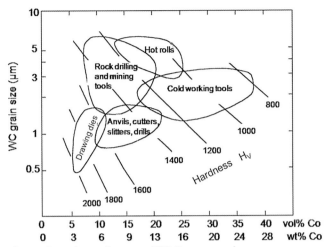

Figure 7.8 Hardness of cemented WC-Co compounds as a function of the WC grain size and the volume and weight contents of cobalt. Several application areas are indicated.

ring test; ASTM G77-05, 2005). A sintered hardmetal (WC-15Co, TRIBO Hartmetall GmbH, Immelborn, Germany) was used as a control. From the results of these tests and the systematic study of the microstructure of the worn coating surfaces a comparison was made among the various coating type, and their wear performance evaluated in response to the different test procedures (Henke, 2001). Table 7.1 shows the composition and particle sizes of spray powders, and the Vickers microhardness and volume losses in $mm^3 km^{-1}$ wear path lengths of the resulting coatings.

Figure 7.9 shows that the surfaces subjected to the *Taber®Abraser test* are characterized by microchipping (weak abrasion) and/or microploughing (strong abrasion) (Wielage *et al.*, 1999; DIN 50320, 1979). Samples 1 (Figure 7.9b) and 3 show weak abrasion ($2.13 \pm 0.12\ mm^3 km^{-1}$), samples 4 and 8 medium abrasion ($2.37 \pm 0.59\ mm^3 km^{-1}$) predominantly by microchipping whereas samples 6, 7 and 9 suffered rather strong abrasion ($3.53 \pm 1.95\ mm^3 km^{-1}$) involving both microchipping and microploughing wear mechanisms. Sample 2 exhibited the most severe abrasion with formation of deep furrows and grooves owing to pronounced microploughing (Figure 7.10b).

The *dry sand/rubber wheel abrasion test* resulted in all samples in an apparently more intense abrasion compared to the Taber® Abraser test. Most samples showed strong parallel grooving related to microchipping. The only samples without grooving were samples 3 and the control sample 9. Weak grooving occurred in samples 1 (Figure 7.9c), 4 and 8 ($1.04 \pm 0.04\ mm^3 km^{-1}$), strong grooving in samples 2 (Figure 7.10c), 6, 7 and 9 ($3.25 \pm 0.86\ mm^3 km^{-1}$). Hence from the volume loss per km wear path length it appears that the integral abrasion losses for most samples are less than those found in the Taber® Abraser test (Table 7.1).

The *block-on-ring test* similar to ASTM G77-05 (2005) to simulate 3-body sliding wear (Figure 10.27) showed the maximum coating surface deterioration. In the case

Table 7.1 Composition and grain size distribution (in µm) of WC-based spray powders (1–8), sintered hardmetal (9) and tool steel substrate (10) and microhardness (in GPa) and volume losses (in mm^3 km^{-1}) of coatings subjected to Taber® Abraser (ΔV_A), dry sand/rubber wheel (ASTM G65-04) (ΔV_B) and 3-body sliding wear tests (ASTM G77-05) (ΔV_C) (Henke et al., 2004).

	Composition	Grain size	Microhardness	ΔV_A	ΔV_B	ΔV_C
1	WC-12Co	−35 + 10	17.20 ± 2.64	2.05	1.00	12.06
2	'75-25'	−45 + 11	13.30 ± 3.72	3.77	4.23	103.25
3	WC-12Co	−45 + 11	19.70 ± 3.25	3.22	1.44	9.74
4	WC-12Co	−35 + 10	19.60 ± 4.35	1.95	1.05	11.35
5	WC-17Co	−45 + 5.5	18.40 ± 5.75	2.33	3.40	36.23
6	WC-17Co	−45 + 11	18.70 ± 3.99	2.33	2.12	20.62
7	WC-10Co4Cr	−45 + 11	20.70 ± 3.52	2.79	1.07	10.15
8	'73-20-7'	−40 + 10	13.20 ± 4.12	5.71	3.27	45.38
9	WC-15Co			1.44	2.27	51.43
10	Tool steel X210Cr12			6.35	2.65	537.00

^{a} Vickers hardness HV(0.05) in GPa;
^{b} A 2-body Taber® Abraser test, B 2-body dry sand/rubber wheel test (ASTM G65-04); C 3-body sliding wear test (block-on-ring, ASTM G77-05).
^{c} 75(WC-12Co)/25NiCrMo;
^{d} 73WC20Cr$_3$C$_2$7Ni

of more moderate abrasion large scale microchipping was generally observed whereas increasing abrasive sliding wear resulted in massive surface wreckage with pitting and cracking features characteristic of this type of wear. In all samples tested formation of highly polished plateaus separated by lateral deep, narrow grooves (Figure 7.9d) was observed. The formation of these polished areas appears to be related to shifting of lattice planes of the WC grains under extreme peak pressures and temperatures up to 500 °C (Suh, 1973).

However, with increasing wear the grooves widened to extreme values of up to 100 µm (Figure 7.11). This allowed abrasive grains to penetrate into the grooves inducing microchipping. On the other hand the high pressure exerted on the surface may result in densification of the hard material grains and associated work hardening thus increasing wear resistance. Branched cracks frequently appeared perpendicular to the direction of slide loading. The only sample investigated that did not show grooving was the control sample (sintered hardmetal WC-15Co) even though its overall wear was high. Pronounced pitting was observed on sample 2 (Figure 7.10d) and sample 4. Sample 2 showed the highest volume loss of all samples tested of 103 mm^3 km^{-1} (Table 7.1).

In conclusion, the dominating abrasive wear mechanism that these WC-based coatings underwent during a Taber® Abraser and a dry sand/rubber wheel (ASTM G65-04) test was sliding wear with microchipping and microploughing. The very high forces exerted on the coating surfaces during a 3-body sliding wear (block-on-ring) test lead to pronounced surface deterioration characterized by pitting as well as severe cracking and grooving. The high frictional forces generate local temperatures

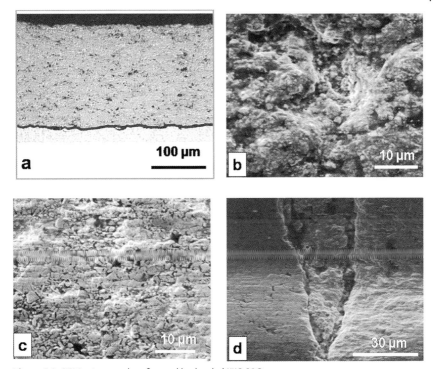

Figure 7.9 SEM micrographs of a weakly abraded WC-12Co coating (sample 1), (a) as-sprayed (HVOF-DJH), (b) after Taber® Abraser test, (c) after dry sand/rubber wheel test, (d) after block-on-ring test.

in excess of 1000 °C (Uetz *et al.*, 1986) may induce plastic deformation of the hard ceramic grains and adhesive wear, respectively. Coatings with a maximum content of WC (WC-12Co) showed the highest resistance against rolling and sliding wear by abrasion with hard mineral grains (quartz). Hence such coatings will be beneficial to protect feeding nozzles and rollers of forming machines for concrete roof tiles and other applications.

Studies by Marx *et al.* 2006 showed that the wear performance of thermally sprayed cemented tungsten carbide coatings depends not only on the chemical and phase compositions of the coatings but also to a large extent on the type of (sliding) wear test applied. Figure 7.12 illustrates the very different wear rates obtained for the same type of coating subjected to different wear tests. In general, the least destructive test was found to be the dry sand/rubber wheel test (ASTM G65-04, 2004) followed by the Taber® Abraser test (ASTM F1978, 1978). The most severe test is the block-on-ring abrasion test performed according to a modified ASTM G77-05 (2005) designation (see also Section 10.2.2). The Ni-based coatings (WC-10Ni, WC-12Ni and WC-17Ni) performed superior over Co-based ones (WC-8Co, WC-10Co4Cr) in the ASTM G65-04 test whereas the Co-based WC coatings were mechanically more stable in the modified ASTM G77-05 test and hence outperformed the Ni-based coatings. The

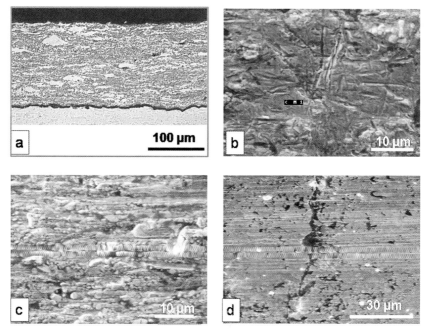

Figure 7.10 SEM micrographs of a strongly abraded coating of sample type 2 (75(WC-12Co)25NiCrMo), (a) as-sprayed (HVOF-DJH), (b) after Taber® Abraser test, (c) after dry sand/rubber wheel test, (d) after block-on-ring test.

70Cr$_3$C$_2$30NiCr coatings showed under more severe conditions (Taber® Abraser ASTM G77-05) the highest wear rates.

The relative wear resistance of various WC17Co, WC12Co and Cr$_3$C$_2$25NiCr coatings were compared by applying a simple Knoop microhardness indentation

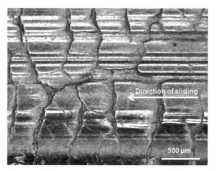

Figure 7.11 Light optical micrograph of the surface of a WC-10Co4Cr coating (sample 7) subjected to a 3-body sliding wear test showing deep grooving, polished plateaus and branching cracks perpendicular to the direction of slide loading (Henke et al., 2004; Marx et al., 2006).

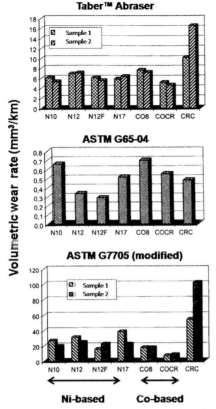

Figure 7.12 Volumetric wear rates ($mm^3\,km^{-1}$ wear path length) of cemented WC coatings (Ni-based; N10, N12, N12F, N17; Co-based: CO8, COCR) and 70Cr$_3$C$_2$-30NiCr (CRC) coatings (control) measured by 2-body Taber® Abraser test (SiC F80, 1 kp load, 5000 revolutions), 2-body dry sand/rubber wheel test (ASTM G65-04; 45 N load, 2000 revolutions corresponding to 4309 m wear path length) and 3-body block-on-ring test (modified ASTM G77-05, 5355 N load, wear path length 2835 m) (Marx et al., 2006).

test (Factor and Roman, 2002a, 2002b). It was found that, when correctly performed, this technique provides a reasonable indication of abrasion and sliding wear resistances, and that the measurements also correlate well with cavitation resistance in Cr$_3$C$_2$25NiCr coatings. However, the technique is less useful for predicting erosion resistances for all coatings investigated and for abrasion resistance when WC-Co coatings were ground against SiC. In the latter case the contribution of microfracturing and brittle failure is larger than indicated by the measured hardness since this is essentially a measure of resistance to plastic deformation under equilibrium conditions.

7.1.2.2 Titanium Carbide-based Coatings

Under conditions of simultaneous mechanical stress and chemical corrosion, in particular in a steam environment, commercially available WC-Co coatings frequently do not perform well at high working temperatures. However, WC coatings frequently outperform TiC coatings in terms of sliding wear resistance (Economou et al., 1998, 2000).

For example, the quest for an increased energy output of power plants and various kinds of heat engines necessitates an increase in the working temperature. In these cases, TiC-Ni is an alternative system (Fuji et al., 1995) that because of higher temperature stability, lower coefficient of thermal expansion, higher hardness and lower specific gravity may outperform other coating systems. If the TiC phase is stable, and well bonded and dispersed in the matrix, a hard, low-friction surface is produced. WC or Cr_xC_y phases tend to decompose at high temperatures thus losing their friction properties.

Problems exist, however, in plasma spray operations that result in decarburization of TiC towards TiC_{1-x}. Therefore, TiC has been processed as 'cermet coating clad' in a metal system (Brunet et al., 1986). In order to suppress decarburization of TiC, a reduction of the dwell time of Ni-plated TiC particles (50TiC50Ni to 30TiC70Ni) using VPS or HVOF techniques led to coatings on mild steel with low porosity (<2%), high microhardness (900–1000 $HV_{0.2}$) and high blast erosion resistance (<0.1 $mm^3 g^{-1}$ at a blast angle of 90°; alumina grit 60 mesh, air pressure 490 kPa) (Fuji et al., 1995). The VPS coatings outperformed the HVOF coatings also in a reciprocating plane wear test.

The wear properties of TiC-based coatings are a strong function of the powder preparation method. For example, 40TiC60(Ni20Cr) powder produced by the PMRS (plasma melted rapidly solidified) method and sprayed with VPS produced a coating with a wear resistance ten times that of a coating of the same composition whose starting powder was produced by physical blending (Smith et al., 1988). The sliding wear performance of five different hard coatings (TiC reinforced alloys: Resistic™ CS-40 (45 vol% TiC, stainless steel matrix); Resistic™ CM (45 vol% TiC, tool steel/Fe, Cr,Mo matrix; Resistic™ HT-6A (40 vol% TiC, Ni20Cr matrix); 88WC12Co composite powder was compared to that of Stellite 6 (Co28Cr4.5W3Fe3Ni1.5Si1Mo1Mn1.15C) sprayed with APS onto steel. The TiC reinforced composite coatings had bond strengths exceeding 58 MPa and sliding wear resistances from 20% higher than the WC/Co coatings, and up to 100% higher than Steelite 6 coatings, depending on the deposit matrix of the TiC reinforced materials (Figure 7.13a). Also, the kind of the auxiliary plasma gas (helium or hydrogen) played an important role for the wear resistance of WC/Co and Stellite 6 but did not influence that of the TiC-based coatings. The kinetic coefficient of friction during the sliding wear test was minimized for CS-40 (Figure 7.13b).

A different way to circumvent the risk of thermal decarburization of TiC under high temperature conditions in a plasma jet is *reactive plasma spraying*. TiC is formed from titanium particles during their flight along the plasma jet. A reactor is being added to a conventional DC plasmatron into which carbon-containing precursor gases (methane, ethane, propane, acetylen etc) are being fed (Figure 7.14). Into the plasmatron titanium powder will be introduced in the conventional manner. Shroud gas (Ar, N_2) led into the reactor tube is thought to prevent build up on the reactor walls of Ti/TiC materials and carbon produced by thermal cracking of the hydrocarbon precursor gases. The modified plasmatron was

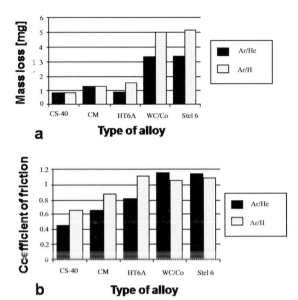

Figure 7.13 Mass loss (a) and coefficient of friction (b) during sliding wear resistance tests of TiC-metal matrix coatings (CS-40, CM, HT6A alloys). Values for WC/Co and Stellite 6 are given for comparison (Smith et al., 1988).

operated in a controlled, sub-atmospheric pressure environment. The amount of TiC formed and retained in the coating is a function of several parameters such as substrate temperature, chamber pressure, type of the reactive gas, flow rate of the reactive gas and the method of introduction of the reactive gas (Smith et al., 1993; Smith and Mutasim, 1992).

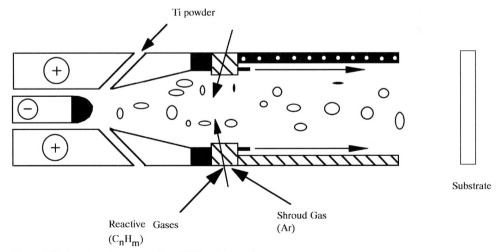

Figure 7.14 Reactive plasma spraying of Ti in a hydrocarbon environment to form TiC-containing coatings (Smith et al., 1993).

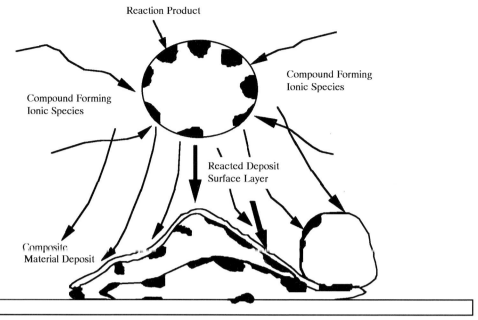

Figure 7.15 Model of TiC formation during reactive plasma spraying of Ti metal powder (Smith et al., 1993).

The hard TiC phases were observed predominantly in the intersplat region. The formation and distribution of the hard phases in the coatings suggest a two-stage mechanism illustrated in Figure 7.15. In these experiments the content of the hard phase is still very low (around 8 vol%) and must be considerably improved by process optimization until reactive plasma spraying of TiC will become an economically competitive technology.

A more promising way appears to be reactive spraying of Ti/Ni20Cr powder with propylene as a precursor gas (Smith and Mutasim, 1992). In the coatings, the hard phases Cr_7C_3 and TiC were deposited in very small crystals ($<1\,\mu m$). From these results it was concluded that reactively plasma-sprayed prealloyed NiCr/TiC powders would show improved wear resistance owing to the better carbide/matrix cohesion of the *in situ* formed carbides Cr_xC_y. Indeed, the reactively sprayed NiCr/TiC material displayed harder coatings (840 $VH_{0.3}$) than nonreactively sprayed pre-alloyed coatings that showed a microhardness of only 700 $VH_{0.3}$ (Smith et al., 1981).

Several other accounts of TiC-based coatings may be consulted to obtain information on applications and performance optimization (Smith et al., 1989; Veilleux et al., 1987; Jungklaus et al., 1992; Cliché and Dallaire, 1990). The properties of the simple TiC/Ni system can be improved by alloying with other elements, such as molybdenum in the carbide and cobalt in the binder phase. Investigations on (Ti,Mo)C-NiCo coatings sprayed by APS and D-Gun techniques (Vuoristo et al., 1995) as well as VPS (Thiele et al., 1996a) indicate that with proper parameter optimization very hard, highly wear-resistant coatings can be produced for a variety of high temperature

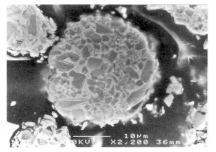

Figure 7.16 Dense as-sintered (Ti,Mo)C-NiCo alloy granule with typical core (nearly pure TiC)-rim (Ti,Mo)C_{1-x} structure (Thiele *et al.*, 1996).

applications including HT erosion protection. It was demonstrated that the typical microstructure of the powders, produced by agglomeration and subsequent sintering, in particular the core (nearly pure TiC) –rim (Ti,Mo)C_{1-x} structure (Figure 7.16) can be transferred to the coating without significant changes. Comparison of VPS- and HVOF-sprayed (Ti,Mo)(C,N) – 45 vol% Ni20Co cermet coatings revealed that the former possess much higher abrasive wear resistance, owing to the absence of oxide scale between the spray splats (Qi *et al.*, 2006). In particular, microcrystalline VPS coatings show high wear resistance at low loads when the hard phase particles can still carry the load and thus support the counterbody. With increasing load a transition from predominant removal of the soft Ni20Co binder phase to a wear mechanism involving fracture of the hard (Ti,Mo)(C,N) grains occurs, resulting at still tougher wear conditions in removal of the fractured hard particles together with the binder phase. In nanocrystalline VPS coatings the dominant mechanism of wear consists of a cutting and ploughing type of wear, displacing simultaneously the nanocrystalline hard particles and the binder phase. This eventually leads to formation of ridges in which the wear debris lodges.

While in the past TiC grains have been used as reinforcement of metallic coatings for wear protection under harsh environmental conditions as described above, novel developments are under way using TiC and SiC in conjunction with an oxide binder phase such as TiC-Al_2O_3/Y_2O_3 or SiC-Al_2O_3/Y_2O_3 and a total carbide content above 65% (Wielage *et al.*, 2007).

7.1.2.3 Chromium Carbide-based Coatings

Although chromium carbide is somewhat softer than tungsten carbide at room temperature (HRC 52 versus HRC 62) it shows excellent sliding-wear resistance (Fessenden *et al.*, 1990) as well as superior oxidation and high temperature solid-particle erosion (SPE) resistance (Hintermann, 1984) in steam environment. However, in normal atmosphere resistance of chromium carbide coatings against sliding wear using the 2-body Taber® Abraser and the ASTM G77.97 3-body block-on-ring tests has been found to be distinctly lower than that of WC-based coatings (Marx *et al.*, 2006).

Because of their high wear resistance in a steam environment chromium carbide coatings offer promising opportunities to protect steam paths surfaces and turbine components from corrosion and erosion (Buchanan, 1987). High-temperature erosion protection is imparted to boiler tubes and fire chambers of coal-fired power plants by Cr_3C_2-25NiCr coatings. Hence several attempts have been made in the past to combat corrosion damage, stress corrosion cracking and corrosion fatigue in steam turbines (Ortolano, 1983) as well as solid-particle erosion (SPE) in boilers, tubing and steam lines including superheaters (Buchanan, 1987). These attempts include the development of plasma-sprayed chromium carbide coatings and more recently HVOF-generated coatings (Guilemany *et al.*, 2002; Morimoto *et al.*, 2006).

An alternate way to protect steam path surfaces in boiler tubes involves *boride diffusion coatings*. Their advantage is that they do not have a line-of-sight limitation, as do plasma spray coatings but this is more than counterbalanced by the disadvantage that boride coatings can result in a fatigue loss in the ferritic base metal of boiler tubes of as much as 50% (EPRI, 1983). This fatigue loss originates from a combination of lack of coating ductility and the fact that the coating forms a metallurgical bond with the base metal. Strain damage occurs in cyclical thermal loading since the higher Young's modulus of the coating causes it to be put under tension that will be amplified in service and leads finally to cracks perpendicular to the stress direction.

Plasma-sprayed *chromium carbide coatings* can considerably reduce the steam path erosion rate. Comparative erosion tests on chromium carbide sprayed with a D-Gun™ technique and chromium boride/iron boride diffusion coatings on AISI 403 and AISI 422 martensitic stainless steels showed that at 538 °C the former are clearly superior in terms of SPE with chromite particles (75 μm, 152 m s^{-1}, 30 min) at impingement angles of 30° whereas the latter show higher SPE resistance at impingement angles of 90° (Qureshi *et al.*, 1986). During deposition of plasma-sprayed coatings the heat input into the ferritic base metal is much lower than for boride diffusion coatings. Therefore boride diffusion coatings require a post-coating treatment to restore the strength and ductility to the coated component (Buchanan, 1987). The austenitic phase formed at the high diffusion temperature creates internal stresses in the base metal that result in distortion and thus the need for time-consuming post-coating machining or grinding operations to restore flatness or critical clearances.

While tungsten carbide coatings were introduced in the early 1960s by General Electric for the protection of steam turbine buckets against SPE, it was subsequently found that such coatings have not been consistently effective in preventing erosion damage. Thus new solutions were sought and chromium carbide-NiCr cermet coatings on 12Cr martensitic stainless steel AISI 422 cast) were developed and tested by the General Electric Turbine Technology Laboratory in a fully-ducted, dynamic burner rig heated by natural gas (Sumner *et al.*, 1985).

The so-called Holdgren practice (Buchanan, 1987) used for protection of steam path surfaces is similar to coating practices specified by General Electric, and Pratt and Whitney for rotating aircraft components. An exothermic nickel aluminide bond coat, consisting of spherical aluminum powder particles overcoated with shells of nickel, is applied to the base metal. During spraying the two elements react to form a nickel aluminide intermetallic compound. Since this reaction results in the evolution of heat that under optimum spraying conditions continues after impact and coating formation, microdiffusion may occur between base metal and bond coat thereby enhancing the adhesion strength by a factor of two compared to NiCr bond coats.

The reason for the outstanding erosion resistance of chromium carbide coatings seems to be based on precipitation-strengthening of Cr_3C_2 by the formation of Cr_7C_3 due to decarburization (oxidation) of the former (secondary carbide precipitation) (Menne et al., 1993). The as-sprayed, carbon deficient Cr_3C_2 transforms in service to Cr_7C_3, thereby more than doubling the hardness and greatly enhancing the SPE resistance (Wlodek, 1985). Since this precipitation-strengthening mechanism requires an air atmosphere environment, good erosion resistance is promoted by APS rather than VPS. Also, fine chromium carbide powders and pre-aging to optimum hardness further increase SPE resistance, as does the replacement of the NiCr or NiCrMo matrix by FeCrAlY or CoCrNiW. Alloys such as FeCrAlY have also been tested in their own right as a barrier layer to high temperature erosion on coal combustion and conversion processes including fireside erosion, erosion of steam turbines and corrosion-assisted wear in flue gas desulfurization equipment (Wright, 1987). The protection mechanism is based on the formation of Al_2O_3 scale (Wright et al., 1986).

Another widespread problem exists in the petrochemical industry. High temperature steam cracking of ethane to produce ethylene causes several deleterious effects to the steel tubing in the convectively and radiantly heated sections of a typical cracking furnace. These effects include external oxidation and internal oxidation/deposition of coke that is thought to be formed under the catalytic action of nickel oxide. This oxide is formed from oxidation of nickel as an alloying element for steel (AISI 304, AISI 410) or nickel superalloys (Inconel 800, Hastelloy X) added to impart high temperature resistance. The coke deposited at the internal surfaces of tubing tends to form carbides with other alloying metals in the steel that can migrate into the metal. As a result, blistering and cracking of the steel can occur. Also, reduction of the heat conduction capability of the tubing may lead to local overheating in the convection section of ethylene steam cracker. Since the surface roughness of the internal tubing walls seems to promote the rate of coke deposition (Albright et al., 1978a; 1978b) mechanical polishing or application of a smooth coating can reduce the coke layer formed. Also, introduction of trace amounts of sulfur compounds or antifoulants into the process gas stream inhibits the catalytic effect of the metal surface and may offer some moderate measure of protection (Swales, 1980). Finally, metallurgical remedies tested were either related to higher alloyed steels to reduce the carbon gradient or to the introduction into the metal of carbide-stabilizing additions such as W, Mo, Nb or V. Unfortunately, these metal additions have adverse effects on the high temperature oxidation resistance and their oxides may form low melting eutectics with nickel or chromium oxide.

In order to reconcile the materials requirements with the severity of the environmental attack, functional chemical barrier coatings (CBC) must be developed with the following properties:

- high temperature resistance;
- high corrosion resistance;
- chemical inertness;
- high hardness;
- high wear, abrasion and erosion resistance;
- creep resistance;
- good adhesion;
- high fracture toughness;
- good thermal conductivity;
- thermal shock resistance;
- smooth surface/low porosity;
- thermal fatigue cracking resistance.

The first five criteria are best met by ceramic coatings, the remaining seven by metal or alloy coatings. To develop a coating material that would meet all requirements, ideally a type of coating should be selected that combines the advantages of both materials classes synergistically. This, however, is extremely difficult to achieve since many of the desired properties listed above are incompatible or even mutually exclusive. For example, partially stabilized zirconia (PSZ) would have all the advantages of ceramic coatings including good adhesion but would fail to meet the important criteria of good thermal conductivity and low porosity. In general, the most difficult problem is to match the coefficients of thermal expansion and the thermal conductivities of the coating and the metal substrate. Therefore the composite coating performance will always have to be compromised. Attempts have been made to plasma spray composite chromium carbide-based coatings such as $62Cr_3C_225W_2C5TiC4Ni3Mo1Cr$ that show excellent erosion resistance in a high temperature steam environment. While there seems to be controversy about the role chromium plays as a possible catalyst for coke formation, there is also evidence that increasing the chromium content and decreasing the nickel content of the alloy metal reduces the coke deposition rate (Brown and Albright, 1976). There is, however, no general solution available yet to the problem of coke deposition and chromium carbide coatings may be but a small step towards such approach.

7.1.2.4 Other Hard Carbide Coatings

Although the borides and diborides of the transition metals of the fourth to sixth group of the Periodic Table have a metallic character (Figure 11.1), extreme hardness, high melting points and high chemical stability they are difficult to process in plasma spray operations (Dallaire and Champagne, 1985). Early attempts to develop boride coatings used the deposition of CrB_2 from a Cr-B-Si-Ni alloy by a welding torch (Schwarzkopf and Glaser, 1953). Modern approaches rely on the so-called 'auxiliary metal bath process' (Menstruum process) (Dallaire and Champagne, 1985; Champagne and Dallaire, 1985) that promotes the reaction of elements by dissolving them in a liquid metal. To deposit TiB$_2$, a ferrotitanium alloy is reacted with elemental boron in an iron bath according to:

$$a\{FeTi\} + b\{Ti\} + c\{B\} = d(TiB_2) + e\{Fe\} \qquad (7.1)$$

Also, the synthesis route can start with melting ferrotitanium and ferroboron mixtures in an iron auxilliary bath (Kuijpers and Zaat, 1974) according to

$$e\{Fe\} + f\{FeTi\} : \{B\}/\{Ti\} < 2 \qquad (7.2a)$$

$$a\{FeTi\} + b\{Ti\} + c\{FeB\} \rightarrow d\{Fe\} : \{B\}/\{Ti\} = 2 \qquad (7.2b)$$

$$h\{Fe\} + i\{Fe_2B\} : \{B\}/\{Ti\} > 2 \qquad (7.2c)$$

Obviously, the synthesized product depends on the {B}/{Ti} atomic ratio and on the temperature. In the case of Equation 7.2a the reaction product contains Fe and FeTi compounds, in the case of Equation 7.2b, Fe and Fe$_2$B occur. Only when the {B}/{Ti} ratio is 2 the reaction product consists of titanium diboride in an iron matrix. Thus it is possible to produce such coatings by plasma spraying micropellets of ferrotitanium and ferroboron. These micropellets may act as a 'microbath' in which the TiB$_2$ synthesis takes place. It should be emphasized that this technique mimics the commercial electrolytic production of borides in a bath of fused borax with highly concentrated boron depositing along the cathode at 900 °C. With additional metal oxides present in the bath, metal borides are deposited in well crystallized agglomerates by reaction of the boron with the reduced metals (Equation 7.1) (Baumgart, 1984).

Aluminum-based composite SiC coatings with concentrations between 20 and 75 vol% SiC were developed for producing plasma sprayed coatings on Al and other metallic substrates. The composite powders (44–140 μm grain size) were sprayed using an axial feed plasma torch. Adhesion strength of the coatings to their substrates were found to decrease with increasing SiC content and with decreasing SiC particle sizes. However, increasing SiC content and decreasing particle size improved the erosive wear resistance of the coatings whereas the abrasive wear resistance was found to improve with increasing SiC particle size and SiC content (Ghosh et al., 1998).

7.2
Nitride Coatings

Compared to carbide coatings plasma-sprayed ceramic nitride coatings clearly take second place at present. This is based on (i) the rather low thermal decomposition temperatures of, for example silicon nitride or titanium nitride; (ii) high reactivity with oxygen at elevated temperature; (iii) the high reactivity of nitrides with the metallic binder phase at elevated temperatures and (iv) the preferential use of chemical and/or physical vapor deposition methods including magnetron sputtering and electron beam deposition to deposit very thin films of nitrides. For example, CrN and TiN coatings were deposited on light metal alloys such as Ti6Al4V, AlSi7Mg an AlMgSi0.5 to improve surface abrasion resistance (Fuchs *et al.*, 1998).

7.2.1
Titanium Nitride-based Coatings

TiN coatings are a popular means of increasing the wear performance of titanium-based components of aerospace engines including fan blades, compressor blades and rotating parts. However, TiN coatings rarely exceed a thickness of 6 μm when deposited by CVD or PVD techniques. Attempts to increase this thickness are generally met with buildup of excessive residual stresses that cause debonding. Consequently an interfacial layer is required that alleviates steep gradients of stresses, hardness and modulus across the coating, and that provides mechanical support the hard top coat.

A combined reactive plasma spray (RPS)/PVD technique was recently employed to deposit Ti/Ti$_x$N$_y$-TiN duplex coatings with excellent mechanical and adhesive properties onto Ti6Al4V substrates (Casadei *et al.*, 2006). In a first step a graded Ti$_x$N$_y$ coating was generated, by reaction of a Ti4.5Al3V2Mo2Fe powder in nitrogen plasma (Valente and Galliano, 2000), that consisted of a Ti matrix with dispersed Ti$_x$N$_y$ phases and phases related to the other elemental species contained in the starting powder. Onto this rather porous layer a dense stoichiometric TiN coating was deposited by PVD arc technique that effectively filled the pores of the underlying RPS layer. The average hardness of the RPS layer was measured to be 762 HV (300g) and that of the complete coatings system (RPS + PVD) to be 1097 HV (300g). The absolute hardness value of the PVD TiN top layer was calculated according to the Chicot–Lesage model (see Equation 10.18) to be about 2500 HV. The elastic modulus of the RPS coatings was 81 GPa.

Reactive plasma-sprayed TiN coatings were deposited over a NiAl bond coat on steel surfaces showing high toughness and an appreciable adhesion strength of about 26 MPa (Zou *et al.*, 2007).

7.2.2
Silicon Nitride-based Coatings

Silicon nitride ceramics show excellent *mechanical* (high bending strength, elastic modulus and fracture toughness up to 1400 °C, high abrasion wear and solid particle erosion resistance, low density), *thermal* (low coefficient of thermal expansion, high thermoshock resistance) and *chemical* (stability against most acids and bases, corrosive gases and liquid metals) properties that are exploited in diverse technologically challenging fields (Michalowski, 1994; Riley, 1996). These include, but are not limited to, applications as roller and ball bearings, coil and disc spring material, cutting tools for high speed machining of hard steels and superalloys, heat exchangers and heat pumps, inert gas welding and brazing fixtures and pins, stationary blades and burner nozzles of gas turbines, moulds for pressure casting of light metals, and ladles and tundishes for horizontal casting of steel tubes (Heimann et al., 1997). Particular promising areas of application exist in gasoline and diesel engines including exhaust valves, valve spring retainers, bucket tappets, rocker arm pads, pistons of internal combustion engines and turbocharger rotors (Mörgenthaler and Bühl, 1994). However, major barriers to the incorporation of these parts into standard power trains and engines of passenger cars relate to the cost of processing, and the severe technological problem of mass-producing complex ceramic components with a very high degree of reproducibility and long term reliability. Hence in many cases the use of monolithic ceramic parts poses problems both during manufacturing and in service. For example, experimental turbocharger rotors and pistons of internal combustion engines are still prone to failure in service owing to their inherently brittle nature. If these parts could be made from tough high-temperature resistant metals that provide superior mechanical stability and in turn are being protected against the attack of highly corrosive combustion gases by a mechanically, thermally and chemically compatible silicon nitride coating than a longer service life as well as increased combustion temperature and hence environmentally beneficial engine performance could be expected.

Deposition of pure silicon nitride coatings by conventional thermal spraying has been considered impossible since Si_3N_4 dissociates and in turn sublimates above 1900 °C. It is also subject to oxidation in the presence of an oxygen-containing atmosphere at elevated temperature. Thin amorphous silicon nitride films are frequently deposited by chemical vapor deposition (CVD) and applied as masking layers for semiconductor integrated circuits during profile etching, diffusion barriers in VLSI production lines, for damage protection of optical fibers, as gate dielectrics for specific metal insulator semiconductor (MIS) memory devices and as moisture barrier for OLED displays (Wu et al., 2005). However, attempts to deposit mechanically stable thick silicon nitride coatings by thermal spraying using metallic (Lugscheider and Limbach, 1990; Lugscheider et al., 1990; Limbach, 1992) or silicate glass binders (Kucuk et al., 2000), in situ nitridation in flight (Eckardt et al., 1994) or conversion in the as-deposited state through a reactive spray process (Eckardt et al., 1996) were not met with resounding success as such coatings contain only little silicon nitride but instead substantial amounts of embrittling metal silicides. More successful were endeavors to prepare high-Si_3N_4 coatings starting from ß'-$Si_{6-z}Al_zO_zN_{8-z}$ powders with different degrees of substitution, z (Sodeoka et al., 1992) or clad-type powder consolidation using alloy bond coats (Tomota et al., 1988).

Recently it was recognized that high particle velocities generated by detonation spraying (DS), Top Gun™ technology, high-frequency pulse detonation (HFPD) and APS with axial powder injection were conducive to deposit dense and well-adhering silicon nitride coatings (Thiele et al., 1996b; Berger et al., 1998; Heimann et al., 1998; Heimann et al., 1999; Thiele et al., 2002; see Figure 7.17). In addition, since optimization of heat transfer into the powder particles was found to be one of the most decisive factors special powder preparation procedures are required (Berger et al., 2000). The development of high grip/high friction coatings based on Si_3N_4 by

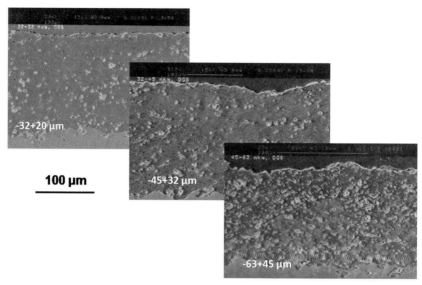

Figure 7.17 SEM micrographs of cross-sections of detonation-sprayed (D-Gun) silicon nitride-based coatings consisting of 68 mass% α-Si_3N_4 + 16 mass% Al_2O_3 + 16 mass% Y_2O_3 with different starting powder grain size distributions. The porosity of the coatings is ≤ 2%, the Vickers indentation microhardness varies between 500 and 600 $HV_{0.05}$ and the mass loss during an ASTM G65-04 wear test was found to be dependent on the powder grain size being 140 mg (−32 + 20 μm), 170 mg (−45 + 32 μm) and 200 mg (−63 + 45 μm), respectively (Berger et al., 1998).

electromagnetically accelerated plasma spraying (EMAPS) will be described in Chapter 8.4 (Usuba and Heimann, 2006; see also Bangert, 2006).

7.3
Oxide Coatings

As frequently pointed out in the preceding paragraphs cemented carbides do not stand up well to chemical degradation at high temperature, in particular a steam environment, due to decarburization, oxidation and matrix-alloying under formation of η-carbides. In these cases the material of choice may be an oxide ceramic coating, most frequently alumina and chromia, and their modifications and composite materials. However, the advantage of considerable increase in chemical and thermal resistance is counterbalanced by the disadvantage of generally low values of the coefficient of thermal expansions, thermal conductivity, mechanical strength and fracture toughness. Also, the adhesion of such oxide coatings to a metallic substrate is compromised by the nonmetallic bonding character of oxides. Indeed, the high proportion of ionic bonds (cp. Figure 11.1) prevents the formation of compatible lattice planes at the interface. In such a case, thin mediating transition layers, for example TiC (cp. Figure 11.2) may be useful. As will be shown later (Section 8.1), thermal barrier coatings of stabilized zirconia require metallic alloy bond coats to alleviate the gradient in the coefficients of thermal expansion between metal substrate and ceramic oxide coating. On exceeding a limiting coating thickness the residual tensile stresses built up in the coating layer will exceed the yield strength of the ceramic material, and delamination and catastrophic cracking will occur (cp. Section 5.7, Figure 5.26b). Therefore, careful control of the residual tensile stresses in ceramic coatings by application of an appropriate bond coat, well-designed temperature schedule including substrate preheating, or reinforcing measures by addition of other oxides (Lugscheider et al., 1995) is mandatory for oxide coatings that are supposed to stand up to the severe in-service conditions in a high-temperature corrosive environment.

7.3.1
Alumina-based Coatings

During plasma spraying, alumina transforms from its α-modification, stable at room temperature, to the γ modification with a defect spinel lattice. Thus plasma-sprayed alumina coatings contain substantial amounts of the essentially metastable γ-phase. The ratio α/γ can be used to characterize the degree of melting of the starting powder with the assumption that high amounts of residual α-phase signal incomplete melting (see Figure 5.27). Also, high amounts of γ-phase appears to lower the wear resistance of alumina coatings even though the microhardness is apparently little affected (Kamachi et al., 1990).

Due to its hardness and good electrical insulation, plasma-sprayed alumina is used as a top coat for insulated metal substrates in automotive applications, as about 50 µm thick coatings for aluminum heat-sinks. On these top coats electronic circuitry is then built up. A basic requirement for this application is maximum dielectric breakdown strength in the range of several hundred volts per 25 µm (Herman et al., 1993). This, however, is compromised by the formation of the γ-phase that tends to absorb water (Brown et al., 1986) that will have a significant effect on the dielectric properties of the coating.

In many cases alumina is used in conjunction with titania: increasing TiO_2 content increases coating fracture toughness but reduces hardness and friction coefficient thus leading to decreased wear resistance. A compromise is to use a ceramic material that contains only a small amount of titania. Alumina/titania (97/3) coatings (called 'grey alumina') are used, for example, to prevent wear, cavitation erosion and chemical corrosion of plungers from reciprocating and centrifugal pumps (Heimann et al., 1990b). Such coatings are known to be very dense, and produce the smoothest surface of any as-sprayed ceramic coatings. This reduces the amount of post-spray grinding and polishing treatment required. The coatings must be thin to conform to available clearances between plunger and barrel of the pump. On the other hand, they should be hard (70–80 HRC) and show high adhesion strength (>50 MPa).

Figure 7.18 shows the cross-section of a thick alumina/titania (97/3) APS coating on a commercial-grade, hot-rolled sheet steel (A 569). The coating appears to be very dense since the apparent porosity is mostly related to grain plug-out during preparation of microscopic samples. Two systems of cracks appear that are due to

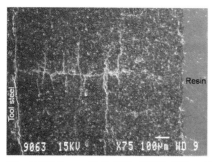

Figure 7.18 Dense Al_2O_3/TiO_2 (97/3) APS coating on A 569 steel (left) (Heimann et al., 1990).

(i) residual tensile stresses owing to the exceptionally thick (0.9 mm) coating (horizontal cracks) and (ii) delamination (vertical cracks). The microhardness of such coatings depends on the thickness and obeys a linear law, $HV_{0.3} = a \times d + b$ [MPa] with a = 12 MPa, d = coating thickness in mm and b = 97.4. Abrasion wear test mass losses (ASTM G65-04, 2004) of the coatings show an inverse proportionality to the thickness ($\Delta m = c/d$ [mg], with c = 45.3 mg mm^{-1} and d = coating thickness in mm) and to the microhardness ($\Delta m = A\, HV_{0.3}^{-B}$ [kg], with $A = 4.3 \times 10^{24}$, $HV_{0.3}$ = microhardness in kg mm^{-2} and B = 9.5. Similarly, dense coatings with SPE-resistance approaching that of bulk alumina were produced by Kingswell et al. (1990) and Chon et al. (1988). Recent research on the tribological behavior of alumina–titania coatings has been reported by Liu et al. (2003), Bounazef et al. (2004), Guessasma et al. (2006), Ctibor et al. (2006) and Yilmaz et al. (2007).

Very dense alumina coatings can be produced by suspension plasma spraying (SPS) of extremely fine powders with grain diameters in the nanometer range. Figure 7.19 shows a high magnification SEM micrograph of the surface of an SPS-alumina coating whose dense structure can be attributed to the homogeneity and similarity of the individual splats that appear as flat disks (see also Section 6.4.2).

To further reduce porosity of alumina-based coatings to below about 5%, laser glazing can be used to thicken the topmost layer of the coating and thus to seal it

Figure 7.19 Alumina coating deposited on a silicon nitride wafer substrate by suspension plasma spraying of nano-sized powder (grain size < 100 nm) (Courtesy: Prof. Francois Gitzhofer, CREPE, U of Sherbrooke, Québec, Canada).

against penetration of corrosive agents (see Section 8.1.3.1). For example, laser treatment and TIG (tungsten inert gas) welding were used to remelt the top surface of plasma-sprayed Al_2O_3-13% TiO_2 coatings (Iwaszko, 2006; see also Dubourg et al., 2007). Remelting resulted in substantial reduction of porosity and surface roughness, coating homogenization, transformation of spray-generated γ- to α-Al_2O_3 and formation of Al_2TiO_5. Inevitably the ceramic layer shows creasing or even severe cracking due to the induced thermal stresses when subjected to a high power-laser beam. These cracks frequently not only propagate through the laser-melted top layer but further extend into the underlying material. A solution to the severe cracking problem may be to introduce into the laser beam ceramic compositions with lower melting points that would develop smaller solidification stresses and fewer differential strains with respect to the unheated alumina material. Such attempts have been made by selecting Al_2O_3/TiO_2 (60/40) that transforms on laser-melting to a mixture of Al_2TiO_5, TiO_2 and α-Al_2O_3 (Hannotiau et al., 1988) and Al_2O_3/ZrO_2 (Petitbon et al., 1989). In both cases the cracking could be considerably reduced.

An instructive example is the optimization of WC/Co-tipped saw teeth for ripsawing wood in the unseasoned state as performed in sawmilling operations. These cermet saw tips were CVD-coated with Al_2O_3-TiO_2 to obtain self-sharpening characteristics of the tool (Kirbach and Stacey, 1989). This requires that the abrasive and erosive/corrosive wear resistance of the coating be synchronized with the wear resistance and microstructure of the substrate so that a cutting edge can be produced that maintains a sharply pointed profile, i.e. that is self-sharpening. As shown in Figure 7.20 this is achieved by coating 60WC40Co tips (E) with alumina/titania on the rake face (rake angle 30°, clearance angle 7°). This materials system maintains a reasonable cutting edge during most of the cutting period for up to 60 km cutting

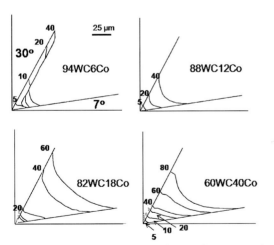

Figure 7.20 Self-sharpening wear behavior of WC/C tipped saw teeth coated with alumina/titania. The numbers refer to kilometer of cutting path. The rake face angle of the sawteeth is 30°, the clearance angle 7°. (Kirbach and Stacey, 1989).

path since the wear-induced retreat of the surface of the coating and that of the substrate roughly coincide. The species cut was unseasoned Western Red Cedar (*Thuja plicata Donn*) of British Columbia, Canada, and the numbers next to the wear profiles indicate km of cutting path. At this point, a self-sharpening behavior up to 60 km is not sufficient for industrial application that require a minimum cutting path of 200 km between mechanical sharpening operations. This approach will become economically competitive to current technology only if a modified substrate material could be designed that ensures such high cutting paths.

Alumina and alumina/zirconia (4:1) coatings have been used for oxidation protection of RSiC kiln aids used in fast firing of China ware. Vacuum plasma-sprayed alumina/zirconia coatings of 400 µm thickness performed best: the RSiC material showed an SiO_2 content of less than 2% after exposure to an O_2/N_2 (1:1) gas mixture for 2500 h (Rump and Möhler, 1995).

7.3.2
Chromium Oxide-based Coatings

Chromia coatings are being applied when corrosion resistance is required in addition to abrasion resistance, *i.e.* in the case of strongly synergistic corrabrasion. Such coatings adhere very well to most metal substrate surfaces and show exceptional hardness of 2300 $HV_{0.05}$ (Gößmann *et al.*, 1990). Applications abound in the chemical industry such as coatings of joints in movable parts, in water pumps, steel rollers for ore classification and smooth top coats for printing rolls (Beczkowiak and Mundinger, 1989). Chromia coatings are also gastight as shown by the occurrence of protective oxide films against sulfidation and carburization on high chromium steel and alloys at high temperatures (Wright *et al.*, 1986). While during APS operations only the ditrigonal α-modification (corundum structure) occurs, at reduced pressure Cr_2O_3 decomposes forming metallic chromium as well as a metastable Cr_3O_4 phase (Gößmann *et al.*,1990).

Chromia coatings perform particularly well in ship and stationary diesel engines where a corrosive environment is created due to the use of less expensive but lower quality diesel fuel (Kvernes and Lugscheider, 1992). Serious corrosion occurs through the impurities of the fuel such as sulfur, vanadium, sodium etc. Erosion is also of concern because catalytic cracking of fuel precursors introduces Al_2O_3-SiO_2 particles into the diesel fuel. Thus chromia coatings are required, for example to protect valve stems of diesel engines from wear and corrosion. Such coatings must be optimized for good adherence (>40 MPa) and maximum fracture toughness. Better coating quality can be obtained by HVOF spraying (Gansert *et al.*, 1990). Here microhardnesses can be routinely achieved exceeding 2000 HV and a porosity of less than 1%. The hardness and the coefficient of friction can be improved by adding other components such as MoO_3 even though the wear performance did not significantly change (Lyo *et al.*, 2003). Friction coatings on braking units of transportation machines have been suggested that consist of very rough plasma-sprayed $60Cr_2O_3$ $40TiO_2$ coatings with a very high friction coefficient of 0.8 and a porosity <3.5% (Seliverstov *et al.*, 1993).

In the automotive industry significant weight savings are obtained by changing the engine crank case material from customary cast iron to aluminum. However, in this case the worse tribological and thermomechanical performance of low alloyed aluminum cylinder liners and piston rings require wear-resistant low friction coatings operating under mixed and boundary lubrication conditions. Cr_2O_3 and $60Cr_2O_3 40TiO_2$ coatings were studied by Woydt *et al.* (2004). APS coatings were found to have higher Vickers indentation hardness, slightly higher porosities and higher elastic moduli compared to HVOF coating. However, coating surface roughness and the type of lubricant applied appeared to influence the coefficients of friction and the wear behavior more strongly than the ceramic coating materials themselves.

In the power industry surface protection is required against corrosive wear of *e.g.* ducts, electrostatic precipitators, flue gas desulfurization installations and chimneys in power and incineration plants. Protective coatings fulfilling these requirements have been researched based on thermally sprayed Cr_2O_3 and Al_2O_3, sealed with aluminum phosphate ($AlPO_4$) (Formanek *et al.*, 2005). These coatings are supposed to be stable up to 1900 °C, resistant to thermal shock and abrasive wear, and show very high thermal emissivities that suggest application for heat consuming and processing installations.

7.3.3
Other Oxide Coatings

Information on design, deposition and properties of other plasma-sprayed oxide coatings can be found in the following chapters for titanium oxide (8.3; 9.4.1.4), zirconium oxide (8.1.1; 9.4.1.2) and $2Al_2O_3 \cdot SiO_2$ (mullite; 8.1.4).

7.4
Metallic Coatings

Metallic coatings are, in general, easy to apply to other base metal surfaces by a variety of thermal spray techniques, usually flame- and HVOF- spraying. However, refractory metals with very high melting points such as W, Mo, Ti, Cr, Nb or Ta and superalloys such as Inconel, Hastelloy and NiCoCrAlY-type alloys require APS- or VPS-plasma spray techniques. However, oxidation-sensitive metals such as Al and NiAl5 suggested as coatings for magnesium alloys such as AZ91 and AE42 developed for lightweight construction materials in the aerospace industry and engine blocks in the automotive industry were also sprayed by APS (Parco *et al.*, 2007).

7.4.1
Refractory Metal Coatings

Problems with Mo and W wear- and corrosion-resistant coatings occur due to the pronounced tendency of the molten metals to absorb gases during the spraying process (Kuijpers and Zaat, 1974). For example, liquid molybdenum exposed to air at about

2730 °C absorbs approximately 13 wt.% oxygen (Kozina and Revyakin, 1970). Oxygen will be entrained from the surrounding air into a turbulent argon plasma jet and will react with the molten Mo particles forming MoO_2 that segregates during solidification along the grain boundaries of the splats. Thus VPS technology has been used to deposit Mo and W coatings (Varacalle et al., 1992; Varacalle et al., 1993; Buchanan and Sickinger, 1994; McKechnie et al., 1993; Cai et al., 1994). Mo coatings show very good adhesion to steel substrates presumably due to the formation of a thin metallurgically bonded Fe-Mo alloy at very high contact temperatures (Houben, 1988). However, with proper spray parameter optimization using statistical experimental design methodology it appears to be feasible to use APS technology (Varacalle et al., 1995) (see Section 11.2.2.2). Water-stabilized atmospheric plasma spraying has been used to study the potential application of W coatings for nuclear fusion application (Matějiček et al., 2005). Because of its extremely high melting point, very low vapor pressure, good thermal conductivity, high temperature strength and high-energy threshold for physical sputtering (Davis et al., 1998) tungsten is among the candidate materials for plasma facing components for ITER and other fusion devices. Exposure of experimental coatings to high temperature plasma in a small tokamak showed no erosion and a high tolerance for thermal shocks up to $0.5\,\text{GW}\,\text{m}^{-2}$.

Nb/Hf and Ta plasma-sprayed coatings are being applied by VPS technology to the interior surface of advanced gun barrels because of their high temperature strength, creep resistance and toughness (Prichard et al., 1990). Addition of 15 vol% TiC increased the ultimate tensile strength (UTS) of 89Nb10Hf1Ti alloy from 469 MPa for the nonreinforced material to 523 MPa at room temperature. However, addition of 45 vol% TiC to 90Ta10W reduced the UTS from 606 MPa for the nonreinforced material to 284 MPa at room temperature. The UTS values at 1093 °C (2000 °F) were 209 MPa (89Nb10Hf1Ti) and 197 MPa (89Nb10Hf1Ti + 15 vol% TiC) and 191 MPa (90Ta10W) and 156 MPa (90Ta10W + 45 vol% TiC), respectively.

Ti and Ta plasma-sprayed coatings require VPS technology since at high temperatures considerable amounts of oxygen and nitrogen can be interstitially absorbed that compromise the properties of the coatings (Steffens et al., 1980; Lugscheider et al., 1987). With VPS, dense (< 1% porosity), well adhering (72 MPa for 92.5Ti5Al2.5Fe) titanium coatings could be produced.

Cr- and NiCr-based coatings are being used to protect waterwall tubes in boilers in paper, power and chemical process industries (for example Sidhu and Prakash, 2006). Plasma-sprayed high chromium-nickel-titanium alloy (TAFALOY 45CT) coatings were developed to combat degradation by chemical corrosion and erosion of unprotected waterwall tubes made from carbon steel in steam generating boilers (Thorpe and Unger, 1986). Recently APS 50Ni50Cr undercoats with Al topcoats were studied as steam oxidation-resistant components for ultra supercritical (USC) power plants operating at 650 °C and 30 MPa in which existing 9Cr1Mo steel (ASME T91) suffers from severe oxidation and creep (Sundararajan et al., 2005).

Thick Cr coatings on copper substrates were developed as PVD sputter targets (Müller et al., 1995; Müller et al., 2000) and their adhesion strengths measured using a tensile adhesion test/ultrasonic c-scan correlation function (Müller et al., 1996) (see Figure 10.18).

7.4.2
Superalloy Coatings

So-called superalloys are nickel-based γ'-phase precipitation hardened alloys of nickel aluminide, NiCoCrAlY alloys and even more complex high-nickel alloys with additions of Cr,Co,Mo,W,Ti,Nb,B,Zr and C (Rairden et al., 1983). They show excellent high temperature-corrosion resistance and are predominantly applied to gas turbine engine blades and vanes. One of the main advantages of such overlay coatings is their inherent compositional flexibility that permits tailoring of the coating composition for both oxidation resistance and coating substrate compatibility (Hebsur and Miner, 1986; McKee and Luthra, 1993). This compositional flexibility allows the addition of active elements such as yttrium that improves the adherence of the oxide scale during in-service thermal cycling (Barrett and Lowell, 1977). Currently, two competing techniques are used for applying NiCoCrAlY overlay coatings: electron-beam physical vapor deposition (EB-PVD) and low-pressure plasma spraying (LLPS, VPS). Increasingly, plasma spraying becomes economically more attractive owing to advantages such as lower cost, tighter compositional control and greater process flexibility. Recently plasma-sprayed NiCrAlY coatings were applied to improve the high-temperature erosion resistance of Fe-based superalloy used in steam boilers, furnaces equipment, heat exchangers and piping in chemical plants as well as reformers and tubing in fertilizer plants (Mishra et al., 2006).

One of the most important properties to be optimized is the low cycle fatigue (LCF) (Gayda et al., 1986) and stress rupture and creep (Hebsur and Miner, 1987) behavior of plasma-sprayed NiCoCrAlY coatings. The coatings can significantly affect the mechanical superalloy component, for example oriented single-crystalline PWA 1460. The higher ductility of the coating compared to the diffusion aluminide coatings (see Figure 1.3) is beneficial to the thermomechanical fatigue life of the coated superalloy component. However, at higher temperature the coatings are significantly weaker than the superalloy substrate and are thus considered to be non-load-bearing. The LCF and creep behavior is essentially connected with the complex microstructure of plasma-sprayed overlay coatings. The fine grained two-phase microstructure consists of a NiAl-phase (β) and a Ni-rich solid solution (γ). This γ-phase contains extremely fine-grained Ni_3Al (γ') precipitates (Merchant and Notis, 1984). A more recent study (Noguchi et al., 1995) dealt with the precipitation and transformation kinetics of the structurally reinforcing γ'-phase. According to this TEM work, in an as-sprayed coating γ'- and β-phase were detected whereby the spherical γ'-phase was surrounded by β-phase. It was therefore concluded that the γ'-phase crystallized primarily from the liquid, and subsequently the β-phase during solidification of the plasma-sprayed splats. During heat treatment at 1273 K for 4 h to homogenize the coating, most of the γ'-phase was transformed into γ-phase by solid state diffusion of Co and Cr from the β- to the γ'-phase. This results in a decrease in microhardness and low cycle fatigue strength. In aged coatings at 873 and 973 °K the γ'-phase was reformed and α-Co precipitates were observed in the β-phase. The reverse diffusion of Co to the β-phase from the γ-phase might contribute to ordering transition from the γ-phase to the γ'-phase. This is accompanied by a hardening process of the aged coating. In conclusion, to retain the hardness, and LCF and creep resistance of the NiCOCrAlY coatings in service, elements should be added that stabilize the γ'-phase such as Ti,V,W,Ta or Mo. This sequence explains the observations by Gayda *et al.* (1986) that NiCoCrAlY coatings show significant softening in LCF runs at 650 °C but show improved fatigue life and cyclic hardening at 1050 °C that may be related to a slower transgranular crack growth rate due to precipitation-hardening by the reformed γ'-phase.

In addition to LCF- and creep-resistant coatings NiCoCrAlY is also used extensively as a bond coat or graded intermediate layer for partially stabilized zirconia (PSZ) thermal barrier top coatings (TBCs) (see Section 8.1), where it has a two-fold function. First, it provides a gradient of the coefficient of thermal expansion (CTE), i.e. it bridges the gap between the CTE of the Inconel or Hastelloy superalloy structure and that of the PSZ. Secondly, it protects the superalloy from hot corrosive gases penetrating the porous PSZ coating. To maintain structural stability and protective performance of LPPS NiCoCrAlY coatings close control of the oxygen content is required. Recently, a quantitative relationship has been established between the oxygen content in the LPPS atmosphere and the as-sprayed bond coats (Mauer *et al.*, 2007). It could be shown that it suffices to measure the stationary oxygen concentration to determine the oxygen content in the coating as there is a strong almost linear correlation between them.

7.5
Diamond Coatings

The overwhelming majority of industrial thin diamond films and coatings are produced by CVD, plasma-assisted CVD (PA-CVD), ion beam deposition and eximer laser ablation techniques (Spear, 1989; Bachmann *et al.*, 1991). Work in this area is directed towards electronic high-temperature devices, high-power switches, blue light-emitting diodes, radiation-hardened electronics and ultra wear-resistant overlays. While qualitatively superior diamond films could be grown with these well-established techniques, their pitifully low deposition rates of the order of only a few $\mu m\, h^{-1}$ precluded any large scale industrial applications. This changed with the invention of a DC plasma jet process (DIA-JET) by Fujitsu Laboratories (Kurihara *et al.*, 1988). The method relies on argon as a plasma gas and a reactive mixture of 5–20 L min^{-1} of hydrogen and 0.01–0.2 L min^{-1} of methane. Deposition rates were as high as 0.25 mm^{-1} and values exceeding 1 mm h^{-1} have been reported (Bachmann *et al.*, 1991). The short distance between nozzle and substrate (<20 mm) requires very effective cooling of the substrate since temperatures exceeding 1300 °C lead to graphite deposition. Here lies one of the major shortcomings of the technique: the high substrate temperature does not allow for the use of steel since instead of diamond carbides will be deposited. This limitation can be somewhat overcome by an intermediate inert molybdenum or tungsten bond coat. However, diamond films on tungsten-coated steel show rather weak adhesion strength (Lugscheider *et al.*,1993b). While silicon, alumina and vitreous silica substrate have been coated in the past

(Heimann, 1991), modern work was concerned with developing ultra wear-resistant coatings for WC/Co (Lugscheider et al., 1993a, 1993b; Lugscheider and Müller, 1992) and silicon nitride cutting tools (Lugscheider and Müller, 1992) and silicon carbide ceramics (Matsumoto et al., 1995).

At relatively high methane concentrations the polycrystalline diamond coatings show cube-shaped crystals whereas low methane concentrations show octahedral crystals. The crystal morphology is also a function of the deposition temperature: {1 1 1} morphology is typical for low substrate temperatures (700 °C), {100} morphology appears at higher temperatures (1100 °C) (Matsumoto, 1988).

The mechanism behind the process is still somewhat elusive. It is thought that the source gas dissociates in the high temperature plasma jet and forms activated atoms, ions and radicals such as H, CH, C_2H^*, CH_3^+, or CH_3^*. Thermodynamic self-cooling occurs in the plasma jet when it exits the nozzle at very high speed. This process is augmented by entrainment of eddies of external cold gas that break up the potential core of the plasma jet, thereby shifting the initially laminar into a turbulent flow regime (Figure 3.29). This will further reduce the mean temperature and, to a certain extent, the velocity of the jet. Thus the plasma enters a nonequilibrium state with an abundance of activated radicals remaining in the jet which transports them quickly to the substrate surface before they can recombine.

Figure 7.21 shows a typical setup of a DC plasma jet-CVD process (Lugscheider et al., 1993a). With such an apparatus optimized fine-grained diamond coatings could be produced with a high deposition rate at a substrate temperature of about 1000 °C,

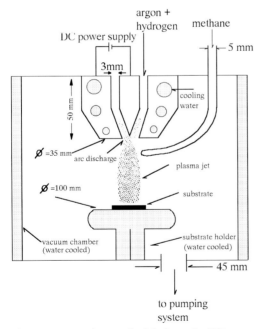

Figure 7.21 Typical setup of a DC plasma jet-CVD reactor to deposit diamond film (Lugscheider et al., 1993a).

chamber pressure of 300 mbar, CH_4/H_2 ratios of less than 5% and stand-off distance of 70 mm. Best results in terms of coating density were reached with surface roughness values of 0.25 µm (Lugscheider et al., 1993b).

Quick and easy production of medium quality diamond films has been reported using a simple oxyacetylene welding torch (Murakawa et al., 1989; Snail, 1991) in which acetylene provides both the carbon source and a high hydrogen flux. Calculations of the sticking probability of various species occurring in a thermal plasma, fed by methane and hydrogen, showed that gas-phase precursors to diamond are acetylene or acetylene-like radicals (Goodwin, 1989). A sticking probability of 10^{-3} for acetylene gives calculated growth rates that compare favorably with experimental growth rates found by Kurihara et al. (1988) and Matsumoto (1988). Acetylene-like radicals provide for much increased diamond growth rates in DC plasma jet deposition but for poorer quality films (Weimer et al., 1991) compared with CH_3 radicals whose gas phase concentration is thought to be the rate-determining step in a typical slow CVD process. On the other hand there is a suggestion that the high growth rates in a d.c. plasma jet are related to high gas phase temperatures. Diamond growth species, whatever their nature might be, may be very effectively produced in 'hot spots' and subsequently rapidly quenched to temperatures below 1000 °C on the substrate.

References

L.F. Albright, C.F. McConnell, K. Welther, in: Proc. ACS 175th National Meeting, Anaheim, CA, March 12–17, 1978a, 175.

L.F. Albright, C.F. McConnell, in: Proc. ACS 175th National Meeting, Anaheim, CA, March 12–17, 1978b, 205.

ASTM F1978, *Standard test method for measuring abrasion resistance of metallic thermal spray coatings by using the Taber Abraser*, 1978.

ASTM G65-04, *Standard test method for measuring abrasion using the dry sand/rubber wheel apparatus*, 2004.

ASTM G77-05,, *Standard test method for ranking resistance of materials to sliding wear using block-on-ring wear test*, 2005.

Fr.-W. Bach, K. Möhwald, B. Drößler, L. Engl, *Mat.-wiss.u.Werkstofftech.*, 2005, **36** (8), 353.

P.K. Bachmann, D. Leers, H. Lydtin, *Diamond Relat. Mat.*, 1991, **1**, 1.

D.S. Bangert, US 2006 0 248 985A1 (2006).

C.A. Barrett, C.E. Lowell, *Oxid. Met.*, 1977, **11**, 199.

W. Baumgart, *Hard Materials*, in: *Process Mineralogy of Ceramic Materials* (eds. W. Baumgart, A.C. Dunham, G.C. Amstutz,), Stuttgart: Enke, 1984, 177.

J. Beczkowiak, K. Mundinger, in: Proc.13th ITSC, London, UK, 1989, 94.

L.-M. Berger, M. Herrmann, M. Nebelung, S. Thiele, R.B. Heimann, T. Schnick, B. Wielage, P. Vuoristo, in: *Thermal Spray: Meeting the Challenges of the 21st Century*, C. Coddet, ed., ASM International, Materials Park, OH, 1998, 1149.

L.-M. Berger, M. Herrmann, M. Nebelung, R.B. Heimann, B. Wielage, 'Modified Composite Silicon Nitride Powders for Thermal Coatings and Process for their Production', *U.S. Patent 6,110,853* (August 23, 2000).

G. Bolelli, R. Giovanardi, L. Lusvarghi, T. Manfredini, *Corr. Sci.*, 2006, **48**, 3375.

M. Bounazef, S. Guessasma, G. Montavon, C. Coddet, *Mater. Lett.*, 2004, **58**, 2451.

S.M. Brown, L.F. Albright, in: *Industrial and Laboratory Pyrolysis* (eds. L.F. Albright,

B.L. Crynes), Chapter 17, Am. Chem. Soc.: Washington, 1976.

L. Brown, H. Herman, R.K. MacCrone, in: *Advances in Thermal Spraying*, Proc. ITSC 1986, Welding Inst. of Canada, Pergamon: Toronto, 1986, 507.

C. Brunet, S. Dallaire, J.G. Sproule, in: *Advances in Thermal Spraying*, Proc. ITSC 1986, Welding Inst. of Canada, Pergamon: Toronto, 1986, 129.

E.R. Buchanan, *Turbomachinery Int.*, 1987, **28** (1), 25.

E.R. Buchanan, A. Sickinger, in: Proc.1994 NTSC, Boston, MA, June 20–24, 1994.

W. Cai, H. Liu, A. Sickinger, E. Muehlberger, D. Bailey, E.J. Lavernia, *J. Thermal Spray Technol.*, 1994, **3** (2), 135.

F. Cacadoi, R. Piloggi, R. Vallo, A. Matthews, *Surf. Coat. Technol.* 2006, **201**, 1200.

B. Champagne, S. Dallaire, *J. Vac. Sci. Technol.*, 1985, **A3** (6), 2373.

P.E. Chandler, A.R. Nicoll, in: Proc.2nd Int. Conf.on Surface Eng., Stratford-on-Avon, UK, June 16–18, 1987, 403.

T. Chon, G.A. Bancke, H. Herman, H. Gruner, in: *Thermal Spray-Advances in Coatings Technology*, ASM International: Cleveland, 1988, 329.

S.G. Cliche, S. Dallaire, in: *Thermal Spray Research and Applications*, Proc.3rd NTSC, Long Beach, CA, May 20–25, 1990, 761.

P. Ctibor, P. Boháč, M. Stranyánek, R. Čtvrtlik, *J. Eur. Ceram. Soc.*, 2006, **26**, 3509.

S. Dallaire, B. Champagne, in: *Modern Development in Powder Metallurgy* (eds. E.N. Aqua, Ch.I. Whitman), 1985, 17, 589.

C. D'Angelo, H. El Joundi, *Adv. Mat. and Processes*, 1988, **12**, 41.

J.W. Davis, V.R. Barabash, A. Makhanov, L. Ploechl, K.T. Slattery, *J. Nucl. Mater.*, 1998, **258–263**, 308.

DIN 50320, Verschleiss, 1979.

L. Dubourg, R.S. Lima, C. Moreau, *Surf. Coat. Technol.*, 2007, **201**, 6278.

T. Eckardt, W. Malléner, D. Stöver, in: *Thermal Spray Industrial Applications*, C.C. Berndt and S. Sampath, eds., ASM International, Materials Park, OH, 1994, 515.

T. Eckardt, W. Malléner, D. Stöver, in: *Thermische Spritzkonferenz: TS96*, E. Lugscheider, ed., DVS Berichte 175, Deutscher Verlag für Schweisstechnik, Düsseldorf, 1996, 309.

S. Economou, M. DeBonte, J.P. Celis, R.W. Smith, E. Lugscheider, *Wear*, 1998, **220**, 34.

S. Economou, M. DeBonte, J.P. Celis, R.W. Smith, E. Lugscheider, *Wear*, 2000, **244**, 165.

EN 657, Euronorm '*Thermal Spraying Technology*', Classification, 1994.

EPRI *Development of Low-Pressure Coatings Resistant to Steam-Born Corrodents*, EPRI Project CS-3139, Vol. 1, June, 1983.

M. Factor, I. Roman, *J. Thermal Spray Technol.*, 2002a, **11** (4), 468.

M. Factor, I. Roman, *J. Thermal Spray Technol.*, 2002b **11** (4), 482.

K.S. Fancondon, C.V. Cooper, L.H. Favrow, A.R. Matarese, T.P. Slavin, in: Proc.3rd NTSC, Thermal Spray Research and Applications, Long Beach, CA, May 20–25, 1990, 611.

B. Formanek, K. Szymánski, B. Szczuka-Lasota, A. Wlodarczyk, *J. Mater. Proc. Technol.*, 2005, **164/165**, 850.

O. Fuchs, C. Friedrich, G. Berg, E. Broszeit, A. Leyland, A. Matthews, *Mat.-wiss. u. Werkstofftech.*, 1998, **29**, 141.

S. Fuji, T. Tajiri, A. Ohmori, in: Proc.14th ITSC, Kobe, May 22–26, 1995, 839.

D. Gansert, E. Lugscheider, U. Müller, in: *Thermal Spray Research and Applications*, Proc. 3rd NTSC, Long Beach, CA, May 20–25, 1990, 517.

J. Gayda, T.P. Gabb, R.V. Miner, *J. Fatigue*, 1986, **8** (4), 217.

K. Ghosh, T. Troczynski, A.C.D. Chaklader, *J. Thermal Spray Technol.*, 1998, **7** (1), 78.

T. Gößmann, H.G. Schütz, D. Stöver, H.P. Buchkremer, D. Jäger, in: *Thermal Spray Research and Applications*, Proc. 3rd NTSC, Long Beach, CA, May 20–25, 1990, 503.

D.G. Goodwin, Mat. Res. Soc. Symp. Proc. EA-19, 1989, 153.

S. Guessasma, M. Bounazef, P. Nardin, T. Sahraoui, *Ceram. Intern.*, 2006, **32**, 13.

J.M. Guilemany, J. Fernández, J. Delgado, A.V. Benedetti, F. Climent, *Surf. Coat. Technol.*, 2002, **153**, 107.

G. Hägg, *Z. Phys. Chem.*, 1929, **6**, 221;1931, **12**, 33.

K. Hajmrle, M. Dorfman, *Modern Dev. Powder Metallurg.*, 1985, **15/15**, 609.

H. Hannotiau, J. Leunen, J. Sleurs, S. Heusdains, H. Tas, in: 1st Int. Conf. Plasma Surface Eng., Garmisch-Partenkirchen, Germany, September 19–23, 1988.

M.G. Hebsur, R.V. Miner, *Mater. Sci. Eng.*, 1986, **83**, 239.

M.G. Hebsur, R.V. Miner, *Thin Solid Films*, 1987, **147**, 143.

R.B. Heimann, D. Lamy, T. Sopkow, *J. Can. Ceram. Soc.*, 1990a, **59** (3), 49.

R.B. Heimann, D. Lamy, T.N. Sopkow, in: *Thermal Spray Research and Applications*, Proc.3rd NTSC, Long Beach, CA, May 20–25, 1990b, 491.

R.B. Heimann, *Proc. Adv. Mater.*, 1991, **1**, 181.

R.B. Heimann, S. Thiele, B. Wielage, M. Zschunke, M. Herrmann, L.-M. Berger, in: *Freiberger Forschungshefte (Maschinenbau)*, 1997, A848, 166 (in German)

R.B. Heimann, S. Thiele, L.-M. Berger, M. Herrmann, M. Nebelung, B. Wielage, T.M. Schnick, P. Vuoristo, in: *Microstructural Science: Analysis of In-Service Failures and Advances in Microstructural Characterization*, **26**, E. Abramovici, D.O. Northwood, M.T. Shehata and J. Wylie, eds., ASM International, Materials Park, OH, 1998, 389.

R.B. Heimann, S. Thiele, L.-M. Berger, M. Herrmann, P. Vuoristo, in: *Proc. Coatings for Aerospace and Automotive Industries* (CAAI-6), Toronto, October 20–22, 1999, 14.

R.B. Heimann, H.D. Lehmann, *Recent Patents on Materials Science*, 2008, **1**, 41.

H. Henke, *Surface Behavior of Sprayed Coatings of Hard Materials During Mineral Sliding Wear*, Unpublished Master thesis, Technische Universität Bergakademie Freiberg, 2001.

H. Henke, D. Adam, A. Köhler, R.B. Heimann, *Wear*, 2004, **256**, 81.

H. Herman, C.C. Berndt, H. Wang, Plasma Sprayed Ceramic Coatings, in: *Ceramic Films and Coatings* (eds. J.B. Wachtman, R.A. Haber,), Noyes: Park Ridge, N.J., 1993, Chapter 5, 131.

H.E. Hintermann, *J. Vac. Sci. Technol.*, 1984, **2**, 816.

J.M. Houben, *Relation of the Adhesion of Plasma Sprayed Coatings to the Process Parameters Size, Velocity and Heat Content of the Spray Particles*, Ph.D.Thesis, Technical University Eindhoven, The Netherlands, 1988.

J. Iwaszko, *Surf. Coat. Technol.*, 2006, **201**, 3443.

W. Jeitschko, H. Nowotny, F. Benesovsky, *Mh. Chemie*, 1964, **94**, 672.

H. Jungklaus, E. Lugscheider, R. Limbach, R.W. Smith, in: Proc.13th ITSC'92, Orlando, FL, 1992, 679.

K. Kamachi, S. Goda, S. Oki, M. Magome, K. Ueno, S. Sodeoka, G. Ueno, T. Yosioka, in: *Thermal Spray Research and Applications*, Proc. 3rd NTSC, Long Beach, CA, May 20–25, 1990, 497.

R. Kingswell, D.S. Rickerby, K.T. Scott, S.J. Bull, in: *Thermal Spray Research and Applications*, Proc.3rd NTSC, Long Beach, CA, May 20–25, 1990, 179.

E.D. Kirbach, M. Stacey, Personal communication, FORINTEK Canada Corp., Western Laboratory, Vancouver, B.C., 1989.

K. Kirner, *Schweißen und Schneiden*, 1989, **41**, 11.

J. Kitamura, S. Usuba, Y. Kakudate, H. Yokoi, K. Yamamoto, A. Tanaka, S. Fujiwara, *J. Thermal Spray Technol.*, 2003, **12** (1), 70.

O. Knotek, E. Lugscheider, H. Reimann, *J. Vacuum Sci. Technol.*, 1975, **12** (4).

H. Koiprasert, S. Dumrongrattana, P. Niranatlumpong, *Wear*, 2004, **257**, 1.

H. Kolaska, K. Dreyer, *DIMA*, 1989, 11.

L.N. Kozina, Yu.P. Revyakin, *Izvest. AN SSSR, Metally*, 1970, 56.

H. Kreye, D. Fandrich, H.H. Müller, G. Reiners, in: Advances in Thermal Spraying, Proc. ITSC 1986, Welding Inst. of Canada, Pergamon: Toronto, 1986, 121.

A. Kucuk, R.S. Lima, C.C. Berndt, *J. Mater. Eng. Perform.*, 2000, **9**, 603.

T.W. Kuijpers, J.H. Zaat, *Metals Technol.*, March 1974, 142.

K. Kurihara, K. Sasaki, M. Kawarada, N. Koshino, *Appl. Phys. Letters*, 1988, **52**, 437; also *Fujitsu Sci. Technol. J.*, 1989, **25**, 44.

I. Kvernes, E. Lugscheider, *pmi*, 1992, **24** (1), 7.

W.J. Lenling, M.F. Smith, J.A. Henfling, in: *Thermal Spray Research and Applications*,

Proc.3rd NTSC, Long Beach, CA, May 20–25, 1990, 451.

M.M. Lima, C. Godoy, P.J. Modenesi, J.C. Avelar-Batista, A. Davison, A. Matthews, *Surf. Coat. Technol.*, 2004, **177/178**, 489.

R. Limbach, *Techn.-Wiss. Berichte Lehr- und Forschungsgebiet Werkstoffwissenschaften der RWTH Aachen*, Nr. 37.03.12.92 E. Lugscheider, ed., 1992.

Y. Liu, T.E. Fischer, A. Dent, *Surf. Coat. Technol.*, 2003, **167**, 68.

E. López Cantera, B.G. Mellor, *Mater. Lett.*, 1998, **37**, 201.

H.L.d.V. Lovelock, *J. Thermal Spray Technol.*, 1998, **7** (3), 357.

E. Lugscheider, M. Kuelkens, in: *Advances in Thermal Spraying*, Proc. ITSC 1986, Welding Inst. of Canada, Pergamon: Toronto, 1986, 137.

E. Lugscheider, P. Lu, B. Hauser, D. Jäger, *J. Surf. Coatings Technol.*, 1987, **32**, 215.

E. Lugscheider, R. Limbach, in: DVS Berichte, 130, Deutscher Verlag für Schweisstechnik, Düsseldorf, 1990, 224.

E. Lugscheider, R. Limbach, A. Liden, J. Lodin, in: Proc. Conf. High Temp. Mater. Powder Eng., Liege, Belgium, 1, F. Bachelet ed., Kluwer, Dordrecht, The Netherlands, 1990, 877.

E. Lugscheider, U. Müller, *Ingenieur-Werkstoffe*, 1992, **4** (7/8), 42.

E. Lugscheider, U. Müller, F. Deuerler, W. Schlump, Proc. TS'93, Aachen, 1993a, DVS 152, 19.

E. Lugscheider, U. Müller, F. Deuerler, W. Schlump, P. Jokiel, P. Remer, in: Proc.13th Intern. Plansee Seminar, Reutte, 1993b, 3, 287.

E. Lugscheider, H. Jungklaus, P. Remer, J. Knuuttila, in: Proc.14th ITSC, Kobe, May 22–26, 1995, 833.

S. Lyckx, C.N. Machio, *Int. J. Refract. Metals & Hard Mater.*, 2007, **25**, 11.

I.W. Lyo, H.S. Ahn, D.S. Lim, *Surf. Coat. Technol.*, 2003, **163/164**, 413.

W. Malléner, D. Stöver, in: Proc. TS'90, Essen 1990. DVS 152, 3.

S. Marx, R.B. Heimann, J.M.F. Orenes, *Tribol. Schmierungstechn.*, 2006, **53** (1), 20.

J. Matějiček, Y. Koza, V. Weinzettl, *Fusion Eng. Design*, 2005, **75–79**, 395.

F. Matsumoto, S. Kato, Y. Tomii, in: Proc.14th ITSC, Kobe, May 20–26, 1995, 821.

S. Matsumoto, in: *Diamond and Diamond-Like Materials Syntheses*, (eds. G.H. Johnson, A.R. Badzian, M.W. Geis), Mat. Res. Soc. Symp. Proc. EA-15, Pittsburgh, PA, 1988, 119.

G. Mauer, R. Vaßen, D. Stöver, *Surf. Coat. Technol.*, 2007, **201**, 4796.

D.W. McKee, K.L. Luthra, *Surf. Coat. Technol.*, 1993, **56**, 109.

T. McKechnie, et al., Proc. 1993 NTSC, Orlando, FL, June 7.11, 1993.

U. Menne, A. Mohr, M. Bammer, C. Verpoort, K. Ebert, R. Baumann, in: Proc. TS'93, Aachen 1993 DVS 152, 280.

C.M. Merchant, M.R. Notis, *Mater. Sci. Eng.*, 1984, **66**, 47.

L. Michalowski, ed.,: *New Ceramic Materials*, Deutscher Verlag für Grundstoffindustrie, Leipzig, Stuttgart, 1994 (in German).

S.B. Mishra, K. Chandra, S. Prakash, *J. Tribol.*, 2006, **128**, 469.

K.D. Mörgenthaler, H. Bühl, in: *Tailoring of Mechanical Properties of Ceramics*, M.J. Hoffmann, G. Petzow, eds., NATO ASI Series E. Applied Sciences, Kluwer, Dordrecht, The Netherlands, 1994, 429.

J. Morimoto, Y. Sasaki, S. Fukuhara, N. Abe, M. Tukamoto, *Vacuum*, 2006, **80**, 1400.

M. Müller, F. Gitzhofer, R.B. Heimann, M.I. Boulos, in: Proc.NTSC'95, Houston, TX, Sept 11–15, 1995, 567.

M. Müller, R.B. Heimann, C. Schurig, K. Schwarz, N. Sommer, in: Proc. Thermal Spraying Conf., TS'96, Essen, Germany, E. Lugscheider (ed.), 1996, DVS 175, 359.

M. Müller, R.B. Heimann, F. Gitzhofer, M.I. Boulos, K. Schwarz, *J. Thermal Spray Technol.*, 2000, **9** (4), 488.

M. Murakawa, S. Takeuchi, Y. Hirose, Mat. Res. Soc. Symp. Proc. EA-19, 1989, 63.

Z.Z. Mutasim, R.W. Smith, L. Sokol, in: *Thermal Spray Research and Applications*, Proc.3rd NTSC, Long Beach, CA, May 20–25, 1990, 165.

S. Nanobashvili, J. Matějiček, F. Žáček, J. Stöckel, P. Chráska, V. Brožek, *J. Nucl. Mater.*, 2002, **307.311**, 1334.

M.C. Nestler, E. Müller, D. Hawley, H.-M. Höhle, D. Sporer, M. Dorfman, *Sulzer Tech. Rev.*, 2007, **2**, 11.

K. Noguchi, M. Nishida, A. Chiba, J. Takeuchi, Y. Harada, in: Proc.14th ITSC, Kobe, May 22–26, 1995, 459.

R.J. Ortolano, *Turbomachinery Int.*, 1983, April, 19.

M. Parco, L. Zhao, J. Zwick, K. Bobzin, E. Lugscheider, *Surf. Coat. Technol.*, 2007, **201**, 6290.

J.M. Perry, A. Neville, T. Hodgkiess, *J. Thermal Spray Technol.*, 2002, **11** (4), 536.

A. Petitbon, D. Guignot, U. Fischer, J.-M. Guillemot, *Mat. Sci. Eng.*, 1989, **A121**, 545.

P.D. Prichard, R.L. McCormick, Z.Z. Mutasim, R.W. Smith, in: Proc. Composite Forming Symp., AIME Winter Meeting, Anaheim, CA, February 1990.

X. Qi, N. Eigen, E. Aust, F. Gärtner, T. Klassen, R. Bormann, *Surf. Coat. Technol.*, 2006, **200**, 5037.

J. Qureshi, A. Levy, B. Wang, *J. Vac. Sci. Technol.*, 1986, **A4** (6), 2638.

R. Rairden, M.R. Jackson, M.F. Henry, in: Proc.10th ITSC, Essen 1983, DVS 80, 205.

S. Rangaswamy, H. Herman, in: *Advances in Thermal Spraying*, Proc. ITSC 1986, Welding Inst. of Canada, Pergamon: Toronto, 1986, 101.

A. Raveh, I. Zukerman, R. Shneck, R. Avni, I. Fried, *Surf. Coat. Technol.*, 2007, **201**, 6136.

F.L. Riley, Application of Silicon Nitride Ceramics, in: *Advanced Ceramic Materials*, Key Engineering Materials, H. Mostaghaci, (ed.), Trans Tech Publ. Ltd., Zurich, Switzerland, 1996, **122/124**, 479.

H. Rump, W. Möhler, *Keram. Z.*, 1995, **47** (4), 284.

H.R. Salimijazi, T.W. Coyle, J. Mostaghimi, L. Leblanc, *J. Thermal Spray Technol.*, 2005, **14** (3), 362.

A. Sandt, *Schweißen und Schneiden*, 1986, **38**, 4.

S. Sasaki, H. Takahashi, JP 2006 177 697A2 (2006).

P. Schwarzkopf, F.W. Glaser, *Z. Metallkunde*, 1953, **44**, 353.

A. Scrivani, S. Ianelli, A. Rossi, R. Groppetti, F. Casadei, G. Rizzi, *Wear*, 2001, **250**, 107.

N.F. Seliverstov, V.A. Ryabin, M.Ya. Berezhneva, A.B. Chudinov, Yu.S. Borisov, in: Proc. TS'93, Aachen 1993, DVS 152, 442.

S. Shimizu, K. Nagai, *Welding Intern.*, 1991, **5** (1).

B.S. Sidhu, S. Prakash, *J. Thermal Spray Technol.*, 2006, **15** (1), 131.

R. Sivakumar, B.L. Mordike, *Surf. Coat. Technol.*, 1989, **37**, 139.

R.W. Smith, M.R. Jackson, J.R. Rairden, J.S. Smith, *J. Met.*, 1981, **33**, 23.

R.W. Smith, D. Gentner, E. Harzenski, T. Robisch, in: *Thermal Spray Technology: New Ideas and Processes.* Proc. NTSC'88, Cincinnati, OH, October 24–27, 1988, 299.

R.W. Smith, E. Harzenski, T. Robisch, in: Proc.12th ITSC, London, UK, June 1989.

R.W. Smith, R. Novak, *pmi*, 1991a, **23** (3), 147.

R.W. Smith, R. Novak, *pmi*, 1991b, **23** (4), 231.

R.W. Smith, Z.Z. Mutasim, *J. Thermal Spray Technol.*, 1992, **1** (1), 57.

R.W. Smith, E. Lugscheider, P. Jokiel, U. Müller, J. Merz, M. Wilbert, Proc.5th NTSC, Anaheim, CA, June 1993.

K.A. Snail, *Inside R&D*, 1991, **10** (23), 1.

S. Sodeoka, K. Ueno, Y. Hagiwara, S. Kose, *J. Thermal Spray Technol.*, 1992, **1**, 153.

K.E. Spear, *J. Am. Ceram. Soc.*, 1989, **72**, 171.

H.-D. Steffens, H.-M. Höhle, E. Ertürk, *Thin Solid Films*, 1980, **73**, 19.

G. Stehr, J. Bast, K. Meltke, *Mat.-wiss. u. Werkstofftech.*, 2000, **31**, 780.

N.P. Suh, *Wear*, 1973, **25**, 111.

W.J. Sumner, J.H. Vogan, R.J. Lindinger, in: Proc. Am. Power Conf., April 22–24, 1985, p. 196.

T. Sundararajan, H. Haruyama, S. Kuroda, F. Abe, *Surf. Eng.*, 2005, **21** (3), 243.

G.L. Swales, in: *Behaviour of High Temperature Alloys in Aggressive Environments* (eds. I. Kirman, J.B. Marriott, M. Merz), The Metals Society: London, 1980, 45.

S. Thiele, R.B. Heimann, L.-M. Berger, M. Nebelung, K. Schwarz, *J. Mat. Sci. Letters*, 1996a, **15** (8), 683.

S. Thiele, R.B. Heimann, M. Herrmann, L.-M. Berger, M. Nebelung, M. Zschunke, B. Wielage, in: *Thermal Spray: Practical*

Solutions for Engineering Problems, C.C. Berndt, ed., ASM International, Materials Park, OH, 1996b, 325.

S. Thiele, R.B. Heimann, L.-M. Berger, M. Herrmann, M. Nebelung, T. Schnick, B. Wielage, P. Vuoristo, *J. Thermal Spray Technol.*, 2002, **11** (2), 218.

M.L. Thorpe, R.H. Unger, in: *Advances in Thermal Spraying*, Proc. ITSC 1986 Welding Inst. of Canada, Pergamon: Toronto, 1986, 3.

T. Tomota, N. Miyamoto, H. Koyama: *Formation of Thermal Spraying Ceramic Layer*, Patent JP 63 169371, A. Int. Cl.^{4}: C23C 4/10, Filing date 29.12.1986, Publication date 13.07.**1988**, Patent Abstracts of Japan, 12, No. 416, 24.11.1988.

D. Tu, S. Chang, C. Chao, C. Lin, *J. Vac. Sci. Technol.*, 1985, **A3** (6).

H. Uetz, J. Föhl, K. Sommer, in: *Abrasion und Erosion*, H. Uetz,(ed.), München: Carl Hanser, 1986.

S. Usuba, R.B. Heimann, *J. Thermal Spray Technol.*, 2006, **15** (3), 356.

T. Valente, F.P. Galliano, *Surf. Coat. Technol.*, 2000, **127**, 96.

D.J. Varacalle, R.A. Neiser, M.F. Smith, in: *Thermal Spray: International Advances in Coatings Technology*, Proc.13th ITSC, Orlando, FL, May 1992.

D.J. Varacalle, L.B. Lundberg, M.G. Jacox, J.R. Hartenstine, W.L. Riggs, H. Herman, G.A. Bancke, *Surf. Coatings Technol.*, 1993, **61**, 79.

D.J. Varacalle, L.B. Lundberg, B.G. Miller, W.L. Riggs, in: Proc.14th ITSC, Kobe, May 22–26, 1995, 377.

G. Veilleux, R.G. Saint-Jacques, S. Dallaire, *Thin Solid Films*, 1987, **154**, 91.

M.E. Vinayo, F. Kassabji, J. Guyonnet, P. Fauchais, *J. Vac. Sci. Technol.*, 1985, **A3** (6), 2483.

P. Vuoristo, K. Niemi, B. Jouve, T. Stenberg, T. Mäntylä, L.-M. Berger, M. Nebelung, W. Hermel, in: Proc.14th ITSC, Kobe, May 22–26, 1995, 699.

W.A. Weimer, F.M. Cerio, C.E. Johnson, *J. Mater. Res.*, 1991, **6**, 2134.

D.W. Wheeler, R.J.K. Wood, *Wear*, 2005, **258**, 526.

B. Wielage, S. Steinhäuser, T. Schnick, D. Nickelmann, *J. Thermal Spray Technol.*, 1999, **8** (4), 553.

B. Wielage, T. Grund, M. Nebelung, S. Thiele, A. Wank, *Mat.-wiss.u.Werkstofftech.*, 2007, **38** (2), 139.

S.T. Wlodek, in: EPRI Workshop on Solid Particle Erosion of Steam Turbines, Chattanooga, TN, 1985.

M. Woydt, N. Kelling, M. Buchmann, *Mat.-wiss.u.Werkstofftech.*, 2004, **35** (10/11), 824.

A. Wozniewski, in: Proc. Int. Conf. Metallurg. Coatings, ICMC, San Diego, April 1–6, 1990.

I.G. Wright, V. Nagarajan, J. Stringer, *Oxid. Met.*, 1986, **25** (3/4), 175.

I.G. Wright, *Mat. Sci. Eng.*, 1987, **88**, 261.

D.S. Wu, W.C. Lo, C. Chiang, H.B. Lin, L.S. Chang, R.H. Horng, C.L. Huang, Y.J. Gao, *Surf. Coat. Technol.*, 2005, **198**, 114.

R. Yilmaz, A.O. Kurt, A. Demir, Z. Tatli, *J. Eur. Ceram. Soc.*, 2007, **27**, 1319.

D.L. Zou, D.R. Yan, J.N. He, X.Z. Li, Y.C. Dong, J.X. Zhang, *Int. J. Iron Steeel Res.*, 2007, **14** (5), 71.

8
Solutions to Industrial Problems II: Functional Coatings

8.1
Thermal (TBC) and Chemical (CBC) Barrier Coatings

8.1.1
Partially Yttria-stabilized Zirconia Coatings (Y-PSZ)

Thermal barrier coatings (TBCs) for components of stationary and aerospace gas turbines allow an increase of the turbine inlet temperature (TIT) by several hundred degrees (100–300 °C) over noncoated parts thus boosting the fuel economy.

TBCs consist of a dense MCrAlY (M = Ni, Co) alloy bond coat adjacent to the metallic substrate (Inconel, Hastelloy) and a porous stabilized zirconia top coat in the thickness range of 300–1000 μm (Cernuschi *et al.*, 2004) as schematically shown in Figure 8.1. During operation oxygen diffuses through the porous zirconia layer and reacts with the bond coat to form a thermally grown oxide (TGO) interface consisting of alumina that is intended to inhibit further oxygen diffusion down to the substrate level. The growth of the TGO layer generally follows a parabolic rate law and is influenced by the surface roughness of the LPPS MCrAlY coating (Seo *et al.*, 2007).

Research in the field of TBCs based on partially stabilized zirconia (PSZ) for aerospace applications is driven by (i) increase of the thermal efficiency of the turbine due to higher gas temperatures; (ii) increase of the compressor efficiency due to a reduced air flow for cooling, and (iii) longer service life of metallic turbine components due to decreased thermal fatigue load (Cosack and Kopperger, 2001; Streibl *et al.*, 2006). To achieve these goals, development and improvement of TBCs are carried on in two main areas.

The first area relates to the aerospace industry and is concerned with coating of austenitic superalloy blades and vanes of gas turbine engines, combustor cans and turbine shrouds. These efforts provide an excellent example of the sophistication now reached by ceramic coatings. Such coatings increase the lifetime of components subjected to a variety of degrading processes. As shown in Figure 8.2, the lifetime of an uncoated superalloy turbine component part is limited by mechanical fatigue at temperatures below 800 °C, by hot corrosion between 800 and 900 °C, and by thermal fatigue between 900 and 1050 °C. Above that temperature, oxidation and creep are

Plasma Spray Coating: Principles and Applications. Robert B. Heimann
Copyright © 2008 WILEY-VCH Verlag GmbH & Co. KGaA, Weinheim
ISBN: 978-3-527-32050-9

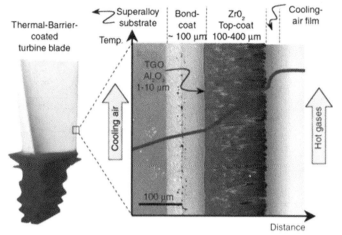

Figure 8.1 Design of thermal barrier coatings (TBCs) applied to the high-Ni superalloy substrate of a gas turbine blade. In this schematic rendering the zirconia top coat shows the typical microstructure of an electron beam-deposited coating (EBC).

lifetime controlling (Spera and Grisaffe, 1973). Thermal barrier coatings fulfill a vital function by increasing turbine blade cooling efficiency, and in conjunction with MCrAlY (M = Ni,Co,Fe) (see Section 7.4.2) bond coats, prevent hot corrosion of the superalloy by molten sodium sulfate and vanadate salts, and corrosive gases. It is thus said that a modern jetliner would not make a single transatlantic flight, but for the coatings (Sivakumar and Mordike, 1989).

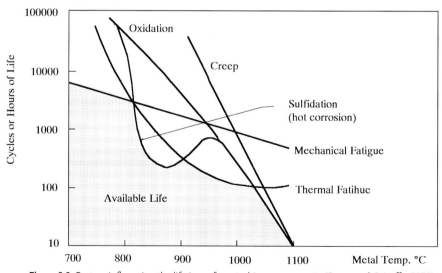

Figure 8.2 Factors influencing the lifetime of gas turbine components (Spera and Grisaffe, 1973).

The second area of application is as a potential material for inclusion in reciprocating internal combustion engines, principally diesel engines. The goals are to insulate components such as pistons, valves, intake and exhaust ports, and to protect moving parts from wear and corrosion. The quest for increasing fuel economy and decreasing levels of hydrocarbons in exhaust gases necessitates an increase in combustion temperatures. A joint venture between Adiabatic Inc. and the US army has been carried out to develop thick TBCs and other ceramic wear coatings for a military adiabatic diesel engine. The thick (0.75 mm) zirconia-based TBC (called 'TTBC') was applied to the recess in the combustion face, to the intake and exhaust ports, and to the piston combustion bowl. The intermediate piston rings were coated with alumina-titania wear ceramic, and the cylinder liner with a proprietary tribological coating. Fuel economy improved from 16 to 37%, and the engine has survived a 400-h durability test with very low lubricating oil consumption.

Currently 90% of zirconia precursor powders produced are used for gas turbine applications, but this market segment is estimated to decrease to 40% by the year 2010, whereas diesel engine applications will increase from 10% today to 40% (Heimann, 1991). Figure 8.3 shows a NASA scenario of development trends in in-service temperatures of gas turbines for aerospace applications. The first step in the development cycle has already been reached by increasing the turbine blade surface temperatures to 1100 °C. The second, somewhat visionary step, requires the development of completely new ceramic structural materials and systems that can withstand surface temperatures of up to 1400 °C required to burn the fuel with considerably higher efficiency with concurrent reduction in polluting gases. Bridging these two steps is the current coating technology that reaches its temperature limit at approximately 1200 °C (BMFT, 1994).

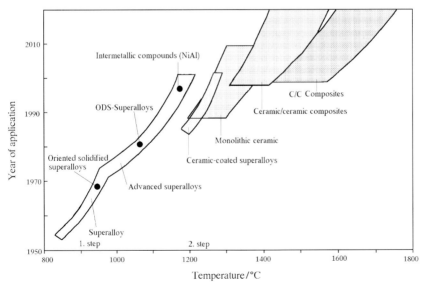

Figure 8.3 NASA scenario of development trends in in-service temperature of aerospace gas turbines (BMFT, 1994).

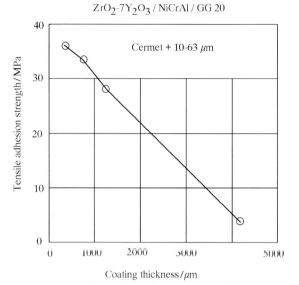

Figure 8.4 Dependence of the adhesion strength of a ZrO_2 (7 mass% Y_2O_3)/NiCrAl TBC on GG20 on the coating thickness (Gramlich and Steffens, 1994).

This limit must be overcome in highly loaded blades and vanes of high-pressure gas turbines and should potentially reach 1500 °C. The resulting increase in efficiency could come from reduced cooling and/or increased inlet temperatures (Peters *et al.*, 1997).

Thick thermal barrier coatings (TTBCs) satisfy requirements of thermal efficiency but pose problems with residual stresses that are maximized in such thick coatings (Figure 8.4). Therefore, optimization of the ceramic material has been performed in two areas: *chemical modifications* and *microstructural modifications* (Miller, 1987).

As a result of *chemical modification*, the lifetime of a TBC has been maximized by varying the yttria content in zirconia. An yttria concentration of 6–8 mass% coincides with the maximum amount of a nonequilibrium 'nontransformable' tetragonal zirconia polymorph (t′-zirconia) (Figure 8.6).

State-of-the-art technology consists of applying zirconia coatings, partially stabilized with 7–8 mass% yttria, over a metallic bond coat of Ni- or CoCrAlY alloy (Figure 8.5). The bond coat must be applied by VPS technology to prevent formation of alumina scale that is thought to reduce adhesion strength to the superalloy substrate. Bond coat adhesion also appears to be controlled by the concentration of yttrium as yttrium decreases the interface stress by minimizing the difference in valence electron densities between substrate and bond coat. The addition of yttrium can also increase the valence electron density ρ_{hkl} of the interface and so increase the interface cohesion force. The most effective yttrium content is at 0.4 mass% or so (Li *et al.*, 2007).

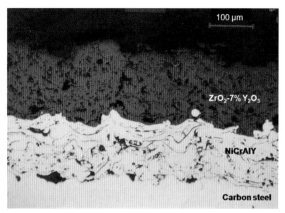

Figure 8.5 Porous Y-stabilized zirconia TBC on top of a dense NiCrAlY bond coat (Heimann et al., 1991).

The PSZ top layer is generally sprayed in the APS mode since VPS experiments have shown that in this case the PSZ layer develops Young's moduli three times larger than those observed in APS coatings (Yajima et al., 1995). Because of this larger thermal stresses are expected to be generated in VPS-PSZ coatings. Figure 8.6 shows the phase relationships of the binary system ZrO_2-Y_2O_3 with several compositions whose grain sizes of the t-phase, bending strengths and fracture toughness are displayed. In general, fracture toughness increases with decreasing amount of yttria, whereas the thermal stability increases with increasing yttria content. This requires trade-offs depending on the application of the material. For 8 mass% (2.65 mole%) yttria the phase assembly at ambient temperature should consist of a mixture of cubic and monoclinic zirconia (Figure 8.6). However, extensive X-ray diffraction work on as-sprayed coatings have shown a tetragonal phase and only a small amount of monoclinic phase (Hodge et al., 1980; Butler, 1985; Scott, 1975; Jasim et al., 1992a). This tetragonal (t′) phase does not transform to the monoclinic (m) phase because of the rapid splat cooling during plasma spraying. It is a tetragonal solid solution with high yttria content that on annealing at high temperatures (1300 °C) transforms slowly by a diffusion-controlled mechanism to equilibrium tetragonal (t) zirconia with low yttria content, and cubic (c) zirconia (Wu et al., 1989).

A second aspect of chemical modification concerns the replacement of the stabilizing yttria by other elements. Work in this area has been relatively scarce. Using ceria instead of yttria, the microhardness and crack resistance could be markedly improved for a PSZ stabilized with 15 mol% CeO_2 and beyond (Iwamoto et al., 1990; Langjahr et al., 2001). To reduce the thermal conductivity of ZrO_2-based TBCs, they were doped with Al_2O_3, NiO, Nd_2O_3, Er_2O_3 and Gd_2O_3 (An et al., 1999; Nicholls et al., 2002) as well as La_2O_3 (Matsumoto et al., 2006).

The *microstructural modifications* relate to either porosity control or development of graded coating systems (for example Leushake et al., 1997). Porosity and microcrack distribution impart to the material a tolerance to thermal and residual stress. Thus the lifetime of a TBC is sensitive to density variations that can be controlled by proper

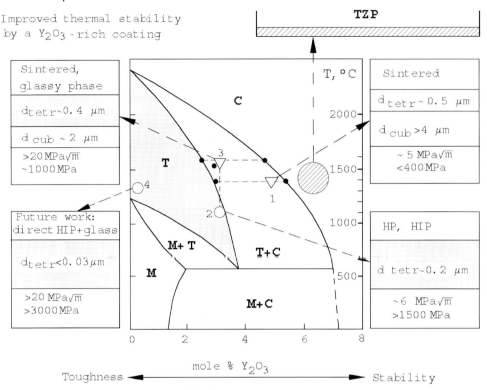

Figure 8.6 Phase diagram of the binary system ZrO_2-Y_2O_3. Different compositions and coating properties for different applications are indicated (TZP: tetragonal zirconia polycrystal).

adjustment of the plasma spray parameters by statistical experimental design methodology. Recently the effect of porosity and splat interfaces on effective properties of zirconia coatings have been studied by small-angle neutron scattering (SANS) and scanning electron microscopy using an object-oriented finite (OOF) element method (Wang et al., 2003) as well as various NDT methods (Rogé et al., 2003).

Not surprisingly, most modern developments in the field of TBCs are carried out by manufacturers of aerospace gas turbine engines, i.e. Rolls-Royce (Restall, 1984; Rhys-Jones, 1989), General Electric (Comassar, 1991; Smith et al., 1981), NASA (Stecura, 1982; Chang et al., 1987; Miller, 1984; Miller, 1988) and Pratt & Whitney (Ruckle, 1980) but also by plasma spray equipment manufacturers (Feuerstein et al., 1986; Longo and Florant, 1976) and superalloy suppliers (Cushnie et al., 1990).

The future performance profile of TBCs will include improved thermal barrier function, oxidation resistance as well as corrosion and erosion resistance. With improved TBCs economic and environmental requirements will be addressed such as higher combustion temperatures resulting in better fuel efficiency, lower cooling air requirements and hence higher compressor effectivity, and lower thermal stresses on materials that result in longer service life and maintenance cycles. There are, however, problems with TBCs that manifest themselves in increased spalling and chipping at higher operational temperatures and times. This is caused by higher residual stresses in the coatings, bond coat oxidation as well as volume changes induced by phase transformations. Here improvements are urgently needed to attain performance levels anticipated by NASA (BMFT, 1984) for airborne gas turbines (10 000 h at turbine blade surface temperatures up to 1400 °C), stationary gas turbines (25 000 h at up to 1000 °C) and large stationary diesel engines (5000 h at cycling loads). Quality control protocols are needed to more accurately predict the lifetime and reliability as well as obtaining deeper insight into failure mechanisms of TBCs by applying various techniques such as acoustic emission (AE) (Andrews and Taylor, 2000) and implementation of real-time and nondestructive tests for on-line characterization of coatings (Nadeau et al., 2006).

Research has recently been performed to test detonation spraying to deposit suitable TBCs based on yttria-stabilized zirconia over a NiCrAlY bond coat on a Ni-base superalloy (M38G) (Ke et al., 2005). The coatings had low thermal conductivities between 1.0 and 1.4 $W\,m^{-1}\,K^{-1}$ (200–1200 °C) and showed excellent resistance against thermal shock. Thermal cycling up to 400 times (heating to 1050 °C, cooling to room temperature by forced air cooling) followed by 200 cycles to 1100 °C and cooling to room temperature by forced water quenching introduced extensive cracking and eventual failure by spalling. Early during cycling a continuous alumina layer was formed at the ceramic/bond coat interface and eventually a Ni/Co- rich TGO layer developed as well as transformation of t-ZrO_2 to m-ZrO_2 occurred that resulted in final, i.e. catastrophic, spallation.

8.1.2
Stress Development and Control

Since coated turbine section parts are subject to steep temperature gradients, high thermal shock resistance of the metallic bond coat/ceramic top coat is generally required. However, residual stresses in the as-sprayed PSZ coating arising from the difference in the coefficient of thermal expansion (CTE) between the Inconel 617 substrate (CTE: 16 ppm at 850 °C) and the 7% yttria-PSZ coating (CTE: 10 ppm at 850 °C) decrease the thermal shock resistance, and thus the life expectancy of the TBC.

Figure 8.7 shows how, in PSZ, strong shrinkage by cooling from the melting temperature at 2680 °C to the temperature of the martensitic tetragonal-monoclinic phase transition at 1170 °C introduces large tensile stresses in the coating (Steffens et al., 1987; Trice et al., 2002). The phase transition in PSZ with approximately 8% of monoclinic phase increases the volume only slightly, and thus adds some compressive component that, however, is too small to offset the strong tensile stresses introduced. Further temperature decrease to room temperature again introduces shrinkage, i.e. tensile stresses in the ceramics. Since the combined tensile

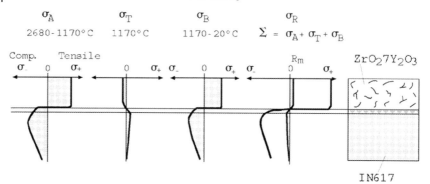

Figure 8.7 Distribution of residual stresses in Y-PSZ coatings (σ_A: residual tensile stresses due to strong shrinkage by cooling from the melting point to the temperature of the t-m transition, σ_T: compressive stress due to volume increase during t-m transition, σ_B: shrinkage during cooling to room temperature) (Steffens et al., 1987).

stresses exceed the yield strength of the material many times over, stress relief occurs through microcracking and/or delamination.

The NiCrAlY bond coat provides a good mechanical bond between the substrate and the ceramic coating, and also oxidation and hot corrosion (sulfidation) resistance for the substrate. Due to the routinely applied grit-blasting procedure before plasma spraying, the interface between the metal substrate and the coating is rough, and can be modeled as a quasi-sinusoidal surface. Finite element analyses (Chang et al., 1987; Liu et al., 2004) have shown that compressive axial and hoop stresses are located in the valleys of the quasi-sinusoidal surfaces but tensile radial stresses at the peaks of the asperities (Figure 8.8 left, a). Thus, a microcrack originating at the point of maximum

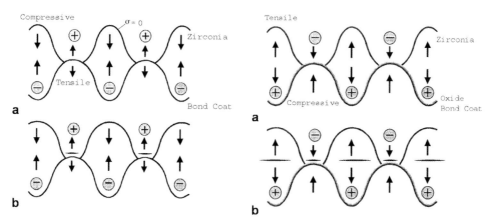

Figure 8.8 Calculated stress states in a PSZ-NiCrAlY bond coat system with sinusoidally modeled interface. Left: a: radial stress due to thermal expansion mismatch, b: development of microcracks. Right: a: reversal of sign of radial stresses due to bond coat oxidation, b: crack extension and delamination. (Miller, 1988).

tensile loading (Figure 8.8 left, b) will be arrested once it enters the realm of compression.

It has been amply confirmed that the key failure mechanism of TBCs is bond coat oxidation (Miller, 1987). If the NiCoCrAlY bond coat is oxidized by penetration through the porous PSZ layer of hot corrosive gases forming an alumina TGO (thermally grown oxide) layer, the signs of the axial and radial stresses are reversed (Figure 8.8 right, a). As a result, crack extension can occur from the region of the peaks, now under compression, into the region of the valley, now under tension. Hence, complete delamination or exfoliation will eventually take place (Figure 8.8 right, b) (Chang et al., 1987).

Tackling the problem of cracking involves two aspects: controlling the residual tensile stresses in the coating, and preventing bond coat oxidation. The first task involves the, at least theoretical, possibility of reversing the stresses in the ceramic layer on an Inconel 617 substrate from tensile to compressive by substrate preheating. Figure 8.9 shows the stress distribution during thermal shock testing of (a) not preheated and (b) preheated substrates (Steffens et al., 1987). The preheating of the IN 617 substrate induces large compressive stresses in the coating that should offset the tensile stresses occurring during cooling. As a result, a slight compressive stress component remains at room temperature thus producing a crack-free coating.

However, coatings cannot be treated as a continuum bulk ceramic but are built up layer by layer in several traverses of the plasma jet. Thus every new layer acts as a heat treatment for the previously deposited one, and heat flow from the substrate decreases. It should be emphasized that this dynamic scheme of *in situ* annealing leads to degrading of the coating, and thus largely foils any attempt to model

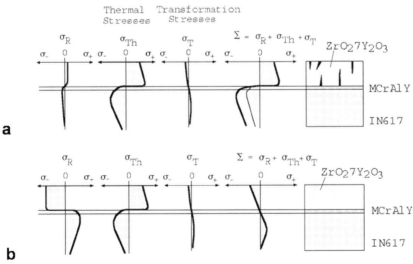

Figure 8.9 Distribution of residual stresses in Y-PSZ coatings on (a) not preheated and (b) preheated substrates (Steffens et al., 1987).

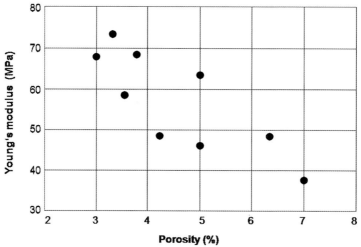

Figure 8.10 Modulus of elasticity as a function of coating porosity in Y-PSZ coatings (Gramlich and Steffens, 1994).

its complete thermal history by finite element analysis. Work in this area (Chang et al., 1987) was based on the assumption of homogeneous, isotropic and linearly elastic behavior of the PSZ layer, the bond coat and the substrate. The modulus of elasticity was assumed to be constant and time-invariant, whereas in reality it is a strong function of the porosity of the coating (Figure 8.10, Gramlich and Steffens, 1994) and shows nonlinearity that is strongly dependent on processing conditions (Liu et al., 2007). The porosity of the coatings will also affect the value of the thermal conductivity: a small change from 1.00 to 0.75 W m^{-1} °C^{-1} can shift the residual stress from overall compressive to overall tensile (Eckold et al., 1987; see also Cernuschi et al., 2004).

A rough estimate of the preheating temperature required to alleviate residual tensile coating stresses can be obtained from Equation 6.26 that determines the critical temperature difference ΔT_c between the lowest temperature of plastic flow and the preheating temperature:

$$\Delta T_c = \sigma (1-\nu)/E\alpha \tag{8.1}$$

with $\sigma = 600$ MPa, $\nu = 0.26$, $E = 200$ GPa, and $\alpha = 10^{-5}$ mm^{-1} °C^{-1}. With these values, $\Delta T_c = 533$ K, i.e. if the critical temperature difference between the lowest temperature of plastic flow (Nabarro–Herring creep) and the preheating temperature is less than 260°, than there is no microcracking. The creep limit is approximately $0.6 T_m = 0.6 \times 3133$ K $= 1880$ K. Thus the minimum preheating temperature becomes $T_p = 1880 - 533 = 1347$ K! For a finite quench rate, the Biot number is $Bi = dh/k$, where $d =$ thickness of the ceramic layer, $h = c_p \times a \times \rho/d = 3.3 \times 10^4$ J s^{-1} m^{-2} K^{-1}, and $k = 1.0$ J s^{-1} m^{-2} K^{-1}. Thus $Bi = 1.16$, $\Delta T_c = 533$ [K] $\times 1.16 = 618$ K; $T_p = 1262$ K! Therefore, the minimum preheating temperature required to alleviate microcracking is close to the range of the melting temperature of Ni-based superalloys. This is to be expected since stress relief in ceramic coatings generally requires very high temperatures owing to the nondislocation type mechanism of lattice relaxation by rearrangement.

It has been observed that during thermal exposure of a TBC at the interface MCrAlY bond coat/stabilized zirconia top coat, not only does a TGO layer develop but also crystalline clusters of $Ni(Cr,Al)_2O_4$ spinel and NiO (Haynes et al., 1996; Lee et al., 2000; Chen et al., 2006a) that tend to provide preferred sites of crack nucleation and hence hasten coating delamination. To counteract this detrimental effect Chen et al. 2006 suggested a heat treatment of the as-sprayed TBC coatings in a low-pressure oxygen environment that will promote formation of a nearly continuous alumina layer and thus suppress formation of deleterious spinel and NiO.

To counteract destruction by cracking and delamination of TBCs for SiC/SiC composite substrates (Zhu et al., 2002) plasma-sprayed multilayer coatings of nanostructured feedstock are being researched that show promising thermomechanical properties (Racek et al., 2006). Layers of yttria-stabilized nanosized zirconia coatings and conventional YSZ were deposited on mullite substrates using an in-house triple torch plasma reactor (TTPR), and on NiCrAlY coated steel substrates using a customary plasma torch (Praxair SG-100). Multilayer coatings consisting of a succession of (i) SiC substrate- (ii) BSAS (barium-strontium aluminum silicate) chemical barrier bond coating- (iii) mullite coating- (iv) YSZ thermal barrier coating, were deposited that employ a strain accommodating BSAS interlayer (Zhu et al., 2003). It could be shown that the strain compliant nanostructured YSZ top coat with reduced Young's modulus also reduced the stress-intensity factor K by up to 26% (Michlik and Berndt, 2006). Likewise the effect of cracking in the coating reduced the driving force for vertical cracks.

Management of crack propagation and hence improvement of the in-service performance and longevity of ZrO_2-7%Y_2O_3 TBCs was attempted by application of suspension plasma spraying (SPS; see Figure 8.22) by injecting a liquid precursor solution into a plasma jet generated by a conventional DC plasma torch (Metco 9MB) (Xie et al., 2006). It was found that vertical cracks, i.e. cracks perpendicular to the coating surface develop preferentially during SPS using acetates or nitrates of zirconium and yttrium as liquid precursors. Since it is widely thought that such vertical cracks impart strain tolerance during thermal cycling their purposeful generation may be advantageous for the performance of TBCs for gas turbine components. For example, so-called DVC (dense vertically cracked) APS-TBCs were formulated by Taylor 1991 that are characterized by cracks oriented normal to the ceramic/metal interface running through more than half of the coating thickness. Xie et al. 2006 associated formation of vertical cracks with the generation of in-plane tensile stresses of the order of 400 MPa originating from the large volume shrinkage occurring during pyrolysis of hitherto unpyrolyzed precursor particles physically incorporated into the growing TBC coating. These precursor particles start to decompose when the coating temperature exceeds their decomposition temperature. Subsequently a point will be reached when the tensile stresses around those decomposing particles will be larger than the tensile yield strength of the coating, and vertical cracks form that on further increase in coating thickness or post-deposition heat treatment extend towards the interface coating/NiCrAlY bond coat. On the other hand, generation of vertical cracks was associated with quenching stress caused by the density difference of molten and solidified particles (Michlik and Berndt, 2006).

8.1.3
Sealing of As-sprayed Surfaces

To prevent bond coat oxidation, sealing the surface of the porous PSZ coating can be performed in several ways.

- spraying of the bond coat under vacuum or low-pressure inert gas conditions prevents *a priori* oxidation, and thus increases drastically the number of cycles-to-failure as compared with air plasma spraying;
- effective sealing can be achieved through infiltration of the PSZ coating with hot corrosion resistant material such as nickel aluminide or CVD-SiO_2;
- reduction of surface porosity by plasma spraying a final top layer of very fine zirconia;
- treatment of plasma-sprayed PSZ with a 0.1 mm slurry layer that is subsequently densified by a heat treatment;
- control of porosity with adjusted powder morphology;
- addition of Si to the MCrAlY bond coat to improve its oxidation/corrosion resistance;
- plasma spraying of low yttria PSZ whose expansion during the martensitic t–m transition offsets the thermal contraction;
- laser surface remelting ('laser glazing');
- reactive laser treatment with alumina and silica to produce dense, low melting eutectic compositions;
- hot isostatic pressing (Abdel-Samad *et al.*, 2006).

8.1.3.1 Laser Surface Remelting of Y-PSZ Coatings

Laser glazing provides a smooth, dense top layer but tends to introduce cracks owing to nonuniform cooling after the point-like introduction of heat through the focused laser beam. Figures 8.11, a–c show scanning electron micrographs of Y-PSZ (Plasmalloy grade AI-1075)/NiCoCrAlY (Plasmalloy grade AI-1065-2) duplex coatings, plasma sprayed with an argon/helium plasma jet at an arc current of 800 A (NiCoCrAlY) and 900 A (Y-PSZ), respectively.

Figure 8.11a shows the as-sprayed TBC with a typically rough and detailed surface, large pores close to the surface as well as a network of radial (transverse) and layer-parallel (longitudinal) microcracks delineating splat boundaries. Laser treatment was performed with an industrial Mitsubishi CO_2-CW laser with a power of 10 kW, a beam diameter of 3 mm, and a wavelength of 10.6 µm (Heimann *et al.*, 1991). Figure 8.11b shows the result of laser glazing using a laser beam scanning speed of 23 cm s^{-1} that translates to a laser irradiance (energy density) of 1.45 J mm^{-2}. The surface is smooth and essentially free of pores down to a depth of 30–40 µm.

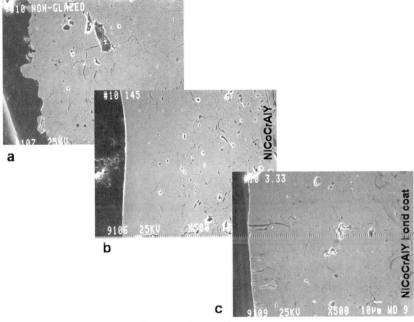

Figure 8.11 SEM micrographs of as-sprayed (a) and laser-remelted Y-PSZ coatings (b: laser irradiance 1.45 J mm^{-2}; c: laser irradiance 3.33 J mm^{-2}) (Heimann et al., 1991).

An increase of the laser energy density to 3.33 J mm^{-2} (Figure 8.11c) results in pronounced cracking. This, however, is not necessarily deleterious to the performance of the TBC since laser-induced *radial (vertical) cracks* will enhance the mechanical stability and the fracture toughness, respectively by a crack-arresting mechanism thus leading to strain-tolerant toughened ceramics. They also constitute sinks for thermal stresses introduced by the differences in the coefficients of thermal expansion. On the other hand, transverse cracks are oriented parallel to the heat flux, and this results in a maximum value of the thermal diffusivity. Effective thermal diffusivity values a_{eff} are

$$a_{eff} = a_0(1 + 8N\,b^3/3)^{-1} \text{ for cracks perpendicular to the heat flux} \quad (8.2a)$$

$$a_{eff} = a_0(1 + 8N\,b^3/9)^{-1} \text{ for randomly oriented cracks} \quad (8.2b)$$

$$a_{eff} = a_0 \text{ for cracks parallel to the heat flux,} \quad (8.2c)$$

where a_0 = bulk thermal diffusivity, N = number of cracks, and b = f(crack radius) (Hasselman, 1978). These equations consider the effects of so-called penny-shaped pores on the thermal conductivity of coatings and have been derived from the

Maxwell model (Maxwell, 1881) extended to rotational ellipsoids (spheroids; see also Section 6.4.4). With increasing layer thickness, both radial and longitudinal cracks will be formed. Longitudinal cracks appear to delineate splat boundaries and are thus undesirable since they tend to increase the risk of delamination and exfoliation of the coatings. On the other hand, they also limit the transverse crack propagation. It was found that despite such cracking the lifetime of laser-densified coatings in corrosive thermally cycled burner rig tests can increase by a factor of three to ten, depending on the coating thickness (Zaplatynsky, 1982). A problem, however, exists since the surface cracks are known to act as nuclei for subsequent cavitational erosion damage (Adamski and McPherson, 1986).

It should be noted that the specific laser energy density required to melt a ceramic material is much lower than that needed to melt a metal. This can be related to (i) the lower reflectivity and (ii) the lower thermal conductivity of ceramics compared with a typical metal. The Cline–Anthony model for a moving Gaussian laser beam to calculate the depth of the melt pool is not applicable (Anthony and Cline, 1977) since the high scanning rate (15 to 30 cm s^{-1}) of the laser beam renders the temperature distribution asymmetric in the direction of the beam traverse.

Figure 8.12a shows the temperature profile that can be approximated by a Gaussian distribution, $T = T_0 \exp(-x^2/2R^2)$, where R = radius of the laser beam. Due to the low thermal conductivity of the material, the molten ceramic pool does not cool instantaneously, and a pronounced asymmetry develops. Figure 8.12a also shows the onset of surface rippling associated with surface tension gradients (Cline, 1981). Also, depressions develop in the ceramic layer under the center of the laser beam (Figure 8.13b). Their origin is related to such surface tension gradients induced by a temperature difference between the center of the beam and the regions further away from it. In the center, the temperature of the melt is at its maximum and thus the surface tension at a minimum. As the temperature decreases away from the center, the surface tension increases. This pulls away the melt from the center thereby

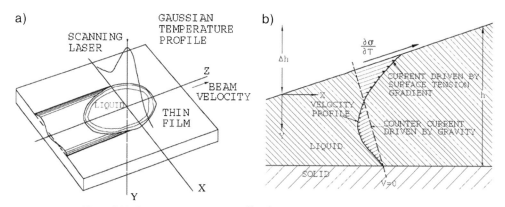

Figure 8.12 Gaussian temperature profile of a scanning laser beam (a) and explanation of the onset of surface 'rippling' due to a strong gradient of surface tension (b) (Cline, 1981).

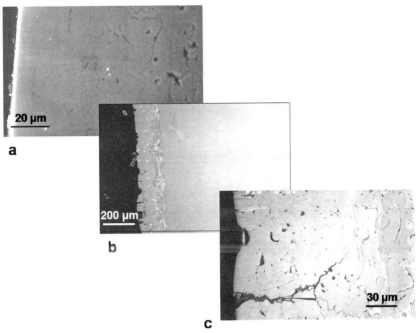

Figure 8.13 Increase of radial cracking with increasing laser irradiance (a: 0.34 J mm^{-2}, b: 1.45 J mm^{-2}, c: 3.33 J mm^{-2}) (Heimann et al., 1991).

depressing the surface under the beam and raise the melt level at the periphery of the melt pool (surface 'rippling', Figure 8.12a). As the height difference increases, a pressure head develops that will eventually induce a gravity-driven counterflow (Figure 8.12b). At steady state, the liquid flow away from the beam center, driven by the surface tension gradient, will be exactly balanced by the gravity-driven counterflow towards the beam center.[1]

Increase of the laser irradiance from 0.34 (Figure 8.13a) to 1.45 (Figure 8.13b) to 3.33 J mm^{-2} (Figure 8.13c) leads to more pronounced cracking (see also Ilyuschenko et al., 2002). Also, as shown in Figure 8.13b, the depths of the laser-induced depressions increase with decreasing scanning speed (top to bottom). It is therefore mandatory to optimize the specific laser energy for sealing of coatings. Experiments have shown (Shieh and Wu, 1991) that at high laser irradiance two solidification fronts exist in the laser-melted coating, one originating at the melt/substrate interface and one at the melt/air interface. The former is related to conductive heat losses, the latter to radiative heat losses. Which front will overtake the other depends on the

---

[1] This mechanism is physically identical to the mechanism responsible for the formation of wine 'tears' where a surface-tension gradient, produced by evaporation of alcohol, exerts a shear stress on the surface of the liquid that drags it up the sides of the wine glass.

laser power. Also, the microstructures of the remelted layers are different: equiaxed grains in case of radiative heat loss, columnar grains in case of conductive heat loss.

Other treatments have been explored with the objective of reducing coating cracking including pulsed laser sealing, reactive laser treatment, and spark plasma sintering. Experiments with *pulsed lasers* (Jasim et al., 1992) revealed that, although microcracking could only be slightly reduced, the resulting coatings showed fewer depressions, smoother surfaces, and improved thermal shock resistance, particularly when laser glazing was preceded by sample preheating. The extremely rapid solidification stabilizes the nontransformable tetragonal zirconia (t′) phase (Jasim et al., 1988). The amount of the cubic equilibrium phase decreases. Nontransformable tetragonal zirconia contains a high amount of yttria in solid solution, and during laser remelting there is not enough time available to redistribute yttria by diffusion to form the equilibrium t-phase with low yttria content. On annealing such a laser remelted coating, the t′-phase transforms indeed to (t + c) phase. The maximum rate of transformation has been found to occur at 1400 °C as represented in a TTT-diagram (Jasim et al., 1992). Comparable results were recently obtained by Batista et al. 2006 using very high energy densities between 6 and 14 J mm^{-2} produced by an industrial 6 kW CO_2-CW laser to seal APS ZrO_2-8%Y_2O_3 coatings. Laser glazing was found to remove local inhomogeneities in the yttria content and resulted in elimination of residual m-ZrO_2 so that only nontransformable t′-ZrO_2 remained in the melted surface layer.

A promising *reactive laser sealing* approach was put forward by Petitbon et al. (1989) by feeding alumina powder directly into a laser beam through a modified nozzle of a 3 kW CO_2 laser. A eutectic alumina-zirconia top layer was formed that produced a smooth, dense and very hard (20 GPa) surface layer with reduced microcracks that shows greatly improved wear- and erosion-resistance and may thus be applied to improve coatings for turbine airfoils and diesel engine valves.

Spark plasma sintering under compressive loading has been explored as a novel way to densify porous TBCs (ZrO_2-25 mass% MgO). The as-sprayed porosity of the coating of 22% was reduced to below 5% while bond strength and microhardness increased (Prawara et al., 2003).

In conclusion, surface engineering of plasma-sprayed TBC systems is a necessary requirement to improve their performance in terms of high-temperature corrosion (Oksa et al., 2004), wear- and thermoshock-resistance for demanding applications including those in the aerospace and automotive industries. Laser surface remelting and reactive laser treatment are modern techniques that need more work and, in particular, the synergistic interaction of various scientific and engineering disciplines. Since laser equipment is predominantly operated by physicists, materials scientists must become more conversant in laser physics, and vice versa, physicists must better understand the specifics of ceramic engineering. Only then can the tremendous challenges be overcome, that are being posed today by ceramic TBCs.

8.1.4
Other Thermal and Chemical Barrier Coatings

The search for other ceramic materials to be used as TBCs was triggered by the large difference in the coefficients of thermal expansion between base metal and ceramic coating that tend to introduce large thermal and residual stresses. Coatings consisting of dicalcium silicate (C_2S) over a NiCrAlY bond coat on stainless steel (SUS-304) revealed a higher coefficient of thermal expansion compared with PSZ (13 ppm vs. 9 ppm), good thermal shock resistance, excellent hot corrosion resistance against V_2O_5 and Na_2SO_4 as well as a small thermal conductivity comparable to or even less than PSZ (Uchikawa et al., 1988).

A second line of investigation deals with the ternary system CaO-SiO_2-ZrO_2, i.e. C_2S-CZ. The high temperature oxidation behavior up to 1100 °C of these coatings is characterized by a lack of exfoliation and severe cracking that has been observed in PSZ TBCs under comparable conditions. Instead, a reaction layer develops, presumably by involvement of Al and/or Cr from the bond coat underneath. In hot corrosion tests involving molten salts, the new coating behaved quite poorly, and showed high reactivity at temperatures around 1000 °C (Yoshiba et al., 1995).

Since zirconia coatings tend to age significantly by undesired densification at temperatures exceeding 1100 °C alternative materials such as lanthanum hexaaluminate were investigated as a novel TBC for gas turbine application (Gadow and Lischka, 2002). Owing to the hexagonal close-packed structure of the magnetoplumbite-type lanthanum hexaaluminate, ion diffusion, and hence densification, is strongly suppressed at potential operating temperature above 1300 °C.

New materials considered for the next generation of plasma-sprayed TBCs with low thermal conductivities include complex ceramics such as perovskite- and pyrochlor-type compounds, for example $BaZrO_3$ and $La_2Zr_2O_7$ (Vaßen et al., 2001) but also Sc-doped zirconia (Stöver et al., 2004; Biedermann, 2002). These materials have been identified as potential successors to YSZ because they are more stable at high temperatures. For example, fluorite-type $La_2Zr_2O_7$ and $Gd_2Zr_2O_7$ exhibit lower thermal conductivities (1.6 W m^{-1} °K^{-1} and 1.14 W m^{-1}K^{-1}, respectively) and better thermal stability > 1200 °C compared with YSZ (2.45 W m^{-1}K^{-1}) and hence were developed for novel TBC systems by several major engine manufacturers (Vaßen et al., 2000; Michael, 2001; Ramesh, 2001; Clarke, 2003; Padture et al., 2000). Zirconia co-doped with rare earth element oxides, including yttrium and gadolinium, was plasma sprayed with liquid precursor solution to yield coating with exceptionally low thermal conductivity of 0.55 − 0.6 W m^{-1} °K^{-1} (RT to 1300 °C) associated with a microstructure characterized by ultrafine splats, high volume porosity, and vertical cracks (Ma et al., 2006).

According to Pitek and Levi 2007 the system ZrO_2-Y_2O_3-Ta_2O_5 may hold great promise for next generation TBCs as it shows improved phase stability, resistance to corrosion by vanadate/sulfate melts, and fracture toughness at least comparable to state-of-the-art ZrO_2-7%Y_2O_3 (7YSZ) material. A tetragonal 16.6%Y_2O_3/16.6% Ta_2O5 stabilized zirconia composition is thermally stable to at least 1500 °C,

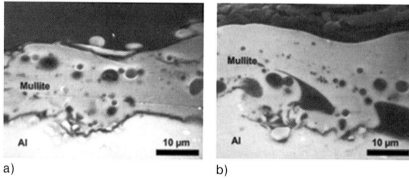

Figure 8.14 Cross-sectional back-scattered electron images of highly porous plasma-sprayed mullite coatings on an aluminum substrate produced by (a) fine and (b) coarse powders. (Seifert et al., 2006a).

and insensitive to the tetragonal/monoclinic phase transition during thermal cycling. It shows remarkable corrosion stability in a Na_2SO_4-30% $NaVO_3$ melt at 900 °C over 500 hours in contrast to severe destabilization of the nontransformable t'-ZrO_2 phase of conventional 7YSZ in which selective depletion of Y by formation of YVO_4 (and $Y_2(SO_4)_3$ dissolved in the melt) occurs.

Plasma-sprayed mullite coatings have been considered viable TBCs and some work has been done to optimize their properties for a variety of high temperature applications (Henne and Weber, 1986; Butt et al., 1990; Lee et al., 1995; Braue et al., 1996; Ramaswamy et al., 1998; Seifert, 2004; Schneider and Komarneni, 2005; Heimann et al., 2007). Plasma-sprayed mullite coatings usually show severe cracking and high porosity (Figure 8.14) and hence will grant easy access of corrosive gases or fluids/melts to the surfaces to be protected (Lee and Miller, 1996; Fritze et al., 1998; Lee et al., 1995; Lee, 1998).

However, in other applications formation of microcracks may even be beneficial as gases trapped in pores and microcracks appear to strongly affect the apparent thermal conductivity of the coatings (Litovsky and Shapiro, 1992; Seifert et al., 2006a). Likewise the occurrence of pores will account for improved relaxation of residual thermal stresses and thus superior thermoshock resistance compared with dense material (Lutz, 1994). Such porous, microcracked mullite coatings may be suitable candidates for advanced TBCs as well as thermal protection systems (TPS) for space re-entry vehicles and could also potentially be applied as thermal control coatings (TCC, Reed and Wright, 1999) of solar reflector type to protect space-bound structures from the effect of heating by solar radiation in a low earth orbit (LEO) (Seifert et al., 2006b).

In a recent study (Heimann et al., 2007) a series of mullite coatings were deposited by APS on aluminum substrate coupons with coarse (median grain size: 47 μm) and fine (median grain size: 17 μm) 2:1-mullite powders (Seifert, 2004). All coatings were deposited with only a single traverse of the plasma torch across the sample surface since it was observed that subsequent traverses would remove some of the previously deposited coating material by solid particle erosion (SPE). This will happen preferentially when rigid solid particles flying on a trajectory outside the hot central core of the plasma jet impact the as-deposited coating surface. Figure 8.14 shows typical cross-sections of APS mullite coatings with substantial porosity of around 30%. In Figure 8.14b two semi-solid splats are shown that have folded over a largely unmelted porous spherical particle to create a train of large voids as well as a collection of spherical pores. The average thicknesses of the mullite coatings deposited during one traverse of the plasmatron were 23 ± 6 µm for coarse powders and 17 ± 6 µm for fine powders. The thickness of the coating produced with coarse powder does not match the median powder grain size of 47 µm. This suggests that the coarser particles suffering from large thermal stresses in the hot plasma jet may disintegrate in-flight before melting as established elsewhere (Reisel and Heimann, 2004). The average coating surface roughness varied between 6.5 and 8.1 µm for coarse powders, and 5.6 and 6.6 µm for fine powders thus mimicking the standard deviation of the coating thickness.

Porous mullite coatings were found to have very low thermal conductivities ranging between 0.19 W m^{-1} °K^{-1} at -196 °C and 0.25 W m^{-1} °K^{-1} at 100 °C (Seifert et al., 2006a). In comparison coatings of 200 µm thickness produced by plasma electrolytic oxidation (PEO) of aluminum yielded values of about 0.5 W m^{-1} °K^{-1} (Curran et al., 2006). These low thermal conductivities suggest that mullite coatings may be suitable TBCs for a variety of applications including abrasion resistant coatings for space re-entry vehicles. Since the thermal conductivity of the aluminum substrate is much higher at around 22 W m^{-1} °K^{-1} rapid quenching of molten splats on impact occurs.

Figure 8.15 shows that the coating close to the Al substrate (a, far left) is essentially glassy and shows exsolution of α-Al$_2$O$_3$ crystals (c). TEM dark field imaging using the (100) reflex at d = 0.475 nm reveals the ordered texture of the precipitate (d). This exsolution shifts the overall composition of the 2:1 mullite coating slightly towards that of 3:2 mullite.

Thin mullite coatings were also developed and tested as chemical barrier coatings for oxidation protection of SiC-based heat exchanger tubes (Butt et al., 1990), SiC/SiC composites (Lee and Miller, 1996), SiC-C/C composites (Rüscher et al., 1997; Fritze et al., 1998), and Si/SiO$_2$ electronic substrates (Meyer et al., 2005; Gutmann et al., 2005). Future applications could include chemical barrier coatings (CBCs) to protect internal surfaces of metal casting moulds for light metals and their alloys, as well as passive thermal control coatings for space structures and thermally protective coatings for re-entry vehicles (Seifert et al., 2006a; Seifert et al., 2006b; Heimann et al., 2007). Because the coefficients of thermal expansion of tantalum oxide and silicon nitride are similar, the former has been suggested as an environmental barrier coating for silicon nitride-based ceramics for high temperature application (Moldovan et al., 2004). Graded alumina/alumina-mullite coatings were deposited by APS on sintered alumina-mullite refractory bricks to protect them against attack by molten glass. The alumina-mullite layer adjacent to the surface of the bricks acted as bond coat to match the low thermal expansion of the porous substrate (Bolelli et al., 2006).

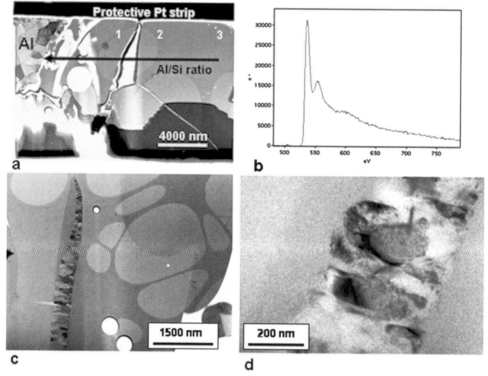

Figure 8.15 Scanning transmission electron micrographs of focused ion beam (FIB) generated cross sections of APS 2:1 mullite coatings (a, c, d) and an EEL spectrum showing the smooth O–K absorption edge at 532 eV (b). (Heimann et al., 2007).

Thermonuclear fusion research relies on high-power, high-frequency millimeter-wave generators such as gyrotrons used as a source to heat plasmas. The power (1–2 MW) and energy (power density, 100 to 200 W cm^{-2}) outputs of a gyrotron can be measured by a bolometer, a hollow copper sphere whose inner surface is coated by APS with a ceramic material. Thermal energy supplied by the thermonuclear plasma is absorbed by the coating and transmitted to circulating cooling water. Recording the temperature difference ΔT between inlet and outlet water permits evaluation of the instantaneous power delivered. The temperature of the ceramic coating measured by a pyrometer yields the thermal conductivity, k, of the coating material. Table 8.1 shows k values of ceramic APS coatings determined for two different power densities (Spinicchia et al., 2005).

Hence plasma-sprayed chromium oxide was found to be a candidate material for absorbing microwave radiation in the 120–160 GHz range.

A new TBC material was proposed by Dai et al. (2006) consisting of neodymium cerium oxide ($Nd_2Ce_2O_7$) plasma sprayed over a 120 μm thick NiCoCrAlY bond coat onto a Ni-base superalloy substrate. The linear thermal expansion coefficient (CTE) of

Table 8.1 Thermal conductivities of ceramic materials used as coatings for the inner wall of a spherical copper bolometer designed for a gyrotron to deliver high-power, high-frequency millimeter waves (Spinicchia et al., 2005).

Ceramic material	Thermal conductivity, k/W m^{-1} °C^{-1}	
Power density	100 W cm^{-2}	200 W cm^{-2}
Cr$_2$O$_3$	2.8	2.3
Al$_2$O$_3$-13%TiO$_2$	1.9	1.6
ZrO$_2$-8%Y$_2$O$_3$ (VPS)	1.3	1.0
ZrO$_2$-8%Y$_2$O$_3$ (APS)	1.0	0.7

Nd$_2$Ce$_2$O$_7$ was measured to be about 13×10^{-6} K^{-1} between 100 and 1250 °C, substantially higher than that of YSZ but very close to that of the substrate. Even more important is the fact that the temperature gradient of the CTE is very similar to that of the NiCoCrAlY bond coat. The thermal conductivity is 1.57 W m^{-1} °K^{-1} at 700 °C, about 30% lower than that of YSZ. Although the materials does not show a phase transition at elevated temperature it appears to show incongruent melting during plasma spraying and associated substantial losses of cerium oxide owing to its rather high vapor pressure. Starting with stoichiometric Nd$_2$Ce$_2$O$_7$ led to a coating with a chemical composition of about Nd$_2$Ce$_{1.25}$O$_{5.5}$. Stoichiometric coatings were produced with a starting powder with surplus cerium oxide such as Nd$_2$Ce$_{3.25}$O$_{9.5}$. Bond coat oxidation was observed as evidenced by the appearance of a TGO suggesting transport of oxygen through a network of interconnected pores in the coating.

8.2
Superconducting and Electrocatalytic Coatings

8.2.1
HT-Superconducting Coatings

Plasma-sprayed HTSC coatings were developed with a variety of materials, among them Y-Ba-Cu-oxide (Welch et al., 1987; Pawlowski et al., 1990) and Bi-Sr-Ca-Cu-oxide (Lugscheider and Weber, 1990). One of the first attempts was undertaken by IBM's Yorktown Heights Laboratory to use thermal spraying to create simple integrated circuits (Welch et al., 1987). The technique involved ionizing a jet of a powdered mixture of the starting oxides in an electric arc close to the substrate surface. Spraying through a mask or template resulted in thin strips of HTSC ceramics that were intended to provide loss-free links between chips in computers cooled with liquid nitrogen. Major problems to overcome included stringent oxygen control to assure compositional integrity, and dimensional control to prevent over-spraying and to guarantee the accuracy of the sprayed pattern.

Oxygen control is indeed one of the major weak points in plasma spraying HTSC coatings. Precursor Y-Ba-Cu-oxide material of a nominal composition of $YBa_{2.06}Cu_{4.13}O_x$ ($6.6 < x < 7$) resulted in oxygen-deficient coatings with a bulk composition of cubic $YBa_{2.04}Cu_{2.98}O_{5.6}$ (Lugscheider and Weber, 1990) pointing to a considerable loss of Cu_2O during plasma spraying. Heat treatment in oxygen at 875 °C for 20 h led to the formation of superconducting orthorhombic $YBa_2Cu_3O_{7-\delta}$ ($\delta = 0.1$) with a very fine-grained, relatively dense microstructure. AC susceptibility data for this sample showed an onset of superconductivity at 90.4 K and a critical current density (shielding current), j_c, at 77 K of 460 A cm^{-2}. This low value is expected since the splat structure of a plasma-sprayed coating provides for many weak links. Moreover, it is surmised that the splats are far from being uniformly conductive throughout their granular structure but that only a rather thin shell of a few micrometer thickness around each 20–50 μm splat is superconducting. Higher critical current densities of $YBa_2Cu_3O_{7-\delta}$ coatings were reported by Karthikeyan *et al.* (1989), Wang *et al.* (1990), Pawlowski *et al.* (1991) and Georgiopoulos *et al.* (2000). In particular, Wang *et al.* (1990) succeeded in producing HTSC coatings with j_c as high as 5000 A cm^{-2} under an applied magnetic field of 1 T at 77 K.

In parallel with the flurry of efforts directed during the late 1980s and 1990s towards development of monolithic HTSC materials with aliovalent replacement of Y and/or Ba to achieve high critical current densities, j_c, and as high as possible critical temperatures, T_c, many potential HTSC formulations were tested as suitable coatings. These included $Bi_2Sr_2CaCu_2O_x$ (Harada *et al.*, 1997), lead-doped Bi-Ca-Sr-Cu oxide coatings (Mangapathi *et al.*, 2006), and $GdBa_2Cu_3O_{7-\delta}$ (Meek *et al.*, 2002). In particular, a recent contribution by Mangapathi *et al.* (2006) described the deposition of $Bi_{1.4}Pb_{0.6}Ca_2Sr_{1.9}Cu_3O_y$ coatings on silver-coated steel substrates with critical current densities as high as 700 A cm^{-2}.

A somewhat easier way to produce HTSC coatings with the required stoichiometry and oxygen deficiency to ensure higher critical current densities may be to use liquid precursors fed into the reaction chamber of an r.f. inductively-coupled plasma system (Shah *et al.*, 1990; Zhu *et al.*, 1989; Zhu *et al.*, 1991; Shah *et al.*, 1989; see also Danroc and Lacombe, 1993).

Despite the initially promising results of these efforts, work on high-temperature superconducting materials including coating has considerably abated mostly owing to the dichotomy of highly exaggerated promises and expectations, and the failure by researchers to deliver tangible results. Indeed, the overly optimistic forecasts that heralded HTSCs as materials that would revolutionize the energy market were far from realistic. Hence industrial interest in HTSC coatings has diminished today.

8.2.2
Electrocatalytic Coatings for Solid Oxide Fuel Cells (SOFC)

Fuel cells convert chemical energy stored in the fuel directly to electrical energy through an electrochemical reaction and are thus considered by many one of the most advantageous energy conversion systems of the future. They combine high energy conversion efficiency (> 60% for SOFCs with bottoming cycle), flexibility in fuel use,

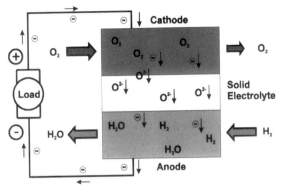

Figure 8.16 Working principle of a solid oxide fuel cell (SOFC).

and cogeneration capability with very low chemical and acoustic pollution of the environment.

An SOFC consists of several layers with specific functions. Figure 8.16 shows the working principle. The anode (fuel electrode) consists of Ni or a Ni-Al$_2$O$_3$ or NiAl-ZrO$_2$ composite material. It is separated from the cathode (air electrode), made for example from Sr-doped lanthanum composite oxide (LaMnO$_3$, LaCoO$_3$, LaSr-MnO$_3$) with perovskite structure, by the Y-stabilized zirconia electrolyte. The arrangement of these three layers ('PEN', Positive electrode-Electrolyte-Negative electrode) can be tubular, monolithic or in a flat plate configuration (Figure 8.17).

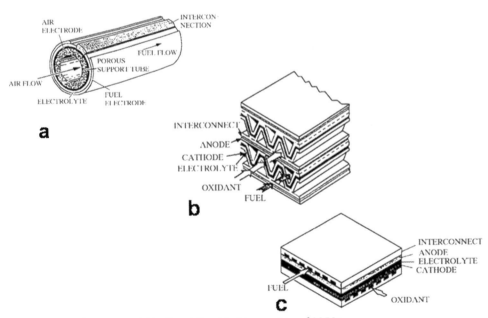

Figure 8.17 Tubular (a), monolithic (b) and flat plate (c) geometries of SOFCs.

The PENs are separated by grooved bipolar plates or interconnects composed of a chromium-iron alloy, Ni alloy-Al_2O_3 (Notomi and Hisatome, 1995) or Sr-doped $LaCrO_3$ (Tai and Lessing, 1991) that transport the fuel gas and the air, respectively as well as heat and current.

At the air electrode (cathode), oxygen combines with electrons, and the negatively charged oxygen ions are transported through the electrolyte via oxygen vacancies in the stabilized zirconia. To obtain reasonable transport rates, the temperature of the electrolyte must be at least 950 °C. At the fuel electrode (anode) a mixture of hydrocarbons and water reacts to hydrogen and carbon monoxide by internal reforming. The oxygen ions carried through the electrolyte oxidize the hydrogen and carbon monoxide to carbon dioxide and water releasing electrons that are transported back via an external circuit to the cathode (Figure 8.16).

From this working principle of an SOFC it follows that the electrodes must have a high proportion of open porosity that will ensure fast and effective gas transport. On the other hand, the electrolyte transports only charged species, i.e. electrons and must thus be gastight. Increasingly work has been concentrated on producing the intricate shapes and compositions required for the PEN by plasma spraying technology. This includes the PSZ electrolyte (Scagliotti et al., 1988; Curtis et al., 1993; Nicoll et al., 1992), the cathodes (Li et al., 1993; Tai and Lessing, 1991), cathodes and anodes (Okiai et al., 1989) as well as the total PEN structure (Kaji et al., 1991; Henne et al., 1994; Malléner et al., 1992). The interconnects of the planar SOFC, i.e. the bipolar plates, up to now hot pressed and rolled to a thickness of 3 mm and machined by ECM (electro-chemical machining), can also be plasma sprayed. Preliminary experiments were conducted with r.f. inductively-coupled plasma-sprayed $94Cr5Fe1Y_2O_3$ powders. The major challenge of producing bipolar plates by plasma spraying economically is concerned with producing a system of orthogonal grooves to carry the natural fuel/water mixture and air, respectively. Orthogonality is required to ensure effective cooling and a uniform temperature distribution that would considerably reduce thermal stresses across the PEN. A second problem is poisoning of the active cathode with nickel or chromium diffusing from the bipolar plate at the working temperature of 950 °C. To prevent this, a protective layer of $(La,Sr)CrO_3$ could be sprayed onto the cathode by VPS.

The function of this protective ceramic layer includes

- electrical insulation of the bipolar plates;
- decrease of the thickness of the soldering gap between the ZrO_2 electrolyte and the bipolar plates;
- prevention of chromium oxide formation;
- barrier function against solid state diffusion of chromium oxide into the active internal cell volume.

These engineering challenges are not trivial and are compounded in monolithic SOFCs owing to the trapeziform gas channels (Figure 8.17). For these reasons the tubular geometry of SOFCs is somewhat easier to realize and is thus closer to mass production (Minazawa et al., 1991). A modern prototype of a tubular SOFC produced entirely by APS/VPS technology was developed in Japan (Notomi and

Hisatome, 1995). Figure 8.18 shows the schematic illustration of the transaxial section of this tubular SOFC. On the surface of the outer support tube multiple cells are arranged: the anode consisting of Ni alloy/PSZ, followed by the Y-PSZ electrolyte, and on top the cathode consisting of a lanthanum composite oxide with perovskite structure, for example $LaCoO_3$. Integrated in this design are the interconnects of Ni alloy/Al_2O_3 coated by an Al_2O_3 protective coating to prevent oxidation of the Ni alloy. The electrodes and the interconnects were sprayed by APS, the very dense electrolyte by VPS. The support tube consisted of fully-stabilized zirconia (CSZ). This design has a very high fuel utilization of 87% and a high power generation efficiency of 38%. A 1 kW prototype module, composed of 48 individual cells of the type shown in Figure 8.18, was successfully operated for 3000 hours without component degradation.

Alternatively the anode layer may be formed by co-spraying an agglomerated NiO/Y-PSZ powder, and subsequent reduction of the NiO by 7 vol% hydrogen/93 vol% argon at 800 °C (Hwang and Yu, 2007). The hydrogen reduction produces nanostructured Ni/Y-PSZ anode coatings that provide increased triple phase boundaries for more efficient fuel gas reduction in an SOFC and presumably also have lower polarization losses.

More recently an approach was described for an integrated manufacturing process for SOFCs using plasma deposition processes in a controlled-atmosphere chamber. The cell performance has been acceptable, with open circuit voltages around 1 V and power densities between 325 and 460 mW cm^{-2} (Chen et al., 2000; see also Stöver et al., 2006). Experiments are under way to manufacture support layers for SOFCs made from dense $La_{0.3}Sr_{0.7}TiO_3$ (60 mass%)-yttria-stabilized zirconia (40 mass%) composite ceramics with an electronic conductivity of 7.0 S cm^{-1} at 700 °C. These properties will make the materials attractive for current collectors in anode supported cells (Ahn et al., 2007).

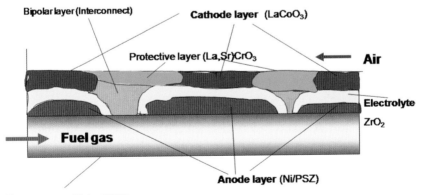

Figure 8.18 Prototype of an entirely plasma-sprayed tubular SOFC (modified after Notomi and Hisatome, 1995).

Figure 8.19 PEN assembly (Anode: Ni, electrolyte: Y-stabilized zirconia, cathode: (La,Sr)MnO$_3$) deposited by r.f. induction plasma spraying on a porous Ni substrate (Courtesy: Dr. Matthias Müller, Bosch AG, Stuttgart, Germany).

Typical perovskites sprayed by VPS are (La,Sr)(Ni,Co)O$_3$ and, using an optimized deLaval nozzle of low Mach number to generate a long laminar plasma jet, also the very attractive but highly parameter-sensitive Co$_3$O$_4$ (Henne *et al.*, 1989; Schiller *et al.*, 1995; Müller, 2001).

Figure 8.19 shows an attempt to manufacture all components of an SOFC by r.f. induction plasma spraying (Müller, 2001).

Alternate coating methods were found in thermal plasma chemical vapor deposition (TPCVD) and suspension plasma spraying (SPS) to deposit perovskite cathode layers (Figs. 8.20; Müller, 2001; Müller *et al.*, 2002)

Figure 8.20 shows that the yttria dopant is homogeneously distributed in the zirconia matrix deposited by TPCVD as well as SPS. This is in contrast to VPS coatings produced from identical precursor powders (Müller *et al.*, 2002). Additional advantages of TPCVD and SPS are deposition rates in excess of 25 µm min^{-1} and 100 µm min^{-1}, respectively, orders of magnitude higher than most of the comparable values found in the literature for TPCVD/SPS of oxide layers (for example Pfender, 1991; Wang *et al.*, 2000). Work was performed to improve the properties of zirconia electrolyte layers deposited by thermal plasma chemical vapor deposition (TPCVD) by adding up to 12 mol% Sc$_2$O$_3$ (Figure 8.21; Biedermann, 2002; Vaßen *et al.*, 2007). The aim of this research was to obtain electrolyte layers with increased electric conductivity by maintaining its low porosity and hence gas tightness. Indeed the specific conductivity of Sc-doped zirconia was found to be around 5 S m^{-1}, more than twice the value found for Y- and Ca-stabilized zirconia (Haering, 2001). Even though attempts to develop better electrolyte layers failed due to the high degree of open porosity as well as strong degradation of the electric conductivity at an operating temperature of 900 °C, such fully Sc-stabilized zirconia material in conjunction with a dense NiCrAlY bond coat could be a suitable solution for thermal barrier application even though the high cost of scandium oxide would presumably counteract its widespread use.

Since these novel TPCVD and SPS process technologies may command a higher degree of attention in the future their working principles will be briefly described

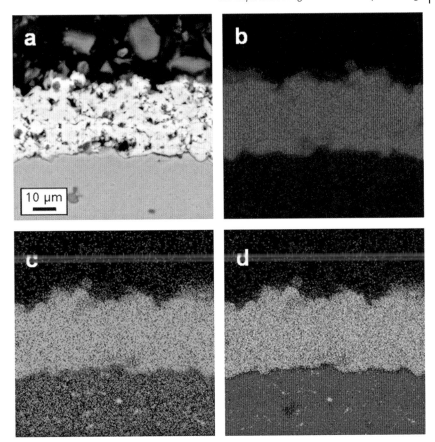

Figure 8.20 SEM cross-sectional micrographs of a TPCVD Y-stabilized zirconia coating (7 mole% Y_2O_3) on a 94Cr5Fe1Y_2O_3 substrate (a) and EDX maps of Zr (b), Y (c) and Cr (d). The uniform yellow color of the coating in panel d originates from the superposition of b and c, and hence attests to the homogenous distribution of Y (Courtesy Dr. Matthias Müller, Bosch AG, Stuttgart, Germany; Müller, 2001).

here. Figure 8.22 summarizes the complex physical processes and the changes to particle geometry during suspension plasma spraying (SPS).

The evaporation behavior and the morphology of aerosol droplets will be determined by the time constants of vapor diffusion, droplet shrinkage, diffusion of dissolved solids, heat conduction in the surrounding gas, and heat conduction within the droplets. Beyond the threshold of the critical supersaturation surface precipitation sets in, i.e. particles will accumulate at the surface of the ever shrinking droplet before they can reach the center thus forming hollow spheres (Figure 8.23b).

These hollow spheres predominantly form in the upper region of the plasma jet. Formation of dense primary spheres is only possible if the solid content of the suspension is very low (Figure 8.22, left). During their flight within the plasma jet the hollow spheres will be further heated and eventually the vapor pressure of the

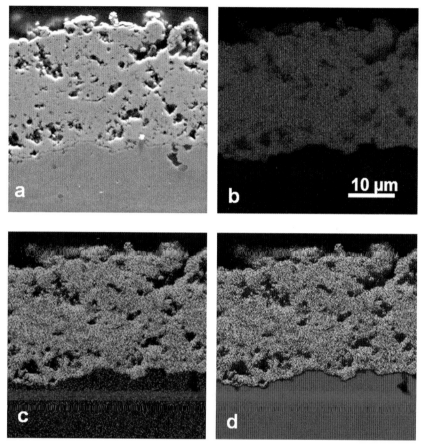

Figure 8.21 SEM cross-sectional micrographs of a TPCVD Sc-stabilized zirconia coating (12 mole% Sc_2O_3) on a 94Cr5Fe1Y_2O_3 substrate (a) and EDX maps of Zr (b), Sc (c) and Cr (d). The uniform yellow color of the coating in panel d originates from the superposition of b and c, and hence attests to the homogenous distribution of Sc (Biedermann, 2002).

residual liquid enclosed in their centers will increase. In case of high wall permeability the vapor can safely escape and closed hollow spheres form as final product. If the wall permeability is low the hollow spheres will crack or may even explode (Figure 8.22, lower right). Figure 8.23 shows typical examples of closed and open hollow cobalt oxide spheres formed by spheroidization of cobalt spinel/ water + methanol suspensions in an r.f. inductively coupled plasma (Müller, 2001).

The physico-chemical processes occurring during TPCVD are somewhat different. Here the distance of the locus of particle formation from the substrate surface plays a decisive role. If the precursor solution completely vaporizes in the plasma jet, on exceeding a critical supersaturation particles will be formed in the bulk volume of the plasma rather far away from the cool substrate surface, a process that subsequently builds up a coating with globular microstructure. However, if the

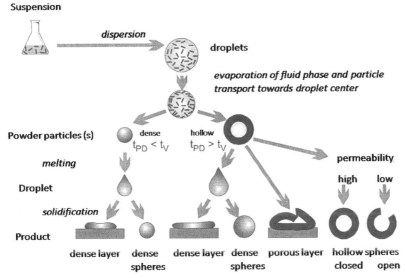

Figure 8.22 Summary of physical processes occurring during suspension plasma spraying (SPS) (after Müller, 2001).

critical saturation occurs only close to the substrate surface, a true CVD process will take place that eventually results in formation of coatings with columnar structure. Figure 8.24 shows that globular structures form when the impinging particles have already been formed in the volume of the plasma rather far away from substrate.

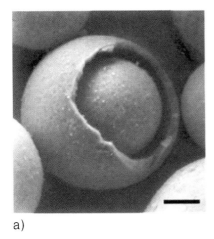

a)

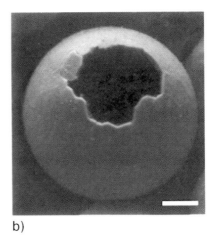

b)

Figure 8.23 Hollow spheres of cobalt oxide formed during SPS spheroidization of a suspension of 57.5 mass% Co_3O_4 and 42.5 mass% water + methanol + Na alginate. Powder feed rate 20 g min^{-1}, chamber pressure 53 kPa, plate power 50 kW; sheath gas O_2 (120 slpm), central gas Ar (30 slpm), suspension feeder gas Ar (5 slpm). The bars correspond to 10 μm. (Müller, 2001).

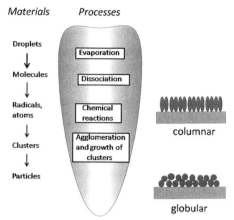

Figure 8.24 Physico-chemical processes during thermal plasma chemical vapor deposition (TPCVD) (after Müller, 2001).

However, if particle formation occurs at the surface directly, columnar coatings will be generated. Examples of columnar coating structures are shown in Figure 8.25. Such a microstructure provides the open porosity required for gas migration in SOFC electrodes (Müller et al., 2002).

Using magnetite ore concentrates, magnetite-Co_3O_4 mixtures, or cobalt oxide, electro-catalytically active coatings for electrodes used in installations for water and sewage disinfection were produced by plasma spraying (Borisov et al., 1995). Even though the current efficiency is lower by 25–30% than for oxide-ruthenium-titanium anodes (ORTA), newly developed oxide-cobalt-titanium anodes (OCTA) have a higher chemical stability, an 8–12% higher productivity as well as other economic advantages.

Prototype HT-SOFCs produced by Siemens in Germany operate between 800 and 1000 °C, and use as cathode gases oxygen or air, as anode gases hydrogen, $H_2/CO/CO_2$ or $H_2/CO/CH_4$ gas mixtures. The planar fuel cells are gathered together in

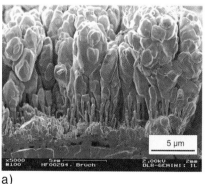
a)

b)

Figure 8.25 SEM micrographs of the fracture surfaces of (a) TPCVD $(La_{0.8}Sr_{0.2})(Co,Fe)O_3$ perovskite columnar coating deposited onto a stationary 94Cr5Fe 1Y_2O_3 substrate and (b) $La_{0.9}Sr_{0.1}MnO_3$ (LSM) coating deposited by scanning the plasma jet across the substrate surface (Müller et al., 2002).

stacks reaching an electrical power output of $1\,MW/m^3$ installed stack. Present problems to be addressed include:

- requirement of varying coating thicknesses on one bipolar plate;
- control of tolerances;
- solderability of the layers to obtain a stack;
- corrosion protection to $1100\,°C$;
- chemical resistance again the soldering material (CaO-SiO_2-B_2O_3 based);
- requirement for high thermoshock resistance.

The market potential for SOFCs appears to be huge: 300–800 $MW\,a^{-1}$ in Europe, 1800 $MW\,a^{-1}$ in the USA, and 1100 $MW\,a^{-1}$ in Asia including Japan. Recently cooperation has been established between Sulzer Metco Inc. and Plansee (Austria) to fabricate SOFC parts using a novel Triplex Pro™-200 plasmatron. Interconnects are being coated with lanthanum-strontium manganate (LSM) as a chemical barrier layer to counteract evaporation of chromium oxide. The performance requirements for such coatings are high as they have to withstand the impact of temperature in excess of $800\,°C$ for > 40 000 hours (Nestler et al., 2007).

Plasma-sprayed oxide electrocatalytic materials with spinel or perovskite structure are also being used as anode materials for water electrolyzers (see also Section 8.3). Here an enormously important energy technology is being addressed as the production of hydrogen gas by electrolysis of water may be one of the mainstays of primary hydrogen energy technology of the future. Electrocatalytic materials must conform to a complex set of stringent requirements (Henne et al., 1989; Müller, 2001) including

- chemical and electrochemical stability under operating conditions in alkaline water electrolysis;
- resistance to atomic oxygen;
- resistance to concentrated electrolyte solutions;
- resistance to working temperatures up to $150\,°C$;
- availability;
- processability as layer or coating on metal electrodes.

8.3
Photocatalytic Coatings

Titanium dioxide (TiO_2) in its anatase modification has demonstrated excellent photocatalytic performance when applied to water and gas purification (Fujishima and Honda, 1972), coupled with chemical inertness, low cost, nontoxicity and the ability to decompose most organic substances contaminating industrial and domestic waste water. While for two decades titania powder suspended in aqueous solution has been used for these purposes, this technique has also revealed several limiting factors, related, among others, to problems with filtration and recycling or regeneration of titania powder after use. However, an alternative way is being found in immobilization of photoactive titania powder on different types of technological support, e.g. metal sheet, ceramics, glass, polymers etc. A variety of deposition techniques have been applied to achieve well-adhering, porous coatings with high photocatalytic activity. Thermal spraying appears to be a particular appropriate process to produce coatings demonstrating such characteristics with additional promising prospects for industrial scale-up. Since the photocatalytic reactions occur only in the very thin top layer of the titania coating much attention must be paid to the morphological and microstructural quality of the coating surface.

Work has been performed by Burlacov *et al.* 2006 to clarify the influence of important induction plasma spray parameters, i.e. plasma power, argon carrier gas flow rate and powder feed rate, on the quality of titania coatings using a statistical experimental design tool (Mawdsley *et al.*, 2001; Lee *et al.*, 2003). To assess the photocatalytical performance based on the relative amounts of anatase and rutile, surface-sensitive analytical methods for characterization of the sprayed coatings are required with high spatial resolution such as Raman microprobe technique using a laser light source. This surface characterization method is very close to the photocatalytic process itself in which the UV light interacts with the coating surface and initiates photochemical reactions. Raman spectroscopic methods are used to monitor the distribution of anatase and rutile phases in coatings with qualitative estimation of their crystallinity.

Induction plasma-sprayed titania coatings were deposited from agglomerated anatase powder. The coatings were mainly composed of crystalline anatase and rutile with traces of oxygen-deficient phases. The concentration of the latter can be controlled by injection of surplus oxygen into the plasma jet as well as adjusting the powder carrier gas flow rate.

Figure 8.26 shows typical Raman spectra taken from the precursor feedstock titania powder (Figure 8.26a) and from plasma sprayed coatings (Figure 8.26b–d). The Raman active modes for anatase, $A_{1g} + 2B_{1g} + 3E_g$ at 147, 197, 396, 515 and 638 cm^{-1}, and rutile, $A_{1g} + B_{1g} + B_{2g} + E_g$ at 144, 238, 447 and 611 cm^{-1} were used as fingerprints. All Raman spectra presented are accompanied by a microscopic view of the measured characteristic objects which can be described as unmelted particles with different composition of anatase and rutile phases (Figure 8.26c, b), and molten (or partly molten) quasi-amorphous splats (Figure 8.26d) emitting very low Raman signals and thus demonstrating their low crystallinity. A comparison of the Raman spectrum of nano-crystalline anatase (crystallite size < 10 nm) of the agglomerated powder with that of anatase particle in the coatings provides a visible difference between the spectra. Red-shifting and sharpening of the Raman bands of anatase is an indication of increasing crystallite size as the results of thermal treatment in the plasma jet, the so called finite-size effect. Another effect produced by thermal annealing of anatase powder is an irreversible transformation of the anatase to the rutile structure above 600 °C. This conversion occurs without a noticeable change of shape of the powder particles (see Figure 8.26c). The essential advantage of Raman microscopy is the possibility of selectively studying microscopic objects. In addition, the automated scan mode of the Raman spectrometer provides high level statistical information on the spatial distribution of titania phases as well as the morphology of

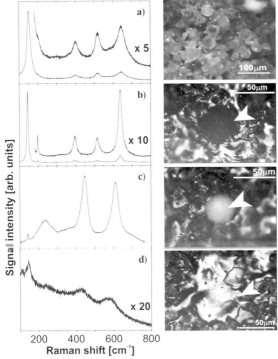

Figure 8.26 Typical micro-Raman spectra from different microscopic objects in induction plasma-sprayed titania coatings. (a) Anatase precursor powder, (b) anatase-rich coating particle, (c) rutile-rich coating particle, (d) molten splat (Burlacov et al., 2006).

the coating surface. Monitoring the signal intensity of two characteristic Raman modes of anatase ($147\,cm^{-1}$) and rutile ($447\,cm^{-1}$) phases has been performed in 560 spectra from a comparatively large coating surface of $50\,\mu m^2$. The information obtained includes spatial distribution of anatase and rutile phases, relative intensity of Raman signals, spectroscopic characteristics of Raman active modes (such as position, width and integrated intensity of Raman bands), estimation of crystallinity of coatings, etc.

Figure 8.27 shows false-color renderings of the spatial distribution of Raman signal intensities selectively measured at two characteristic Raman modes of anatase at $147\,cm^{-1}$ (left panel) and rutile at $447\,cm^{-1}$ (center panel). Surface regions essentially devoid of either mode were assigned to Raman-inactive material (dark blue in Figure 8.27, left and center panels). Their microstructure correspond to the short range ordered (SRO) nature of the molten or semimolten splats (black in Figure 8.27, right panel). The plasma power increases from 20 kW (top row) to 24 kW (center row) to 28 kW (bottom row) resulting in a monotonous decrease of the anatase content with increasing plasma power as measured by the relative intensity of the Raman band at $144\,cm^{-1}$.

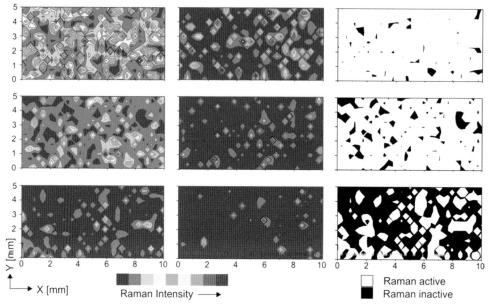

Figure 8.27 Intensity maps of the spatial distribution of characteristic Raman signals of anatase (147 cm^{-1}, left panel), rutile (447 cm^{-1}, center panel) and a Raman-inactive SRO titania phase (right panel) (Burlacov et al., 2006).

Photocatalytic degradation of 4-chlorophenol in aqueous solution showed encouraging performance of some coating compositions. In spite of the rather low anatase content of the coatings, the presence of titania multiphase heterostructures acting as coupled semiconductor systems appears to be a reasonable explanation for this effect. It was found that a high flow rate of the argon carrier gas has a profound positive influence both on the amount of anatase in the coating and its crystallinity. However, the major drawback of using a high-temperature plasma for coating deposition is the transformation of the photocatalytically active anatase modification of titania to the much less active rutile phase. Hence efforts are under way to use alternative low temperature deposition techniques such as cold gas dynamic spraying (CGDS) (Li and Li, 2003; Burlacov et al., 2007; Ballhorn et al., 2007).

The photocatalytic efficiency of agglomerated nanosized (average: 7 nm) anatase (TiO$_2$) deposited by conventional atmospheric plasma spraying (APS) and suspension plasma spraying (SPS) was compared in a study by Toma et al. (2006). The fact that SPS coatings provided a noticeably better efficiency in removal of NO and NO$_x$ than APS coatings was related to a high degree of conversion of the anatase phase to the thermodynamically stable rutile phase in the APS plasma jet. In contrast to that a part of the plasma energy was consumed to evaporate the liquid of the droplets in SPS, so that less energy was available for the thermally controlled phase transformation.

Plasma-sprayed TiO_2 coatings doped with Fe_3O_4 develop microstructures containing anatase, rutile and also pseudobrookite (Fe_2TiO_5) and ilmenite ($FeTiO_3$) whose relative proportions significantly influence the photocatalytic behavior (Ye et al., 2007). Increasing $FeTiO_3$ content appears to increase the photocatalytic efficiency of decomposing acetaldehyde according to a two-step electron transfer model. On the other hand, larger amount of pseudobrookite in the coating were found to reduce the photocatalytic performance owing to an impaired photo-induced electron-hole transfer process.

8.4
Coatings with High Friction Coefficient

Si_3N_4 coatings deposited by electromagnetically accelerated plasma spraying (EMAPS) onto stainless steel surfaces show excellent adhesion strength in excess of 70 MPa, high abrasion and sliding wear resistance, indentation microhardness of around 550 $HV_{0.025}$, and extremely high friction coefficients in air and water exceeding values of 1.0 (Usuba and Heimann, 2006). Such coatings could be used to provide tribological surfaces for automotive brake pads actuating against grey cast iron rotors (Eriksson and Jacobson, 2000), improved grip-and-release surfaces for printing rolls, or high friction coatings between sliding interfaces to damp vibrations (Yu et al., 2005) of positioning equipment operating in outer space. Likewise, silicon nitride coatings could be applied to protect internal surfaces of tool steel pressure casting moulds from chemical attack by molten light metals and their alloys (see Section 7.2.2), and thus to provide a semi-permanent release interface. The excellent coating performance notwithstanding, the batch-type mode of EMAPS with the required replacement of the powder and gas feeding vessel after each shot renders the novel deposition technique uneconomical at present. Hence scale-up and optimization work have to be carried out during future development cycles.

Figure 8.27 shows the principle of this novel deposition technique (Kitamura et al., 2003).

The spraying system consists of an evacuated accelerating channel that comprises a pair of parallel electrodes connected to a high current power source by a switch, and a vessel that contains a pressurized gas and the powder source. The process is being initiated by activating a fast opening valve at the nozzle of the vessel (Figure 8.28a). After introducing the powder and the working gas into the accelerating channel (Figure 8.28b), an arc discharge is initiated at the desired position in the accelerating channel while the powder remains suspended in the accelerating channel (Figure 8.28c). The plasma at the arc initiating point then receives an electromagnetic force (Lorentz force) and forms an electromagnetically accelerated plasma of the working gas that heats and propels the powder towards the substrate (Figure 8.28d and e).

The composite powders used for electromagnetically accelerated plasma spraying (EMAPS) were synthesized by mixing commercially available α-Si_3N_4 powder (Silzot 7038, SKW Trostberg, Germany) with sintering aids such as alumina,

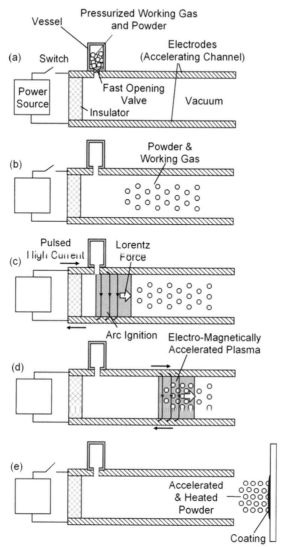

Figure 8.28 Schematic of electromagnetically accelerated plasma spraying (EMAPS) (Kitamura et al., 2003; Usuba and Heimann, 2006).

yttria, silica, and AlN, agglomerating by spray drying using an organic binder, and subsequent sintering at 1450 °C in a nitrogen-containing atmosphere (Berger et al., 2000; Thiele et al., 2002). After sintering the powders were mechanically treated by a mild milling process, and finally fractionized by dry sieving. Three types of powder composition were tested: 68 mass% Si_3N_4 + 16 mass% Al_2O_3 16 mass% Y_2O_3 (powder 1); 65 mass% Si_3N_4 + 11 mass% Al_2O_3 + 12 mass% Y_2O_3

4 mass% AlN, + 8 mass% SiO_2 (powder 2), and as-received powder 1 ball-milled for 2 hours to reduce its grain size (powder 3). The addition of SiO_2 decreases the viscosity of the alumina-yttria binder matrix during sintering while addition of AlN promotes the formation of SiAlON. The grain size fraction $-45 + 20\,\mu m$ was used for EMAP spraying of powders 1 and 2.

The silicon nitride coatings obtained are very dense and adhere well to the stainless steel (SUS 304) surface as shown in Figure 8.29.

The friction coefficients of the as-EMAP sprayed silicon nitride-based coatings were measured by a pin-on-disk test in water and air, showing surprisingly steady and significantly higher friction coefficient values of 1.0–1.1 in water, and even higher values of 1.3–1.4 in air (Figure 8.30) compared with friction coefficients below 0.1 obtained by sliding a polished Si_3N_4 ball against a polished sintered monolithic Si_3N_4 slab (Tomizawa and Fischer, 1987). After being scratched by the Si_3N_4 ball during the

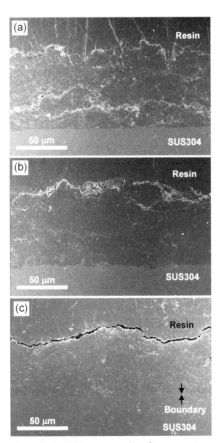

Figure 8.29 SEM micrographs of cross-sections of silicon nitride-based coatings obtained from powders 1 (a), 2(b) and 3(c) (Usuba and Heimann, 2006).

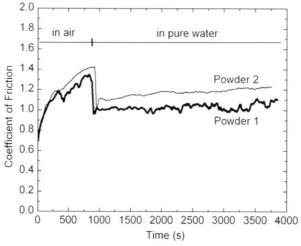

Figure 8.30 Friction coefficient of EMAPS silicon nitride-based coatings measured by a pin-on-disk test (ball: Si_3N_4, load: 1N, sliding speed: 10 mm s^{-1}) (Usuba and Heimann, 2006).

friction test all coatings remained tightly attached to the substrate except for minor localized chipping that resulted in the exposure of the substrate surface. Correspondingly, the surface of the sliding ball showed significant wear by grooving, indicating that at least some parts of the coatings had a hardness and wear resistance comparable to or even exceeding those of the sintered Si_3N_4 ball. Almost identical friction and wear properties of coatings produced from powders 1 and 2 suggest that there was no significant difference in hardness and adhesion strength of the two types of coating, at least not under the conditions of the friction tests performed.

References

A. Abdel-Samad, E. Lugscheider, K. Bobzin, M. Maes, *Surf. Coat. Technol.*, 2006, **201**, 1224.

A. Adamski, R. McPherson, in: *Advances in Thermal Spraying*, Proc. ITSC 1986, Welding Inst. of Canada, 555.

K. Ahn, S. Jung, J.M. Vohs, R.J. Gorte, *Ceramics Intern.*, 2007, **33** (6), 1065.

K. An, K.S. Ravichandran, R.E. Dutton, S.L. Semiatin, *J. Am. Ceram. Soc.*, 1999, **82**, 399.

D.J. Andrews, J.A.T. Taylor, *J. Thermal Spray Technol.*, 2000, **9** (2), 181.

T.R. Anthony, H.E. Cline, *J. Appl. Phys.*, 1977, **48** (9), 3888.

R. Ballhorn, H. Kreye, T. Stoltenhoff, J. Jirkovský, I. Burlacov, F. Peterka, L. Kavan, US Patent 2007 0 110 919A1 (2007).

C. Batista, A. Portinha, R.M. Ribeiro, V. Teixeira, M.F. Costa, C.R. Oliveira, *Surf. Coat. Technol.*, 2006, **200**, 2929.

L.-M. Berger, M. Herrmann, M. Nebelung, R.B. Heimann, B. Wielage, US Patent 6,110,835 (2000).

K. Biedermann, *Synthese von Schichten aus kubisch-stabilisiertem Zirkonoxid mit Hilfe plasmaunterstützter Abscheidung aus der Dampfphase*, Unpublished Master Thesis, Technische Universität Bergakademie Freiberg, June 2002.

BMFT, *Neue Materialien für Schlüsseltechnologien des 21.Jahrhunderts*, MaTech, Bundesmin. für Forschung und Technologie, Bonn, August 1994.

G. Bolelli, V. Cannillo, C. Lugli, L. Lusvarghi, T. Manfredini, *J. Eur. Ceram. Soc.*, 2006, **26**, 2561.

Yu. Borisov, A. Murashov, A. Ilienko, V. Balakin, V. Slipchenko, A. Slipchenko, V. Maksimov, in: Proc. 14th ITSC, Kobe, May 22–26, 1995, 141.

W. Braue, G. Paul, R. Pleger, H. Schneider, J. Decker, *J. Eur. Ceram. Soc.*, 1996, **16**, 85.

I. Burlacov, J. Jirkovský, M. Müller, R.B. Heimann, *Surf. Coat. Technol.*, 2006, **201**, 255.

I. Burlacov, J. Jirkovský, L. Kavan, R. Ballhorn, R.B. Heimann, *J. Photochem. Photobiol. A: Chemistry*, 2007, **187**, 285.

E.P. Butler, *Mat. Sci. Technol.*, 1985, **1** (6), 417.

D.P. Butt, J.J. Mecholsky, Jr., M. van Roode, J.R. Price, *J. Am. Ceram. Soc.*, 1990, **73** (9), 2690.

F. Cernuschi, S. Ahmaniemi, P. Vuoristo, T. Mantyla, *J. Eur. Ceram. Soc.*, 2004, **24**, 2657.

G.C. Chang, W. Phucharoen, R.A. Miller, *Surf. Coat. Technol.*, 1987, **30**, 13.

H.C. Chen, J. Heberlein, R. Henne, *J. Thermal Spray Technol.*, 2000, **9** (3), 348.

W.R. Chen, X. Wu, D. Dudzinski, P.C. Patnaik, *Surf. Coat. Technol.*, 2006a, **200**, 5863.

W.R. Chen, X. Wu, B.R. Marple, P.C. Patnaik, *Surf. Coat. Technol.*, 2006b, **201**, 1074.

D.R. Clarke, *Surf. Coat. Technol.*, 2003, **163–164**, 67.

H.E. Cline, *J. Appl. Phys.*, 1981, **52** (1), 443.

D.M. Comassar, *Metal Finishing*, March 1991, 39.

T. Cosack, B. Kopperger, *Mat.-wiss.u. Werkstofftech.*, 2001, **32**, 678.

J.A. Curran, H. Kalkanci, Yu. Magurova, T.W. Clyne, *Surf. Coat. Technol.*, 2006, doi:10.1016/j.surfcoat.2006.06.050

C.L. Curtis, D.T. Gawne, N. Priestnall, in: Proc. 1993 NTSC, Anaheim, CA, June 1993, 519.

K. Cushnie, J.E. Bell, G.D. Smith, in: Proc. 3rd NTSC, Thermal Spray Research and Applications, Long Beach, CA, May 20–25, 1990, 539.

H. Dai, X. Zhong, J. Li, J. Meng, X. Cao, *Surf. Coat. Technol.*, 2006, **201**, 2527.

J. Danroc, J. Lacombe, in: *Plasma Spraying: Theory and Applications*, R. Suryanarayanan, (ed.), World Scientific, 1993.

G. Eckold, I.M. Buckley-Golder, K.T. Scott, in: Proc. 2nd Intern. Conf. Surface Eng., Stratford-on-Avon, UK, June 16–18, 1987, 433.

M. Eriksson, S. Jacobson, *Tribol. Intern.*, 2000, **33** (12), 817.

A. Feuerstein, W. Dietrich, E. Muehlberger, H. Moyor, in: Proc. Conf. High Temp. Alloys for Gas Turbines and Other Appl., Liege, Belgium, Oct 6–9, 1986, 1227.

H. Fritze, J. Jojic, T. Witke, C. Rüscher, S. Weber, S. Scherrer, R. Weiß, B. Schultrich, G. Borchardt, *J. Eur. Ceram. Soc.*, 1998, **18**, 2351.

A. Fujishima, K. Honda, *Nature*, 1972, **238**, 37.

R. Gadow, M. Lischka, *Surf. Coat. Technol.*, 2002, **151–152**, 392.

E. Georgiopoulos, A. Tsetsekou, C. Andreouli, *Supercond. Sci. Technol.*, 2000, **13** (11), 1539.

M. Gramlich, H.-D. Steffens, Working Group Session 'Ceramic Coatings', German Ceramic Society (DKG), Heilbronn, June 8.9, 1994.

E. Gutmann, A.A. Levin, L. Pommrich, D.C. Meyer, *Cryst. Res. Technol.*, 2005, **40**, 114.

C. Haering, *Degradation der Leitfähigkeit von stabilisiertem Zirkoniumdioxid in Abhängigkeit von der Dotierung und den damit verbundenen Defektstrukturen*, Unpublished Ph.D. thesis, Universität Erlangen, 2001.

N. Harada, T. Kameyama, H. Kawano, K. Kuroda, K. Osaki, N. Tada, *IEEE Appl. Supercond.*, 1997, **7** (2), 1719.

D.P.H. Hasselman, *J. Comput. Mater.*, 1978, **12** (19), 403.

J.A. Haynes, E.D. Rigney, M.K. Ferber, W.D. Porter, *Surf. Coat. Technol.*, 1996, **86–87**, 102.

R.B. Heimann, *Process. Adv. Mater.*, 1991, **1**, 181.

R.B. Heimann, D. Lamy, V.E. Merchant, in: Trans.17th CUICAC Workshop, R.B. Heimann, ed., Laval University, Québec City, Québec, Canada, Oct 2, 1991.

R.B. Heimann, H.J. Pentinghaus, R. Wirth, Eur.J.Mineral., 2007, **19**, 281.

R. Henne, W. Weber, *High Temp. High Pressure*, 1986, **18**, 223.

R. Henne, M. v.Bradke, G. Schiller, W. Schnurnberger, W. Weber, in: Proc. 12th ITSC'89, London, UK, June 4–9, 1989, 175.

R. Henne, E. Fendler, M. Lang, in: Proc. 1st Europ. Solid Oxide Fuel Cell Forum, Lucerne, Switzerland, Oct 3–7, 1994, 2, 617.

P.E. Hodge, R.A. Miller, M.A. Gedwill, *Thin Solid Films*, 1980, **73**, 447.

C. Hwang, C. Yu, *Surf. Coat. Technol.*, 2007, **201**, 5954.

A.P. Ilyuschenko, V.A. Okovity, N.K. Tolochko, S. Steinhäuser, *Mater. Manuf. Processes*, 2002, **17** (2), 157.

N. Iwamoto, N. Umesaki, M. Kamai, G. Ueno, in: Proc. TS'90, Essen 1990, DVS 130, 99.

K.M. Jasim, D.R.F. West, W.M. Steen, R.D. Rawlings, in: Proc. 7th Intern.Congr.on Applic. Lasers and Electrooptics (ICALEO '88), ed. G. Bruck, Springer, 1988, 17.

K.M. Jasim, R.D. Rawlings, D.R.F. West, *Mat. Sci. Technol.*, 1992a, **8** (1), 83.

K.M. Jasim, R.D. Rawlings, D.R.F. West, *J. Mat. Sci.*, 1992b, **27**, 3903.

I. Kaji, S. Yoshida, N. Nagata, T. Nakajima, Y. Seino, in: 2nd Int. Symp. on SOFC, Nagoya, Japan, 1991, 221.

J. Karthikeyan, K.P. Sreekumar, N. Venkatramani, M.B. Kurup, D.S. Patil, V.K. Rohatgi, *Appl. Phys. A.: Mater. Sci. Process.*, 1989, **48** (5), 489.

P.L. Ke, Y.N. Wu, Q.M. Wang, J. Gong, C. Sun, L.S. Wen, *Surf. Coat. Technol.*, 2005, **200**, 2271.

J. Kitamura, S. Usuba, Y. Kakudate, H. Yokoi, K. Yamamoto, A. Tanaka, S. Fujiwara, *J. Thermal Spray Technol.*, 2003, **12** (1), 70.

P.A. Langjahr, R. Oberacker, M.J. Hoffmann, *Mat.-wiss.u.Werkstofftech.*, 2001, **32**, 665.

C.H. Lee, H.K. Kim, H.S. Choi, H.S. Ahn, *Surf. Coat. Technol.*, 2000, **124**, 1.

C.H. Lee, H. Choi, C. Lee, H. Kim, *Surf. Coat. Technol.*, 2003, **17**, 192.

K.N. Lee, *J. Am. Ceram. Soc.*, 1998, **81** (12), 3329.

K.N. Lee, R.A. Miller, N.S. Jacobsen, *J. Am. Ceram. Soc.*, 1995, **78** (3), 705.

K.N. Lee, R.A. Miller, *J. Am. Ceram. Soc.*, 1996, **81** (12), 620.

U. Leushake, T. Krell, U. Schulz, *Mat.-wiss.u. Werkstofftech.*, 1997, **28**, 391.

C.J. Li, W.Y. Li, *Surf. Coat. Technol.*, 2003, **167**, 278.

Z. Li, W. Mallener, L. Fuerst, D. Stöver, F.-D. Scherberich, in: Proc. 1993 NTSC, Anaheim, CA, June 1993, 343.

Z. Li, W. Liu, Y. Wu, Mater. Chem. Phys., 2007, doi:10.106/j.matchemphys.2007.04.062

E.Yu. Litovsky, M. Shapiro, *J. Am. Ceram. Soc.*, 1992, **75** (12), 3425.

Y. Liu, C. Persson, J. Wigren, *J. Thermal Spray Technol.*, 2004, **13** (3), 415.

Y. Liu, T. Nakamura, V. Srinivasan, A. Vaidya, A. Gouldstone, S. Sampath, Acta Mater., 2007, doi:10.1016/j.actamat.2007.04.037.

F.N. Longo, H. Florant, in: Proc. 8th ITSC, Miami, Sept 1976

E. Lugscheider, T. Weber, in: Proc. 3rd NTSC, Thermal Spray Research and Applications, Long Beach, CA, May 20–25, 1990, 635.

E.H. Lutz, *J. Am. Ceram. Soc.*, 1994, **77**, 1274.

X. Ma, F. Wu, J. Roth, M. Gell, E.H. Jordan, *Surf. Coat. Technol.*, 2006, **201**, 4447.

W. Malléner, K. Wippermann, H. Jansen, Z. Li, D. Stöver, in: Proc. ITSC'92, Orlando, FL, May 28.June 5,1992, 835.

D. Mangapathi, D. Mishra, Y.C. Venudhar, *Cryst. Res. Technol.*, 2006, **28** (2), 245.

M. Matsumoto, H. Takayama, D. Yokoe, K. Mukai, H. Matsubara, Y. Kagiya, Y. Sugita, *Scripta Mater.*, 2006, **54** 2035.

J.R. Mawdsley, Y.J. Su, K.T. Faber, T.F. Bernecki, *Mater. Sci. Eng.*, 2001, **A308** (1/2), 189.

J.C. Maxwell, *A Treatise on Electricity and Magnetism*. 2nd edn, Vol. 1. Clarendon Press: Oxford. 1881.

T.T. Meek, A. Kobayashi, E. Herbert, R.L. White, *Vacuum*, 2002, **65** (3), 409.

D.C. Meyer, A.A. Levin, P. Paufler, *Thin Solid Films*, 2005, **489** (1–2), 5.

M. Michael, US Patent 6,284,323 (2001).

P. Michlik, C. Berndt, *Surf. Coat. Technol.*, 2006, **201**, 2369.

R.A. Miller, *J. Am. Ceram. Soc.*, 1984, **67**, 517.

R.A. Miller, *Surf. Coat. Technol.*, 1987, **30**, 1.

R.A. Miller, NASA Technical Memorandum TM 100283, 1988.

K. Minazawa, K. Toda, S. Kaneko, N. Murakami, A. Notomi, *MHI Technical Bulletin*, 1991, **28** (1), 41.

M. Moldovan, C.M. Weyant, D.L. Johnson, K.T. Faber, *J. Thermal Spray Technol.*, 2004, **13** (1), 51.

M. Müller, *Entwicklung elektrokatalytischer Oxidschichten mit kontrollierter Struktur und Dotierung aus flüssigen Prekursoren mittels thermischer Hochfrequenzplasmen*, Unpublished Ph.D. Thesis, Technische Universität Bergakademie Freiberg, Germany, June 2001.

M. Müller, E. Bouyer, M. v.Bradke, D.W. Branston, R.B. Heimann, R. Henne, G. Lins, G. Schiller, *Mat.-wiss.u.Werkstofftech.*, 2002, **33**, 322.

A. Nadeau, L. Pouliot, F. Nadeau, J. Blain, S.A. Berube, C. Morceau, M. Lamontagne, *J. Thermal Spray Technol.*, 2006, **15** (4), 744.

M.C. Nestler, E. Müller, D. Hawley, H.-M. Höhle, D. Sporer, M. Dorfman, *Sulzer Techn. Rev.*, 2007, **2**, 11.

J.R. Nicholls, K.J. Lawson, A. Johnstone, R.S. Rickerby, *Surf. Coat. Technol.*, 2002, **151–152**, 383.

A.R. Nicoll, G. Barbezat, A. Salito, SOFC-Seminar, Yokohama, Feb 1992.

A. Notomi, H. Hisatome, in: Proc. 14th ITSC, Kobe, May 22–26 1995, 79.

R. Okiai, S. Yoshida, I. Kaji, in: 1st Int. Symp. on SOFC, Nagoya, Japan, Nov 13–14, 1989, 191.

M. Oksa, E. Turunen, T. Varis, *Surf. Eng.*, 2004, **20** (4), 251.

N.P. Padture, M. Gell, P.G. Klemens, US Patent 6,015,630 (2000).

L. Pawlowski, A. Hill, R. McPherson, D. Garvie, Z. Przelozny, T. Finlayson, in: Proc. 3rd NTSC, Thermal Spray Research and Applications, Long Beach, CA, May 20–25, 1990, 641.

L. Pawlowski, A. Gross, R. McPherson, *J. Mater. Sci.*, 1991, **26** (14), 3803.

M. Peters, K. Fritscher, G. Staniek, W.A. Kaysser, U. Schulz, *Mat.-wiss.u.Werkstofftech.*, 1997, **28**, 357.

A. Petitbon, D. Guignot, U. Fischer, J.-M. Guillemot, *Mat. Sci. Eng.*, 1989, **A121**, 545.

E. Pfender, *Mat. Sci. Eng.*, 1991, **A139**, 352.

F.M. Pitek, C.G. Levi, *Surf. Coat. Technol.*, 2007, **201**, 6044.

B. Prawara, H. Yara, Y. Miyagi, T. Fukushima, *Surf. Coat. Technol.*, 2003, **162**, 234.

O. Racek, C.C. Berndt, D.N. Guru, J. Heberlein, *Surf. Coat. Technol.*, 2006, **201**, 338.

P. Ramaswamy, S. Seetharamu, K.B.R. Varma, K.J. Rao, *J. Thermal Spray Technol.*, 1998, **7** (4), 497.

S. Ramesh, US Patent 6,258,467 (2001).

C.K. Reed, J.M. Wright, *Thermal Control Coatings for High Thermal Conductivity (K) Substrates*. Report AFRL-ML-WP-TR 1999-4031, Wright-Patterson Air Force Base, OH, 275 pp.

G. Reisel, R.B. Heimann, *Surf. Coat. Technol.*, 2004, **185**, 215.

J.E. Restall, in: Proc. 5th Intern. Symp. on Superalloys, Champion, PA, Oct 7–11 1984, 721.

T.N. Rhys-Jones, *Corrosion Science*, 1989, **29** (6), 623.

B. Rogé, A. Fahr, J.S.R. Griguère, K.I. McRae, *J. Thermal Spray Technol.*, 2003, **12** (4), 530.

D.L. Ruckle, *Thin Solid Films*, 1980, **73**, 455.

C.H. Rüscher, H. Fritze, G. Borchardt, T. Witke, B. Schultrich, *J. Am. Ceram. Soc.*, 1997, **80** (12), 3225.

M. Scagliotti, F. Parmigiani, G. Samoggia, G. Lanzi, D. Richon, *J. Mat. Sci.*, 1988, **23**, 3764.

G. Schiller, R. Henne, V. Borck, *J. Thermal Spray Technol.*, 1995, **4** (2), 185.

H. Schneider, D. Komarneni, *Mullite*, John Wiley & Sons Ltd., 2005.

H.G. Scott, *J. Mat. Sci.*, 1975, **10** (9), 1527.

S. Seifert, *Development of Ceramic Lightweight Thermal Protection Coatings for Reusable Space*

Launch Vehicles, Unpublished Master thesis, Dept. of Mineralogy, Technische Universität Bergakedemie Freiberg, June 2004.

S. Seifert, E. Litovsky J.I. Kleiman, R.B. Heimann, *Surf. Coat. Technol.*, 2006a, **200** (11), 3404.

S. Seifert, J.I. Kleiman, R.B. Heimann, *J. Spacecraft Rockets*, 2006b, **43** (2), 439.

D. Seo, K. Ogawa, M. Tanno, T. Shoji, S. Murata, *Surf. Coat. Technol.*, 2007, **201**, 7952.

A. Shah, S. Patel, E. Narumi, D.T. Shaw, *Appl. Phys. Lett.*, 1990, **57** (14), 1452.

A. Shah, T. Haugan, S. Witanachi, S. Patel, T. Shaw, *Proc. Mat. Res. Soc. Symp.*, Boston, MA, 1989, 747.

J.-H. Shieh, S.-T. Wu, *Appl. Phys. Lett.*, 1991, **59** (12), 1512.

R. Sivakumar, B.L. Mordike, *Surf. Coat. Technol.*, 1989, **37**, 139.

R.W. Smith, W.F. Schilling, H.M. Fox, *Trans. ASME*, 1981, **103**, 146.

D.A. Spera, S.J. Grisaffe, NASA Technical Memorandum TMX-2664. National Technical Information Service, Springfield, VA, 1973.

N. Spinicchia, G. Angella, R. Benocci, A. Bruschi, A. Cremona, G. Gittini, A. Nardone, E. Signorelli, E. Vassallo, *Surf. Coat. Technol.*, 2005, **200**, 1151.

S. Stecura, *Ceram. Bull.*, 1982, **61** (2), 256.

H.-D. Steffens, Z. Babiak, U. Fischer, in: Proc. 2nd Intern. Conf. on Surface Eng., Stratford-upon-Avon, UK, June 16–18, 1987, 471.

D. Stöver, G. Pracht, H. Lehmann, M. Dietrich, J.E. Döring, R. Vaßen, *J. Thermal Spray Technol.*, 2004, **13** (1), 76.

D. Stöver, D. Hathiramani, R. Vaßen, R.J. Damani, *Surf. Coat. Technol.*, 2006, **201**, 2002.

T. Streibl, A. Vaidya, M. Friis, V. Srinivasan, S. Sampath, *Plasma Chem. Plasma Proc.*, 2006, **26** (1), 73.

L.-W. Tai, P.A. Lessing, *J. Am. Ceram. Soc.*, 1991, **74** (3), 501.

T.A. Taylor, US Patent 5,073,433 (1991).

S. Thiele, R.B. Heimann, L.-M. Berger, M. Herrmann, M. Nebelung, T. Schnick, B. Wielage, P. Vuoristo, *J. Thermal Spray Technol.*, 2002, **11** (2), 218.

F.L. Toma, G. Bertrand, S.O. Chwa, C. Meunier, D. Klein, C. Coddet, *Surf. Coat. Technol.*, 2006, **200**, 5855.

H. Tomizawa, T.E. Fischer, *ASLE Trans.*, 1987, **30** (1), 41.

R.W. Trice, Y.J. Su, R. Mawdsley, K.T. Faber, A.R. de Arellano-López, H. Wang, W.D. Porter, *J. Mater. Sci.*, 2002, **37**, 2359.

H. Uchikawa, H. Hagiwara, M. Shirasaka, H. Yamane, in: Proc. Surf. Eng. Intern. Conf., Tokyo, Japan, Oct 18. 22, 1988, 45.

S. Usuba, R.B. Heimann, *J. Thermal Spray Technol.*, 2006, **15** (3), 356.

R. Vaßen, X.Q. Cao, F. Tietz, D. Basu, D. Stöver, *J. Am. Ceram. Soc.*, 2000, **83** 2023.

R. Vaßen, M. Dietrich, H. Lehmann, X. Cao, G. Pracht, F. Tietz, D. Pitzer, D. Stöver *Mat.-wiss.u.Werkstofftech.*, 2001, **32**, 673.

R. Vaßen, D. Hathiramani, J. Mertens, V.A.C. Haanappel, I.C. Vinke, *Surf. Coat. Technol.*, 2007, **202** (3), 499.

H. Wang, H. Herman, H.J. Wiesmann, Y. Zhu, Y. Xu, R.L. Sabatini, M. Suenaga, *Appl. Phys. Lett.*, 1990, **57** (23), 2495.

H.B. Wang, G.Y. Meng, D.K. Peng, *Thin Solid Films*, 2000, **368**, 275.

Z. Wang, A. Kulkarni, S. Deshpande, T. Nakamura, H. Herman, *Acta Mater.*, 2003, **51**, 5319.

D.O. Welch, V.J. Enery, D.E. Cox, *Nature*, 1987, **327**, 278.

B.C. Wu, E. Chang, S.E. Chang, D. Tu, *J. Am. Ceram. Soc.*, 1989, **72** (2), 212.

L. Xie, D. Chen, E.H. Jordan, A. Ozturk, F. Wu, X. Ma, B.M. Cetegen, M. Gell, *Surf. Coat. Technol.*, 2006, **201**, 1058.

H. Yajima, Y. Kimura, T. Yoshioka, in: Proc. 14th ITSC, Kobe, May 22–26, 1995, 621.

F.X. Ye, T. Tsumura, K. Nakata, A. Ohmori, Mater. Sci. Eng. B, doi:10.1016/j.mseb.2007.09.057.

M. Yoshiba, K. Abe, T. Aranami, Y. Harada, in: Proc. 14th ITSC, Kobe, May 22–26, 1995, 785.

L. Yu, Y. Ma, C. Zhou, H. Xu, *Mater. Sci. Eng. A*, 2005, **408**, 42.

I. Zaplatynsky, NASA Technical Memorandum 82830, Proc. Int. Conf. Metall. Coatings and Process Technology, San Diego, CA, April 5–8, 1982.

H. Zhu, Y.C. Lau, E. Pfender, in: Proc. 9th Int. Symp. Plasma Chem., Pugnochiuso, 1989, 876, 1497.

D. Zhu, K.N. Lee, R.A. Miller, *ASME Int. Gas Turbine Inst. Publ. IGTI*, 2002, vol. **4**, 171.

D. Zhu, S.R. Choi, J.I. Eldridge, K.N. Lee, R.A. Miller, *Eng. Sci. Proc.*, 2003, **24** (3), 469.

H. Zhu, Y.C. Lau, E. Pfender, *J. Appl. Phys.*, 1991, **69** (5), 3404.

9
Solutions to Industrial Problems III: Bioceramic Coatings

Research and development on bioceramic materials have reached a level of involvement and sophistication comparable only to electronic ceramics. The reason is obvious: large proportions of an aging population rely increasingly on bone replacement ranging from alveolar ridge augmentation to hip endoprostheses. Ceramics based on biocompatible calcium phosphates such as hydroxyapatite ($Ca_{10}(PO_4)_6(OH)_2$, HAp; $C_{10}P_3H$) or tricalcium phosphate ($Ca_3(PO_4)_2$; C_3P) are prime candidates for both replacement of bone subjected to low loading conditions and osseoconductive coatings of the femoral shafts of hip endoprostheses and dental root replacement parts (Søballe, 1993; see also Heimann, 2007).

9.1
Classification of Biomaterials and Mechanism of Bone Bonding

Bioconductive monolithic ceramics and coatings made from calcium phosphates and bioglasses not only interact with the body by stimulating osseointegration but will sometimes also be resorbed and transformed to calcified osseous tissue if the coating thickness is low. The dominating biorelevant mechanism is *bonding osteogenesis* characterized by a chemical bond between implant and bone. Bioinert materials such as alumina, zirconia, carbon but also some metals like titanium or tantalum react to the host bone by *contact osteogenesis* characterized by a direct contact between implant and bone. Biotolerant materials such as bone cement (PMMA) but also stainless steel and Co-Cr alloys will be accepted by the body and develop an implant interface characterized by a layer of connective tissue between implant and bone thus resulting in a *distance osteogenesis*.

Bone bonding generally occurs throughout life at bony remodeling sites. The bone-resorbing osteoclasts provide a three-dimensionally complex surface with which the cement line, the first matrix elaborated during de novo bone formation, interdigitates and is interlocked. The structure and composition of this interfacial bony matrix has been conserved during evolution across species. This interfacial matrix can be recapitulated at a biomaterial surface implanted in bone, and no

Plasma Spray Coating: Principles and Applications. Robert B. Heimann
Copyright © 2008 WILEY-VCH Verlag GmbH & Co. KGaA, Weinheim
ISBN: 978-3-527-32050-9

evidence has ever emerged to suggest that bone bonding to artificial materials is any different from this natural biological process. In particular, bone bonding capability is not restricted to calcium phosphate-based bioconductive materials. Indeed, without sufficient surface porosity, calcium phosphate biomaterials are not bone bonding but behave in a bioinert fashion (DeGroot, 1991). On the other hand, nonbonding materials can be rendered bone bonding by modifying their surface topography. As suggested by Davies (2007) the driving force for bone bonding is bone reconstruction by *contact osteogenesis* that has to occur on a sufficiently stable recipient surface which has micron-scale surface topography with undercuts in the sub-micron range.

9.2
Properties of Bioceramic Coatings

The application of bioactive calcium phosphate ceramics is generally restricted to non-load-bearing areas because of their low fracture toughness and limited modulus of elasticity. To counteract this drawback, calcium phosphate coatings are applied to load-bearing metallic parts (Figures 9.2, 9.3) to create a synergistic composite of a tough and strong metallic substrate providing strength, and a porous bioactive hydroxyapatite (HAp) coating providing osseointegration function. In fact, a multitude of *in vivo* investigations have shown that a coated implant is readily accepted by the body and within a few weeks develops a strong bond with the surrounding bone. In contrast uncoated implants react by a distance osteogenetic mechanism developing a connective tissue layer at the interface (for example Rivero *et al.*, 1988; Klein *et al.*, 1991; Heimann *et al.*, 2004a; Heimann *et al.*, 2004b; Heimann, 2006a). Figure 9.1 demonstrates that in the absence of a bioconductive HAp layer the body reacts to an experimental Ti6Al4V implant by developing a protective capsule of connective tissue thus preventing solid integration of the implant to living bone tissue. A recent clinical survey of over 6600 cases concluded that in patients younger

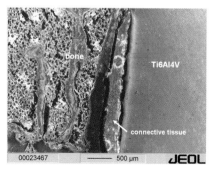

Figure 9.1 Formation of a connective tissue capsule separating an uncoated Ti6Al4V implant surface (right) from the surrounding bone (left) after implantation into the femoral condyle of an adult dog for 6 months (Heimann *et al.*, 2004b).

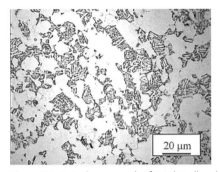

Figure 9.2 Optical micrograph of Ti6Al4V alloy showing the hcp α-phase (light grey) and the lamellar bcc β-phase (dark grey).

than 60 years, only 2% of HAp-coated hip endoprostheses failed within 10 years after implantation whereas more than 25% of uncoated did (Havelin et al., 2000).

Coatings of only a few micrometer thickness have limited application for stimulating bone in-growth but may be beneficial to their selfrepair function in contact with body fluid. Thick plasma-sprayed HAp coatings (DeGroot et al., 1987) are now applied routinely to the shafts of hip endoprosthetic devices and to dental root implants. Their preparation, however, requires a high degree of skill, and general principles of total quality management must be strictly adhered to.

Figure 9.2 shows a micrograph of a polished section of Ti6Al4V ubiquitously used[1] as bioinert material for medical implants, most notably as stems for hip endoprostheses and dental roots. The microstructure of the Ti6Al4V alloy consists of a composite $(\alpha + \beta)$ phase with approximately 80 vol% of primary hcp α- and 20 vol % bcc β-phase.

Figure 9.3 shows the surface (left) and the cross-section (right) of an as-sprayed HAp coating of about $180 \pm 20\,\mu m$ thickness with some microcracks and pores present. The pores appear to have formed as a result of poor bonding between adjacent splats whereas microcracking arises from shrinkage of the splats during quenching and subsequent differential thermal contraction between substrate and coating (Figure 9.3b). The surface morphology of as-sprayed sample revealed only partially molten HAp particles and some porosity (Figure 9.3a).

The plasma spray parameters used to deposit the HAp coating shown in Figure 9.3 are collected in Table 9.1 (Topić et al., 2007).

In the past many attempts have been made to optimize the essential properties of bioconductive plasma-sprayed HAp coatings. These properties include coating cohesion and adhesion, phase composition, crystallinity, porosity and surface roughness, residual coating stresses, and coating thickness (Fazan and Marquis, 2000; Heimann, 2006b). A guideline for a performance profile of plasma-sprayed bioconductive HAp coatings has been given by Wintermantel and Ha (1996) (Table 9.2).

---

[1] As there is some concern about metabolically negative effects, e.g. cytotoxicity of vanadium in current development, this element is being replaced by niobium, tantalum or iron, for example Ti6Al7Nb, Ti5Al12Ta or Ti5Al2.5Fe. Also, aluminum-free Ti alloys are under scrutiny.

Figure 9.3 SEM micrographs of the surface (a) and the cross-section (b) of an as-sprayed HAp coating on a Ti6Al4V substrate (Topić et al., 2007).

Table 9.1 Plasma spray parameters used to deposit the HAp coating shown in Figure 9.3.

Parameter	Value
Argon plasma gas [slpm]*	45
Hydrogen plasma gas [slpm]	6.5
Argon powder carrier gas [slpm]	5
Rotation of powder feeder plate [% of maximum]	20
Relative hopper stirrer rate [% of maximum]	20
Spray distance [mm]	90
Plasma power [kW]	30
Traverse speed [m s^{-1}]	6

*Standard liters per minute

Table 9.2 Performance profile of plasma-sprayed HAp coatings (Wintermantel and Ha, 1996).

Property	Optimum value	Function
Coating thickness	< 50 µm	Easy resorption
	< 200 µm	Long-term stability
Roughness/porosity	> 75 µm	Optimum cell ingrowth
HAp content	> 95%	Chemical stability
Crystallinity	> 90%	Resorption resistance
Adhesion strength	> 35 MPa	Implant integration

In contrast to a desired value of at least 35 MPa (Table 9.2) the *adhesion strength* of plasma-sprayed calcium phosphate layers to the titanium alloy implant surface was found to be notoriously weak.[2] Despite claims that a thin reaction layer of calcium dititanate (CaTi$_2$O$_5$) or calcium titanate (perovskite, CaTiO$_3$) exists that will mediate adhesion (Filiaggi et al., 1991; Ji et al., 1992; Webster et al., 2003) experimental

[2] The ISO 13779.4 standard requires an adhesion strength of hydroxyapatite coatings of only 15 MPa (ISO 13779.4:2002).

evidence of such a reaction layer in as-sprayed coatings is scant (Lu et al., 2004) or absent (Park et al., 1998), and its visualization by electron microscopy even at high magnification (Heimann and Wirth, 2006) is hampered by its exiguity owing to the very short diffusion paths of Ca^{2+} and Ti^{4+} ions, respectively that render any potential reaction zone extremely thin. However, long-time annealing of as-sprayed HAp coatings on a titanium substrate beyond 900 °C resulted in the formation of an interfacial Ca-Ti-Oxide layer of several micrometer thickness (DeGroot et al., 1987; Gross et al., 1998a). On the other hand, biomimetic nucleation of HAp generally observed during incubation of plasma-sprayed coatings in simulated body fluid strongly points to an interfacial molecular recognition by a close 2D-lattice match of (022) of calcium titanate and (0001) of HAp (see Figure 9.11).

To improve adhesion, the degree of melting of the HAp particles in the plasma jet must also be improved by an increase of the plasma enthalpy. However, there is a conundrum. High plasma enthalpies inevitably lead to increased thermal decomposition and thus to a decrease of the resorption resistance, i.e. the *in vivo* longevity of the coatings. As a consequence the plasma spray parameters and the resulting microstructure of the deposited coatings need to be carefully optimized by controlling the heat transfer from the hot core of the plasma jet to the center of the powder particles. Alternatively other solutions have to be sought such as addition of suitable bond coats (see Section 9.4). Moreover, in addition to improvement of coating adhesion to the substrate surface, coating cohesion may be improved in HAp-ZrO_2 (for example Kumar et al., 2003; Rapacz-Kmita et al., 2006) and HAp-TiO_2 composite coatings (for example Li et al., 2003; Lu et al., 2004) that show enhanced cohesion due to particle reinforcement as well as gradient coatings (Ning et al., 2005) with reduced stress gradients.

Not only is the *phase composition* of the coatings of vital importance for their in-service performance, but also their *crystallinity* that to a large extent controls the *in vivo* dissolution behavior (Ducheyne et al., 1993; DeBruijn et al., 1994). Well-crystallized HAp is very stable at pH values above 4.5 showing essentially bioinert characteristics (DeGroot, 1991) and an inhibiting effect on cell proliferation as measured by decreased levels of alkaline phosphatase (ALP) activity (Frayssinet et al., 1994; Leali Tranquilli et al., 1994) and osteocalcin secretion (DeSantis et al., 1996). However, amorphous calcium phosphate (ACP), thermal decomposition products such as tricalcium phosphate (TCP, $Ca_3(PO_4)_2$), tetracalcium phosphate (TetrCP, $Ca_4O(PO_4)_2$) and calcium oxide (CaO), and dehydroxylation products with short range order (SRO) structure such as oxyhydroxyapatite (OHAp) and/or oxyapatite (OAp, $Ca_{10}(PO_4)_6O$) show enhanced solubilities in human blood serum and simulated body fluid (SBF) that follow the order (Klein, 1990; Ducheyne et al., 1993):

$$CaO \gg TCP > ACP > OHAp/OAp \gg HAp.$$

While moderately enhanced levels of Ca^{2+} and HPO_4^{2-} ions in the biofluid space at the implant–tissue interface are desired to assist in bone remodeling, excessive amounts of these ions released from the dissolving decomposition products of HAp drive up the local pH values with concurrent cytotoxic effects on living bone cells

(LeGeros *et al.*, 1991; Wang *et al*, 1993; Chou *et al.*, 1999). Consequently short-term release of ions from dissolving calcium phosphate phases must be kept at bay and optimized by adjusting the amount of well-crystallized HAp in the as-sprayed coatings (Gross *et al.*, 2004). This can be achieved by several measures including optimizing the set of plasma spray parameters that significantly influence the plasma enthalpy and in turn control the thermal history of HAp (Heimann *et al.*, 1997; Heimann *et al*, 1998; Graßmann and Heimann, 2000; see also Section 9.3.2). Also, the presence of a bioinert bond coat appears to improve the adhesion between HAp coating and metal substrate (Kurzweg *et al.*, 1998a; Heimann, 1999a; Heimann, 1999b; Heimann *et al.*, 2004b; Lu *et al.*, 2004; Ng *et al.*, 2005; see Section 9.4). In addition, such bond coats are thought to act as thermal barriers that may aid in enhanced crystallization of HAp at the expense of ACP (Heimann, 1999a).

Besides phase composition, crystallinity and adhesion of the coatings, the *porosity* and the *surface roughness* play decisive roles in the quest for enhancing the biomedical performance of endoprosthetic implants. While optimum coating porosity and roughness (Wintermantel and Ha, 1996; Table 9.2) are mandatory for the ingrowth of bone cells, accumulation of macropores at the substrate–coating interface leads to an intolerable weakening of the coating adhesion as well as cohesion strengths. The denser the microstructure of the bioceramic coating the lower is the risk of bonding degradation by cracking, spalling and delamination during *in vivo* contact with aggressive body fluids (Yang *et al.*, 1995). Since the integrity and continuity of the substrate–coating interface is of paramount importance for implants, the two conflicting requirements of the need of porosity for bone cell ingrowth and the need of high coating density for superior adhesion have to be carefully considered and controlled (Graßmann and Heimann, 2000). This is particularly important considering the risk of release of coating particles that will be distributed by the lymphatic system and is known to lead to inflammatory responses with formation of undesirable giant cells and phagocytes ('particle disease'; Lemons, 1994).

The occurrence of *residual stresses* at the biomaterial coating–substrate interface as well as within the coating will lead to weakening of the adhesion by delamination and crack formation, depending on the sign of the stresses (Yang, 2007). Residual stresses originate from the large temperature gradients experienced during the spraying process (see Section 5.7). When the molten particles strike the cold substrate, they will be rapidly quenched while their contraction is restricted by adherence to the thick, rigid substrate (Fauchais *et al.*, 2004). This leads to accumulation of high levels of tensile stresses both in the coating and at the coating–substrate interface, commonly referred to as 'quenching stresses' (Kuroda *et al.*, 1995; Matejicek and Sampath, 2003; Topić *et al.*, 2007). The first layer to form at the very interface, found to be amorphous (see for example Heimann and Wirth, 2006), will to a large extent control the occurrence of residual stresses in terms of magnitude as well as sign (see Section 5.7). As shown above in Figure 5.31 the transformation of this ACP to crystalline calcium phosphate phases during *in vitro* contact with simulated body fluid and *in vivo* contact with biofluid, respectively will lead to stress relaxation (Heimann *et al.*, 2000; Topić *et al.*, 2007).

While the substrate is usually at some elevated temperature during deposition, post-deposition cooling to room temperature generates additional stress by thermal mismatch proportional to the differences in the thermal expansion coefficients of the coating and the substrate as well as the intrinsic elastic moduli (see Section 5.7). The principal equation (Equation 5.52c, Section 5.7) governing the generation of thermal coating stress, σ_c has been derived by Dietzel and expressed by

$$\sigma_c = \{E_c(\alpha_c - \alpha_s)\Delta T\}/(1 - v_c) + [(1 - v_s)/E_s]d_c/d_s\},$$

where E = Young's modulus, α = coefficient of thermal expansion, T = temperature, v = Poisson's number, and d = thickness, and the subscripts c and s refer to coating and substrate, respectively. Since for any given values of v and E the thermal coating stress σ_c increases with increasing coating thickness d_c the risk of spalling is much higher in thick coatings than in thin ones. Depending on the sign of the difference $(\alpha_c - \alpha_s)$, the so-called 'thermal stress' can be tensile or compressive. Quenching and thermal stresses, combined with the complicated solidification process of the coating, are the two main contributors to the overall residual stress. Hence control of residual stresses is important for the integrity of the deposit–substrate system and in turn its mechanical performance (Clyne and Gill, 1996; Tsui *et al.*, 1998a, 1998b) since high residual stresses can lead to cracking and delamination of the coating, shape changes of thin substrates, and in general can undermine the performance of the entire part. Tensile stresses exceeding the elastic limit cause cracking in coatings perpendicular to the direction of the tensile stress. While in general some degree of compressive stress is considered to be desirable as it closes cracks originating at the surface and thus improves fatigue properties, excessive compressive stress can cause cohesive (spallation) and adhesive failure (Pina *et al.*, 2003). In biomedical service, the existing residual stresses superpose with the applied loading stress during movement of the patient and failure may occur or fatigue life be shortened if the residual stress is sufficiently high.

It is somewhat puzzling that in the literature concerned with residual stresses in plasma-sprayed HAp coatings highly contradictory results were reported in terms of sign and magnitude of the measured stresses. For example, while Sergo *et al.* (1997), Tsui *et al.* (1998a) and Han *et al.* (2001) found the stresses to be tensile using respectively, the specimen curvature method (see Section 5.7.3) and the borehole method (see Section 5.7.1), Brown *et al.* (1994), Millet *et al.* (2002) and Cofino *et al.* (2004) reported compressive stresses throughout the coating depth measured by the $\sin^2\Psi$ method (see Section 5.7.2). The reasons for these discrepancies may be manifold. The $\sin^2\Psi$ method is possibly unreliable owing to the sensitivity of the measured interplanar spacings to the impurity content of the crystalline HAp and the fact that the measured deformation ε is dependent on intrinsic materials parameters such as the Poisson ratio v and the modulus of elasticity E that are not well defined for plasma-sprayed HAp. In addition, E is strongly dependent on porosity. Though it is possible to determine E experimentally on free-standing coatings (Tsui *et al.*, 1998a; Yang *et al.*, 2000) the inevitably cracked deposits and always existing pores (see Figure 9.3) will alter the modulus of the coating compared with dense bulk samples, and hence the values of the measured deformation ε and the calculated stress state will become questionable. The existence of pores, surface roughness, and a network of microcracks render invalid the simplifying assumption of a two-dimensional plane stress model and a real 3D stress field distribution must be invoked (Cofino et al., 2004). Such a 3D stress state has been determined by depth modeling of the full stress tensor using a seminumerical method (Härting, 1998). For all these reasons it is widely believed that the specimen curvature method (see Section 5.7.3) may yield a more accurate description of the sign and magnitude of the residual stress field of plasma-sprayed HAp coatings (see for example Topić et al., 2007; Yang, 2007).

The last property in need of control is the *coating thickness*. A thin HAp layer (<50 μm) yields better adhesion compared with thicker coatings (Wintermantel and Ha, 1996; Table 9.2) owing to reduced residual coating stresses (Heimann et al., 2000) but will be quickly resorbed in the course of bone integration. Thicker coatings (± 150 μm) show substantially reduced adhesion strength, but may be required in some instances to ensure a more permanent bond to guarantee implant stability by a lasting biological effect (Dörre, 1989, 1992; Heimann et al., 2004a; Heimann et al., 2004b). This situation occurs during an endoprosthetic replacement operation involving an exchange of the implant, when the cortical bone matter has been previously damaged, often in concurrence with an undesirable geometric configuration of the implant-supporting bone. In this special case a thin, quickly resorbed calcium phosphate coating will not be sufficient to sustain the required large scale bone regeneration. Hence thicker coatings will be needed to stimulate bone reconstruction over longer times.

9.3
Plasma Spraying of Osseoconductive Hydroxyapatite

Atmospheric plasma spraying (APS) constitutes the state-of-the-art procedure to improve the biological integration of implants into the body. Competing techniques applied include flame spraying (FS), high velocity oxyfuel (HVOF) spraying (Sturgeon and Harvey, 1995), pulsed laser deposition (Kim et al., 2007), micro-arc oxidation (MAO) (Sun et al., 2007) and occasionally, low pressure plasma spraying (LPPS) and suspension plasma spraying (SPS) (Bouyer et al., 1997). Both HAp and fluorapatite (FAp) can potentially be deposited by plasma spray techniques. FAp in particular does not easily decompose in the plasma jet compared with HAp that tends to form C_3P or C_4P by thermal decomposition (see below). Also, plasma spray-deposited FAp coatings show a high degree of crystallinity (DeGroot, 1991). Histological and histomorphometric investigations of coatings implanted in animals show, however, that in contrast to HAp the FAp coatings are intensely attacked by soft tissue (Zimmermann, 1992). In addition, the ingrowth of bone cells into the porous FAp coatings is much suppressed (Lugscheider et al., 1994). Obviously fluoride ions released from the coatings cause cytotoxic responses and thus hamper implant integration. This suggests an exclusion of FAp from the list of potential biocompatible ceramics. It should be mentioned that in the vicinity of uncoated titanium implants significant titanium ion concentrations were measured owing to corrosion in the extremely aggressive body environment. Even though at this time this accumulation of titanium ions is considered nontoxic, an implant coated with HAp seems to provide a more benign environment in which only Ca and P ions concentrate.

Spraying of HAp is performed worldwide under widely varying conditions hence resulting in coatings very different in their phase purity, adhesion strength and crystallinity. The control of these coating properties is crucial: HAp shows a significantly higher stability than C_3P and C_4P during both *in vitro* and *in vivo* tests. Since the resistance to resorption also increases with increasing crystallinity, and proper adhesion of the coating to the metal implant prevents the invasion of acellular connective tissue leading potentially to a loosening of the bond to the bone, optimization of the three responses is mandatory.

9.3.1
Phase Composition

When the molten HAp particles impinge at the metal surface the idealized and neat phase separation shown in Figure 5.10 will be lost. The result is an extremely inhomogeneous calcium phosphate layer in which HAp, OHAp/OAp, TCP (α-TCP; β-TCP, whitlockite), TetrCP (hilgardite), CaO (oldhamite), and ACP (amorphous calcium phosphate) of various compositions are interspersed on a nano- to microcrystalline scale (Götze *et al.*, 2001; Li *et al.*, 2004). At the immediate interface to the metal substrate a thin layer of ACP exists formed by rapid quenching of the outermost shell with heat transfer rates beyond $10^6\,K\,s^{-1}$ (see also Figure 9.8, Figure 9.9).

Figure 9.4b shows patches of ACP scattered throughout the crystalline calcium phosphate phases attesting to the inhomogeneous structure of the coating. While ACP blends into the assembly of other phases in SEM-BSE mode (Figure 9.4a) it can easily be identified by cathodoluminescence imaging (Figure 9.4b).

The thin ACP layer at the coating interface takes on a specific significance as its high solubility *in vivo* and pronounced cracking by shrinking on transformation to HAp by rehydroxylation are considered one of the leading causes of coating delamination observed in clinical application. Hence much research is directed towards deposition of well-crystallized HAp layers with a minimum content of ACP.

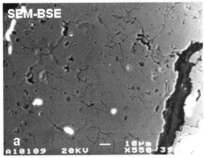 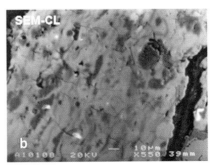

Figure 9.4 SEM micrographs in backscattered electron (BSE) (a) and cathodoluminescence (CL) modes (b) of an APS HAp coating deposited on Ti6Al4V (Götze *et al.*, 2001).

However, there is a caveat: too high a crystalline content may compromise coating performance in hip and dental prostheses owing to reduced adhesion strength and enhanced dissolution through a crack network (Gross et al., 1998a).

The morphology of the deposited particle splats follows the laws of fluid dynamics under supersonic conditions (Fauchais et al., 2004; Heimann, 2006b). The complex evolutionary steps a molten HAp particle undergoes after impacting the solid substrate has already been schematically shown above (Figure 5.11). While the first particle layer of about 5 μm thickness solidifies very quickly forming an ACP layer (Gross et al., 1998b), the following particles arriving at the substrate surface will splash over the previously deposited, already cooled particles compressing them further. Since the arrival time and the very short solidification time of an individual particle differ by several orders of magnitude a molten particle will never meet a molten pool at the surface. As a consequence any particle stresses will be retained and accumulate as the coating grows in thickness.

A useful technique to determine the approximate spray parameter settings in terms of plasma energy and spray distance is the so-called 'wipe' test during which a thin flat surface, e.g. a glass slide is moved quickly across the loaded plasma jet to collect individual particle splats (see Section 10.2.2.1). Figure 9.5 shows as-sprayed HAp splats deposited under different spray conditions.

While at a comparatively long spraying distance of 260 mm even a high plasma power of 45 kW does not melt the HAp particle completely (Figure 9.5A), shortening of the distance by only 20 mm to 240 mm leads at lower energy (30 kW) to a molten circular splat whose viscosity is still rather high so that in its center part many small pores persist (Figure 9.5B). Further narrowing the spray distance to 220 mm results in close to ideal melting with only a few pores remaining (Figure 9.5C). When at this distance the plasma power is again increased to 45 kW overheating occurs and an 'exploded' splat geometry results (Figure 9.5D) as described above. Quantification of the spreading behavior of molten particles is an important prerequisite of numerical simulation of deposit profiles (Kang and Ng, 2006).

Figure 5.11 above suggests that a macroscopic coating will consist of an extremely inhomogeneous jumble of various amorphous and crystalline calcium phosphate phases in several stress states, separated by pores, gas-filled voids, and spherical unmelted oversized particles. It is intuitively obvious that any attempt to unravel the complex thermal history of such coatings by integral analytical methods such as X-ray diffraction (XRD) or vibrational spectroscopic methods such as infrared spectroscopy (FTIR) will be doomed. In spite of this, virtually every contribution on the composition of plasma-sprayed HAp coatings compares the XRD pattern of the starting powder with that of the coating and concludes hastily that the latter consist of essentially homogeneous, more or less monophasic and stoichiometric HAp. This is far from reality. Since the presence of ACP to a large extent controls the mechanical, chemical and biological performance of a bioconductive coating, analytical methods able to distinguish between ACP and crystalline CP phases have to be applied, such as XRD using nonlinear least square fitting (Keller and Dollase, 2000) or standard-free (Bufler and Trettin, 2000) and standard-based Rietveld refinement (Gualtieri et al., 2006), micro-Raman spectros-

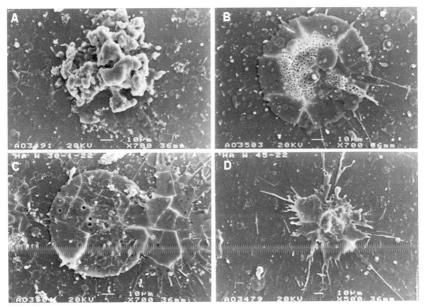

Figure 9.5 Splat configurations of HAp obtained by a 'wipe' test.

copy (Wen *et al.*, 2000; Hartmann *et al.*, 2001; Heimann *et al.*, 2003) or spatially highly resolved techniques such as STEM or TEM in conjunction with EDX, ED and EELS (Li *et al.*, 2003; Heimann and Wirth, 2006).

9.3.2
Parametric Study of HAp Coating Properties

A study was conducted applying low pressure plasma spraying (LPPS) technology to deposit HAp coatings onto Ti6Al4V substrates that led to new insight into the deposition mechanism. The high temperature of the plasma jet promotes, even during the very short residence time of the HAp powder particles in its hot zone, severe decomposition towards tricalcium (C_3P) and tetracalcium (C_4P) phosphates, or even CaO. This fact has discouraged commercial suppliers of medical implants from using LPPS to deposit osseoconductive HAp coating to Ti alloy stems of hip endoprostheses. The calcium phosphate phases occurring in the sprayed samples are compatible with the phase diagram shown in Figures 9.6 and 10.36 that indicate incongruent melting and hence decomposition of HAp above 1360 °C (deGroot, 1988). The stability of HAp is also a function of the water vapor partial pressure, $p(H_2O)$ of the surrounding atmosphere (Figure 5.10). The temperature representing the equilibrium

$$HAp + CaO \rightarrow HAp + C_4P + H_2O$$

decreases with increasing water vapor partial pressure, i.e. at a given temperature and a higher $p(H_2O)$ is the phase assembly HAp + CaO more stable than HAp + C_4P.

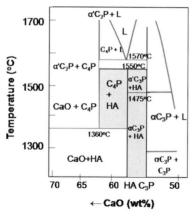

Figure 9.6 Binary phase diagram CaO-P_2O_5 in the range of stability of HAp (deGroot et al., 1990). The shaded areas refer to the processing range to obtain HAp coatings at a water partial pressure of 6.7×10^4 Pa (modified after Hench, 1991).

Parameters studied includes the plasma energy η, the powder feed rate m, the stand-off distance X, the substrate preheating temperature T_s, and the deposition chamber pressure P. The following trends were observed.

The powder feed rate m influences the heat transfer from the plasma to the powder particles, i.e. at the same plasma energy η the thermal decomposition of the HAp decreases with increasing powder feed rate. By the same token HAp decomposition is minimized at low substrate temperature T_s, increasing chamber pressure P and increasing stand-off distance X. On the other hand, optimizing the coating adhesion requires a reduction of residual stresses σ in the coating. This can be achieved at low plasma energies η, high substrate temperature T_s and low stand-off distance X.

The high substrate temperature promotes diffusion of Ti into the coating and leads to an increase in thickness of the reaction layer assumed to consist of CT (perovskite) (DeGroot et al., 1987) and/or CT_2 (Ji et al., 1992). This thicker reaction layer in turn reduces the thermal decomposition of subsequently deposited HAp layers (Ducheyne et al., 1990; Weng, 1994). Under these conditions the sprayed HAp layer attains a brown color thus supporting the ideas of increased Ti diffusion. The brown color disappears after heat treatment at 800 °C in air for 2 hours, possibly by oxidation of Ti^{3+} originally formed in the reducing argon/hydrogen plasma (Heimann et al., 1997).

Increasing the chamber pressure P produces a shorter plasma jet so that at a constant stand-off distance X lower surface temperatures occur. This in turn suppresses the thermal decomposition of HAp.

Hence the following coating optimization strategies should be applied:

- To suppress thermal decomposition of HAp the enthalpy supplied to the powder and the coating, respectively must be reduced. This can be done effectively by minimizing the plasma power η and the substrate temperature T_s but maximizing the powder feed rate m, the stand-off distance X and the chamber pressure P.

- To reduce the coating porosity but retain its biological functionality the enthalpy supplied to the powder must be increased by both increasing the plasma power η and the substrate temperature T_s but minimizing the stand-off distance X.

- Finally, to optimize the coating adhesion both the plasma power η and the stand-off distance X should be minimized but the substrate be preheated to alleviate the gradient of the coefficients of thermal expansion between coating and substrate, and hence the residual stresses introduced during cooling.

As it turns out the simultaneous fulfillment of these three conditions is clearly impossible and the production of HAp coatings on Ti6Al4V implants optimized in terms of phase purity, porosity and adhesion requires a parameter and hence property trade-off. It is not surprising then that a large number of studies can be found in the literature attempting to optimize HAp coating properties. Since the temperature and velocity of HAp particles play a dominant role the main parameters influencing their in-flight properties are the plasma power, plasma gas flow rate, and stand-off distance whereas the effects of powder feed rate and auxiliary gas flow rate are much less significant (Cizek et al., 2007).

9.4
Bioinert Bond Coats

In clinical application, coating failure by chipping, spalling, delamination and dissolution is observed on explanted endoprostheses consistently close to the Ti implant–coating interface (Spivak et al., 1990; Yang et al., 1995). This has frequently been attributed to the existence of a layer of amorphous calcium phosphate (ACP) at the coating–Ti interface formed during rapid quenching of molten or semi-molten droplets of calcium phosphate with exceptionally high cooling rates ($10^5 - 10^6$ K s^{-1}) (Ji et al., 1992) (see Figure 5.11). A continuous ACP layer is thought to act as a low energy fracture path (Park and Condrate, 1999) and, owing to its comparatively high solubility, will preferentially dissolve *in vivo* (Ducheyne et al., 1993; DeBruijn et al., 1994; Gross et al., 1998c) thus further weakening the mechanical integrity of the interface (Gross et al., 1998a; Tsui et al., 1998a, 1998b).

Clinical studies of retrieved hip prostheses coated with HAp clearly showed that micromorphological features of plasma-sprayed coatings play a decisive role in implant longevity as they critically influence adhesion, and cracking and fragmentation of the coatings, chemical dissolution of the amorphous phase of the coatings in areas of high loading as well as secondary carbonate apatite precipitation, and osteoclastic resorption (see for example Gross et al., 2004). In particular, failure of coating adhesion to the implant metal causes formation of a gap into which acellular connective tissue will inevitably invade. This pliable tissue layer will prevent solid attachment of the implant to the bone tissue with eventual aseptic loosening that will require a remediation operation. Hence this gap formation must be suppressed to guarantee long-term functionality and performance of the biomedical implant.

One way to achieve this goal is the application of stable bioinert bond coats. The use of bond coats to increase the adhesion of thermally sprayed ceramic wear- and corrosion-resistant coatings as well as thermal barrier coatings to a metallic substrate is well documented in the relevant literature. Hence, biocompatible bond coats may offer a means to further enhance the adhesion of HAp coatings to metallic substrates (Heimann, 1999b; Chou and Chang, 2002). In addition to improving coating adhesion such bond coats should also exhibit the following properties:

- The bond coat should prevent direct contact between Ti and HAp since metallic Ti is thought to catalyze thermal decomposition of HAp (Ji *et al.*, 1992; Weng *et al.*, 1994).

- The bond coat should reduce or even completely prevent the release of metal ions from the Ti6Al4V substrate to the surrounding living tissue. Such ion release has been found to cause massive hepatic degeneration in animals (Pereira *et al.*, 1995) as well as impaired development of human osteoblasts in *in vitro* tests (Tomas *et al.*, 1996) since the heavy metal ions are thought to affect negatively the transcription of RNA in cell nuclei, and also influence the activity of enzymes by replacing calcium or magnesium ions at binding sites (Schroeder, 1965).

- The bond coat should reduce the thermal gradient at the substrate–coating interface induced by the rapid quenching of the molten particle splats that leads to deposition of amorphous calcium phosphate (ACP) with a concurrent decrease in resorption resistance (Heimann *et al.*, 1997) and hence to reduced *in vivo* performance, i.e. longevity of the implants.

- The bond coat should prevent a steep gradient in the coefficients of thermal expansion between substrate and coating that will otherwise induce large residual tensile stresses leading to cracking and delamination of the coatings.

- The bond coat should cushion damage of the coating initiated by cyclic micromotions of the implant during movement of the patient in the initial phase of osseointegration (Søballe, 1993).

There is evidence that even without a bond coat acting as a sink for metal ions, HAp itself effectively absorbs titanium ions released from Ti6Al4V implants (Ducheyne and Healy, 1988; Ribeiro *et al.*, 1995; Ribeiro *et al.*, 2006), chromium and nickel ions released from stainless steel implants (Sousa and Barbosa, 1995), or nickel ions released from porous NiTi shape memory alloy (Jiang and Rong, 2006). According to a study by Ribeiro *et al.* (1995), calcium atoms in HAp were partially replaced by titanium atoms from a saline physiological solution of 0.9% NaCl to which titanium ions were added to simulate the release of Ti from Ti6Al4V. FTIR and FT–Raman (FTRS) investigations suggested the formation of a Ti-substituted hydroxyapatite, $Ca_{10-n}Ti_{n/2}[(PO_4)_6(OH)_2]$.

9.4.1
Composition of Bioinert Bond Coats

9.4.1.1 Calcium Silicate Bond Coats
Studies by Lamy (1993) and Lamy *et al.* (1996) confirmed the feasibility of applying dicalcium silicate-based (β-Ca_2SiO_4) bond coats to enhance the adhesion strength

and resorption resistance of the bioceramic HAp top coat by up to 30%. These bond coats should be biocompatible since calcium silicate glasses and ceramics were found to bond easily to living bone forming a biomimetic apatite surface layer when exposed to simulated body fluid (Zheng et al., 2005; Long et al., 2006). In fact, even P_2O_5 free CaO-SiO_2 glasses were shown to bond easily to living bone mediated by a thin apatite layer formed adjacent to the bone tissue (Ohura et al., 1991). The cyclic silicate trimer (Si_3O_9) group of pseudowollastonite was thought to act as an active site for heterogeneous, stereochemically promoted nucleation of HAp in simulated body fluid (Sahai and Anseau, 2005). In addition, histological investigations provided evidence that silicon may be allied to the initiation of mineralization of pre-osseous tissue in periosteal or endochondral ossification (Carlisle, 1970), presumably through Si-OH functional groups that are known to induce apatite nucleation (Cho et al., 1996; Kokubo et al., 2003) by providing bonding sites for cation-specific osteonectin attachment complexes on progenitor cells (Hench and Ethridge, 1998).

Wollastonite ($CaSiO_3$) was plasma sprayed onto Ti6Al4V substrates and its dissolution kinetics studied by immersion in TRIS solution (tris-hydroxymethyl aminomethane, $(CH_2OH)_3CNH_2$) buffered to a pH value of 7.4 with HCl (Xue et al., 2005). The coating showed low crystallinity with a large amount of easily soluble amorphous phase as well as CaO that appeared to promote apatite nucleation but led to coating disintegration. In contrast to that heat-treated coatings with increased crystallinity showed much higher stability in contact with SBF but apatite formation was greatly retarded.

9.4.1.2 Zirconia Bond Coats

Several attempts have been made to use zirconia as a bond coat material to improve the performance of bioconductive HAp coatings for hip endoprostheses. Since monolithic sintered HAp shows insufficient mechanical properties in terms of low bending strength, fracture toughness, modulus of elasticity and microhardness, that makes it undesirable for load bearing applications, research has been performed to improve these properties by the addition of fine zirconia particulates (Tamari et al., 1988a, 1988b). Generally, an increase of the bending strength and fracture toughness by a factor of two to three was observed, attributed to formation of reaction phases such as calcium zirconate and transition of tetragonal to cubic zirconia.

While HAp coatings on titania alloy are notoriously weak in cohesion as well as adhesion to the substrate surface, work was performed to strengthen these coatings by adding reinforcing particles of zirconia to form composite coatings deposited by either radio frequency suspension plasma spraying (Kumar et al., 2003), atmospheric plasma spraying (Chou and Chang, 2002), or low pressure plasma spraying (LPPS) (Heimann et al., 1998). The idea was to further mechanical interlocking between bond coat layer and substrate as well as establishing a chemical bond between bond coat and HAp. This, however, has not met with resounding success. Despite claims to the contrary zirconia-reinforced HAp coatings neither significantly increased the bond strength in as-sprayed coatings nor did they slow down the resorption in simulated body fluid (Chang et al., 1997a, 1997b). Moreover $CaZrO_3$ thought to be formed as a reaction product at the substrate–coating interface is suspected to deteriorate the mechanical properties of the coating system (Caetano-Zurita et al., 1994). Although a calcium zirconate bond coat appears to adhere well to the substrate it tends to exhibit lateral cracks parallel to the coating interface when subjected to even low tensile forces (Heimann et al., 1998). Also, stresses introduced into the HAp coating by thermally induced tetragonal-monoclinic phase transformation within the partially Ca-stabilized zirconia bond coat was found to lead to extensive scaling and concurrent leaching during treatment in simulated body fluid (HBSS). These findings were disputed by Chou and Chang (2002) who claimed an increase of the adhesive bond strength of a HAp/ZrO_2 composite coating from 28.6 ± 3.2 MPa for a pure HAp coating to 36.2 ± 3.0 MPa, owing presumably to diffusion of calcium ions from the HAp matrix into the zirconia bond coat. However, the peel adhesion strength[3] of as-sprayed coatings in the presence of a zirconia bond coat was measured by Kurzweg et al. (1998a, 1998b) to be significantly lower (probability point of a double-sided t-test of 2.54 compared with a tabulated value of 2.07 for 22 degrees of freedom, 95% confidence interval) at $18\,\text{N}\,\text{m}^{-1}$ compared with the peel strength of $22\,\text{N}\,\text{m}^{-1}$ of a HAp coating without a bond coat. While the reason for this discrepancy is not clear a different thermal history of the coatings as well as the use of fully stabilized zirconia by Chou and Chang (2002) as opposed to partially Ca-stabilized zirconia used by Heimann et al. (1998) has been suspected. In particular, the existence of massive residual stresses at the bond coat–HAp coating interface related to a mismatch in the coefficients of thermal expansion (thermal stress, see above) will be relieved during leaching for 28 days in simulated body fluid (HBSS) resulting in strong coating delamination (Heimann et al., 1998).

9.4.1.3 Zirconia/Titania Bond Coats

Some limited improvement was observed after application of a mechanical mixture of nonstabilized zirconia (27 mol%)/titania (73 mol%) as a bond coats with eutectic composition (Heimann et al., 1998; Kurzweg et al., 1998a, 1998b). Figure 9.7 shows an SEM micrograph of a cross-section Ti6Al4V substrate–ZrO_2/TiO_2 bond coat–HAp (Kurzweg et al., 1998a).

While the bond coat composition corresponds to a eutectic ratio, the extremely short dwell time (microseconds) of the only mechanically mixed powder particles in the hot core of the plasma jet is not sufficient to produce features akin to a eutectic structure. Instead, the bond coat consists of seemingly unrelated streaks of intertwined titania-rich (dark) and zirconia-rich (light) segments, possibly related to separation already occurring during flight on different trajectories in the hot plasma jet as the result of different particle densities. Also, very small spherical zirconia-rich particles appear to delineate the interface with the HAp top layer as the result of

---

[3] While the numerical values of a tensile adhesion test according to ASTM C633-01 and a peel adhesion test according to a modified ASTM D3167-03 designation (Sexsmith and Troczynski, 1994, 1996) cannot directly be converted into each other their relative values are proportional. The conventional tensile pull test measures failure stress, expressed as the ratio of applied force to (geometric) coating area in dimension: $\text{N}\,\text{m}^{-2}$, the peel test measures the energy required to separate the coating and the substrate along a line in dimension: $\text{N}\,\text{m}^{-1}$ (Kurzweg et al., 1998b).

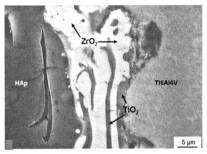

Figure 9.7 SEM micrograph of a cross-section Ti6Al4V substrate–ZrO$_2$/TiO$_2$ bond coat–HAp (Kurzweg et al., 1998a).

splashing during relaxation of the outmoving shock wave as shown in the schematic of Figures 5.11e and 5.11f. The peel adhesion strength (Kurzweg et al., 1998a) is noticeably and significantly higher (probability point of a double-sided t-test of 2.15 compared with a tabulated value of 2.07 for 22 degrees of freedom, 95% confidence interval) than in the presence of a zirconia bond coat, showing values around 32 N m^{-1} compared with the peel strength of 22 N m^{-1} of a HAp coating without a bond coat.

9.4.1.4 Titania Bond Coats and TiO$_2$/HAp Composite Coatings

Considerable work has been expended in the past towards studying the biological behavior and potential application as a bioconductive implant coating material of TiO$_2$/HAp composite coatings and TiO$_2$ bond coat/HAp top coat systems. These composite and 'duplex' coatings were deposited by a large range of techniques including (i) electrochemical anodization to form a TiO$_{2-x}$ layer that acted as a template for a biomimetically formed HAp layer on immersion in SBF (Ng et al., 2005); (ii) micro-arc oxidation (MAO) of an electron beam-evaporated HAp layer on Ti (Wei et al., 2007a); (iii) coating of a Ti substrate in a sol consisting of colloidal particles of titania and submicron HAp particles (Milella et al., 2001); (iv) deposition of TiO$_2$ and HAp layers by a sol-gel process using spin coating and subsequent heat treatment at 500 °C (Kim et al., 2004); (v) a combination of MAO and electrophoretic deposition of HAp (Nie et al., 2000); (vi) electrochemical codeposition of HAp and titania under hydrothermal conditions (Xiao et al., 2006); (vii) HVOF spraying of TiO$_2$/HAp composite powders (Li et al., 2003); (viii) plasma spraying of a TiO$_2$/HAp bond coat underneath a HAp top coat (Lu et al., 2004); (ix) a combination of plasma spraying of TiO$_2$ nanoparticles and hydrogen implantation into the resulting coating, and subsequent immersion in SBF to form a biomimetical HAp top layer (Liu et al., 2005), and (x) atmospheric plasma spraying of HAp on top of a TiO$_2$ bond coat (Heimann et al., 1997; Heimann et al., 1998; Kurzweg et al., 1998a; Heimann, 1999a; Heimann et al., 2004a; Heimann et al., 2004b; Heimann and Wirth, 2006).

Titania coatings were deposited in lieu of HAp to exploit their bioconductive and bioadhesive properties (Lima et al., 2005; Liu et al., 2006; Zhao et al., 2006). In fact they will act as a natural extension of the existing passivating oxide layer. In particular, recent work by Zhao et al. (2006) showed that ASP titania layers on Ti6Al4V substrates develop a biomimetical carbonate-apatite ('bone-like' apatite) layer when incubated in SBF after treatment with NaOH (Kokubo et al., 2003). Without activation by NaOH no apatite was formed. The chemical composition of plasma-sprayed titania coatings varies widely and is usually thought to consist of a succession of substoichiometric Ti_nO_{2n-1} Magnéli-type phases or a mixture of rutile and anatase. However, the crystalline stoichiometric brookite polymorph of titania has also been found (Figures 9.8, 9.9; Heimann and Wirth, 2006) as well as near stoichiometric (n = 1.04) TiO (Park and Condrate, 1999).

Figure 9.8 shows an STEM image of the cross-section of a titania bond coat and a calcium phosphate top coat whose portion adjacent to the bond coat consists of ACP (Heimann and Wirth, 2006).

Figure 9.9 depicts the interface titania (brookite) bond coat–calcium phosphate top coat at higher magnification as well as electron diffraction pattern confirming the amorphous nature of the ACP layer (Heimann, 2007).

Studies of the *in vivo* behavior of titania bond coat–HAp coating 'tandem' assemblies revealed (Itiravivong et al., 2003; Heimann et al., 2004a, 2004b; Heimann, 2006a) that the coating adhesion strength was noticeably improved in the presence of a titania bond coat. Figure 9.10 shows a cross-section of an explanted 'tandem'-coated Ti6Al4V cube that had been implanted for 6 months in the femoral condyle of dog. Statistical evaluation applying Kruskal–Willis and Mann–Whitney tests, respectively (Riffenburgh, 1999) confirms that a solid and continuous bone apposition occurred whose osseointegration and coating adhesion reached average values of $82.2 \pm 1.0\%$ and $94.3 \pm 2.2\%$, respectively (Itiravivong et al., 2003; Heimann et al., 2004b).

More details on the properties, functionality, and *in vitro* and *in vivo* behavior of titania bond coat-HAp top coat duplex layers can be found in Heimann (2007).

9.5
Other Thermal Coating Techniques

Despite some attempts to improve HAp coatings by applying HVOF (Ogushi et al., 1992; Sturgeon and Harvey, 1995) and r.f. inductively coupled-plasma techniques

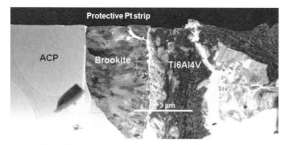

Figure 9.8 STEM image of a cross-section of a plasma-sprayed titania (brookite) bond coat-HAp top coat assembly. On the Ti6Al4V substrate side embedded α-alumina (corundum) particles are visible. The sample was prepared by focused ion beam (FIB) cutting (Heimann and Wirth, 2006).

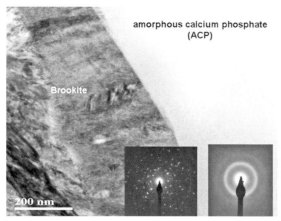

Figure 9.9 Bright field STEM image of the interface between an oriented polycrystalline titania (brookite) bond coat (left) and an amorphous calcium phosphate (ACP) top coat (right). The insets show electron diffraction pattern of both phases (Heimann and Wirth, 2006; Heimann, 2007).

(Kameyama *et al.*, 1993) the results were not very convincing. Even though HVOF thermal spraying reduces powder losses in the flame due to the short residence time and produces coatings with high crystallinity it leads to coatings with rather low adhesion (<18 MPa) and some decomposition as evidenced by the presence of C_3P and C_4P. Inductively coupled-plasmas avoid contamination of the coating by W and Cu used as electrode materials in the DC plasmatrons but lead even at low plasma powers between 3 and 12 kW to pronounced decomposition of HAp owing to the long residence time of the particles in the plasma with free-stream velocities typically around only 10–30 m s^{-1}. The long residence times of the powder compared with a

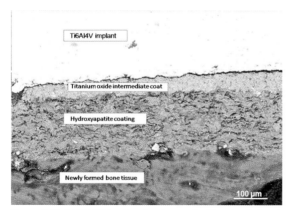

Figure 9.10 A Ti6Al4V implant (top) coated with a titania bond coat-HAp top coat assembly shows solid and continuous bone apposition after implantation for 6 months in the femoral condyle of dog (Itiravivong *et al.*, 2003; Heimann *et al.*, 2004b).

DC plasma or HVOF system leads to intense particle heating and thus decomposition. Also, low adhesion strength of only 5.14 MPa may be attributed to the low plasma velocities that impart only low momentum to the particles.

Suspension plasma spraying (SPS) may yield better prospects. The introduction of the powder material into the plasma jet in form of a solid–liquid suspension allows the use of very fine powders coupled with a substantial increase of the deposition efficiency and rate of coating deposition. The possibility of depositing very thin coatings (<1 µm) provides advantages in terms of minimization of the residual coating stresses. The technique of SPS was developed in the 1990s (Gitzhofer et al., 1997; Bouyer et al., 1997; Bouyer, 1997), using an inductively coupled plasma source to deposit HAp coatings. The morphology of the powder formed by evaporation of the liquid will be determined by the time constants of the vapor diffusion, the shrinking of the droplets, the diffusion of the suspended solids, the heat conductivity in the surrounding gas and the heat conduction within the droplet (see Section 8.2.2; Figure 8.21). Beyond a critical supersaturation solid material will be precipitated at the droplet surface that on further evaporation of the liquid moves towards the center of the ever shrinking droplet. This process results inevitably in the formation of hollow spheres that will impart a high porosity to the coatings. However, if the solids content of the suspension is low precipitation takes place in the volume of the droplet and dense spheres will be generated that on melting will behave exactly as a solid powder during conventional atmospheric plasma spraying. Hence the porosity of the coatings can be engineered throughout a wide range by selecting the appropriate values of solid/liquid ratio, viscosity of the suspension, suspension feed rate, injection point into the plasma jet, and plate power, i.e. plasma enthalpy.

Flame assisted chemical vapor deposition (FACVD)[4] of HAp has been reported recently to yield highly crystalline coatings with a network of open and interconnected pores (Trommer et al., 2007). A precursor solution consisting of a stoichiometric mixture of calcium acetate and diammonium hydrogenphosphate in ethanol was radially injected through an atomization device into a propane–air pilot flame produced by a Bunsen burner. When the atomized solution intercepts this pilot flame another flame, the so-called main flame is ignited by the combustion of the ethanolic precursor solution, and propagates perpendicular to the pilot flame towards the target to be coated. Temperature is managed by using the output of a thermocouple attached to the rear of the substrate that controls cooling of the substrate. Hence the deposition temperature can be controlled at $500 \pm 5\,°C$. The atomization gas (compressed air) pressure was kept at 4 MPa, the precursor solution flow rate at $12\,ml\,min^{-1}$, the deposition time at 10 min, and the distance between the tip of the atomizer and the substrate at 140 mm. Using these parameters, the highly porous HAp coating formed had a thickness of about 410 µm. The authors emphasize that this novel FACVD technique is very cost effective and produces coatings whose tailored pore size distributions may have the potential to enhance in-growth of bone cells.

4) A more appropriate name would be 'solution flame spraying', in line with other well-known techniques such as powder flame spraying and wire flame spraying.

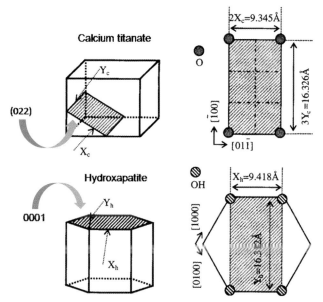

Figure 9.11 Epitaxial relations between (0 2 2) of calcium titanate (perovskite) and (0 0 0 1) of HAp (Wei et al., 2007b).

Plasma electrolytic oxidation (PEO) or MAO (Yerokhin et al., 2000: Gupta et al., 2007) has been applied to deposit bioceramic films and coatings (for example Dong et al., 2007; Paulmier et al., 2007; Wei et al., 2007a). A recent contribution describes the structure of calcium titanate/titania composite films and the mechanism of biomimetic apatite formation during incubation in simulated body fluid. A plausible sequence of events was proposed by interfacial molecular recognition during apatite formation involving epitaxial crystal lattice matching as shown schematically in Figure 9.11 (Wei et al., 2007b).

The lattice plane (022) in the case of perovskite is defined by the position of oxygen atoms; the plane (0001) in the case of HAp by hydroxyl ions. A 2D lattice match exists; $2X_{perovskite} \approx X_{hydroxyapatite}$ (mismatch: 0.8%) and $3Y_{perovskite} \approx Y_{hydroxyapatite}$ (mismatch: 0.09%). This finding is well in accord with current assumptions about the control of apatite nucleation by a calcium titanate surface (Kokubo et al., 2003; Webster et al., 2003; Ohtsu et al., 2006) and may also account for the dramatic increase of the adhesion strength of HAp to a Ti alloy substrate in the presence of a Ti oxide bond coat (Kurzweg et al., 1998a).

9.6
Outlook

The move is on towards third generation biomaterials. The first generation biomaterials relied on bioinert materials that were incorporated into the human body to *replace* missing bone parts or fill bone cavities including dental amalgam, gypsum, ivory or even wood. Second generation biomaterials already provided bioactive and bioresorbable functions typical examples of which are HAp coatings on endoprosthetic implants described above. Now a limit has been reached to medical practice that emphasizes replacement of tissue by a materials science-based approach. In the future more biologically based methods are called for that will concentrate on *repair/ regeneration* of tissue, including development of third generation biomaterials that activate genetic repair, for example through functionalization of HAp surfaces with bone morphogenetic proteins (BMPs) and other osteoinductive proteins. This is reflected in a recent definition that 'biocompatibility is the property of a material to generate in the host an appropriate response'. Since this response can also be negative it is one of the challenging tasks of biomaterials research to minimize and control these negative responses.

Hildebrand *et al.* (2006) reported on surface coatings for biological activation and functionalization of medical devices including *mechanical* functionalization by polishing, machining or irradiation of metallic biomaterials, *physical* functionalization by surface coatings including APS (atmospheric plasma spraying), CVD (chemical vapor deposition), CGDS (cold gas dynamic spraying), PLD (pulsed laser deposition), MAO (micro-arc oxidation or plasma electrolytic oxidation, PEO), sol-gel, and polyelectrolyte coatings and films as well as *chemical and biological* functionalization including DDS (drug delivery systems); biodegradable collagen and polylactic acid molecules; release of biomolecules grafted onto cage molecules such as cyclodextrin, biomimetic HAp coatings, recombinant human bone morphogenetic proteins (rhBMPs), and other materials and processes. Increasingly precursor powder compositions are tailored to the real composition of biological HAp by substituting Ca by Na, Mg, Sr and other metabolic elements (for example Xue *et al.*, 2007) as well as PO_4^{3-} and OH^- by CO_3^{2-}.

Also, stem cell engineering, still a highly contentious issue hotly debated among various segments of the population and politicians, and rapid, predictive *in vitro* test methods for biomaterials–cell responses that might replace costly and ethically dubious animal models, are high up on the agenda of future biomaterials research and development to solve the socio-economic issues of an ageing population in developed countries around the world.

References

E. Bouyer, *Étude de la Préparation et de Dépôts à Partir de Suspension par Plasma Inductif – le Cas de l'Hydroxyapatite Phosphocalcique*. Unpublished Ph.D. Thesis, Université de Sherbrooke, Québec, Canada, 1997.

E. Bouyer, F. Gitzhofer, M.I. Boulos, *J. Mater. Sci. Lett.*, 1997, **49** (2), 58.

S.R. Brown, I.G. Turner, H. Reiter, *J. Mater. Sci.: Mater. Med.*, 1994, **5**, 756.

M. Bufler, R. Trettin, in: *Applied Mineralogy in Research, Economy, Technology, Ecology and Culture*. Proc. 6th Intern. Congress on Applied Mineralogy, D. Rammlmair *et al.* (eds.), ICAM2000, Göttingen, Germany, July 17–19, 2000, 107.

J. Caetano-Zurita, O. Bermudez, I. Lopez-Valero, E.B. Stucchi, J.A. Varella, J.A. Planell, S. Martinez, In: *Bioceramics*, **7**, Proc. 7th

Intern. Symp. Ceramics in Medicine, O.H. Anderson, A. Yli-Urpo (eds.), Turku, Butterworth-Heinemann: Oxford, 1994, 267.

E.M. Carlisle, *Science*, 1970, **167** (January 16), 279.

E. Chang, W.J. Chang, B.C. Wang, C.Y. Yang, *J. Mater. Sci.: Mater. Med.*, 1997a, **8**, 193.

E. Chang, W.J. Chang, B.C. Wang, C.Y. Yang, *J. Mater. Sci.: Mater. Med.*, 1997b, **8**, 201.

S.B. Cho, F. Miyaki, T. Kokubo, K. Nakanishi, N. Soga, T. Nakamura, *J. Biomed. Mater. Res.*, 1996, **32**, 375.

B.Y. Chou, E. Chang, *J. Mater. Sci.: Mater. Med.*, 2002, **13**, 589.

L. Chou, B. Marek, W.R. Wagner, *Biomaterials*, 1999, **19**, 977.

J. Cizek, K.A. Khor, Z. Prochazka, *Mater. Sci. Eng. C*, 2007, **27** (2), 340.

T.W. Clyne, G.C. Gill, *J. Thermal Spray Technol.*, 1996, **5**, 401–416.

B. Cofino, P. Fogarassy, P. Millet, A. Lodini, *J. Biomed. Mater. Res., A*, 2004, **70** (1), 20.

J.E. Davies, *Biomaterials*, 2007, **28** (34), 5058.

J.D. DeBruijn, Y. Lovell, C. van Blitterswijk, *Biomaterials*, 1994, **15**, 543.

K. DeGroot, R.T.G. Geesink, C.P.A.T. Klein, P. Serekian, *J. Biomed. Mater. Res.*, 1987, **21**, 1375.

K. DeGroot, *Ann. N.Y. Acad. Sci.*, 1988, **523**, 227.

K. DeGroot, C.P.A.T. Klein, J.G.C. Wolke, J. de Blieck-Hogervorst, In: *Handbook of Bioactive Ceramics*, T. Yamamuro, L.L. Hench, J. Wilson (eds.), CRC Press: Boca Raton, FL, Vol. II, 1990, 3.

K. DeGroot, *J. Ceram. Soc. Jpn, Int. Edn*, 1991, **99**, 917.

D. DeSantis, C. Guerriero, P.F. Nocini, A. Ungersbock, G. Richards, P. Gotte, U. Armato, *J. Mater. Sci.: Mater. Med.*, 1996, **7**, 21.

E. Dörre, *Biomed. Technik*, 1989, **34**, 46.

E. Dörre, in: *Künstlicher Knochenersatz in der Orthopädie und Traumatologie*, A. Kirgis, W. Noack (eds.), Pontenagel Press: Bochum, 1992, 17.

Y.X. Dong, Y.S. Chen, Q. Chen, B. Liu, Z.X. Song, *Surf. Coat. Technol.*, 2007, **201**, 8789.

P. Ducheyne, K.E. Healy, *J. Biomed. Mater. Res.*, 1988, **22**, 1137.

P. Ducheyne, S. Radin, M. Heughebaert, J.C. Heughebaert, *Biomaterials*, 1990, **11**, 244.

P. Ducheyne, S. Radin, L. King, *J. Biomed. Mater. Res.*, 1993, **27**, 25.

P. Fauchais, M. Fukumoto, A. Vardelle, M. Vardelle, *J. Thermal Spray Technol.*, 2004, **13** (3), 337.

F. Fazan, P.M. Marquis, *J. Mater. Sci.: Mater. Med.*, 2000, **11**, 787.

M.J. Filiaggi, N.A. Coombs, R.M. Pilliar, *J. Biomed. Mat. Res.*, 1991, **25**, 1211.

P. Frayssinet, F. Tourenne, N. Roquet, P. Conte, C. Delga, G. Bonel, *J. Mater. Sci.: Mater. Med.*, 1994, **5**, 11.

F. Gitzhofer, E. Bouyer, M.I. Boulos, US 5,609,921, (1997).

J. Götze, R.B. Heimann, H. Hildebrandt, U. Gburek, *Mat.-wiss.u.Werkstofftech.*, 2001, **32**, 130.

O. Graßmann, R.B. Heimann, *J. Biomed. Mater. Res.*, 2000, **53** (6), 685.

K.A. Gross, V. Gross, C.C. Berndt, *J. Am. Ceram. Soc.*, 1998a, **81** (1), 106.

K.A. Gross, C.C. Berndt, H. Herman, *J. Biomed. Mater. Res.*, 1998b, **39** (3), 407.

K.A. Gross, B. Ben-Nissan, W.R. Walsh, E. Swarts, in: *Thermal Spray. Meeting the Challenges of the 21st Century*, C. Coddet (ed.), Proc. 15th ITSC. May 25.29, 1998c, Nice, France, 1133.

K.A. Gross, W. Walsh, E. Swarts, *J. Thermal Spray Technol.*, 2004, **13** (2), 190.

M.L. Gualtieri, M. Prudenziati, A.F. Gualtieri, *Surf. Coat. Technol.*, 2006, **201** (6), 2984.

P. Gupta, G. Tenhundfeld, E.O. Daigle, D. Ryabkov, *Surf. Coat. Technol.*, 2007, **201**, 8746.

M. Härting, *Acta Mater.*, 1998, **46** (4), 1427.

Y. Han, K. Xu, J. Lu, *J. Biomed. Mater. Res.*, 2001, **55** (4), 596.

P. Hartmann, C. Jäger, St. Barth, J. Vogel, K. Meyer, *J. Solid State Chem.*, 2001, **160**, 460.

L.I. Havelin, B. Engesaeter, B. Espehang, O. Furnes, S.A. Lie, S.E. Vollset, *Acta Orthop. Scand.*, 2000, **71**, 337.

R.B. Heimann, T.A. Vu, M.L. Wayman, *Eur. J. Mineral.*, 1997, **9**, 597.

R.B. Heimann, H. Kurzweg, D.G. Ivey, M.L. Wayman, *J. Biomed. Mater. Res.*, 1998, **43**, 441.

R.B. Heimann, *J. Thermal Spray Technol.*, 1999a, **8** (4), 597.

R.B. Heimann, *Mat.-wiss. u. Werkstofftechn.*, 1999b, **30**, 775.

R.B. Heimann, O. Graßmann, M. Hempel, R. Bucher, M. Härting, in: *Applied Mineralogy in Research, Economy, Technology, Ecology and Culture.* Proc. 6th Intern. Congress on Applied Mineralogy, D. Rammlmair et al. (eds.), ICAM2000, Göttingen, Germany, July 17–19, 2000, 155.

R.B. Heimann, H.V. Tran, P. Hartmann, *Mat.-wiss. u. Werkstofftechn.*, 2003, **34** (12), 1163.

R.B. Heimann, N. Schürmann, R.T. Müller, *J. Mater. Sci.: Mater. Med.*, 2004a, **15**, 1945.

R.B. Heimann, P. Itiravivong, A. Promasa, *BIOmaterialien*, 2004b, **5** (1), 38.

R.B. Heimann, *BIOmaterialien*, 2006a, **7** (1), 29.

R.B. Heimann, *Surf. Coat. Technol.*, 2006b, **201**, 2012.

R.B. Heimann, R. Wirth, *Biomaterials*, 2006, **27**, 823.

R.B. Heimann, in: *Trends in Biomaterials Research*, P.J. Pannone, (ed.), Nova Science Publishers, Inc.: Hauppauge, N.Y., 2007, Chapter 1, 1–80.

L.L. Hench, E.C. Ethridge, *Biomaterials, an Interfacial Approach.* Academic Press: New York, 1998, 87.

L.L. Hench, *J. Am. Ceram. Soc.*, 1991, **74** (7), 1487.

H.F. Hildebrand, N. Blanchemain, G. Mayer, F. Chai, M. Lefebvre, F. Boschin, *Surf. Coat. Technol.*, 2006, **200**, 6318.

ISO 13779.4, *Implants for surgery-hydroxyapatite. Part 4*: Determination of coating adhesion strength., 2002.

P. Itiravivong, A. Promasa, T. Laiprasert, T. Techapongworachai, S. Kuptnirasaikul, V. Thanakit, R.B. Heimann, *J. Med. Assoc. Thai*, 2003, **86** (Suppl. 2), 422.

H. Ji, C.B. Ponton, P.M. Marquis, *J. Mat. Sci.: Mater. Med.*, 1992, **3**, 283.

H.C. Jiang, L.J. Rong, *Surf. Coat. Technol.*, 2006, **201**, 1017.

T. Kameyama, M. Ueda, A. Motoe, K. Ohsaki, H. Tanizaki, K. Iwasaki, in: Proc. 1st Intern. Conf. on Processing Materials for Properties, H. Henein, T. Oki (eds.), The Minerals, Metals and Materials Soc., 1993, 1097.

C.W. Kang, H.W. Ng, *Surf. Coat. Technol.*, 2006, **200** (16/18) 5462.

L. Keller, W.A. Dollase, *J. Biomed. Mater. Res.*, 2000, **49**, 244.

H. Kim, R.P. Camata, S. Lee, G.S. Rohrer, A.D. Rollett, Y.K. Vohra, *Acta Mater.*, 2007, **55** (1), 131.

H.W. Kim, Y.H. Koh, L.H. Li, S. Lee, H.E. Kim, *Biomaterials*, 2004, **25** (13), 2533.

C.P.A.T. Klein, *Biomaterials*, 1990, **11**, 509.

C.P.A.T. Klein, P. Patka, H.B.M. van der Lubbe, J.G.C. Wolke, K. DeGroot, *J. Biomed. Mater. Res.*, 1991, **25**, 53.

T. Kokubo, H.M. Kim, M. Kawashita, *Biomaterials*, 2003, **24**, 2161.

R. Kumar, P. Cheang, K.A. Khor, *Biomaterials*, 2003, **24**, 2611.

S. Kuroda, T. Deudo, S. Kitahara, *J. Thermal Spray Technol.*, 1995, **4** (1), 75.

H. Kurzweg, R.B. Heimann, T. Troczynski, *J. Mater. Sci.: Mater. Med.*, 1998a, **9**, 9.

H. Kurzweg, R.B. Heimann, T. Troczynski, M.L. Wayman, *Biomaterials*, 1998b, **19**, 1507.

D. Lamy, *Development of Plasma-Sprayed Hydroxylapatite Coating Systems for Surgical Implant Devices.* Unpublished Master thesis, 1993, University of Alberta, Edmonton, Alberta, Canada.

D. Lamy, A.C. Pierre, R.B. Heimann, *J. Mater. Res.*, 1996, **11** (3), 680.

P. Leali Tranquilli, A. Merolli, C. Gabbi, A. Cacchioli, G. Gonizzi, *J. Mater. Sci.: Mater. Med.*, 1994, **5**, 345.

R.Z. LeGeros, I. Orly, M. Gregoire, G. Daculsi, in: *The Bone-Biomaterials Interface* J.E. Davies, (ed.), University of Toronto Press, 1991, 76.

J.E. Lemons, in: *Bone Implant Interface*, H.U. Cameron, (ed.). Mosby: St. Louis, Baltimore, Boston, 1994, 307.

H. Li, K.A. Khor, P. Cheang, *Biomaterials*, 2003, **24** (6), 949.

H. Li, K. A. Khor, P. Cheang, *Biomaterials*, 2004, **25** (17), 3463.

R.S. Lima, B.R. Marple, H. Li, K.A. Khor, *Biocompatible Titania Thermal Spray Coating Made from Nanostructured Feedstock*, Patent application, March, 2005.

X. Liu, X. Zhao, R.K.Y. Fu, J.P.Y. Ho, C. Ding, P.K. Chu, *Biomaterials*, 2005, **26**, 6143.

X. Liu, X. Zhao, C. Ding, P.K. Chu, *Appl. Phys. Lett.*, 2006, **88**, 013905.

L.H. Long, L.D. Chen, S.Q. Bai, J. Chang, K.L. Lin, *J. Eur. Ceram. Soc.*, 2006, **26** (9), 1701.

Y.P. Lu, M.S. Li, S.T. Li, Z.G. Wang, R.F. Zhu, *Biomaterials*, 2004, **25** (18), 4393.

E. Lugscheider, M. Knepper, B. Heimberg, A. Dekker, C.J. Kirkpatrick, *J. Mat. Sci.:Mater. Med.*, 1994, **5**, 371.

J. Matejicek, S. Sampath, *Acta Mater.*, 2003, **51** (3), 863.

E. Milella, F. Cosentino, A. Licciulli, C. Massaro, *Biomaterials*, 2001, **22** (11), 1425.

R. Millet, E. Girardin, C. Braham, A. Lodini, *J. Biomed. Mater. Res.*, 2002, **60** (40), 679.

B.S. Ng, I. Annergren, A.M. Soutar, K.A. Khor, A.E. Jarfors, *Biomaterials*, 2005, **26** (10), 1087.

X. Nie, A. Leyland, A. Matthews, *Surf. Coat. Technol.*, 2000, **125** (1–3), 407.

C.Y. Ning, Y.J. Wang, X.F. Chen, N.R. Zhao, J.D. Ye, G. Wu, *Surf. Coat. Technol.*, 2005, **200** (7), 2403.

H. Ogushi, K. Ishikawa, S. Ojima, Y. Hirayama, K. Seto, G. Eguchi, *Biomaterials*, 1992, **13**, 471.

N. Ohtsu, K. Saito, K. Asami, T. Hanawa, *Surf. Coat. Technol.*, 2006, **200**, 5455.

K. Ohura, T. Yamamura, T. Nakamura, T. Kokubo, Y. Ebisawa, Y. Kotoura, M. Oka, *J. Biomed. Mater. Res.*, 1991, **25**, 357.

E. Park, R.A. Condrate, D.T. Hoelzer, G.S. Fischman, *J. Mater. Sci.: Mater. Med.*, 1998, **9** (11), 643.

E. Park, R.A. Condrate, *Mater. Lett.*, 1999, **40**, 228.

T. Paulmier, J.M. Bell, P.M. Fredericks, *Surf. Coat. Technol.*, 2007, **201**, 8761.

M.L. Pereira, A.M. Abreu, J.P. Sousa, G.S. Carvalho, *J. Mater. Sci.: Mater. Med.*, 1995, **6**, 523.

J. Pina, A. Dias, J.L. Lebrun, *Mater. Sci. Eng. A*, 2003, **347** (1–2), 21.

A. Rapacz-Kmita, A. Ślósarczyk, Z. Paszkiewicz, *J. Eur. Ceram. Soc.*, 2006, **26** (8), 1481.

C.C. Ribeiro, M.A. Barbosa, A.A.S.C. Machado, A. Tudor, M.C. Davies, *J. Mater. Sci.: Mater. Med.*, 1995, **6**, 829.

C.C. Ribeiro, I. Gibson, M.A. Barbosa, *Biomaterials*, 2006, **27** (9), 1749.

R.H. Riffenburgh, *Statistics in Medicine*, Harcourt Brace: Orlando, 1999, 277.

D.P. Rivero, J. Fox, A.K. Skipor, R.M. Urban, J.O. Galante, *J. Biomed. Mat. Res.*, 1988, **22**, 191.

N. Sahai, M. Anseau, *Biomaterials*, 2005, **26** (29), 5763.

H.A. Schroeder, *Adv. Int. Med.*, 1965, **8**, 159.

V. Sergo, O. Sbaizero, D.R. Clarke, *Biomaterials*, 1997, **18** (6), 477.

M. Sexsmith, T. Troczynski, *J. Thermal Spray Technol.*, 1994, **3** (4), 404.

M. Sexsmith, T. Troczynski, *J. Thermal Spray Technol.*, 1996, **5**, 196.

K. Søballe, *Acta Orthop. Scand.*; No. 255, 1993, **64**, 58 p.

S.R. Sousa, M.A. Barbosa, *J. Mater. Sci.: Mater. Med.*, 1995, **6**, 818.

J.M. Spivak, J.L. Ricci, N.C. Blumenthal, H. Alexander, *J. Biomed. Mater. Res.*, 1990, **24**, 1121.

A.J. Sturgeon, M.D.F. Harvey, in: Proc. 14th ITSC'95, Kobe, May 22–26, 1995, 933.

J. Sun, Y. Han, X. Huang, *Surf. Coat. Technol.*, 2007, **201**, 5655.

N. Tamari, I. Kondo, M. Mouri, *J. Ceram. Soc. Jpn, Int. Edn*, 1988a, **96**, 108.

N. Tamari, I. Kondo, M. Mouri, M. Kinoshita, *J. Ceram. Soc. Jpn, Int. Edn*, 1988b, **96**, 1170.

H. Tomas, G.S. Carvalho, M.H. Fernandes, A.P. Freire, L.M. Abrantes, *J. Mater. Sci.: Mater. Med.*, 1996, **7**, 291.

M. Topić, T. Ntsoane, R.B. Heimann, *Surf. Coat. Technol.*, 2007, **201** (6), 3633.

R.M. Trommer, F.N. Souza, M.D. Lima, L.A. Santos, C.P. Bergmann, *Surf. Coat. Technol.*, 2007, **201**, 9587.

Y.C. Tsui, C. Doyle, T.W. Clyne, *Biomaterials*, 1998a, **19**, 2015.

Y.C. Tsui, C. Doyle, T.W. Clyne, *Biomaterials*, 1998b, **19**, 2031.

B.C. Wang, E. Chang, C.Y. Yang, D. Tsu, C.H. Tsai, *Surf. Coat. Technol.*, 1993, **58** (2), 107.

T.J. Webster, C. Ergun, R.H. Doremus, W.A. Lanford, *J. Biomed. Mater. Res., A* 2003, **67** (3), 975.

D. Wei, Y. Zhou, D. Jia, Y. Wang, *Surf. Coat. Technol.*, 2007a, **201**, 8723.

D. Wei, Y. Zhou, D. Jia, Y. Wang, *Surf. Coat. Technol.*, 2007b, **201**, 8715.

J. Wen, Y. Leng, J. Chen, C. Zhang, *Biomaterials*, 2000, **21**, 1339.

J. Weng, *J. Mat. Sci. Lett.*, 1994, **13**, 159.

J. Weng, X. Liu, X. Zhang, X. Ji, *J. Mater. Sci. Lett.*, 1994, **13**, 159.

E. Wintermantel, S.W. Ha, *Biokompatible Werkstoffe und Bauweisen. Implantate für Medizin und Umwelt*, Springer: Berlin, Heidelberg, Tokyo, 1996.

X.F. Xiao, R.F. Liu, Y.Z. Zheng, *Surf. Coat. Technol.*, 2006, **200** (14/15), 4406.

W. Xue, X. Liu, X. Zheng, C. Ding, *Surf. Coat. Technol.*, 2005, **200**, 2420.

W. Xue, H.L. Hosick, A. Bandyopadhyay, S. Bose, C. Ding, K.D.K. Luk, K.M.C. Cheung, W.W. Lu, *Surf. Coat. Technol.*, 2007, **201**, 4685.

Y.C. Yang, B.C. Wang, E. Chang, J.D. Wu, *J. Mater. Sci.: Mater. Med.*, 1995, **6**, 258.

Y.C. Yang, E. Chang, B.H. Hwang, S.Y. Lee, *Biomaterials*, 2000, **21**, 1327.

Y.C. Yang, *Surf. Coat. Technol.*, 2007, **201**, 7187.

A.L. Yerokhin, X. Nie, A. Leyland, A. Matthews, *Surf. Coat. Technol.*, 2000, **130**, 195.

X. Zhao, X. Liu, C. Ding, P.K. Chu, *Surf. Coat. Technol.*, 2006, **200** (16/18), 5487.

X. Zheng, X. Liu, W. Xue, C. Ding, *Mater. Sci. Forum*, 2005, **475.479** (III), 2371.

Ch. Zimmermann, *Histologische und histomorphometrische Untersuchungen am Interface Apatit-beschichteter Titan-Implantate*. Unpublished Ph.D. Thesis, 1992, Universität Heidelberg

10
Quality Control and Coating Diagnostic Procedures

10.1
Quality Implementation

An important aspect of the plasma-spray technique is the development and implementation of stringent quality control and assurance procedures to ensure consistency in the properties of the coatings. Because a multitude of spray parameters can potentially influence the coatings properties, parameter optimization involves statistical experimental design procedures. Such procedures provide a maximum of information on the behavior of a system with a minimum number of experiments. Thus there is generally a very favorable experimental economy that can save time and resources, and hence money. Principles of multifactorial analyses and several case studies will be dealt with in Chapter 11.

10.1.1
Total Quality Management (TQM)

Total Quality Management (TQM) is a complex system of several innovative and interacting disciplines including management ('Achieving success through others'), philosophy, psychology, stochastics, engineering and scientific expertise. It involves the concept of continuous improvement (CI) (Japanese: Keizen; see for example Ishikawa and Lu, 1985), and can be divided into three groups and classified as *quality tools, quality philosophy, and management style*. To successfully implement TQM it is mandatory that all three units are linked and interact smoothly. It is generally not sufficient to improve only one or two of these pillars of TQM. Even though some advantages will be gained, the final product will not be optimized. Thus TQM is more than just a 'quality' evolution but a system of continuous improvement of the process (CPI) and the product. The impressive success of Japanese industry in the past decades had its roots in rigorous TQM procedures (Keys and Miller, 1984; Fakuda, 1986).

10.1.1.1 Quality Tools

Statistical Design of Experiments (*SDE*) is the backbone 'quality tool' of Total Quality Management. Using SDE many of the factors can be screened out that vitally control the process and/or the product performance. Closely associated with SDE is Statistical Experimental Strategy (SES) that attempts to answer crucially important questions at the start of a research program, such as the number of experiments needed, the number and ranges of the parameters to be selected, the costs of the program, the equipment and manpower needed, the duration.[1] Likewise important is the initial selection of the quality characteristics that the experimental program is supposed to satisfy, i.e. the customer expectation. This will allow confident declaration as to what exactly the analyses of the experimental data will teach the experimenter about the system under investigation. Linking SDE with SES will establish the only way to plan and execute experiments under conditions that will result in valid and statistically accurate and precise conclusions.

Several other quality tools are also used, mostly based on statistics. These are Statistical Quality Assurance (SQA) (Wadsworth et al., 1991), Statistical Quality Control (SQC) (Montgomery, 2004), Pareto and other distributions (Juran, 1988a), Cause–Effect diagrams, Benchmarking (Camp, 1989), Just-in-time (JIT) concept (Hay and Zonderman, 1988) and more. The most widely applied methods are *Statistical Process Control* (*SPC*) (Owen, 1989) and *Quality Function Deployment* (*QFD*) (Hauser and Clausing, 1988).

After factor screening by SDE/SES and determination of those factors that significantly influence the plasma spraying process and the coating properties, *Statistical Process Control* is used to control the process so that despite the existence of internal and external variations the deposited coating is always within design specification. However, if information about the ranking of the controlling factors, i.e. their importance, is lacking the experimenter risks assuming that he/she is in control of a good process with SPC when in reality a nonoptimized process is being controlled (i.e. a local instead of a global extremum of the response surface), or worse, unimportant factors (Pouskouleli and Wheat, 1991).

An important element of TQM is *Quality Function Deployment*. The customer of the deposited coating defines its 'quality', i.e. a set of properties that must be adhered to. This information supplied by the customer is analyzed by the R&D team and transformed into engineering design and specification requirements. If this is done properly the final coating will have the predefined 'quality' even if the customer as a non-expert cannot explain clearly this desired 'quality' in engineering terms.

10.1.1.2 Quality Philosophy

With QFD entering the picture the dichotomy between non-expert customer and expert designer/engineer can be resolved. In general, the two sources of quality are

---

[1] This is not a trivial task. Starting a research program is rather easy but it takes guts and confidence to kill a program if it does not yield the desired results, i.e. if the program is not on time and within the allocated budget. Managers beware: 'Self-perpetuating' programs usually waste resources and block the execution of other potentially more promising projects!

the research/engineering/technical staff of the coating developer, and the customers who define the term 'quality of the coating'. Since the most important asset of any organization is its staff, empowerment of that staff will be a natural part of any TQM implementation. This means that as much information about the process and the product must flow up and down the hierarchical structures of the company as needed by the staff to understand the customers' needs. This informational empowerment encourages staff participation at all levels of the organization with the result of the creation of a considerable degree of initiative, commitment and motivation: a static and routine custodial organization can be transformed to a dynamic and flexible intra- and entrepreneurial one (Pouskouleli and Wheat, 1991). Because the process and the product quality requirements are fully understood by staff, it will be possible to anticipate, conform to, and also exceed the customers' requirements. Only then 'total quality' can be achieved: doing the right things right, the first time, any time!

10.1.1.3 Management Style

The quality philosophy expounded above can only succeed if the style of management matches the quality philosophy's quest for motivation and challenge. This means that no philosophy can work unless it is applied, and it cannot be applied unless it is encouraged. Products not conforming to quality are subjected to a detailed 'failure/ success analysis' or 'design/process monitoring'. The result of such types of analysis is to decide what needs to be done so that the root causes of the unreliability can be controlled. This is the hallmark of a successful Total Quality Management: to control the process, not the staff. Staff will feel encouraged to strive for continuous process improvement (CPI) that will create a high degree of autonomy of the teams and an increased level of responsiveness. While working under such conditions of 'enlightened participatory management' team members build trust and respect, share vital information and acquire common values. Hence a code of conduct is created that helps to resolve conflicts effectively and rapidly. In the end productivity as well as quality will be maximized. Several theories describe the productivity improvement operations such as 'Theory Z' (Ouchi, 1981), Juran's Quality Trilogy (Juran, 1988b), Deming's Principles (Deming, 1989) and Crosby's 14 Steps (Crosby, 1979).

To conclude, the total quality management (TQM) approach involves a chain of events that should be implemented in order to arrive at a high quality process that will result in a high quality product, in this case a superior plasma-sprayed coating that meets the customers demand and expectations (Ishikawa and Lu, 1985). The quality tools of this chain link research, development and product: *Statistical Design of Experiments* (SDE/SES) identifies those plasma spray parameters that significantly influence coating performance, process-based *Statistical Process Control* (SPC) ensures consistency in the industrial production of coating through Taguchi-type control designs, and the *Quality Function Deployment* (QFD) translates consumer demands into technical reality, i.e. engineering factors.

An excellent example of how to introduce a quality management system based on the ISO 9001 (ISO 9001, 1994) into a thermal spray company was given by Ebert and Verpoort 1995. The ISO 9001 standard is a model for quality assurance in design and development, production and servicing. It contains 20 quality management (QM) elements that were described in detail for a company developing new coatings and application areas.

10.1.2
Qualification Procedures

Qualification procedures of equipment, spray powders, process design and implementation, and operators are a mandatory part of total quality management (TQM).

The success of plasma-sprayed coatings depends on the skill of the operator, the condition of the equipment, and the selection and optimization of the internal and external process variables. Thus it is logical that qualification tests be part of any quality assurance (QA) and total quality management (TQM) programs implemented by an organization required to produce components for severe service. The major purchasers of plasma-sprayed components for aircraft engines, for example, require potential suppliers to demonstrate their capabilities before being approved as vendors. The qualification procedures are intended to establish that the vendor has operators, equipment, and processes capable to produce plasma-spayed coatings of acceptable quality and service life.

As far as plasma spray system operators are concerned, qualification procedures should demonstrate skill and knowledge, the ability to follow process instructions, and eventually to produce acceptable products. As in the widely used welding qualification procedures, a combination of the following exercises is deemed appropriate (Roseberry and Boulger, 1977).

- The operator should take and pass a short written test covering questions pertinent to cleaning, surface preparation and masking procedures, and the general principles of the plasma-spraying equipment and procedures.
- The operator should demonstrate familiarity with appropriate equipment by connecting, setting up, and operating plasma-arc spraying equipment safely and with confidence according to manuals supplied by the manufacturers.
- The operator should demonstrate capability by depositing an acceptable coating to a specified thickness on an appropriate test specimen. The quality of the test coating should be judged by suitable methods such as bond strength measurements.

The qualification of plasma-spraying equipment required by purchasers of critical components is usually directed towards two quality control objectives. First, using a qualified operator, the equipment must be shown to be capable of producing coatings that meet the acceptance quality agreed upon by the vendor and the purchaser. Secondly, all of the control and metering devices governing deposition variables must be shown to be properly calibrated and checked at regular intervals but at least every 30 days (Roseberry and Boulger, 1977).

The purchaser of high quality plasma spray coated parts normally requires that the deposition process proposed for production be qualified by experiments, in general using methods and principles of statistical design of experiments (SDE). In the qualification tests, the coatings are deposited on strips or coupons of standard size

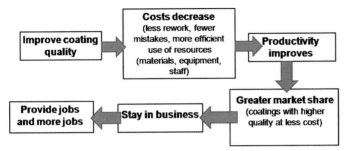

Figure 10.1 The Deming chain reaction.

representing the material and surface characteristics of those to be used in the spray shop. The coatings are deposited to a specified thickness by qualified operators using written process sheets and qualified, calibrated equipment. Then the coatings will be evaluated by test methods mutually agreed upon.

All important tenets of the quality control philosophy and their economic results are best summed up in the Deming chain reaction, an intuitive approach that will help coating companies to compete successfully in the global market (Figure 10.1).

10.2
Characterization and Test Procedures

10.2.1
Powder Characterization

The suppliers of spray powders, and shot and grit used to roughen the surface of the parts to be coated normally provide chemical analyses and information of particle size of the material. When chemical compositions are to be checked, standard analytical methods are employed. When analyzing metals optical emission spectroscopy, X-ray fluorescence spectroscopy or inductively-coupled plasma spectrometry will be used. Gas contents such as oxygen and nitrogen of metal powders, coatings and substrates are determined by vacuum fusion.

Figure 10.2 shows a compilation of testing procedures and quality control measures for plasma-spray powders and coatings (Elvers *et al.*, 1990). For more details on methods of powder production and characterization see Pawlowski (1995).

The particle size distribution of spray powders can be determined by a variety of methods and described by different designations. For particle sizes larger than 45 μm the range is characterized by the minimum and maximum cumulative percentages, by mass, that will pass through or be retained by sieves with different designations (ISO designation: mm and μm, respectively; ASTM designation: mesh size). In general, sieving is done with a set of standard sieves, for example ASTM E11-04 2002 and B214-07 (2007), assembled in suitable order by nesting in a mechanical shaker. The mass of powder retained at the appropriate sieves will be weighed with a balance

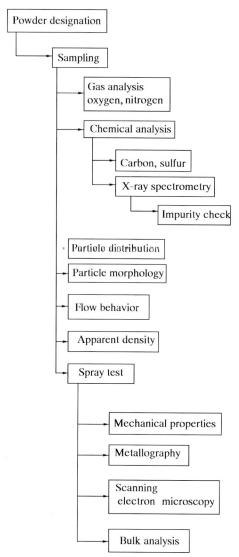

Figure 10.2 Testing procedures and quality control for plasma-spray powders and coatings (Elvers *et al.*, 1990).

to a sensitivity of 0.01 g. The recommended sample masses for sieve analyses are between 50 and 100 g depending on apparent density and grain size of the material. Data obtained from a sieve analysis of a plasma-spray powder should identify the specific mass fractions passing a particular size opening and retained on screens with smaller openings. The classification should be in size steps small enough to be useful for characterizing the material .

Particle sizes smaller than 45 μm are used to produce coatings with a very smooth finish. Their size distributions are normally determined by the elutriation method described in ASTM specification B293-76 (1976). The method is based on the velocity of particles falling in a countercurrent of air or gas. The results are not strictly accurate for porous particles. Better accuracies can be obtained by applying the Coulter Counter that measures the change in electrical resistivity across an orifice as particles suspended in a electrolyte are passed through the orifice. This change in resistivity is related to the volume of the particles. The effective size range of this instrument is 1.0 to 500 μm. Sedimentation techniques, for example the MSA (Mining Safety Appliance) apparatus extend the particle range to be measured to the submicron range (0.1 to 80 μm). Modern laser-operated particle analyzers allow for an automated measuring protocol.

'True' particle and coating densities can be determined by Archimedes' technique by dividing the mass of the sample by the volume of water it displaces. If the powder is representative of the material in the coating after plasma spraying, the density determination can be used to compute the porosity of the sprayed coating. More exact densities can be obtained with a helium-air pycnometer.

Useful information on flow rates and apparent densities of powders can be obtained with the Hall flow funnel (DIN EN ISO 4490, 2002). Data will give information on the ease of handling of the powder during processing as well as the densities to expect when powder is placed in the powder feeding device.

10.2.2
Microstructure of Coatings

To test and qualify coatings, a limited set of procedures are applied to evaluate *microstructural* (splat development, porosity, surface roughness), *mechanical* (cohesive and adhesive bond strength, shear strength, macro- and microhardness, fracture toughness), *tribological* (adhesive, sliding and erosive wear, mechanical fatigue) and *chemical* (corrosion, oxidation) properties.

A few principles should be considered that are important to assess and control the microstructure of ceramic coatings (see for example Fowler *et al.*, 1990):

- Flow and solidification of the molten droplets upon impact is a very complex process.
- Solidification occurs in less than 1 μs.
- Complete cooling of an isolated particle of 50 μm diameter lasts about 100 μs depending on the thermal conductivity of the material.
- the time lag between the arrival of two particles flying in the same trajectory with a velocity of $100 \, \text{m s}^{-1}$ is about 1 ms.
- Wetting and flow properties are very important.
- Substrate roughness will control mechanical properties and residual stresses of coatings.
- Substrate heating or cooling will influence the splat configuration and residual stresses of coatings.
- Coatings possess fractal properties.

10.2.2.1 Splat Configuration

A rather good estimation of the flow characteristics of the molten particles can be obtained by applying the simple 'wipe' test (Gruner, 1984). Here a flat surface is quickly brought into the path of a molten particle trajectory with the intention of capturing only a few particles. The solidified particle splats are investigated with optical or electron microscopy. Figure 10.3 shows typical examples of hydroxyapatite particle splats obtained from an argon/hydrogen plasma jet under low-pressure conditions (Heimann et al., 1997). The plasma enthalpy determined by the interaction of the plasma power and stand-off distance, increases from Figures 10.3A through 10.3D. In Figure 10.3A (plasma power 45 kW, stand-off distance 26 mm) the enthalpy supplied to the particle is not sufficient to achieve melting. Figure 10.3B (plasma power 30 kW, stand-off distance 24 mm) shows a splat pattern of a particle whose outer rim has been melted but its core has remained highly viscous as exemplified by its porous microstructure. In Figure 10.3C (plasma power 30 kW, stand-off distance 22 mm) a well-melted splat is shown whereas Figure 10.3D shows the 'exploded' splat features (see Figures 5.8 and 5.9) of a severely overheated particle (plasma power 45 kW, stand-off distance 22 mm). The significance of these splat properties is further shown in Chapter 9.3.

Figure 10.4 illustrates that substrate preheating can profoundly influence the morphology of splats and in particular is able to consolidate the splat shape by enhancing surface diffusion. Hence it leads to improved coating integrity, microstructure, and other mechanical and thermal properties (Sampath et al., 1999, 2003, 2004).

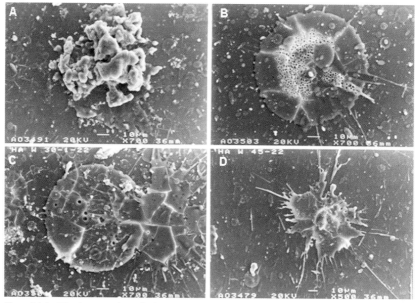

Figure 10.3 'Wipe' test results on hydroxyapatite particles sprayed onto a moving glass slide (VPS argon plasma) (Heimann et al., 1997).

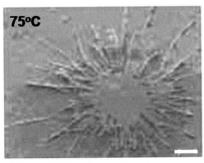 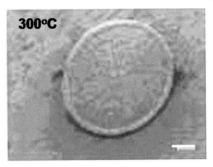

Figure 10.4 Dependence of the splat configuration of suspension plasma-sprayed (SPS) zirconia on substrate temperature. The bar corresponds to 25 µm. (Courtesy Prof. Francois Gitzhofer, U of Sherbrooke, Québec, Canada).

Figure 10.5 confirms by the novel tool of process mapping that increasing substrate temperature and decreasing kinetic energy of particles lead to more contiguous, less fragmented splats. This in turn yields better coating properties in terms of higher hardness, thermal conductivity and adhesion as well as lower oxygen content, porosity and modulus (Sampath *et al.*, 2003).

For the results of modeling splat shapes as a function of substrate preheating see Chapter 6.4.2.

Many modeling studies deal with the behavior of particles impacting on a rigid substrate surface. Bertagnolli *et al.* (1995) performed finite element calculations of the spreading process of ceramic liquid droplets on a flat cold surface and found that there is a correlation between the degree of flattening (Equation 5.8) and the initial process parameters. Since the mechanical performance of coatings depends crucially

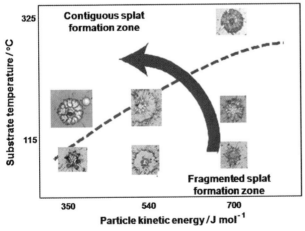

Figure 10.5 Process map showing splat morphologies (molybdenum) as functions of substrate (steel) temperature and kinetic energy of particles (Courtesy: Sampath *et al.*, 2003).

on the way how particles flatten and establish intersplat bonding, such studies are very important to unravel the complex interaction of spray parameters and coating properties. A research group led by Javad Mostaghimi at the University of Toronto has performed a series of studies to simulate splat shapes and solidification pattern of molten particles hitting a surface with high velocity (Pasandideh-Fard *et al.*, 2002; Mehdizadeh *et al.*, 2004; Ghafouri-Azar *et al.*, 2004; Raessi *et al.*, 2005; see also Chapter 6.4.2).

Image analysis has been applied to thermal spray deposits and a microstructural index was defined based on the determination of several stereological and morphological parameters relating to the size–shape distribution of phases and inclusions, the fractal dimension of the coating surface, and the Euclidean distance map of the features of interest (Montavon *et al.*, 1998).

10.2.3
Mechanical Properties

10.2.3.1 Adhesion of Coatings and Determination of Bond Strength

The quality of a thermally sprayed coating is to a large extent determined by the quality of its adhesion to the substrate. While it is generally accepted that the main contribution to the adhesion is a mechanical interlocking of the particle splats with asperities of the grit-blasted substrate surface, increasingly chemisorptive and epitaxial processes are considered important contributors to coating adhesion (Steffens and Müller, 1972). This assumption is partly based on the experimental evidence that the presence of intermediate bond coats, preheating of the substrate and high particle temperatures generated by increased plasma enthalpy and/or the residence time, translate into thermally activated bonding mechanisms.

Results by Gawne *et al.* (1995) indicate that mechanical interlocking may only play a secondary role in coating adhesion. Surface roughening by grit-blasting is then considered only a vehicle to promote disc-shaped particle splats and consequently to suppress exploded splash-type splats. The latter result in voids at the coating–substrate interface because the spaces between individual splashes may be too small for the second lamella to penetrate. Secondly, the flattening on impact of the splash-type splats is more extensive than for the disc-shaped splats. Consequently the lamellae are thinner, cool more rapidly and thus decrease the time available for chemical bonding. Thirdly, during splashing the particles break up into smaller droplets resulting in a loss of continuity of the flowing melt and thus decreased bonding to the substrate.

The requirement of a roughened substrate surface can apparently be relaxed by the application of an initial thin layer (about 25 μm) of ceramic on a smooth metal surface by low-pressure plasma spraying (LPPS) followed by the application of a thermal barrier coating (TBC) by APS technique (Miller *et al.*, 1993). The smooth surface can be an uncoated oxidation-resistant alloy, a metallic diffusion coating, or a plasma-sprayed metallic bond coat ground smooth or even lapped. The LPPS ceramic layer adheres well to the smooth metallic surface but its top surface is sufficiently rough to accommodate the normal thick TBC.

Adhesion of coatings is controlled by three main mechanisms.

1. *Mechanical anchorage.* Surface roughness plays an overriding role. The particles must have sufficient plasticity, high impact velocity, low viscosity and good wettability. The adhesion strength of a ceramic coating is in many cases a linear function of the average surface roughness, R_a deliberately produced by grit blasting. However, as shown in Chapter 5.6 the true influencing parameter appears to be the fractal dimension of the surface roughness (Reisel and Heimann, 2004).

   A study performed on plasma-sprayed Tribaloy 800 on a Ti-base alloy showed that the maximum adhesion strength was reached when the blasting and spraying angles were close to 90°. However, the grit residue reaches its maximum at a 75° blasting angle (see also Figure 5.19c). From the image analysis of the interface in different directions, it was found that the nonperpendicular grit blasting produces an anisotropic surface. The fractal analysis method showed a rather good correlation with the blasting angle. However, no good correlation between the fractal number and the adhesion strength was found (Bahbou et al., 2004).

2. *Physical adhesion.* This mechanism is controlled by diffusive bonding, where the diffusivity increases with increasing contact temperature according to Fick's law. This can be maximized by substrate preheating. Because of the small diffusion depth (produced by the rapid solidification), the diffusive adhesion generally plays only a minor role as an adhesive mechanism.

3. *Chemical adhesion.* Chemical adhesion can be engineered by adjusting the contact diffusivities. Thin reaction layers may be formed that improve the adhesion on a molecular scale by forming a true metallurgical bond (Godoy and Batista, 1999).

In more detail, adhesion mechanisms can be classified as 'microbonding' and 'macrobonding' (Dennis et al., 1965). *Microbonding* refers to the bonding that takes place along very small surface areas the size of an individual particle of sprayed powder. *Macrobonding* refers to areas much larger, by 10 to 100 times. Macrobonding relates to the macroroughness produced by threading and grooving methods or by extremely coarse grit blasting.

As indicated above, the microbonding between sprayed particles and the substrate, and between sprayed particles themselves, is never completely mechanical. There may be considerable bonding among particles at the microbond level, and no bond at all over a macroarea after a substantial coating has been built up. The reason for this is shrinkage. As each particle impinges at the substrate surface it flattens out, sticks to some extent, and then shrinks. An initial shrinkage occurs when the particle changes from a liquid, i.e. plastic to a solid, i.e. rigid state. In addition to this state-change shrink there is normal thermal shrinkage that continues as the particle cools further down. At the individual particle level shrinking may not cause much stress or at least not enough to rupture the microbond. A large body of evidence indicates that there is a strong initial film adhesion between the sprayed particles and the substrate, and between neighboring particles. This microbonding mechanism is still not well understood. It is the same type of adhesion that occurs between an anodized coating and aluminum, or between chrome plate and steel. This adhesion has been variously referred to as 'film adhesion', 'solid-phase bond' or 'solid-state bond'. The bond may be very strong, nonmechanical and occur on a molecular level, but is not usually referred to as a 'metallurgical' or 'chemical bond'. Metallurgical bonds imply some alloying of the materials at the interface that is not normally observed in plasma-sprayed coatings. In the case of nickel aluminide coatings, however, the exothermic reaction of this material causes alloying at the interface, and the result is a true metallurgical bond. Also, if very high contact temperatures occur at the interface as in the case of plasma-sprayed molybdenum on steel (see Chapter 5.4.2, Figure 5.14), metallurgical bonding involving alloy formation is predominant. Various test methods to evaluate coating cohesion and adhesion are described below.

Over the years a large variety of different tests have been devised to measure bond strengths in an accurate and reproducible way. Despite those efforts, however, there is to date no reliable method available that can be generally applied to any kind of coating on any kind of substrate. An older review by Davies and Whittaker (1967) identified the ultracentrifuge and various ultrasonic techniques as those that could presumably successfully meet requirements of routine quality control. However, the tensile pull test is still most widely used to determine bond strength (Pawlowski, 1995) even though there is a rather urgent call to interpret tensile adhesion tests in terms of a more appropriate design philosophy (Frielinghaus et al., 1990).

Tensile Tests with Adhesives The strength of the bond between a plasma-sprayed coating and the substrate is extremely important for most coating applications. For this reason, tensile testing is most commonly applied to evaluate the cohesive and/or adhesive strength as described by the ASTM Specification C633-01 (2001) or DIN EN 582-1994 (1994). The main differences between these tests concern the sample dimensions. The method is limited to rather thick coatings exceeding 380 μm (ASTM C633-01) and 150 μm (DIN EN 582-1994), respectively.

As pointed out be Milewski (1993) the tensile test methods can be subdivided into (a) tests in which the coatings are being pulled off from the substrate with the help of a counter fixture glued or soldered to the coating, and (b) tests in which the coatings are pulled off from an appropriately formed auxiliary fixture without an adhesive (Ollard–Sharivker test, see below).

The properties of the organic adhesives used limit the application of the test to temperatures at or near room temperature. The method is recommended for qualification, quality control, and component or process acceptance testing. It is also frequently applied to compare the adhesive or cohesive strengths of different coatings or different methods of substrate preparation. However, because of complicating factors such as the penetration of the adhesive into a porous coating, the strength data obtained are not suitable for design purposes. Figure 10.6 shows the fixtures for aligning the test specimens. A test specimen consists of a substrate fixture to which the sprayed coating is applied, and a loading fixture. Both fixtures should be round solid cylinders not substantially shorter than their diameters that should be between 23 and 25 mm. The fixtures should be preferably made from the same material that will be used for the production substrates. The material used for

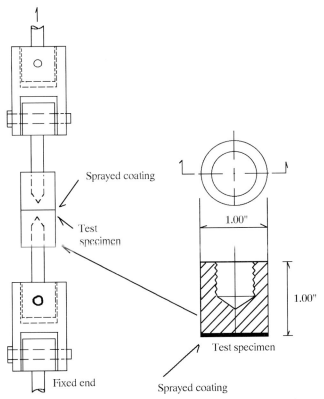

Figure 10.6 Loading fixture used to align the test specimens according to ASTM C633-01.

adhesive bonding the loading fixture to the substrate fixture must have a tensile strength at least as high as the adhesive and cohesive strengths of the coating. The adhesive should be sufficiently viscous not to penetrate through the coating.

The bonding strength or the cohesive strength of the coating is determined by the quotient of the maximum load F in N required to separate the two fixtures subjected to the tensile test, and the cross-sectional area A in mm^2: Strength = F/A [N mm^{-2}]. If the failure occurs entirely at the coating–substrate interface, the value is reported as *adhesion strength*. If the failure occurs entirely in the coating, the measured strength is considered the *cohesive strength* of the coating. Failure in the adhesive can be considered a satisfactory result if the strength value exceeds requirements for quality assurance or qualification tests (Figure 10.7).

This qualitative distinction between adhesive and cohesive failure modes of a coating can be quantified by a modified ASTM C633-01 tensile test jig developed by Berndt 1986 that is able to measure the specimen extension during loading. Since already slight misalignments of the specimen with respect to the pulling axis cause inappropriate errors in determining the failure load any deviations were taken into account by using clip gauges positioned on opposite sides of the jig. For tensile experiments using this modified test set-up on ZrO$_2$–8 wt% Y$_2$O$_3$–NiCrAlY 'tandem'

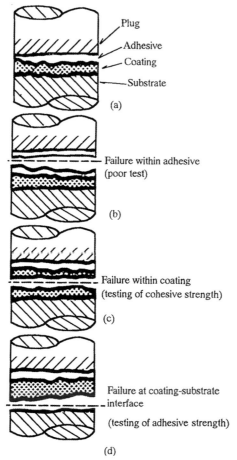

Figure 10.7 Definition of failure surfaces of loading and substrate fixtures.

coatings it was found that the stress/strain data approximately fitted a straight line, i.e. the slope of the stress/strain plot is directly proportional to Young's modulus. The probability plot of the stress/strain gradients (Figure 10.8) reveals a bimodal distribution: large gradients, i.e. large moduli correspond to adhesive failure, small gradients, i.e. low moduli correspond to cohesive failure. Mixed-mode failures are distributed over the entire gradient range.

The German specification DIN 50 160/10.90, analogous to ASTM C633-01, has been replaced by the European Standard DIN EN 582-1994 1994. This specification prescribes loading and coated substrate fixtures (configuration A) that allow for torsion- and momentum-free fixation by using ball- and socket-joints. Also, there is provision for having a coated disk of 25 or 40 mm diameter glued between two loading fixtures (configuration B).

Tensile Tests without Adhesive Tensile tests without using an adhesive avoid the problem of the adhesive penetrating into open porosity of the coatings thus

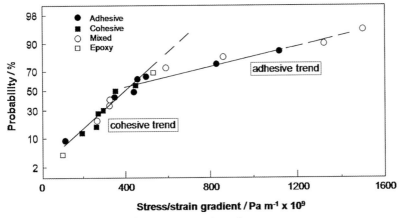

Figure 10.8 Probability plot of stress/strain gradients for a modified ASTM C633-01 test on ZrO_2-8%Y_2O_3/NiCrAlY TBC (Berndt, 1986).

compromising the accuracy of the measurement. Also, interpretative problems related to the influence of the coating thickness on the measured strength values could presumably be alleviated.

The Ollard–Sharivker test (Ollard, 1925; Sharivker, 1967) uses a special jig (Figure 10.9a) consisting of a base and a smoothly fitting washer supported on the shoulder of the base. This device is grit-blasted and inserted into a plasma spray apparatus through which the top face is coated as shown in the figure. The adhesion strength is determined by tearing away the base from the coating. Since the adhesion strength appears to be a function of the coating thickness, it seems feasible to determine the value of adhesion strength unaffected by internal stresses by extrapolating the adhesion strength vs. coating thickness curve to zero thickness as shown in Figure 10.9b. When a certain critical coating thickness δ_c is reached, spontaneous peeling of the coating from the substrate is observed. With decreasing thickness the strength of adhesion increases linearly. Failure takes place in an adhesion-type stripping mode along the coating/substrate interface (zone 1). With further decrease of the thickness a transition to cohesion-type stripping occurs (zone 2), and finally coating rupture is observed (zone 3).

For example, the adhesion strength at 'zero' thickness (P_0) is extrapolated for an alumina coating on a mild steel substrate (coating conditions: 220 A, 85.95 V, argon/hydrogen plasma, argon flow rate 35 L min^{-1}, hydrogen flow rate 4.5 L min^{-1}, stand-off distance 75–100 mm) to be 13 MPa (Figure 10.9b, left). This value is comparable to that of a Mo coating on an Al substrate (17 MPa; Figure 10.9b, right) but much lower than that of a Mo coating on mild steel (60 MPa; Figure 10.9b, center). The geometry of the jig shown in Figure 10.9a may not be optimal since there is a risk that separation occurs not by tensile forces but by bending or shearing. To account for this the original Ollard test was modified by Roehl (1940), Hothersall and Leadbetter (1938), Bullough and Gardam (1947), and Williams and Hammond (1954) (see also Milewski, 1993).

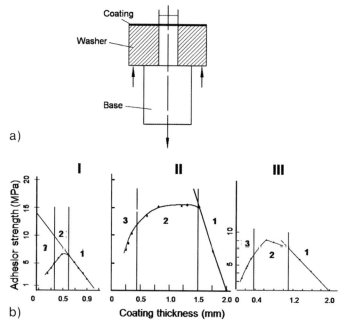

Figure 10.9 Ollard-Sharivker test. (a) Test jig; (b) adhesion strength vs. coating thickness curves for several coating/substrate combinations extrapolated to zero thickness. (I) Alumina on mild steel; (II) molybdenum on mild steel; (III) molybdenum on aluminum (Sharivker, 1967).

A variant of the Ollard test is the 'pin-hole' test (Figure 10.10). A pin with a diameter of 2 mm is fitted into a massive disc so that the end faces form a planar surface onto which a coating can be deposited (Rhyim et al., 1995). The pin is then pulled off the disc, and the bond strength can be determined by the force at which it detaches from the coating.

ShearTests These tests are based on the generation of stresses that act tangentially to the coating/substrate interface. It is quite difficult to localize the stresses exactly in the contact plane when the coated workpiece had undergone a preliminary milling or

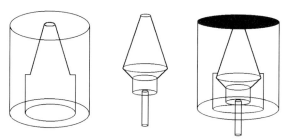

Figure 10.10 'Pin-hole' test arrangement (Rhyim et al., 1995).

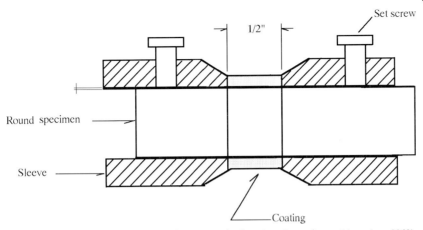

Figure 10.11 Device for measuring the shear strength of coatings (according to Metco Inc., 1963).

threading treatment. But even after customary surface preparation by grit blasting the shear forces occur predominantly in the coating itself and not at the contact plane coating/substrate (Steffens, 1963). Variants of the experimental realization of testing devices including finite-element evaluation have been shown by Milewski (1993) and Zhu et al. (1999). Figure 10.11 shows a device suggested earlier by Metco Inc. (Metco, 1963). The test is performed on a half-inch wide coating band deposited at the grit-blasted surface of a solid round cylinder. The former is subjected to an axial pressure to shear off the coating from the cylinder. After assembling the two sleeves and the round cylinder, grit-blasting, cleaning and depositing the coating, the set screws will be loosened and the cylinder will be pushed in the direction of the arrow to shear off the coating. The accurately measured load is used to calculate the shear force, and thus the shear strength of the coating. Friable or porous coatings should be given a top coat of stainless steel or Ni-Cr alloy to improve the distribution of the shear stress over the test section. A drawback of this technique is that shrinkage stresses induced into the coating during cooling after spraying will affect the measured shear strength to a considerable extent. A shear test suggested by Grützner and Weiss (1991) attempts to circumvent this problem.

However, recently a novel promising shear test was developed that does not require gluing and relies on a well-defined and reproducible mode of loading (Marot et al., 2006). Figure 10.12 shows the principle of the test device. The coated sample will be shear-loaded in a direction parallel to the coating interface by a hard metal plate. The load acting on the coating is increased until failure by delamination and/or cracking occurs. The main advantage of this test consists in its ability to be performed quickly with a minimum of preparation time. Hence it may be adapted to application during the production process of coatings. A disadvantage is that the test is recommended only for thicker coatings in excess of 150 μm.

Peel Test A modified peel test (present designation ASTM D-3167-03) was introduced by Sexsmith and Troczynski (1995) (Figure 10.13). A coating is deposited onto a metal foil soldered to a massive copper block that provides mechanical support and

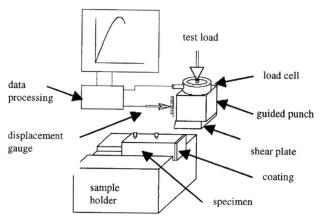

Figure 10.12 Shear test assembly according to EN 15340:2007 (Marot et al., 2006).

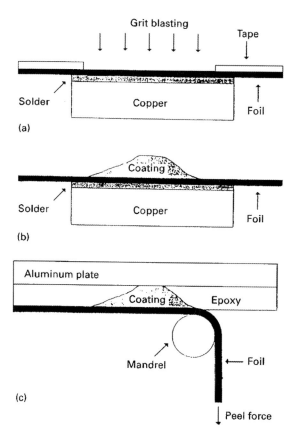

Figure 10.13 Peel test after Sexsmith and Troczynski 1995. (a) Schematic cross section of the sample block during grit blasting before spraying. (b) Coated sample block ready to be soldered to Al plate. (c) Sample sandwich ready for peel test.

acts as a heat sink. The block, foil and coating assembly are glued to a stiff aluminum plate and the copper block is then removed.

Peeling off the foil from the coating causes a crack to propagate precisely along the coating/foil interface in a controllable manner because the sample geometry forces the crack tip to move along the interface (Crocombe and Adams, 1981) where it encounters the local least energy path. Although more work is required to fully characterize and evaluate the potential of this adhesion test, its highly controlled crack tip behavior opens up a new way of coating quality testing (see Kurzweg et al., 1998).

Tape Test This simple test according to ASTM D-3359-02 (2002) involves a qualitative assessment of the adhesion of coatings by applying and then removing a pressure-sensitive tape over cuts made in the coatings by scribing with a sharp point. The cuts are placed approximately 4 mm apart with a length of 20 mm to produce a 2D square or diamond grid. Then the center of a strip of tape is placed over the grid, rubbed on firmly, and rapidly removed within 90 ± 30 seconds. Then the grid will be examined under a stereomicroscope and the areas where the coating was peeled off recorded, and classified and rated according to a scale (Seifert, 2004).

Scratch Test This method uses a Rockwell diamond pressed with increasing load into the coating surface and subsequent pulling away the sample. The ultrasonic signal from breaking of the coating and the interface, respectively as well as the increasing tangential force are measured during loading. Changes in the slope of this force signal changes of the coating properties, and changes in the intensity of the ultrasonic signal point to coating failure through chipping and spalling as well as loss of adhesion. After the test the trace of the scratch can be evaluated microscopically. Figure 10.14 shows the recorded tangential force of a VPS (Ti,Mo)C-NiCo coating (Thiele, 1994) tested within a range of 100 to 200 N with a CSEM Revetest Automatic Scratch Tester. The coating thickness was 70 μm. The slope of the tangential force vs.

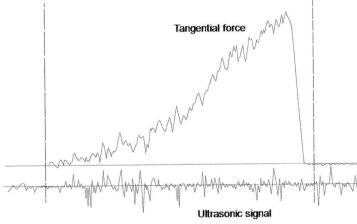

Figure 10.14 Recorded tangential force of an LPPS (Ti,Mo)C-NiCo coating (top) and ultrasonic signal (bottom) obtained with a CSEM Revetest Automatic Scratch Tester (Thiele, 1994).

distance (time) shows a noticeable change. Since the very hard and extremely well-adhering coating did not fail this change in slope may be attributed to heterogeneities in the coating or even the influence of the substrate because of the rather thin coating. The noisy curve is due to the high surface roughness of $R_a = 30\,\mu m$. The lower wiggly line in Figure 10.14 is a recording of the ultrasonic signal that also shows no indication of a coating failure but remains within the range of the instrumental noise.

Scratch tests showed that plasma-sprayed chromium oxide and alumina–titania coatings tend to fracture along thermal cracks and pores rather than splat boundaries indicating that the former are the weakest links in the coatings (Erickson et al., 1998). Scratch tests were found to be a useful tool to determine the critical pressure of biomimetically deposited hydroxyapatite coatings on cp-Ti substrates (Forsgren et al., 2007).

A similar test involves traversing a Rockwell 'C' diamond under a fixed load across a polished cross-section of a coated substrate. Figure 10.15 shows schematically the interfacial fracture processes occurring (Gudge et al., 1990). At a critical distance (cono depth) L_c the cracks formed in the coating propagate to the free surface and form a half cone-shaped chip whose depth R is a function of the applied indenter load and seems to be a measure of the coating cohesion. According to Belzung (Lopez et al., 1989) the cone depth L_c is related to the indentation load by $F_N = A\,L_c^{3/2}$ where A is proportional to the fracture toughness, K_c. However, the relationship for APS tungsten coatings was found to be linear by Gudge et al. (1990).

Similar relationships were developed by Démarcéaux et al. (1996) and Lesage and Chicot (1999) (see also Marot et al., 2006). The interface toughness K_c was found to be a function of the residual coating stresses. Lesage and Chicot (2002) investigated HVOF-sprayed Cr_3C_2-NiCr coatings by Vickers indentation and found that the interface toughness varied with the inverse square of the coating thickness.

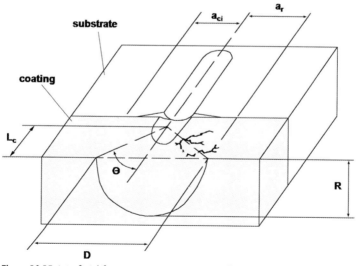

Figure 10.15 Interfacial fracture processes occurring during a scratch test (Gudge et al., 1990).

Moreover, it was postulated that two levels of residual stresses exist, one level that is independent of the coating thickness and one level that depends on thickness. Consequently the total interface toughness value K_{ca} behaves additively and can be separated in a term $K_{ca0} = 0.015(P_{c0} \times a^{3/2})(E/H)^{1/2}$ independent of coating thickness (P_{c0} = critical load at onset of cracking, a = crack length, E = modulus of elasticity, H = Vickers hardness) and a term $p(\sigma)/e^2$ where $p(\sigma)$ is a parameter dependent on the existing stress and e is the coating thickness.

Ultrasonic Tests With high-frequency ultrasonic waves defects at the interface coating/substrate can be detected and qualitatively related to the coating adhesion. In the second technique, low-frequency ultrasound induces stresses at the interface sufficiently high to detach the coating. The third technique is a combination of the first two, low-frequency ultrasonic energy being used to produce interfacial stresses, and a simultaneously applied high-frequency signal to detect any defects generated by the stressing (Davies and Whittaker, 1967). Holographic imaging of ultrasonic waves allows to make visible the defects and to determine their size, geometric form, and position. The best sensitivity is obtained if the penetration depth of the induced surface waves is about eight times larger than the coating thickness (Meyer and Pohl, 1992).

Figure 10.16a shows the geometry of the measuring device. As long as there is good bonding between substrate (copper) and coating (chromium) the transmitted signal

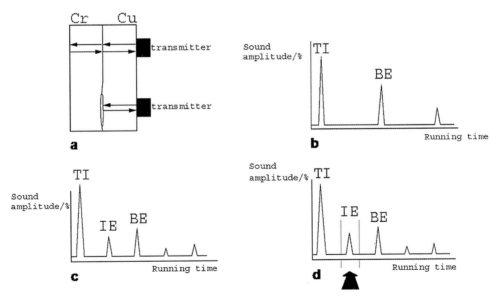

Figure 10.16 High-frequency ultrasonic adhesion test. (a) Geometry of the measuring device, (b) a-scan diagram of well-adhering chromium coating on a copper substrate, (c) a-scan diagram indicating a coating failure by the presence of an interface echo, and (d) setup for obtaining a c-scan diagram by filtering out the interface echo (Müller, 1994).

passes through the interface almost undisturbed. It is first reflected at the back wall and then detected at the copper surface (top). Coating failure is depicted in the bottom part of Figure 10.16a: the sound waves will be totally reflected at the Cu/Cr interface. Figure 10.16b shows the so-called a-scan diagram of the first case with the transmitted impulse TI and the first back wall echo (BE) with its much weaker repetition after twice the original running time. In Figure 10.16c the situation is shown when a small failure in adhesion occurs. The waves will be partially reflected at the interface (IE) and only later at the back wall (BE), again with their weaker repetitions. Since only the amplitude of the interface echo is indicative of a good adhesion it is filtered out of the total spectrum by a narrow strip-shaped window (Figure 10.16d) and processed separately. Within the bounds of the set window the transmitter is scanned across the sample and the so-called c-scan diagram is obtained. A useful scanning row distance and step width of 0.5 mm each results in a pixel area of 0.25 mm^2. The best signal resolution in the case of thick VPS-sprayed chromium coatings on copper was achieved using a sound wave frequency of 5 MHz (Müller, 1994).

A specific color can be assigned to each detected amplitude value. Figure 10.17 shows such a color-coded c-scan map of a thick VPS-sprayed chromium coating on an 8 mm thick copper plate. Dark red colors were assigned to 0% coating failure, i.e. maximum adhesion strength, dark purple colors to 100% coating failure, i.e. no adhesion. It can be deduced from the color map that a broad band with decreased adhesion passes through the center of the electrospark-machined sample (Müller, 1994; Heimann and Müller, 1995).

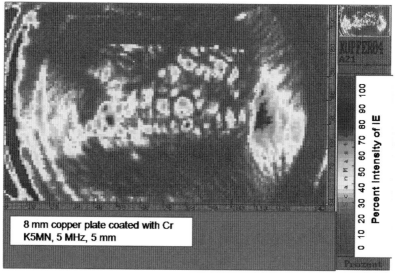

Figure 10.17 Color-coded c-scan map of a LPPS chromium coating on a copper plate of 8 mm thickness (Müller, 1994; Heimann and Müller, 1995).

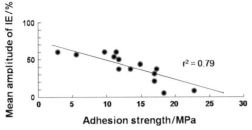

Figure 10.18 Representative results of the quasi-linear correlation between the mean amplitude of the interface echo (see Figure 10.16d) and the adhesion strength measured by ASTM C633-01 and EN 582-1994, respectively for chromium coatings on a 8 mm thick copper substrate (Müller, 1994; Heimann and Müller, 1995).

To quantify the results a combination of the signal amplitude of the interfacial echo (IE) in a c-scan, expressed through the color code, and the adhesion strength obtained destructively by an ASTM C633-01 (2001) tensile test can be used. In several cases a good correlation was found (Suga *et al.*, 1993; Heimann and Müller, 1995; Lescribaa and Vincent, 1996). Figure 10.18 shows representative results of the correlation between the mean amplitude of the interfacial echo in percent calibrated against a coating with no bonding (100%) and the adhesion strength measured by an ASTM C633-01 tensile test for an 8 mm thick substrate (see Figure 10.17).

In order to extend the application of ultrasonic testing to samples with complex geometrical shapes, a robot with five axes was constructed that can change the angle between sample surface and ultrasonic beam automatically (Lian *et al.*, 1995). Echo-impulse techniques with an auxiliary reflector, for example by using water as an immersion medium, measure the adhesion of very thin coatings reliably, since the resolution of the reflected signal is much improved. Also, the coupling of the signal head is more uniform (Frielinghaus *et al.*, 1990; Suga *et al.*, 1995). Several ways to use the ultrasonic c-scan technique to characterize thermal fatigue damage of TBCs were explored by Mesrati *et al.* (2004).

Spallation Techniques An advanced spallation technique was applied by Baumung *et al.* (2001) to probe the adhesive strength of MCrAlY coatings applied by plasma spraying onto Inconel IN 738LC surfaces. An aluminum flyer plate (0.2–2 mm thickness) accelerated by an explosive impacted the sample creating a compressive pulse that will be reflected at the free surface of two composite two-layer sample. Figure 10.19a shows schematically the stress wave moving through the sample. The interaction of the rarefaction tail of the incident stress wave and the reflected rarefaction wave generate tensile stresses within the interface coating/substrate. Stress relief by fracturing causes the tensile stress to decrease to zero and a compressive wave forms instead close to the spall plane that propagates back to the free surface and thus generates a spall signal in the free velocity profile (Figure 10.19b). The velocity of the receding free surface, Δu_{fs} was measured with a VISAR velocimeter (Barker and Hollenbach, 1972) and found to be a function of the

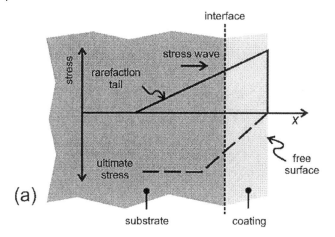

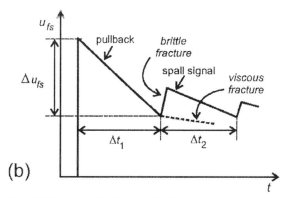

Figure 10.19 Schematic diagram of (a) the tensile stress generated by the reflection of a pulsed compression wave at the free surface and (b) free surface velocity history (Baumung et al., 2001).

spall strength σ^*, i.e. the coating adhesion strength that can be described by the linear approximation

$$\sigma^* = 1/2\rho_0 \times c_0 \times \Delta u_{fs}, \tag{10.1}$$

where ρ_0 = material density and c_0 = sound speed (Davison et al., 1996). Multiple reverberation of the wave within the solid causes oscillations of the surface velocity. The time interval Δt_1 of the first oscillation is proportional to the spall thickness h_s and can be expressed by

$$\Delta t_1 = 2\,h_s/c. \tag{10.2}$$

Brittle materials respond rapidly thus creating a steep and close to instantaneous spall signal whereas in viscous materials fracture develops at a slower rate (Figure 10.16b). For elastic–plastic materials Equation 10.1 should be modified to yield

$$\sigma^* = 1/2\rho_0 \times c_b \times (\Delta u_{fs} + \delta), \tag{10.3}$$

where δ is a correction factor for the elastic–plastic deformation of the velocity profile, and c_b is the bulk sound speed. If the thickness of the coating becomes small compared with the wavelength of the compression pulse, the interfacial tensile stress equals a decelerating inertial force and can be expressed by Newton's law as

$$\sigma = -\rho \times h_c \times (\partial u_{fs}/\partial t). \tag{10.4}$$

There is a major problem associated with the determination of the adhesion strength of coatings by spallation technique: the rupture has to occur within the interface coating/substrate, and must be recorded in real time. Hence the incident stress pulses must possess varying gradients in their unloading parts.

In addition, stress pulses of much shorter duration were generated by a high-power pulsed proton beam (1.5 MeV, 0.15 TW cm^{-2}) which is absorbed within the first 20 μm of the coating free surface, creating compressive waves with peak stresses between 5 and 11 GPa, depending on coating thickness. The velocity histories were simultaneously recorded for many surface points with a line-imaging Doppler velocimeter (Baumung et al., 1996). High adhesion strengths between 1.3 and 1.6 GPa were measured depending on the substrate temperature (600–900 °C) and the degree of annealing (1 h at 1080 °C).

Thermal Wave Interferometry Nonbonded areas perturb the heat flow through the coating into the substrate and thus affect the transient surface temperature (Travis et al., 1986). Bonding defects as small as 300 μm in size can be reliably localized, and such of 150 μm can be detected. Although the test results are affected by scores of parameters the operation does not require a high degree of knowledge of the emissivity of the materials. Thus thermal wave scanning has proved to be a valuable quality assurance tool for thermally sprayed coatings.

Heat is applied to the coating surface by an air-operated heating nozzle. The stream of hot air (500 to 600 °C) is switched between the nozzle and a bypass tube. A computerized infrared radiometer is used to record and store the surface temperature that is affected by the density of bonding defects. Figure 10.20 illustrates a scan of a plasma-sprayed NiCrAlMo bond coat-Al_2O_3/TiO_2 top coat duplex system on a valve stem that failed in service. With time-resolved infrared radiometry (TRIR) thickness variations and debonding of zirconia TBCs have been studied (Spicer et al., 1990). In this case the sample is pulse-heated as opposed to the continuous wave photo-thermal radiometry (CW-PTR) where a continuous modulated heating from a laser source is used. Mathematical modeling of thermal wave NDT of TBCs using numerical finite differences has been attempted by Georgiou et al. (1995).

Thermal wave interferometry using fast infrared scanning technique (Hartikainen, 1989; Bendada et al., 2005), thermal wave imaging systems (Aumüller et al., 1996; Delgadillo-Holtfort et al., 2000; Newaz and Chen, 2005), and pulsed laser thermography (Bendada, 2004) were applied for nondestructive inspection of coatings for various applications including on-line process identification as well as monitoring damage evolution in TBCs (Franke et al., 2005).

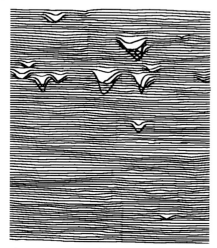

Figure 10.20 Thermal wave ND scan of a NiCrAlMo/Al$_2$O$_3$-TiO$_2$ duplex coating showing bonding defects (Travis et al., 1986).

10.2.3.2 Macro- and Microhardness Tests

Coating hardness values are often reported to compare the performance of coatings in service as well as the effects of spray variables. These hardness values should generally not be considered as a measure of the actual coating strength. For thin and porous coatings with low cohesion strength macrohardness tests are not applicable. Microhardness tests require careful determination and interpretation so that frequently Rockwell Superficial Hardness tests are applied that are simpler to conduct than the standard Vickers or Knoop indentation tests (see below).

Rockwell Hardness Tests Procedures for Rockwell Superficial Hardness tests are laid down in ASTM E18-03 (2007). In contrast to normal Rockwell test this procedure employs smaller loads. The superficial hardness apparatus measures the difference in depth of indentation caused by a minor load (3 kgf) applied first and a major load (15 to 50 kgf). One unit on the superficial hardness scale represents a penetrator movement of 1 µm between minor and major loads. Most commonly used are the diamond penetrator (N scale) and the ball (T scale). There is no reliable general method of converting hardness numbers from one Rockwell scale to another, or to tensile strengths. The choice of an appropriate scale to use for measuring superficial hardness depends on the hardness and thickness of the coating.

The requirements of the plasma-sprayed coatings to be tested are surface cleanliness and absence of gross imperfections. The impressions of the penetrator should be spaced at least three impression-diameters from each other and from a free edge. Five determinations are normally sufficient to obtain reliable results. A surface finish of 1 µm is recommended for using the 15 N (15 kgf, diamond penetrator) scale. While flat specimens are preferred testing of curved specimens requires correction factors.

Microhardness Testing The standard methods to determine microhardness by indentation of a diamond pyramid are described in ASTM standard E384-07 (2007). Depending on the shape of the indenter, Knoop- and Vickers-type diamonds as well as Berkovich indenters can be distinguished. In all systems, the hardness number in $N\,mm^{-2}$ is the force exerted on the specimen by the diamond indenter used to produce the impression. In principle, all systems are less affected by porosity than scratch tests based on measuring the indenter travel caused by a specific increase in load. Microhardness tests are usually made transverse to the coating surface, even though loads occurring in service are usually normal to the surface and hardness may vary because of the anisotropy of the microstructure. It is often convenient to make microhardness indentations on specimens prepared for metallographic studies of coatings.

Even though it appears to be relatively easy to perform the measurements, the microhardness can be affected by very many parameters including residual stresses, grain size, grain orientation and the presence of pores and microvoids (Sundgen and Hentzell, 1986). Therefore, microhardness is not an intrinsic quantity independent of operating conditions but characterized by the fact that the measured value depends on the applied load. The usual formula used for determining the Vickers hardness is

$$H_V = 1.8544(P/d^2), \tag{10.5}$$

where P is the load in kg or Newton, and d the diagonal of the indent in mm. In this equation, however, the nonlinear influence of the load P on H_V is not considered even though it is often observed experimentally that the measured hardness increases as the pyramid load decreases. This nonlinear behavior has been found for SiC (Sargent and Page, 1978), TiO_2 (Perry and Pulker, 1985), Cr_7C_3 (Perry and Horvath, 1978) and VC (Horvath and Perry, 1980). To account for the load dependence of the hardness two approaches are usually taken:

1. *Meyer's approach* (Meyer, 1908):

   $$P = a_m d^n, \tag{10.6}$$

   where n represents the load-hardness dependence. If $n = 2$, than the hardness is independent of the load. Also, a relation was given by Burnett and Page (1984) to describe the hardness in the regime of small loads as $H = qd^{m-2}$, where q is a constant and m the ISE index (indentation size effect).

2. *Thomas' approach* (Thomas, 1987):

   $$H = H_0 + (b/d). \tag{10.7}$$

These relations, however, cannot be directly applied to thin films or coatings unless their thicknesses are several times (typically ten times; Burnett and Rickerby, 1987) greater than the indentation depth so that the subsurface deformation beneath the indenter is not influenced by the proximity of interfaces or free surfaces (Burnett and Rickerby, 1987; Almond, 1984). Those interfaces can be considered for a stratified coating using Bückle's proposal (Bückle, 1971) that the hardness should be expressed as a weighted sum of the hardnesses of the different layers. In this spirit a coating/substrate tandem could be considered a two-layer material whose composite hardness is then

$$H_C = \alpha H_F + \beta H_S = H_S + \alpha(H_F - H_S), \tag{10.8}$$

with $\alpha + \beta = 1$, and where H_F = coating hardness and H_S = substrate hardness. The correlation factor α varies from 1 (coating hardness is not dependent on the substrate, i.e. the coating thickness t is at least ten times the indentation depth D) to 0 (coating thickness t is negligeable compared with the indentation depth D).

The geometrical approach by Jönsson and Hogmark (1984) separates coating and substrate contributions to the measured composite hardness by applying a simple 'area law of mixtures':

$$H_C = (A_F/A)H_F + (A_S/A)H_S, \tag{10.9}$$

where A_F = area of indentation within the coating and A_S = area of indentation within the substrate ($A = A_F + A_S$). Figure 10.21a shows the geometry used by Jönsson and Hogmark for their model, and the definitions of A_F and A_S. From geometrical considerations an equation can be derived that describes the composite hardness as a function of the ratio t/d using a constant C that takes the value $C = 2\sin^2(11°)$ for hard coatings on very soft substrates or $C = \sin^2(22°)$ for coatings whose hardness is comparable to that of the substrate:

$$H_C = H_S + \{2Ct/d - C^2(t/d)^2\}(H_F - H_S). \tag{10.10}$$

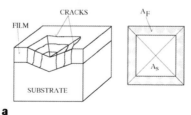

a

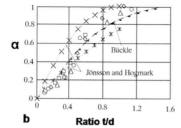

b

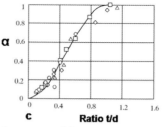

c

Figure 10.21 Jönsson–Hogmark 'area law of mixture' model (see text). (a) Geometry and definitions, (b) data for thick Cr_3C_2-25% NiCr coatings fitted to Bückle's 1971 and Jönsson-Hogmark's (1984) model, (c) data fitted to Jönsson–Hogmark's modified model (Lesage and Chicot, 1995).

In Equation 10.10 the parameter t is the thickness of coatings and d is the diameter of the indent (d ≈ 7D).

Comparison with Equation 10.8 shows that Jönsson and Hogmark's model is identical to Bückle's when $\alpha = A_F/A = 2Ct/d - C^2(t/d)^2$. Figure 10.21b shows experimental data for thick Cr_3C_2-25% NiCr coatings (Lesage and Chicot, 1995) and the predictions obtained from Bückle's and Jönsson-Hogmark's relations when the fitting parameter α is plotted against the ratio t/d. While Bückle's model fits the data quite well even for thick coatings, the Jönsson–Hogmark model does not. However, modifying the original Jönsson–Hogmark model by assuming that the C-value is not constant but varies continuously with t/d according to

$$C = (t/d)^n, \tag{10.11}$$

than the parameter α may be written as

$$\alpha = 2(t/d)^{n+1} - (t/d)^{2(n+1)}. \tag{10.12}$$

Fitting the experimentally measured hardness values to the modified Jönsson–Hogmark model satisfactory results are obtained for a value of $n = 3/4$ (Figure 10.21c). This value, however, is purely arbitrary and there is no physical confirmation of its validity yet.

A different way to account for the load dependence of the measured hardness is to apply a correction factor to the measured diagonal of the indent (d_{cor}) to obtain a constant absolute hardness (Iost and Aryani-Boufette, 1992). Combining Equations 10.5 and 10.7 one obtains

$$H = 1.8544(P/d^2) = H_0 + (b/d)$$

$$H_{cor} = 1.8544(P/d_{cor}^2) = H_0 \tag{10.13}$$

$$d_{cor}^2 = d^2 + (b/H_0)d.$$

Introducing this expression into Jönsson and Hogmark's original equation the following equation is obtained (Lesage and Chicot, 1995)

$$H_C = H_{0,S}[1 - (t/d_{cor})^{n+1}]^2 + H_{0,F}(t/d_{cor})^{n+1}[2 - (t/d_{cor})^{n+1}], \tag{10.14}$$

with the limiting conditions

if $t/d_{cor} = 0$ than $H_C = H_{0,S}$

if $t/d_{cor} = 1$ than $H_C = H_{0,F}$.

The hardness of a material is related to the plastic work done in creating an indentation. Based on this concept, Burnett and Rickerby (1987) developed a model that took into account the relative plastic zone size and the related amount of plastic work. Spherical cavity analysis (Marsh's relation; Marsh, 1964) done on the indentation that creates a hemispherical plastic zone showed that the size of the latter varies with the size of the Vickers indentation according to Lawn et al. (1980)

$$b/a = c(E/H)^{1/2} \cot^{1/3} \varphi, \tag{10.15}$$

where a = indentation semi-diagonal, b = radius of plastic zone, c = constant close to unity, and φ = indenter semi-angle (74°). With this relation a nonlinear 'volume law of mixture' model was suggested by Burnett and Rickerby (1987) that incorporated the so-called ISE term (indentation size effect), i.e. the dependence of hardness on load at small indentations as well as the plastic zone size term to yield

$$H_C = (V_F/V)H_F + (V_S/V)\chi^3 H_S \text{ for } H_S < H_F \tag{10.16a}$$

$$H_C = (V_S/V)H_S + (V_F/V)\chi^3 H_F \text{ for } H_F < H_S, \tag{10.16b}$$

where V_F and V_S are deforming volumes to be calculated using Equation 10.15, and χ is an 'interface parameter':

$$\chi \propto (E_F H'_S / E_S H'_F)^{n/2}, \tag{10.17}$$

where E_F and E_S = modulus of elasticity of coating and substrate, respectively and H'_F and H'_S = characteristic hardness values. The interface parameter is a strong function of the mismatch between the radii of the plastic zone predicted from Equation 10.15 and also includes deviations from the ideal geometry of the indent. Its value is strongly dependent on the ability of an interface to accommodate the shear stress arising from this mismatch. It may therefore be regarded as a measure of the rigidity of the interface, i.e. the coating/substrate adhesion governing the transmission of shear stresses from a deformed layer to an initially undeformed substrate. It should be mentioned that according to Veprek et al. (2003) (see also Fischer-Cripps et al., 2006) the indentation size effect (ISE) is not seen in data calculated on the basis of SEM indentation images of superhard or ultrahard coatings as well as silicon and sapphire.

A more complex expression has been developed by Chicot and Lesage (1995) to account for the interaction effects of substrate and coating such as elastic rebound:

$$H = H_s + \left\{ \left(\frac{3}{2} \sqrt[3]{\tan\xi} \cdot \right) \frac{d_c}{D} \left[\left(\sqrt{\frac{H_c}{E_c}}\right) + \left(\sqrt{\frac{H_s}{E_s}}\right) \right] \right\} (H_c - H_s), \tag{10.18}$$

where H is the composite hardness of substrate and coating, H_s and H_c, and E_s and E_c are the hardness and modulus values of substrate and coating in MPa, respectively, d_c is the coating thickness (μm), D is the indentation diagonal (μm), and ξ is the half angle between opposite edges of the nano-indenter (74°).

10.2.3.3 Fracture Toughness

It is questionable whether toughness measurements on ceramic coatings actually tell much about the material. Data are often inconsistent since strength and toughness do not always respond in the same manner to changes in microstructure of the coating or their interfacial properties. These inconsistencies arise from the sensitivity of the measurements to specimen preparation, i.e. precracking, getting a crack to grow properly in a double-torsion test, and specimen alignment. These problems have been well worked out over the years in more forgiving metals. For example, the fracture toughness K_{Ic} of bulk materials measured with the double torsion technique are around 50 MPa$\sqrt{m}$ for medium-strength steel, 13 MPa$\sqrt{m}$ for Co-bonded tungsten carbide, and only 7 MPa$\sqrt{m}$ for Ca-stabilized zirconia (double cantilever beam test).

Determination of the critical strain energy release rate G_{Ic} can provide insight into the dependence of fracture toughness on plasma spray parameters such as spray distance (Li et al., 2004a). The model proposed uses the lamellar interface mean bonding ratio and the effective surface energy of the bulk ceramic (alumina) to predict G_{Ic} that was found to be in good agreement with the observed G_{Ic} data. It was concluded that the fracture toughness of plasma-sprayed alumina coatings along the coating surface is determined by the strength of the lamellar interface bonding, i.e. coating cohesion.

Work by Lima et al. (2004) showed that fracture toughness of WC-Co coatings designed to be resistant against cavitation erosion can be measured by a Vickers indentation test. Since the toughness K_C is proportional to either $(E/H)^{1/2}$ in the Marshall–Lawn indentation model (Marshall and Lawn, 1977; Equation 10.15) or $(E/H)^{2/5}$ in the Niihara indentation model (Niihara, 1983) a better tribological performance can be achieved by reducing the modulus E and increasing the Vickers hardness H. For a post-melted 50(WC12Co)50NiCr coating a fracture toughness of 32 ± 12 MPa$\sqrt{m}$ was measured and a correlation between K_C and the cavitation erosion resistance could be established.

Malzbender 2006 compared the fracture toughness of ceramic coatings measured by surface and cross-sectional indentation by a Vickers diamond according to the DIN ISO 14577 (2003) procedure. Owing to the high compressive in-plane residual stresses indentation in the surface did not produce radial cracks. Cross-sectional indentation generated cracks parallel to the interface from which extension c the fracture toughness K_{Ic} could be determined according to the well-known Marshall–Lawn equation

$$K_{Ic} = \chi_r \frac{P}{c^{3/2}} + Z(c)\sigma_r \sqrt{C}, \tag{10.19}$$

where $\chi_r = \chi(E/H)^{1/2}$ is related to the size of the plastic zone, σ_r is the residual stress, P is the applied load, c is the crack length, and $Z(c)$ is a shape factor. A reliable assessment of the fracture toughness appears to be feasible for cross-sectional Vickers indentation if the indentation pressure as a measure of the apparent hardness is incorporated into the relationship.

A four-point bending test can provide some information on the toughness of plasma-sprayed coatings (Cox, 1988). The test consists of placing a coated beam in pure four-point bending with the coating in tension and recording cracks by acoustic emission (AE) with a piezoelectric transducer attached to the surface of the coating (see also Miguel et al., 2003). Simultaneously the coating strain is monitored by strain gauges, and the test results are presented as strain to fracture (STF). It was found that for WC/Co coatings the STF (toughness) depends strongly on the residual stresses present (see Chapter 5.7) but neither appreciably on the microhardness nor the metal content of the coating. The acoustic signals picked up by the transducer occur at four different amplitudes thus suggesting different cracking mechanisms. Early in the test, low amplitude events around 50 dB took place that were related to pre-cracking or microcracking. On release of the bend stress no visible damage in the test specimen could be discerned. The second type of noise at greater 100 dB amplitude are true coating cracks as confirmed by a close to one-to-one correlation between high amplitude events and the number of macrocracks in the coating following testing. During macrocracking the number of low amplitude events increase strongly. They are considered reflections of the stress wave developed at the crack front and thus are not related to changes in the material. The last type of AE events around 80 dB may be related to cracks that propagate through fewer lamella than the cracks causing the 100 dB events.

Bend tests can also be used to determine the probability of rupture of ceramic coatings using Weibull analysis (Rigal *et al.*, 1995). Free standing samples of PSZ with a length L and thickness d were subjected to a three-point bend test, and the modulus of elasticity, E and the mechanical strength, σ were determined from the moment of inertia, I:

$$E = FL^3/48 f I, \tag{10.20}$$

where F = force, f = displacement of the center point, and

$$\sigma = FLd/8 I. \tag{10.20a}$$

The probability of rupture Pr of the zirconia ceramic is given by

$$Pr = \exp[-(\sigma/\sigma_0)^m], \tag{10.21}$$

where σ is the applied stress, σ_0 the normalized stress below which 63% of the samples fail, and m the Weibull modulus. Equation 10.21 can be written in the following form

$$\ln[\ln(1/1 - Pr^i)] = m \ln(\sigma_r^i) + \ln(1/\sigma_0^m), \tag{10.21a}$$

where i refers to the experiment number i when all experiments are classified starting from the lowest values σ_r to the highest, and where Pr^i is estimated by

$$Pr^i = (i - 0.5)/n \tag{10.22}$$

with n = total number of experiments. By plotting the left-hand term of Equation 10.21a against ln(σ), the slope of the resulting straight line determines the Weibull modulus m, and the intersection with the ln(σ)-axis results in σ_r.

Using the scratch test described in 10.2.3.1 and data obtained for the coating fracture toughness from the half-cone fracture shown in Figure 10.15, Lopez *et al.* (1990) estimated the coating cohesion for plasma-sprayed alumina, alumina–titania, chromium oxide, chromium carbide-NiCr, and WC-Co coatings.

Acoustic emission (AE) inverse analysis was used by Watanabe *et al.* (2003) to evaluate microfractures originating in APS alumina coatings on stainless steel (SUS304) and by combining it with numerical analysis the *in situ* fracture toughness was determined to be 40–50 J m^{-2}, consistent with the results of a double cantilever beam test (see also Bansal *et al.*, 2006).

Elastic moduli of plasma-sprayed coatings required to determine the coating fracture toughness can be determined by Knoop indentation (Kim and Kweon, 1999; Lima *et al.*, 2005), Brinell-type indentation (Wallace and Ilavsky, 1998) or laser-ultrasonics to determine the velocity of ultrasound waves in the coating material that is thought to be directly proportional to the elastic modulus (Lima *et al.*, 2005). Comparison of the values of the modulus obtained by three-point bending and Knoop indentation led to the conclusion that for porosities around 10% irrespective of the chemical composition the former are always lower than the latter (Kim and Kweon, 1999). Elastic–plastic anisotropic parameters of thermally sprayed NiAl coatings were determined using instrumented nano-indentation with spherically and Berkovitch-type profiled indenter heads, and applying inverse analysis to extract unknown parameters of the elastic-plastic transversely isotropic material (Nakamura and Gu, 2007).

10.2.3.4 Porosity of Coatings

Whereas metallic coatings with nearly theoretical densities can be obtained by plasma spraying, the porosity of plasma-sprayed ceramic coatings is, in general, between 3 and 20%. While in some cases high porosity is advantageous, for example to reduce the thermal conductivity of TBCs, to act as a retaining reservoir of lubricants in some wear-resistant coatings, or to enhance in-growth of bone cells into bioceramic coatings, in most wear applications the wear and corrosion resistances decrease dramatically with increasing porosity. Thus, porosity has to be tightly controlled in order to maximize coating performance.

Common causes of porosity are:

- formation of large, spherical pores around particles that have already solidified prior to impact or were never completely molten due to their large size (Figure 5.3);
- shadow effects when splashing of a second particle over previously arrived ones may lead to a gap within one lamella layer (Figure 5.3);
- narrow planar inter-lamellar pores and/or gas inclusions between the *i*th and the $(i + 1)$th lamella (Figure 10.22);
- exploded particles due to overheating, excessive particle velocities and thus occurrence of disruptive shock waves (Figure 5.9, Figure 5.15D);
- flat planar crack-like pores formed during cooling to relax stresses originating from restricted thermal shrinkage (Figure 6.5);
- gas-filled voids caused by dissolution of gas in molten material;

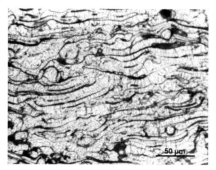

Figure 10.22 Microstructure of splats in an etched cross-section of a chromium coating on copper (LPPS, 43 kW, 30 kPa, grain size $-63 + 20\,\mu m$, powder feed rate 23 g min^{-1}, 220 mm spray distance, porosity 2.8%) (Müller et al., 1995).

- pores originating from condensation of partially evaporated particles;
- pores caused by formation of dendritic structures, preferentially in metallic coatings;
- microcracks formed by solidification, quenching, external loading and so on.

The quest to reduce porosity includes the following measures:

- Preheating the substrate to increase the contact temperature and to reduce the viscosity of the impinging molten droplets.
- Use of low-pressure plasma spraying (LPPS) with increased particle velocities and thus increased kinetic impact energies.
- Post-spraying treatment such as:
  - annealing of coatings to reduce microporosity by solid-state diffusion;
  - hot isostatic pressing ('HIPping') of coatings (Inada et al., 1988);
  - laser surface densification (Khanna et al., 1995; Takasaki et al., 1995; Heimann et al., 1991; Sivakumar and Mordike, 1989; Zaplatynsky, 1986);
  - infiltration of coatings with polymers for low-temperature applications, Ni aluminides or other alloys for high-temperature uses (Ohmori et al., 1995), or by the sol-gel process (Moriya et al., 1995).

Measurement of the porosity of plasma-sprayed coatings can be accomplished by a wide variety of methods that can be divided into those yielding as a result a simple number, the 'porosity' or pore volume related to the total volume of the coating in cm^3/g, and those that yield a pore size distribution function. In many cases the former methods are sufficient to characterize the porosity of a coating.

The following methods can be applied.

- Point-counting using optical microscopy (Glagolev, 1933; Chayes, 1954).
- Electron-optical microscopy in conjunction with discriminant analysis of optical density using automated image analysis (Li and Wang, 2004b; Ctibor et al., 2006; Venkataraman et al., 2006; Shanmugavelayutham and Kobayashi, 2007).
- Mercury pressure porosimetry using stepwise filling of smaller and smaller pore sizes with increasing pressure (Ulmer and Smothers, 1967).

- Bubble pressure method using stepwise squeezing out of liquid from a completely filled pore ensemble (Zagar, 1955).
- Vacuum impregnation with a low-viscosity fluorescent resin and 3D-image reconstruction using confocal laser scanning microscopy (Llorca-Isern et al., 1999).
- Dynamic penetration followed by measuring the electrical conductivity of the porous coating (Dietzel and Saalfeld, 1957).
- Determination of the distribution function of pores by small angle X-ray (Ritter and Erich, 1948) or small angle neutron scattering (Deshpande et al., 2004).
- Measurement of topological and spatial distribution of pores with Voronoi cells and Euclidean distance mapping (Venkataraman et al., 2007).

The volume fraction and size distribution of pores may be used to predict coating properties such as microhardness. However, this does not appear to be straightforward. Instead, it was found that the spatial and topological arrangement of pores as mathematically expressed using Voronoi cells and Euclidean distance mapping could be directly correlated to the microhardness of plasma-sprayed ceramic coatings (Venkataraman et al., 2007). Hardness was found to be increased in those locations which would result in a large number of Voronoi cells and/or longer Euclidean distances characterizing areas with a large number of widely separated pores.

The dominant features of pore orientation in plasma-sprayed coatings are identified as strongly oblate voids either parallel or normal to the substrate. During modeling of the anisotropic elastic properties of coatings (Sevostianov and Kachanov, 2000) it was found that the scatter in pore orientation and the difference between pore aspect ratios have a pronounced effect on the effective modulus. These irregularities may even be responsible for the observed effect of 'inverse' anisotropy, i.e. the fact that Young's modulus is higher in the direction normal to the substrate than that in the transverse direction, as well as for the relatively large value of the Poisson ratio in the plane of isotropy.

Point Counting This approach is based on the quantitative relationships between measurements on the two-dimensional plane of polish and the magnitudes of the microstructural features in three-dimensional materials. The determination of the volume of pores is based on Gauss' principle that the spatial extension of a plane and a volume, respectively are determined by the number of hits that are obtained by randomly or regularly spatially distributed points in the various components of an aggregate. In modern point-counting techniques the points in space are made to points in a plane that are counted optically or electronically. Thus the volume fraction occupied by a microstructural feature such as a void or pore, is equal to the point ratio of the selected feature as seen on random sections through the microstructure. Because point counting is tedious and special instruments for quantitative metallography are expensive, sometimes porosity is estimated by comparing the microstructures with standard photomicrographs. This method is simpler, faster and quite suitable for control and quality acceptance purposes. Depending on their purpose, comparisons are usually based on micrographs taken at magnifications ranging from 50 to 500 times (Roseberry and Boulger, 1977).

Mercury Pressure Porosimetry   The technique requires fully automated equipment for pore size distribution functions ranging from 10 μm to 10 nm. It is based on the Washburn equation (1921) that relates the diameter of a pore r to the pressure p required to fill it with a liquid. Since the ability of a liquid with a contact angle φ exceeding 90° to fill a pore is limited by its surface tension σ, the applied pressure must overcome this surface tension:

$$r[\text{Å}] = 2\sigma \cos\varphi/p, \qquad (10.23)$$

where for mercury σ attains the value of 480 dyn cm^{-1} and $\varphi = 140°$. One of the disadvantages of this method is that Equation 10.23 is strictly valid only for cylindrical pores. For noncylindrical pores correction factors must be applied. This is particularly true for 'ink bottle' pores. Also, on application of high mercury pressures pore walls can be destroyed and higher porosities will be suggested.

Archimedes' Method   This is the classical method to determine the apparent density of a material and from this value the porosity. For plasma-sprayed metallic coatings a procedure has been suggested (Metco Inc., 1963) involving the following steps:

1. Preparation of a solid cylindrical bar (12.7 by 2.2 cm).
2. Application of the coating approximately 2.8 mm thick for a length of about 6.4 cm.
3. Using the center holes, the bar is mounted on a lathe and the coated section is machined to a thickness of 2.54 mm.
4. Specimens of about 2.8 cm length are cut from both the coated and uncoated regions of the bar.
5. Grinding of both ends of the specimens flat and perpendicular to the central axes.
6. Weighing of the specimens to an accuracy of 0.001 g.
7. Calculation of the volume.
8. Determination of the density of the coating using the following equation: (Weight of coated sample − weight of uncoated sample)/(Volume of coated sample − volume of uncoated sample). The volume fraction of porosity can be obtained from 'measured' and 'true' density values by the equation: Pore volume fraction (%) = ('true' density − 'measured' density)/'true' density.

10.2.4
Tribological Properties

There are three primary types of wear: adhesive, abrasive, and erosive. Other composite types of wear include surface fatigue, fretting, gouging, and cavitation erosion. Since there is no universal type of wear, there is also no universal method or machine for testing wear. Laboratory tests are aimed at simulation of service conditions, and consider (i) the position of the fixed or loose abrasive, (ii) the size, shape and hardness of the dominant abrasive, (iii) the direction and speed of relative motion during abrasion, and (iv) the contact pressures or loads in the system.

A rubber-wheel test (ASTM G65-04) simulates low-stress or scratching abrasion with loose abrasive. Gouging abrasion is tested in a jaw crusher (ASTM G 81-97a). Sliding wear tests (ASTM G 77-05, ASTM G 83-96) and erosive wear tests (ASTM G 73-04, ASTM G 76-07) are generally applied to metals and plastics. Tests to evaluate the wear in ceramics are based on the pin-on-disc (POD) concept. In the microwear POD apparatus, a diamond pin with a predetermined applied load rides on the rotating specimen (coating). In the macrowear tester, two freely rotating wheels (Taber™ Abraser apparatus) ride over the rotating specimen assembly that consists of twelve trapezoidal-shape sections held together by a circular ring on the outside edge and a disc in the center. A general discussion of tribological properties of thin films and coatings, thick coatings and hardfacing has been presented by Kelley *et al.* 1988.

Since the wear behavior of coatings is strongly influenced by composition, microstructure including porosity, residual stresses (Kuroda *et al.*, 1995; Matejicek *et al.*, 2003) and surface conditions, tribological properties of coatings must be evaluated under conditions that match as closely as possible the actual in-service conditions (Borel *et al.*, 1990). As this is generally not possible, wear model tests are applied that simulate the very complex wear processes in technical tribosystems under simplifying assumptions at ambient conditions (Steinhäuser and Wielage, 1995). As a consequence, application of the results of such wear model tests to the real world is generally unsuccessful.

To illustrate the degree of complexity, Figure 10.23 shows a methodological approach to wear tests for tribomaterials (Steinhäuser and Wielage, 1995). Wear mechanism maps are particularly useful to reveal the relationships between interaction parameters and dominant wear mechanism (Borel *et al.*, 1990).

10.2.4.1 Simulation of Basic Wear Mechanisms

These quality assurance procedures involve testing for adhesive and abrasive wear as well as for long-term fatigue and erosive wear. The tests permit to investigate the local behavior of the coating/substrate tandem system subjected only to those basic wear mechanisms. Evaluation of existing models for abrasive wear and impact erosion was performed by Dimond *et al.* (1983) to reconcile the results of laboratory wear tests and theoretical models to the true wear performance of a material in service.

Adhesive Wear Friction is generated by local adhesion and subsequent separation of the contact faces of a tribological couple. The contact of the two surfaces does not occur along the entire geometrical surface area A_0 but only with the fraction

$$A/A_0 = (\sigma/H)R, \tag{10.24}$$

where A is the effective contact area. This ratio increases with increasing compressive stress $\sigma = F/A_0$ and surface roughness R, and decreasing hardness H of the

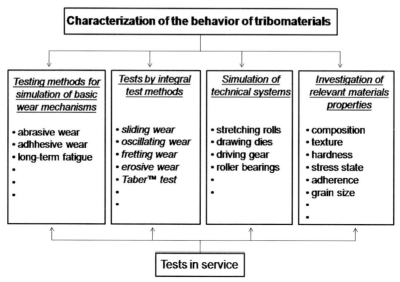

Figure 10.23 Methodology of wear tests for tribomaterials (Steinhäuser and Wielage, 1995).

materials. The friction coefficient μ is given by $\mu = F_R/F$, where $F_R =$ frictional force $= (d\gamma/dx)A$ and $F =$ compressive load. The energy dissipation γ per glide distance x, i.e. $d\gamma/dx$ in the effective contact area A is the main physical reason for dry friction (Hornbogen, 1991).

The test of adhesive wear uses the adhesion tendencies of tribocouples to determine an 'adhesion number' $\mu_{v,ad}$ that is numerically different from the friction coefficient defined above (Steinhäuser and Wielage, 1995). The adhesion number is determined as the ratio of the tangential force F_T to the normal force F_N of a tribosystem consisting of a counterbody attached to a torque rod transferring the torsional momentum to the coating surface that is pressed against the conterbody with the normal force F_N. Since the torsional momentum is only maintained by adhesive forces, the normal force F_N is relaxed when the counterbody slides back to its starting position. From the displacement diagram both F_T and F_N and therefore $\mu_{v,ad}$ can be determined as a measure of the adhesive tendency of the tribocouple.

Abrasive Wear Using the scratch tester mentioned in 10.2.3.1 information can be obtained on the abrasive wear properties of coatings subjected to the indentation load of a Vickers diamond pyramid. The indenter is pressed against the surface of the coating with the normal force F_N and at the same time the sample is being moved relative to the indenter for a distance L_R with a velocity v_R. The measured tangential force F_T is related to the volume of the produced scratch that can be determined by laser beam tracing and the scratch energy density W_R is obtained in J mm^{-3}:

$$W_R = (F_T \, L_R)/(A_R \, L_R), \tag{10.25}$$

where A_R is the so-called scratch square, i.e. the cross-section of the scratch. Note that the expression in the denominator of Equation 10.25 is the volume of the material removed by the scratching operation.

Problems frequently occur due to microploughing, microcutting and -chipping and particle pull-out that tend to obscure the scratch traces. Figure 10.24 shows (a) a scratch produced in a (Ti,Mo)C-NiCo coating on a mild steel substrate and (b) its laser-generated profile. The scratch energy density was calculated to be 5.2 J mm^{-3} (Thiele, 1994).

The most frequently applied abrasive wear test is the dry sand-rubber wheel abrasion test according to ASTM G65-04 (2004). This simple test measures material losses that occur when a coated sample surface is pressed with a defined force against a steel wheel whose circumference is lined with rubber. Into the gap between the sample and the wheel sand (Ottawa sand) or other abrasive materials are fed from a hopper reservoir with a constant flow rate (Figure 10.25). After 2000 revolutions of the wheel the sample is removed and weighed. The loss of material is a measure for the abrasive resistance of the coating.

Other tests to determine abrasive wear exploit 2-body rotating ring-on-flat (reciprocating Cameron–Plint tester, Prchlik and Sampath, 2007) or 3-body block-on-ring

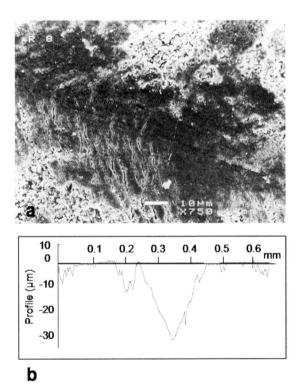

Figure 10.24 (a) Scratch produced with a diamond indenter in a (Ni,Mo)C-NiCo coating and (b) its laser-generated profile (Thiele, 1994).

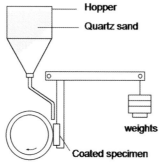

Figure 10.25 Dry sand-rubber wheel abrasion test device according to ASTM G65-04 (2004).

(ASTM G77-05; see also Günther, 1982; Henke et al., 2004) configurations. The latter test generally inflicts much damage (see Chapter 7.1.2.1) and thus is useful to estimate wear performance of coatings subjected to severe in-service conditions as encountered, for example in mining machinery, ore processing, and earth moving equipment. Figure 10.26 shows the functional principle of the Cameron–Plint reciprocating rotating ring-on-flat system (Model TE-77, Compton, Berkshire, UK; Prchlik and Sampath, 2007).

Owing to the combination of the disk rotation and the flat sample oscillation this wear test maintains a well-defined close to ideal cylinder/planar contact geometry throughout the test. The contact width b and the maximum contact load p_{max} are expressed as

$$b = 1.13 \times \sqrt{(P \times d/2L \times E^*)}, \tag{10.26}$$

with $p_{max} = 3P/4Lb$ and $E^* = [(1-v_c^2)/E_c + (1-v_i^2)/E_i)$, where P = load [N], d = ring diameter [mm], and E_c, v_c and E_i, v_i are the elastic moduli and Poisson ratios of coating and indenter, respectively (Prchlik et al., 2000).

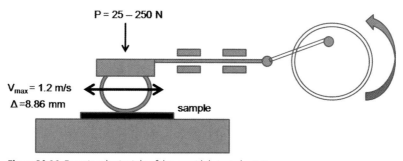

Figure 10.26 Functional principle of the quasi-lubricated rotating ring-on-flat sliding wear test (P: load, v_{max}: velocity of ring perimeter, Δ: stroke of reciprocating motion) (after Prchlik and Sampath, 2007).

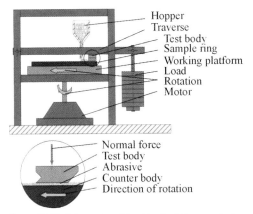

Figure 10.27 Schematics of the 3-body sliding wear test apparatus according to a modified ASTM G77-05 designation (top) and contact area between the loose abrasive, and the test and counter bodies (bottom) (after Günther, 1982; Henke et al., 2004).

Figure 10.27 shows the 3-body block-on-ring sliding wear test configuration according to ASTM G77-05 in a modified version after Günther 1982 (see also Henke et al., 2004). In this modified test the load (normal force) can be varied between 687 and 7×10^3 N by applying dead weights to the sample traverse; with a wire rope a load reduction can be achieved to below 180 N. For a sample area of 50 mm × 35 mm this results in a nominal contact pressure between 12.3 MPa (7×10^3 N) and 0.31 MPa (180 N), resulting in a nominal contact pressure between 0.1 and 4 MPa. As abrasive the quartz sand HB33 was used.

Long-term Fatigue Wear In this test the cylindrical sample to be tested is pressed against a curved counterbody disc whose diameter is ten times that of the cylindrical sample (Figure 10.28).

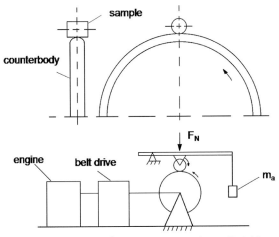

Figure 10.28 Long-term fatigue wear testing device (Steinhäuser und Wielage, 1995).

The curvature of the rotating disc is different in two perpendicular directions in order to minimize the contact surface between disc and cylindrical sample. The load m_A applied generates the normal force F_N. The critical number of load reversals as well as scratch and pit formation are evaluated.

Erosive Wear A widely applied test for erosive wear of ceramics and ceramic coatings is the impact abrasion/solid particle erosion test based on depth of penetration produced by a standard sand or grit blast (ASTM G76-07). Erosion is a mechanism of wear resulting from the impact of abrasive particles on a target material (Ruff and Wiederhorn, 1983; Wiederhorn and Hockey, 1983). The erosion rate of a plasma-sprayed coating is a complex function of many variables including the size, shape, velocity, flux and angle of impact of the impinging particles and such coating properties as hardness, grain size diameter (Orowan–Petch relation), porosity (Ryshkevich–Duckworth equation), ductility or fracture toughness.

Figure 10.29 shows a schematic of the erosion test apparatus according to ASTM G76-07. A screw-feed type metering system releases controlled amounts of an erodent (sand, glass beads, crushed alumina, carborundum etc.) into a flowing gas stream. Particles delivered are picked up in the stream and accelerated through a tungsten carbide nozzle before being directed at the coating surface.

Figure 10.30 shows the schematic sequence of events for impact of a high-energy erodent particle (Wright *et al.*, 1986). The highly accelerated particles (a) are able to penetrate the coating completely and to deform the metallic substrate underneath in a ductile manner by cutting and ploughing (b) . The release of the deformation energy results in a chipping mechanism (c) that removes parts of the coating and thus exposes the substrate to environmental attack by corrosion or abrasive wear. Maximum loss of coating would be expected at shallow impact angles in the 15 to 30° range (Figure 10.31, Sumner *et al.*, 1985) and the velocity dependence exponent of erosive wear would be about 2.0 to 2.5 (Wright *et al.*, 1986).

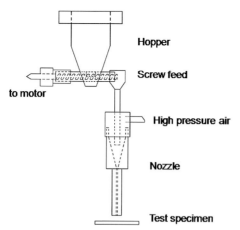

Figure 10.29 Solid particle erosion (SPE) test device according to ASTM G76-07.

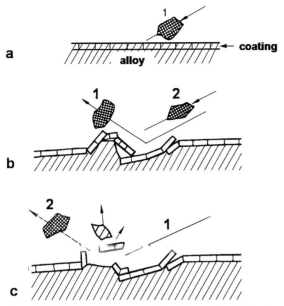

Figure 10.30 Schematic sequence of events occurring during the impact a high-energy erodent particle onto a plasma-sprayed coating surface (Wright et al., 1986).

It should be pointed out that this is an important mechanism of the erosion of high-pressure turbine blades and valve components of fossil fuel-fired power plants (Qureshi et al., 1986; Buchanan, 1987). Hard particles of magnetite scale formed at elevated temperature by reaction of steam with ferritic alloy boiler tube material can exfoliate from the interior surfaces of the boiler tubes during boiler transients (startup and cool down cycles). This solid particle erosion (SPE) attacks the blade

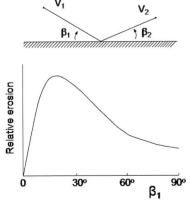

Figure 10.31 Dependence of the relative erosion rate on the impact angle β (Sumner et al., 1985).

airfoil at various impingement angles when passing through the turbine and also erodes any other component of the steam path. By this mechanism protective coatings along the steam path such as plasma-sprayed $80Cr_3C_2/20(NiCrMo)$ coatings (Wlodek, 1986) can be completely destroyed.

A potentially useful noncontact technique has been explored by Gentleman and Clarke 2004 to monitor by luminescence sensing the wear and foreign particle erosion damage (Janos et al., 1999; Zhang and Desai, 2005) inflicted on plasma-sprayed ceramic TBCs during service and hence to establish the 'health' of the TBC before it eventually fails. Trivalent rare earth elements (REE) such as Gd, Er, or Sm forming stable solid solutions with stabilized zirconia (YSZ) or gadolinium zirconate ($Gd_2Zr_2O_7$, GZO) will be excited by an argon-ion laser beam to give off luminescent radiation whose intensity and wavelength will be monitored. Figure 10.32 shows the principle. The TBC consists of a succession of layer, each one doped with a different REE luminescing at a different frequency hence constituting a 'rainbow' sensor (Figure 10.32a). As the coating erodes away, successive doped layers will be removed and the recorded visible luminescence will characterize the remaining layers (Figure 10.32b). If mid-UV excitation above the optical band gap is being used, the outermost layer could be detected in the 'rainbow' sensor and thereby the progress of erosion could be directly assessed. Recently this approach has been extended to REE zirconates, $(REE)_2Zr_2O_7$ (REE = Eu, Gd, Sm) with pyrochlore structure. These materials were proposed as temperature sensing coatings in different configurations such as (i) a thin Eu-doped YSZ coating underneath an undoped YSZ layer to sense the temperature at the bond coat interface, (ii) a thin Eu-doped YSZ or $Eu_2Zr_2O_7$ coating deposited on top of an existing TBC to measure the temperature at the surface of a TBC, and (iii) two doped layers at the inside and outside of a TBC to monitor the temperature difference through the TBC (Gentleman and Clarke, 2005). A reflectance-enhanced luminescence technique was developed to monitor delamination of TBCs that uses the discernible luminescence contrast between unbacked and NiCr-backed sections of Eu-doped YSZ (Eldridge and Bencic, 2006).

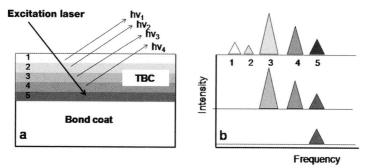

Figure 10.32 Schematic rendering of a 'rainbow' sensor consisting of a succession of REE-doped TBC layers (a) and the change of the luminescence spectra during progressive erosion (b) (Gentleman and Clarke, 2004).

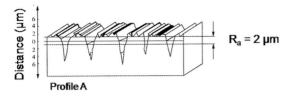
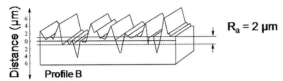

Figure 10.33 Surface roughness profiles of a plasma-sprayed coating according to DIN 4768. Despite identical average roughness two situations arise: a smooth plateau surface with depressions (profile A) and a strongly ragged surface with protrusions (profile B).

10.2.4.2 Surface Roughness of Coatings

Thermally sprayed coatings exhibit on mechanical treatment such as grinding, lapping and polishing very different surface structures compared with homogenous bulk materials. Homogeneous materials show after mechanical treatment a ploughed surface that is characterized by scratches; plasma-sprayed surfaces show an inhomogeneous profile given by the original splat structure, porosity and pulled-out and chipped-off areas. To describe such surfaces according to their functional behavior in service it is necessary to separate the undulation, i.e. the waviness of the surface (macro-roughness) and the roughness *per se* (micro-roughness) that is responsible for the tribological behavior of the coating.

Using a diamond-stylus surface roughness tester the profile obtained can be manipulated by electronic filtering methods so that a cut-off line is obtained that correspond to the long-wave surface profile (Kühn *et al.*, 1990). Applying an RC high-pass filter (DIN 4768, ISO 3274) the information obtained from smooth surfaces with pores and pull-outs may be severely falsified. This is shown in Figure 10.33. The average roughness R_a (DIN 4768)[2] is measured to be identical for the two very different surface roughness profiles (A: smooth plateau surface with depressions, B: strongly ragged surface with protruding asperities). Similar results are obtained for a comparison of the maximum roughness R_{max} and the median roughness R_z.

Obviously, profile A has a much larger 'carrying' surface, i.e. supports the countersurface of a bearing much more effectively than profile B. The depressions of profile A will also much more efficiently act as reservoirs for lubricating materials and thus promote the frictional properties of the coating. In order to describe the surface roughness more properly by determining the plateau-like amount of

---

[2] The value R_a is the arithmetic mean of the deviation of the protrusions and depressions of the roughness profile from the average line, $R_a = (1/lm) \int |f(x)| \, dx$.

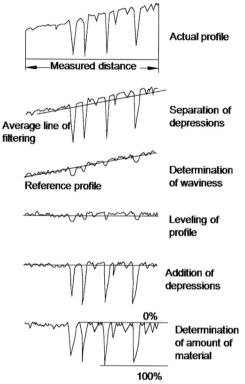

Figure 10.34 Determination of the waviness of a plasma-sprayed surface through Abbot's curve (see text, Kühn et al., 1990).

material, surface roughness should be measured by DIN 4776 (Kühn et al., 1990). Here the waviness of the surface can be determined by (i) cutting off the depressions, and (ii) application of digital phase-true filters to measure Abbot's curve of the amount of material 'carrying' the profile (Figure 10.34). For thermally sprayed coatings the Abbot curve always has an S-shaped character. Approximating this curve by three straight lines the roughness profile can be subdivided into three areas: (i) the core area of the profile, (ii) the spike area of the profile, and (iii) the depression area of the profile. The *reduced spike height* characterizes the height of the spikes protruding from the core area, the *reduced depression depth* characterizes the depth of the depressions below the core area.

10.2.5
Chemical Properties

10.2.5.1 Thermally Induced Chemical Changes

Frequently, the molten droplets react with the plasma gas, or with air pumped by the plasma jet. In particular, oxidation or decomposition of carbides and nitrides takes place. For example, a WC/Co coating may encounter three connected chemical processes:

1. matrix alloying, i.e. solubility of WC in the Co metal matrix under formation of so called 'η-carbides' of the general composition Co_nW_mC,
2. decarburization, and
3. deoxidation (Hajmrle and Dorfman, 1985).

The general equation for matrix alloying is

$$a\{WC\} + b\{Co\} + c\{O_2\} = d\{WC\} + e[Co_3W_3C] + f\{CO_2\} + \Delta H. \tag{10.27}$$

With increasing temperature and residence time in the hottest regions of the plasma jet decarburization reactions take place thought to occur in three stages (Bouaifi et al., 1995; Vinajo et al., 1985):

$$2\,WC \rightarrow W_2C + C \tag{10.28a}$$

$$W_2C + 1/2\,O_2 \rightarrow W_2(C,O) \tag{10.28b}$$

$$W_2(C,O) \rightarrow 2\,W + CO \tag{10.28c}$$

As a consequence, HVOF spraying with lower temperature and shorter residence time considerably suppress tungsten and chromium carbide decomposition. Therefore, highly wear- and corrosion-resistant coatings are increasingly sprayed successfully with HVOF technique (Brandt, 1995).

Figure 10.35 shows schematically the chemical and phase changes during thermal spraying of WC/Co(Ni) powder. It can be seen that the matrix of the coating consists in principle of η-carbide with some residual WC, and a small degree of porosity (Hajmrle and Dorfman, 1985; see also Figure 7.1).

A study by Brunet et al. (1995) showed that during plasma spraying of TiC severe carbon depletion occurs even when argon is used as a shroud gas to protect the sprayed TiC from oxidation. Plasma spraying in air results in the formation of crystalline oxides (Ti_3O_5, TiO_2) and free titanium (Morozumi et al., 1981).

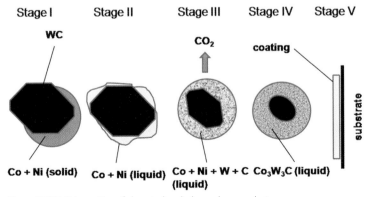

Figure 10.35 Schematics of chemical and phase changes during thermal spraying of WC(Co(Ni)) powder (after Hajmrle and Dorfman, 1985).

A second example of chemical changes that occur during plasma spraying is the thermal decomposition of hydroxyapatite (see also Chapter 9). Figure 10.36 shows the binary phase diagram of the system CaO-P_2O_5 with the region of particular interest around the composition of apatite. Hydroxyapatite melts incongruently in the absence of water at 1843 K to form a mixture of α'-$Ca_3(PO_4)_2$ (high-temperature tricalcium phosphate, α'-C_3P) and $Ca_4O(PO_4)_2$ (tetracalcium phosphate, C_4P) (Riboud, 1973). The actual melting temperature depends on the water partial pressure: at a water partial pressure $p_{H2O} = 6.7 \times 10^4$ Pa hydroxyapatite is stable up to 1633 K and its melting point increases to 1823 K at $p_{H2O} = 1.33 \times 10^2$ Pa. Since HAp melts incongruently, the oxyapatite formed by stepwise dehydroxylation decomposes into tri- and tetracalcium phosphates and the entire transformation sequence can be expressed in four steps (Graßmann and Heimann, 2000; Carayon and Lacout, 2003).

The squares in the formulae of oxyhydroxyapatite and oxyapatite shown in Box 10.1 refer to lattice vacancies in the OH positions along the crystallographic c-axis in the structure of hydroxyapatite. In the completely dehydroxylated oxyapatite (OAp) there exists a linear chain of O^{2-} ions parallel to the c-axis each one followed by a vacancy (x = 1 in step 2). While OAp appears to be stable in the absence of moisture it will readily transform to hydroxyapatite in the presence of water according to the equation (Montel et al., 1980)

$$O^{2-}(s) + \square(s) + H_2O(g) \rightarrow 2OH^-(s). \qquad (10.29)$$

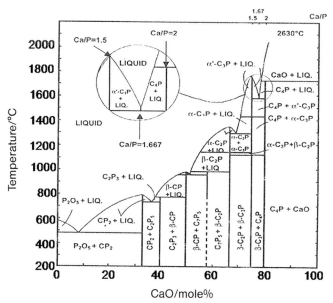

Figure 10.36 Binary phase diagram CaO-P_2O_5 (Kreidler and Hummel, 1967). The inset shows the region of interest, i.e. the incongruent thermal decomposition of hydroxyapatite (Ca/P = 1.667) to α'-tricalcium phosphate (Ca/P = 1.5) and tetracalcium phosphate (Ca/P = 2).

> **Box 10.1**
>
> **Thermal decomposition sequence of hydroxyapatite**
>
> Step 1: $Ca_{10}(PO_4)_6(OH)_2 \rightarrow Ca_{10}(PO_4)_6(OH)_{2-2x}O_x\square_x + xH_2O$
> (hydroxyapatite) (oxyhydroxyapatite)
>
> Step 2: $Ca_{10}(PO_4)_6(OH)_{2-2x}O_x\square_x \rightarrow Ca_{10}(PO_4)_6O_x\square_x + (1-x)H_2O$
> (oxyhydroxyapatite) (oxyapatite)
>
> Step 3: $Ca_{10}(PO_4)_6O_x\square_x \rightarrow 2Ca_3(PO_4)_2 + Ca_4O(PO_4)_2$
> (oxyapatite) (tricalcium phosphate)+(tetracalcium phosphate)
>
> Step 4a: $Ca_3(PO_4)_2 \rightarrow 3\,CaO + P_2O_5$
>
> Step 4b: $Ca_4O(PO_4)_2 \rightarrow 4\,CaO + P_2O_5$

The equilibrium temperature of step 3 is determined by the temperature of incongruent melting of hydroxyapatite at 1843 K (Riboud, 1973), (see Chapter 9.3 for more details).

Performance specifications of coatings for high-temperature applications, for example gas turbine blades, combustor cans, ladles and tundishes for metal casting, require quality testing that must be able to simulate the severe service conditions at which the coated parts are supposed to function. A comprehensive review of production and performance evaluation of high-temperature coatings has been given by Nicoll 1983a. Since the coatings are subjected to synergistic effects of mechanical stresses, temperature and corrosive environment small coating failures can lead to catastrophic destruction of components of engineering systems. Environmental test considerations include the gas temperature, composition, pressure, velocity and temperature cycling. Also, contaminants such as sulfates and vanadates as well as particulate matter can lead to deposition and corrosion, and also destructive erosion effects. Figure 10.37 shows several of such degradation mechanisms that can affect plasma-sprayed stabilized zirconia-NiCrAlY duplex coatings (Kvernes and Forseth, 1987; Levine et al., 1979). While in combustion environment hot corrosion of ceramic components such as SiC, Si_3N_4, and SiAlON occurs in a manner similar to dry oxidation, i.e. under formation of a protective SiO_2 surface layer, gaseous impurity species such as Na_2SO_4 and NaCl can condense at the surface of engine components at temperatures as high as 1100 °C and lead to severe corrosion, pitting and strength reduction (Jacobson and Smialek, 1985; Davies et al., 1986).

10.2.5.2 Tests of Chemical Corrosion Resistance

To select a proper protective coating system for high-temperature applications, three main factors must be considered: the applications, the structural alloys to be protected, and the coatings themselves. Figure 10.38 show schematically the interactions of mechanical properties, coating processes, and the environmental attack the system is subjected to. The design of the component determines the service stresses, maximum operation temperatures, and service environment. Alloy properties are controlled by chemistry, processing and the resulting microstructure that can also control the high temperature stability (Nicoll, 1983a). Environmental considerations include the gas temperature, composition, pressure stream velocity, and temperature cycling. Since materials at high temperatures are subject to corrosion

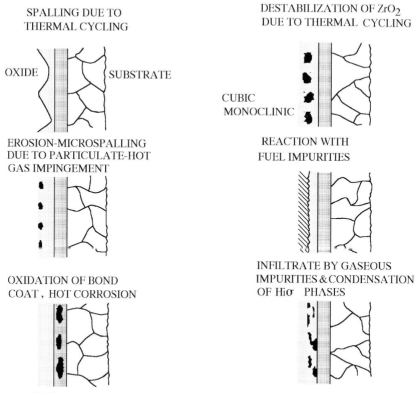

Figure 10.37 Degradation mechanisms of stabilized ZrO2/NiCrAlY TBCs (Kvernes and Forseth, 1987).

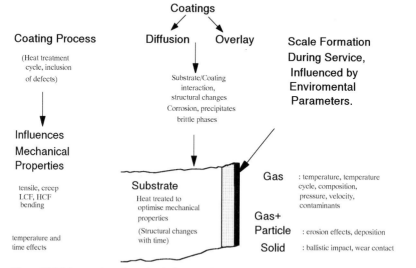

Figure 10.38 Interaction of mechanical properties, coating processes, and environmental attack of high temperature-resistant coatings (Nicoll, 1983a).

phenomena, e.g. oxidation, hot corrosion/erosion or carburization tests to evaluate high temperature performance of coating systems are used to account for these interactions. Degradation of coatings also involves creep (Hebsur and Miner, 1986, 1987), and low-cycle (Gayda et al., 1986) and thermal fatigue (Berndt, 1989).

All evaluation tests at various stages of the development of a high temperature protection coating system can be grouped according to their cost, number of tests required and their extrapolation risk. Thus *screening tests* require many samples at comparatively low cost performed as crucible tests at isothermal exposure. *Bench-scale tests* such as creep testing need fewer samples but the costs are increased. *Component tests* use a simulated service environment such as burner rig testing for gas turbine blades. Finally, tests designed to provide *service life prediction and systems verification* are most expensive, require but a few samples, and are performed in a pilot plant or for 5000 hours in a stationary gas turbine. Figure 10.39 shows a typical coating evaluation program for the evaluation of new coating systems for industrial gas turbines that includes a combination of short-term complex environmental exposure, mechanical tests and long-term laboratory test for structural stability (up to 10 000 hours).

The short-term chemical corrosion tests in their required complexity are shown in Figure 10.40 (Nicoll, 1983b). Since corrosion resistance of a coatings depends to a large extent on the rate of formation of a protective scale, the simplest method is to expose the coating isothermally to either air or oxygen, to atmospheres containing hydrogen sulfide, sulfur dioxide or trioxide, or to 'coal' atmospheres (methane). The weight change of the sample will be measured, and the nature of the scale formed be determined by X-ray diffraction analysis. Cyclic modes of testing are

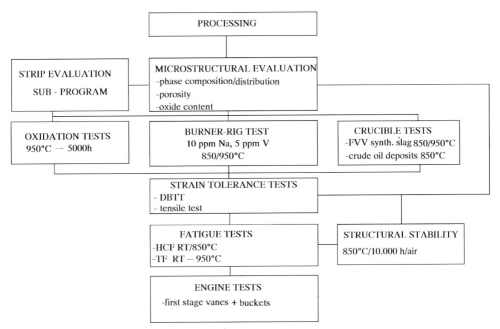

Figure 10.39 Typical coating evaluation program for coatings applied to stationary gas turbine blades (Nicoll, 1983a).

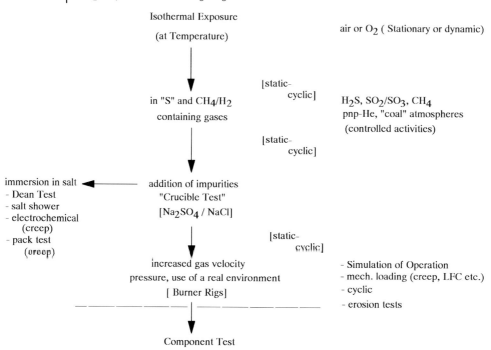

Figure 10.40 Complexity of short-term chemical corrosion tests (Nicoll, 1983b).

utilized to introduce thermal strain between coating and substrate, and coating and scale.

Abrupt weight changes measured with a thermal balance may indicate spalling and chipping of the scale. Addition of impurities to the corrosive environment accounts for the presence of chlorides, sulfates and vanadates in the combustion gases. Such tests are usually performed as immersion tests where the coated sample is immersed in an appropriate salt melt. The extent of attack can be determined from thickness changes on the metallographic cross-section or from weight loss after descaling (Lewis and Smith, 1965). Frequently the attack during such a crucible test would appear to be more severe than that encountered during normal service conditions of the coating. The reasons for this are manifold. For example, the salt composition is often unrealistic, the oxidation potential is low, and the test is static. To overcome these problems other test schedules were devised such as the salt-shower test (Shaw et al., 1979), the synthetic slag test (Bauer et al., 1979), and the modified Dean test (Dean, 1964).

10.2.5.3 Burner Rig Test

This test simulates reasonably well the severe conditions at which a coating has to function in a gas turbine (Janke and Nicoll, 1983). Such a test rig consists of a combustion chamber taken from a small turbine into which fuel and compressed air is fed in the usual manner. Contaminants are supplied either to the fuel or to the air, or can be sprayed directly into the combustion chamber onto specimen coupons to be tested. These coupons are either stationary or rotate. Variables to be tested include coating temperature, gas pressure, velocity, dwell time, contaminant concentration and composition, and also the fuel-to-air ratio (Nicoll, 1983a). The samples are investigated after the test by measuring weight loss and/or penetration depth of the corrosive gases. To increase the realistic evaluation of coating systems the burner rig test can be cycled (Ruckle, 1980; Nicholls and Bennet, 2000) or supplemented by atomized salt water solution to simulate helicopter engine conditions (Raffaitin *et al.*, 2006).

References

E.A. Almond, *Vacuum*, 1984, **35**, 835.

B. Aumüller, A.N. Kirkbride, R. Heller, H.W. Bergmann, *Mat.-wiss.u.Werkstofftech.*, 1996, **27**, 72.

M.F. Bahbou, P. Nylen, J. Wigren, *J. Thermal Spray Technol.*, 2004, **13** (4), 508.

P. Bansal, P.H. Shipway, S.B. Leen, L.C. Driver, *Mater. Sci. Eng. A*, 2006, **430**, 104.

L.M. Barker, R.E. Hollenbach, *J. Appl. Phys.*, 1972, **43**, 4669.

R. Bauer, K. Schneider, H.W. Grünling,in: Proc. DOE Conf. on Adv. Mater. for Alternate Fuel Capable Directly Fired Heat Engines, Castine/Maine, USA, 1979.

K. Baumung, J. Singer, S.V. Razorenov, A.V. Utkin, *Shock Compression of Condensed Matter – 1995* (S.C. Smith, W.C. Tao, eds.), AIP Conf. Proc., 1996, **370**, 1015.

K. Baumung, G. Müller, J. Singer, G.I. Kanel, S.V. Razorenov, *J. Appl. Phys.*, 2001, **89** (11), 6523.

A. Bendada, *J. Adhesion Sci. Technol.*, 2004, **18** (8), 943.

A. Bendada, N. Baddour, A. Mandelis, C. Moreau, *Intern. J. Thermophys.*, 2005, **26** (3), 881.

C.C. Berndt,in: Advances in Thermal Spraying, Proc. ITSC 1986, Welding Inst. of Canada, 149.

C.C. Berndt, *J. Mater. Sci.*, 1989, **24**, 3511.

M. Bertagnolli, M. Marchese, G. Jacucci, *J. Thermal Spray Technol.*, 1995, **4** (1), 41.

M.O. Borel, R.K. Smith, A.R. Nicoll,in: Proc. TS'90, Essen 1990, DVS 130, 68.

B. Bouaifi, U. Draugelates, I. Grimberg, K. Soifer, B.Z. Weiss,in: Proc. 14th ITSC'95, Kobe, May 22–26, 1995, 627.

O. Brandt,in: Proc. 14th ITSC'95, Kobe, May 22–26, 1995, 639.

C. Brunet, S. Dallaire, I.G. Sproule,in: 14th ITSC'95, Kobe, May 22–26, 1995, 129.

E.R. Buchanan, *Turbomachinery Intern.*, 1987, **28** (1), 25–27, 31.

H. Bückle,in: *The Science of Hardness Testing and its Research Applications* (eds. J.W. Westbrook, H. Conrad), ASM, Metal Park, OH, 1971, 453.

W. Bullough, G.E. Gardam, *J. Electrodepos. Tech. Soc.*, 1947, **22**, 169.

P.J. Burnett, T.P. Page, *J. Mater. Sci.*, 1984, **19**, 845.

P.J. Burnett, D.S. Rickerby, *Thin Solid Films*, 1987, **148**, 41.

R.C. Camp, *Benchmarking: The Search for Industry Best Practices that Lead to Superior Performance*, Quality Press and UNIPUB/Quality Resources, 1989.

M.T. Carayon, J.L. Lacout, *J. Solid State Chem.*, 2003, **172**, 339.

F. Chayes, *J. Geol. USA*, 1954, 62.

D. Chicot, J. Lesage, *Thin Solid Films*, 1995, **254**, 123.

L.C. Cox, *Surf. Coat. Technol.*, 1988, **36**, 807.

A. Crocombe, R. Adams, *J. of Adhesion*, 1981, **12**, 127.

P.B. Crosby, *Quality is Free*, McGraw-Hill Co., New York, 1979.

P. Ctibor, R. Lechnerová, V. Beneš, *Mater. Charact.*, 2006, **56**, 297.

D. Davies, J.A. Whittaker, *Metallurg. Reviews*, 1967, **12**, 15.

G.B. Davies, T.M. Holmes, O.J. Gregory, *Adv. Ceram. Mat.*, 1986, **3**, 542.

L. Davison, D.E. Grady, M. Shahinpoor, *High Pressure Shock Compression of Solids II. Dynamic Fracture and Fragmentation*, Springer: New York, 1996.

A.V. Dean, *Investigation into the Resistance of Various Nickel and Cobalt Base Alloys to Sea Salt Corrosion at Elevated Temperature*, NGTE Report, January 1964.

I. Delgadillo-Holtfort, J. Gibkes, D. Dietzel, M. Kaak, B.K. Bein, J. Pelzl, *Mater.-wiss. u. Werkstofftech.*, 2000, **31**, 850.

P. Démarécaux, D. Chicot, J. Lesage, *J. Mater. Sci.*, 1996, **15**, 1377.

W.E. Deming, *Out of the Crisis*, MIT Press: Cambridge, MA., 1989.

P.R. Dennis, C.R. Smith, D.W. Gates, J.B. Bond (eds.) *Plasma Jet Technology* Technology Survey, NASA SP-5033, NASA: Washington, D.C., October 1965, p. 42.

S. Deshpande, A. Kulkarni, S. Sampath, H. Herman, *Surf. Coat. Technol.*, 2004, **187**, 6.

A. Dietzel, H. Saalfeld, *Ber. Deutsche Keram. Ges.*, 1957, **34**, 363.

C.R. Dimond, J.N. Kirk, J. Briggs, *Wear of Materials*, April 1983, 333.

K. Ebert, C.M. Verpoort, in: Proc. 14th ISTC'95, Kobe, May 22–26, 1995, 1191.

J.I. Eldridge, T.J. Bencic, *Surf. Coat. Technol.*, 2006, **201**, 3926.

B. Elvers, S. Hawkins, G. Schulz, (eds.), *Ullmann's Encyclopedia of Industrial Chemistry*, 5th edn, Metals, Surface Treatment, 1990, Vol. A16, p. 433, VCH: Weinheim.

L.C. Erickson, R. Westergård, U. Wiklund, N. Axén, H.M. Hawthorne, S. Hogmark, *Wear*, 1998, **214**, 30.

J. Fakuda, *J. General Managem.*, 1986, **11**, 16.

A.C. Fischer-Cripps, P. Karvánková, S. Vepřek, *Surf. Coat. Technol.*, 2006, **200**, 5645.

J. Forsgren, F. Svahn, T. Jarmar, H. Engqvist, *Acta Biomater.*, 2007, **3** (6), 980.

D.B. Fowler, W. Riggs, J.C. Russ, *Adv. Mater. Process.*, 1990, **141** (1), 41.

B. Franke, Y.H. Sohn, X. Chen, J.R. Price, Z. Mutasim, *Surf. Coat. Technol.*, 2005, **200**, 1292.

R. Frielinghaus, G. Schmitz, U. Wielpütz, in: Proc. TS'90, Essen, 1990, DVS 130, 147.

D.T. Gawne, B.J. Griffith, G. Dong, in: Proc. 14th ITSC'95, Kobe, May 22–26, 1995, 779.

J. Gayda, T.P. Gabb, R.V. Miner, *Int. J. Fatigue*, 1986, **8** (4), 217.

M.M. Gentleman, D.R. Clarke, *Surf. Coat. Technol.*, 2004, **188/189**, 93.

M.M. Gentleman, D.R. Clarke, *Surf. Coat. Technol.*, 2005, **200**, 1269.

G.A. Georgiou, M.B. Saintey, A.M. Lank, D.P. Almond, in: Proc. 14th ITSC'95, Kobe, May 22–26, 1995, 1047.

R. Ghafouri-Azar, J. Mostaghimi, S. Chandra, *Int. J. Comput. Fluid Dynamics*, 2004, **18** (2), 133.

A.A. Glagolev, *Trans. Inst. Econ. Min., Moscow*, 1933, 59.

C. Godoy, J.C.A. Batista, *J. Thermal Spray Technol.*, 1999, **8** (4), 531.

O. Graßmann, R.B. Heimann, *J. Biomed. Mater. Res.*, 2000, **53** (6), 685.

H. Grützner, H. Weiss, *Surf. Coat. Technol.*, 1991, **45**, 317.

H. Gruner, *Thin Solid Films*, 1984, **118**, 409.

M. Gudge, D.S. Rickerby, R. Kingswell, K.T. Scott, in: Proc. 3rd NTSC, Thermal Spray Research and Applications, Long Beach, CA, May 20–25, 1990, p. 331.

H. Günther, *Investigation of the Wear Performance of Metallic Sliding Pairs Considering the Action of Abrasives*, Unpublished Ph.D. thesis, Faculty of Mechanical Engineering, Process and Energy Technology, Bergakademie Freiberg, 1982.

K. Hajmrle, M. Dorfman, *Modern Develop. Powder Metall.*, 1985, **15/17**, 609.

J. Hartikainen, *Rev. Sci. Instrum.*, 1989, **60** (7), 1334.

J.R. Hauser, D. Clausing, *The House of Quality*, Harvard Business Review, May–June, 1988.

E.J. Hay, J. Zonderman, *Just in Time Manufacturing: How the JIT System can Decrease Costs, Increase Productivity and Enhance Quality*, John Wiley & Son: New York, 1988.

M.G. Hebsur, R.V. Miner, *Mater. Sci. Eng.*, 1986, **83**, 239.

M.G. Hebsur, R.V. Miner, *Thin Solid Films*, 1987, **147**, 143.

R.B. Heimann, D. Lamy, V.E. Merchant, in: Trans. 17th Workshop CUICAC, (R.B. Heimann ed.), Québec City, Québec, Canada, October 2, 1991.

R.B. Heimann, M. Müller, in: Proc. 4th Intern. Seminar/Course on Coatings for Aerospace Industry (CAI-4), Toronto, Oct 26–27, 1995.

R.B. Heimann, T.A. Vu, M.L. Wayman, *Eur. J. Miner.*, 1997, **9**, 597.

H. Henke, D. Adam, A. Köhler, R.B. Heimann, *Wear*, 2004, **256**, 81.

E. Hornbogen, *Werkstoffe*. Springer: Berlin, Heidelberg, New York, 1991, 202.

E. Horvath, A.J. Perry, *Thin Solid Films*, 1980, **65**, 309.

A.W. Hothersall, C.J. Leadbetter, *J. Electrodepos. Techn. Soc.*, 1938, **14**, 207.

M. Inada, T. Maeda, T. Shikata, in: Proc. ATTAC'88, Adv. Thermal Spraying Technology and Allied Coatings, Osaka, May, 1988, 211.

A. Iost, J. Aryani-Boufette, *J. Fonct. Mém. Sci., Rev. Mét.*, 1992, **11**, 681.

K. Ishikawa, D.J. Lu, *What is Total Quality Control? The Japanese Way*, Prentice-Hall: Englewood Cliffs, NJ, 1985.

ISO 9001, Quality systems model for quality assurance in design/development, production, installation and servicing, 1994.

N.S. Jacobson, J.L. Smialek, *J. Am. Ceram. Soc.*, 1985, **68**, 432.

B. Janke, A.R. Nicoll, in: Proc. Conf. on Frontiers of High Temperature Materials II, London, 1983.

B.Z. Janos, E. Lugscheider, P. Remer, *Surf. Coat. Technol.*, 1999, **113**, 278.

B. Jönsson, S. Hogmark, *Thin Solid Films*, 1984, **114**, 257.

J.M. Juran, *Juran's Quality Control Handbook*, McGraw-Hill: New York, 1988.

J.M. Juran, *Planning for Quality*, The Free Press: New York, 1988.

J.E. Kelley, J.J. Stiglich, G.L. Sheldon, in: *Surface Modification Technologies* (eds. T.S. Sudarshan, D.G. Bhat), in: Proc. 1st Int. Conf. Surf. Modification, Phoenix, AZ, January 25–28 1988, The Metallurgical Society, 169.

B. Keys, T. Miller, *Acad. Managem. Review*, 1984, **9**, 342.

A.S. Khanna, A.K. Patniak, K. Wissenbach, in: Proc. 14th ITSC'95, Kobe, May 22–26, 1995, 993.

H.J. Kim, Y.G. Kweon, *Thin Solid Films*, 1999, **342**, 201.

E.R. Kreidler, F.A. Hummel, *Inorg. Chem.*, 1967, **6**, 884.

H. Kühn, O. Stitz, R. Letzner, in: Proc. Thermal Spray Conf., TS90, 1990, Essen, August 29–31, p. 38.

S. Kuroda, T. Dendo, S. Kitahara, *J. Thermal Spray Technol.*, 1995, **4** (1), 75.

H. Kurzweg, R.B. Heimann, T. Troczynski, *J. Mater. Sci.: Mater. Med.*, 1998, **9**, 9.

I. Kvernes, S. Forseth, *Mater. Sci. Eng.*, 1987, **88**, 61.

B.R. Lawn, A.G. Evans, D.B. Marshall, *J. Am. Ceram. Soc.*, 1980, **63**, 574.

J. Lesage, D. Chicot, in: Proc. 14th ITSC'95, Kobe, May 22–26, 1995, 951.

J. Lesage, D. Chicot, *Surf. Eng.*, 1999, **15**, 447.

J. Lesage, D. Chicot, *Thin Solid Films*, 2002, **415**, 143.

D. Lescribaa, A. Vincent, *Surf. Coat. Technol.*, 1996, **81**, 297.

S. Levine, P.E. Hodge, R.A. Miller, in: Proc. 1st Conf. on Advanced Fuel Capable Directly Fired Heat Engines, Maine Maritime Acad., Castine, ME; July 31–August 3, 1979; DOE, Washington, D.C., 1979.

H. Lewis, R.A. Smith, in: Proc. 1st Int. Congr. Met. Corr., 1965, 202.

C.J. Li, W.Z. Wang, Y. He, *J. Thermal Spray Technol.*, 2004a, **13** (3), 425.

C.J. Li, W.Z. Wang, *Mater. Sci. Eng. A*, 2004b, **386**, 10.

D. Lian, Y. Suga, S. Kurihara, in: Proc. 14th ISCT'95, Kobe, May 22–26, 1995, 879.

M.M. Lima, C. Godoy, P.J. Modenesi, J.C. Avelar-Batista, A. Davison, A. Matthews, *Surf. Coat. Technol.*, 2004, **177/178**, 489.

R.S. Lima, S.E. Kruger, G. Lamouche, B.R. Marple, *J. Thermal Spray Technol.*, 2005, **14**, 52.

N. Llorca-Isern, M. Puig, M. Español, *J. Thermal Spray Technol.*, 1999, **8** (1), 73.

E. Lopez, F. Belzung, G. Zembelli, *J. Mat. Sci. Lett.*, 1989, **8**, 346.

E. Lopez, G. Zambelli, A.R. Nicoll, in: Proc. TS'90, Essen, 1990, DVS 130, 241.

J. Malzbender, *Surf. Coat. Technol.*, 2006, **201**, 3797.

G. Marot, J. Lesage, Ph. Démarcéaux, M. Hadad, St. Siegmann, M.N. Staia, *Surf. Coat. Technol.*, 2006, **201**, 2080.

D.M. Marsh, *Proc. R. Soc. London, Ser. A.*, 1964, **279**, 420.

D.B. Marshall, B.R. Lawn, *J. Am. Ceram. Soc.*, 1977, **60**, 86.

J. Matejicek, S. Sampath, D. Gilmore, R. Neiser, *Acta Mater.*, 2003, **51**, 873.

N.Z. Mehdizadeh, S. Chandra, J. Mostaghimi, *J. Fluid Mech.*, 2004, **510**, 353.

N. Mesrati, Q. Saif, D. Treheux, A. Moughil, G. Fantozzi, A. Vincent, *Surf. Coat. Technol.*, 2004, **187**, 185.

Metco Inc., *Evaluation Methods and Equipment for Flame-Sprayed Coatings*, Metco Inc., 1963, 15 pp.

E. Meyer, *Phys. Z.*, 1908, **9**, 66.

E.-H. Meyer, K.-J. Pohl, in: *Moderne Beschichtungsverfahren* (eds. H.-D. Steffens, W. Brandl), DGM Informationsgesellschaft Verlag, 1992.

J.M. Miguel, J.M. Guilemany, B.G. Mellor, Y.M. Xu, *Mater. Sci. Eng. A*, 2003, **352**, 55.

W. Milewski, in: Proc. TS'93, Aachen, 1993, DVS 152, 258.

R.A. Miller, W.J. Brindley, C.J. Rouge, G. Leissler, *NASA Tech. Briefs*, 1993.

G. Montavon, C. Coddet, C.C. Berndt, S.-H. Leigh, *J. Thermal Spray Technol.*, 1998, **7** (2), 229.

G. Montel, G. Bonel, J.C. Trombe, J.C. Heughebaert, C. Rey, *Pure Appl. Chem.*, 1980, **52**, 973.

D.C. Montgomery, *Introduction to Statistical Quality Control*, 5th edn, Wiley-VCH: Weinheim, 2004.

K. Moriya, W. Zhao, A. Ohmori, in: Proc. 14th ITSC'95, Kobe, May 22–26, 1995, 1017.

S. Morozumi, M. Kikuchi, S. Kanazawa, *J. Nucl. Mat.*, 1981, **103/104**, 279.

M. Müller, *Haftungsuntersuchungen an vakuumplasmagespritzten Chromschichten*, Unpublished Master Thesis, Technische Universität Bergakademie Freiberg, June 1994.

M. Müller, F. Gitzhofer, R.B. Heimann, M.I. Boulos, in: Proc. NTSC'95, Houston, TX, USA, Sept 11–15, 1995.

T. Nakamura, Y. Gu, *Mech. Mater.*, 2007, **39**, 340.

G. Newaz, X. Chen, *Surf. Coat. Technol.*, 2005, **190**, 7.

J.R. Nicholls, M.J. Bennet, *Mater. High Temp.*, 2000, **17** (3), 413.

A.R. Nicoll, in: *Coatings and Surface Treatment for Corrosion and Wear Resistance* (eds. K.N. Strafford, P.K. Datta, C.G. Googan), Ellis Horwood Ltd.: Chichester, chapt. 13, 180, 1983a.

A.R. Nicoll, in: *Coatings for High Temperature Applications*, E. Lang, ed., London: Applied Science Publishers, 1983b.

K. Niihara, *J. Mater. Sci. Lett.*, 1983, **2**, 221.

E.A. Ollard, *Trans. Faraday Soc.*, 1925, **21**, 81.

A. Ohmori, Z. Zho, K. Inue, K. Murakami, T. Sasaki, in: Proc. 14th ITSC'95, Kobe, May 22–26, 1995, 549.

W. Ouchi, *Theory Z*, Addison-Wesley, MA, 1981.

M. Owen, *SPC and Continuous Improvement*, Springer-Verlag: London, 1989.

M. Pasandideh-Fard, V. Pershin, S. Chandra, J. Mostaghimi, *J. Thermal Spray Technol.*, 2002, **11** (2), 206.

L. Pawlowski, *The Science and Engineering of Thermal Spray Coatings*, John Wiley & Sons: Chichester, New York, Brisbane, Toronto, Singapore, 1995.

A.J. Perry, H.K. Pulker, *Thin Solid Films*, 1985, **124**, 323.

A.J. Perry, E. Horvath, *J. Mater. Sci.*, 1978, **13**, 1303.

G. Pouskouleli, T.A. Wheat, in: *Ceramic Coatings- A Solution Towards Reducing Wear and Corrosion* (R.B. Heimann, ed.), Trans.17th CUICAC Workshop, Quebec City, October 2, 1991.

L. Prchlik, A. Vaidya, S. Sampath, in: Proc. 6th Int. Symp. on FGMs, Estes Park, CO, USA, 2000.

L. Prchlik, S. Sampath, *Wear*, 2007, **262**, 11.

J. Qureshi, A. Levy, B. Wang, *J. Vac. Sci. Technol.*, 1986, **A4** (6), 2638.

M. Raessi, J. Mostaghimi, M. Bussmann, *Thin Solid Films*, 2005, **506–507**, 133.

A. Raffaitin, F. Crabos, E. Andrieu, D. Monceau, *Surf. Coat. Technol.*, 2006, **201**, 3829.

G. Reisel, R.B. Heimann, *Surf. Coat. Technol.*, 2004, **185**, 215.

Y.M. Rhyim, C.G. Park, S.B. Kim, M.C. Kim, in: Proc. 14th ITSC'95, Kobe, May 22–26, 1995, 773.

P.V. Riboud, *Ann. Chim.*, 1973, **8**, 381.

E. Rigal, T. Priem, E. Vray, in: Proc. 14th ITSC'95, Kobe, May 22–26, 1995, 851.

H.L. Ritter, L.C. Erich, *Anal. Chem.*, 1948, **26**, 665.

E.J. Roehl, *Iron Age*, 1940, **146**, 17, 30.

T.J. Roseberry, F.W. Boulger (eds.), *A Plasma Flame Spray Handbook*. Final Report No. MT-043 to Naval Sea Systems Command, Naval Ordnance Station. Louisville, KY, 1977.

D.L. Ruckle, *Thin Solid Films*, 1980, **73**, 455.

A.W. Ruff, S.M. Wiederhorn, *Treat. Mat. Sci. Technol.*, 1983, **16**, 69.

S. Sampath, X.Y. Jiang, J. Matejicek, *Mater. Sci. Eng, A*, 1999, **272**, 194.

S. Sampath, X. Jiang, A. Kulkarni, J. Matejicek, D.L. Gilmore, R.A. Neiser, *Mater. Sci. Eng. A*, 2003, **348**, 54.

S. Sampath, X.Y. Jiang, J. Matejicek, L. Prchlik, A. Kulkarni, A. Vaidya, *Mater. Sci. Eng. A*, 2004, **364**, 216.

P.M. Sargent, T.F. Page, *Proc. Brit. Ceram. Soc.*, 1978, **26**, 209.

S. Seifert, *Development of Ceramic Lightweight Thermal Protection Coatings for Reusable Space Launch Vehicles*, Unpublished Master Thesis, Technische Universität Bergakademie Freiberg, 2004.

I. Sevostianov, M. Kachanov, *Acta Mater.*, 2000, **48**, 1361.

M. Sexsmith, T. Troczynski, in: Proc. 14th ISTC'95, Kobe, May 22–26, 1995, 897.

G. Shanmugavelayutham, A. Kobayashi, *Mater. Chem. Phys.*, 2007, **103** (2–3), 283.

S.Yu. Sharivker, *Poroshk. Metallurgiya*, 1967, **6** (54), 70; Sov. Powder Met. Metal Ceram., 1967, 483.

S.W.K. Shaw, M.S. Starkey, M.T. Cunningham, *High Temperature Corrosion in Salt Shower Test and Oxidation of a Range of Superalloys*, Materials for Gas Turbines: Cost 50, Final Report, March 1979.

R. Sivakumar, B.L. Mordike, *Surf. Eng.*, 1989, **4**, 127.

J.W.M. Spicer, W.D. Kerns, L.C. Aamodt, J.C. Murphy, *Review of Progress in Quant. Nondestruct. Eval.*, 1990, **9**, 1169.

H.-D. Steffens, *Haftung und Schichtaufbau beim Lichtbogen- und Flammspritzen*, Ph.D. Dissertation, Hannover, 1963.

H.-D. Steffens, K.-N. Müller, *Adhäsion*, 1972, **2**, 34.

S. Steinhäuser, B. Wielage, in: Proc. 14th ITSC'95, Kobe, May 22–26, 1995, 693.

Y. Suga, H. Makaba, K. Makabe, in: Proc. TS'93, Aachen, 1993. DVS 152, 201.

Y. Suga, D. Lian, S. Kurihara, in: Proc. 14th ISTC'95, Kobe, May 22–26, 1995, 961.

W.J. Sumner, J.H. Vogan, R.J. Lindinger, Proc. Am. Power Conf., April 22–24, 1985, 196.

J.E. Sundgen, H.T.G. Hentzell, *J. Vac. Sci. Tech. A.*, 1986, **4**, 2259.

N. Takasaki, M. Kumagawa, K. Yairo, A. Ohmori, in: Proc. 14th ITSC'95, Kobe, May 22–26, 1995, 987.

S. Thiele, *Mikrohärte, Mikrostruktur und Haftung vakuumplasmagespritzter TiC-Mo$_2$C-Ni,Co-Verbundschichten*. Unpublished Master Thesis, Technische Universität Bergakademie Freiberg, June 1994, p. 66.

A. Thomas, *Surf. Eng.*, 1987, **3**, 117.

R. Travis, C. Ginther, C. Zanis, in: *Advances in Thermal Spraying*, Proc. ITSC 1986, Welding Inst. of Canada, 309.

G.C. Ulmer, W.J. Smothers, *Am. Ceram. Soc. Bull.*, 1967, **46**, 649.

R. Venkataraman, G. Das, B. Ventakaraman, G.V. Narashima Rao, R. Krishnamurthy, *Surf. Coat. Technol.*, 2006, **201**, 3691.

R. Venkataraman, G. Das, S.R. Singh, L.C. Pathak, R.N. Ghosh, B. Venkataraman, R. Krishnamurthy, *Mater. Sci. Eng. A.*, 2007, **445/446**, 269.

S. Veprěk, S. Mukherjee, P. Karnáková, H.-D. Männling, J.L. Hee, K. Moto, J. Procházka, A.S. Argon, *Thin Solid Films*, 2003, **436**, 220.

M.E. Vinajo, F. Kassabji, J. Guyonnet, P. Fauchais, *J. Vac. Sci. Technol.*, 1985, **A3**, 2483.

H.M. Wadsworth, K.E. Stephens, A.B. Godfrey, *Modern Methods for Quality Control and Improvement*, John Wiley & Sons: New York, 1991.

J.S. Wallace, J. Ilavsky, *J. Thermal Spray Technol.*, 1998, **7** (4), 521.

E.W. Washburn, *Proc. Natl. Acad. Sci. USA*, 1921, **7**, 115.

M. Watanabe, T. Okabe, M. Enoki, T. Kishi, *Sci. Technol. Adv. Mater.*, 2003, **4**, 205.

S.M. Wiederhorn, B.J. Hockey, *J. Mat. Sci.*, 1983, **18**, 766.

C. Williams, R.A.F. Hammond, *Trans. Inst. Metal Finishing*, 1954, **31**, 124.

S.T. Wlodek, EPRI 1885.2, Phase II, Final Report, 1986.

I.G. Wright, V. Nagarajan, J. Stringer, *Oxid. Metals*, 1986, **25** (3/4), 175.

L. Zagar, *Arch. Eisenhüttenwesen*, 1955, **26**, 561.

I. Zaplatynsky, *The effect of Laser Glazing on Life of ZrO_2 TBS's in Cyclic Burner Tests*. NASA Techn. Memorandum 88821, August 1986.

J. Zhang, V. Desai, *Surf. Coat. Technol.*, 2005, **190**, 90.

Y.L. Zhu, S.N. Ma, B.S. Xu, *J. Thermal Spray Technol.*, 1999, **8** (2), 328.

ASTM D214-07, 'Standard Test Method for Sieve Analysis of Metal Powders', 2007.

ASTM B293-76, 'Standard Method for Subsieve Analysis of Granular Metal Powders by Air Classification', withdrawn 1984, no replacement.

ASTM C633-01, 'Standard Test Method for Adhesion or Cohesive Strength of Thermal Spray Coatings', 2001.

ASTM D3167-03a, 'Standard Test Method for Floating Roller Peel Resistance of Adhesives', 2004.

ASTM D3359-02, 'Standard Test Method for Measuring Adhesion by Tape Test', 2002.

ASTM E11-04, 'Standard Specification for Wire Cloth and Sieves for Testing Purposes', 2004.

ASTM E18-03, 'Standard Test Method for Rockwell Hardness and Rockwell Superficial Hardness of Metallic Materials', 2003.

ASTM E384-07, 'Standard Test Method for Microindentation Hardness of Materials', 2007.

ASTM G65-04, 'Standard Test Method for Measuring Abrasion Using the Dry Sand/Rubber Wheel Apparatus', 2004.

ASTM G73-04, 'Standard Practice for Liquid Impingement Erosion Testing', 2004.

ASTM G76-07, 'Standard Test Method for Conducting Erosion Tests by Solid Particle Impingement Using Gas Jets', 2007.

ASTM G77-05e1, 'Standard Test Method for Ranking Resistance of Materials to Sliding Wear Using Block-on-Ring Wear Test', 2005.

ASTM G81-97a(2002)e1, 'Standard Test Method for Jaw Crusher Gouging Abrasion Test', 2002.

ASTM G83-96, 'Standard Test Method for Wear Testing with a Crossed-Cylinder Apparatus', 1996, withdrawn 2005.

DIN EN ISO 14577, 'Metallic materials – Instrumented indentation tests for hardness and materials parameters', 2003.

DIN 50160, 'Ermittlung der Haftzugfestigkeit im Stirnversuch', 1981.

DIN EN 582-1994, European Norm: 'Determination of adhesion strength (Thermal spraying)', 1994.

DIN EN ISO 4490 (4/2002), 'Metal powders; Determination of flow duration by means of a calibrated funnel (Hall flowmeter)'.

DIN 4768 (5/1990), 'Ermittlung der Rauheitskenngrößen R_a, R_z, R_{max} mit elektrischen Tastschnittgeräten'.

DIN 4776 (5/1990), 'Kenngrößen R_K, R_{pK}, R_{vK}, M_{r1}, M_{r2} zur Beschreibung des Materialanteils im Rauheitsprofil'.

EN 15340:2007, 'Thermisches Spritzen. Bestimmung des Scherbeanspruchungswiderstandes bei thermisch gespritzten Schichten'.

11
Design of Novel Coatings

11.1
Property-based Approaches

11.1.1
Coating Design Based on Chemical Bonding

Thermally sprayed and PVD/CVD (Physical Vapor Deposition/Chemical Vapor Deposition) coatings must be carefully designed in order to obtain maximum performance for the required application. An initial example will be given based on the concept of advanced layered coatings for cutting tools. Even though it relates predominantly to thin PVD or CVD coatings the strategy can also be applied, in principle, to thermally sprayed coatings.

Coatings for cutting tools are important to

- improve service life of the tool, i.e. reduce the frequency of tool changes;
- provide more favorable operating conditions, i.e. higher speed and feed rates and thus improve the economy of the cutting and milling operations;
- provide higher product quality, i.e. better surface finish and tighter tolerances;
- open up new areas of applications, i.e. machining of difficult to handle nonmetallic materials such as wood, plastic and composites;
- achieve resource conservation, i.e. reduced materials losses and machining of more economical materials.

There are, however, conflicting basic requirements for coating performances. The hard coating materials in question are frequently oxide ceramics such as Al_2O_3 and Cr_2O_3, or non-oxide ceramics such as transition metal carbides (TiC, TaC, WC, Mo_2C), nitrides (TiN, Si_3N_4, SiAlONs) and borides (TiB_2).

> *Problem I:* Ceramic coatings must have good adhesion to the metallic substrate, but little tendency to react with the material of the metal chip removed by the tool from the workpiece. This is particularly important when at high cutting speeds and therefore high frictional temperatures the material of the coating dissolves in the chip.

Plasma Spray Coating: Principles and Applications. Robert B. Heimann
Copyright © 2008 WILEY-VCH Verlag GmbH & Co. KGaA, Weinheim
ISBN: 978-3-527-32050-9

Problem II: The ceramic constitutes a hard layer with a high melting point, but there should be little crack propagation, i.e. high fracture toughness of the coating is required.

Problem III: The ceramic layer must be hard and high melting, but good adhesion is needed under widely changing temperature conditions imposed by friction and intense shear forces at the worked interface, and the associated high thermal stresses and high thermal expansion (Holleck, 1991).

A way to look at the compatibility of different ceramic materials is through their differing chemical bonding type (Figure 11.1). Metallic hard materials, such as WC, TiC or TiN are characterized by their high proportion of metal–metal bonds in the fcc closed-packed arrangement of metal atoms, whereas the small carbon or nitrogen atoms occupy the octahedral interstitial sites. On the other hand, covalent ceramic materials have a high proportion of highly directional covalent bonds, and

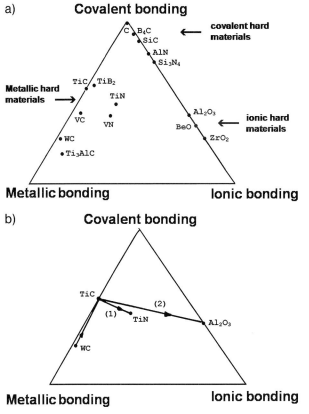

Figure 11.1 Chemical bonding types of different hard ceramic materials (a) and possible designs of gradient-layer coatings WC–TiC–TiN (1) and WC–TiC–Al$_2$O$_3$ (2) (b) (Holleck, 1991).

heteropolar (ionic) ceramic materials possess simple sublattices of large close-packed oxygen anions with the small metal atoms at interstitial sites, opposite to the metal bonding structure.

In general, all materials used for protective layers are characterized by mixed bonding types. Their properties change with the position of the material in the bonding triangle (Figure 11.1). TiN close to the center of the triangle has a particularly favorable combination of metallic, covalent and ionic bonding types. Carbide, nitrides and borides of the transition metals (Ti, V, W) crystallize in simple, densely packed lattices of large metal atoms with the nonmetallic atoms joined to the metal lattice by covalent or mixed covalent/ionic bonds. If metallic and ionic structures are combined, i.e. TiC–Al_2O_3, there are no corresponding metal planes at the interface but the Ti-planes of TiC will match the oxygen planes of Al_2O_3 in position and size. Because the covalent structures have highly directional, saturated bonds, there is only a small tendency to interact with other materials at the interface.

11.1.2
Design of Novel Advanced Layered Coatings

The concept of 'advanced layers' includes the design of gradient layers (Sampath *et al.*, 1995) and multilayers.

11.1.2.1 **Gradient Layers**
Functionally gradient materials (FGMs) display continuously (or discontinuously) varying compositions or microstructures over definable geometrical distances (Sampath *et al.*, 1995).

Very thin layers can be deposited by physical vapor deposition (PVD) techniques. Such coatings are designed in a way that to the substrate is first added an adhesive layer with a large fraction of metal bonds. These bonds tend to form a stable external layer with little tendency to interact. Examples are given in Figure 11.1(b) of a gradient layer combination WC–TiC–TiN (1) and WC–TiC–Al_2O_3 (2). Such gradient layer coatings for cutting tools will exhibit superior flank wear resistance. In fact, the wear depth of uncoated WC during interrupted cutting is reduced by using a PVD TiC–TiN gradient coating from >160 µm to a minimum wear depth of 20 µm (Holleck, 1991). These data were obtained from interrupted cutting conditions of CK 45 (speed: 125 m min^{-1}, feed: 0.2 mm per revolution, depth of cut: 2 mm, cutting time: 5 min; corresponding to 3500 impacts).

Plasma-spray processing offers an alternative way to deposit functionally gradient materials (FGMs) in a flexible and economical manner. It is possible to deposit multiple constituents simultaneously, for example to produce thermal barrier graded layers with enhanced survivability for gas turbines and diesel engines. Gradient layers are primarily employed to reduce discontinuities in the coefficients of thermal expansion to avoid mismatch-related failure in service. Such discontinuities frequently result in fatal cracks and spallation at the sharp interface between the substrate and a thermal barrier coating. FGMs are capable of spreading out this mismatch stress and thus reduce crack initiation. Equipment to deposit FGMs can be classified as (i) single plasmatron–multiple powder feeders with blended or composite powders, (ii) multiple plasmatron–independent feeding systems for each components, and (iii) process combinations–wire/powder feed systems combinations. Using a single plasmatron–dual feeder combination, a graded NiCrAl/Y-PSZ coating has been produced with an almost linear increase of Y-PSZ as a function of the distance from the interface to the substrate. The microstructure of this 2 mm thick gradient coating was composed of 70 μm thick discrete layers of varying compositions (Sampath *et al.*, 1995). Likewise, the elastic modulus of such a gradient coating decreased nearly linearly with increasing Y-PSZ content, i.e. away from the interface (Marshall *et al.*, 1982). Despite these advances the benefits of grading for stress relaxation need to be studied in detail in the context of other coating characteristics such as environmental stability and manufacturing complexity.

Today the list of research and development effort in the area of FGMs has spread to the following products and applications (ASM, 2004):

- thick, multilayer thermal barrier coatings (TTBCs) for heavy-duty diesel engine pistons;
- wear-resistant coatings for diesel engine piston rings;
- oxidation-resistant coatings for high-temperature operation of machinery;
- ceramic seals in aircraft gas turbines;
- clearance control coatings in rotating machinery;
- thermal protection of lightweight polymeric insulating materials in aircraft components;
- graded metallic/oxide/intermetallic advanced batteries and solid oxide fuel cells;
- oxide/metal/air-type electrode/electrolyte systems;
- dielectric and electromagnetic shielding coatings for electronic devices;
- biomedical implants for enhanced bone–tissue apposition.

11.1.2.2 Multilayers

Fine-grained multiphase structures with many phase boundaries frequently produce structures with increased toughness. Limited crack propagation takes place in the modulated layer structure because of deflection and arrest of cracks at the interfaces, as there are many oscillations between tensile and compressive stresses. However, great care must be taken to assure the integrity and solid adhesion of the boundary layer next to the substrate. PVD methods are ideally suited to produce TiC/TiB_2, TiC/TiN or TiN/TiB_2 composite coatings with up to 500 individual layers with a total layer thickness not exceeding 5 μm. The advantage of the multilayer concept lies in the reduction of grain growth, the introduction of many interfaces, and the formation of modulated layer material that changes the mechanical properties, and leads to effective stopping of crack propagation. Materials with a large difference in the shear modulus (ΔG) should be considered for synthesizing superlattice coatings (Veprek, 1999).

11.1.3
Coating Design Based on Materials Informatics

The design of novel coatings involves rational selection of material constituents and development of manufacturing processes to produce desired microstructures. This approach can be guided by investigating the effect of chemical bonding and microstructure on coating properties and performance (see Section 11.1). To design nanostructured hard coatings layer materials should be chosen with a maximum difference in shear modulus, ΔG, a property that would counteract dislocation propagation from the softer to the hard layers (Chu and Barnett, 1995). Design criteria for super- and ultrahard nanocomposite coatings may also involve combinations of transition metal nitrides such as TiN, W_2N and VN with amorphous Si_3N_4 or BN as grain boundary phases (Veprek, 1999). The physical mechanism to provide such coatings with hardness exceeding 50 GPa is effective retardation of grain boundary sliding by the pinning effect of the amorphous phase. Since fracture strength and hardness are proportional to Young's modulus coating development based on evaluation of elastic data is an important design tool. For example, hard coatings were developed using information on elastic properties of transition metal nitrides calculated from first-principle density functional theory (DFT) (Zhao et al., 2005). This particular approach is consistent with the principles of materials informatics (Rodgers, 2001) and combine (i) calculation of the elastic coefficients c_{ij} of single crystals and elastic moduli of polycrystalline single-phase materials by DFT, (ii) their validation by FPLMTO (full-potential linear muffin-tin orbital) and FLAPW (full-potential linearized augmented plane-wave) methods, and experiments, (iii) trends identified by 'data mining' in appropriate property databases, and (iv) selecting rationally materials from the property trends. Using these techniques, hardness and ductility trends were identified for binary nitrides TiN, ZrN, CrN, VN, NbN and AlN, and a multitude of ternary transition metal nitrides, and their interaction mapped as a function of the shear modulus, G and the Cauchy pressure, $(c_{12}-c_{44})$ of the B1 structure.

However, since the data derived from the first-principles calculation do not contain microstructural information additional knowledge of microstructural effects is required.

11.1.4
Process Mapping

A process map is an integrated set of relationships that link processing parameters to coating properties, and ultimately to performance. Process maps allow a more intelligent control of the plasma spraying process in an industrial environment and are thus closely related to the quest for economically and environmentally beneficial coating deposition. During APS of molybdenum on a steel substrate an empirical model was developed based on experimental data that link the plasmatron input parameters (arc current, auxiliary helium gas flow rate, powder carrier gas flow rate) to the properties of the plasma jet (Sampath et al., 2003). This is

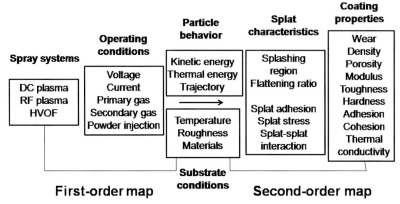

Figure 11.2 Philosophy of process mapping for plasma spraying (after Sampath et al., 2003).

called a first-order process map (Figure 11.2, left). Based on these results those parameters were identified that control splat characteristics as well as coating properties to establish a correlation between process microstructures and properties (second-order process map; Figure 11.2, right). The combination of these maps eventually yields an understanding of the complex interaction of intrinsic and extrinsic plasma parameters and coating properties, and hence provides a useful tool for use of the acquired in-depth knowledge of the process in the manufacturing sector.

Figure 11.3 illustrates a typical process map developed for APS Mo coatings on a steel substrate (Sampath et al., 2003). The map shows the responses of elastic modulus in GPa, residual stress in MPa, and splat morphology as functions of the process parameters substrate temperature and kinetic energy of particles in

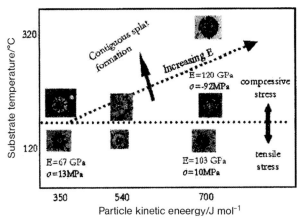

Figure 11.3 Second-order process map for plasma spraying of Mo on a steel substrate (Courtesy: Sampath et al., 2003).

J mole^{-1}. The dotted line corresponds to zero stress condition indicating the balance of the tensile residual stress generated during splat quenching and the compressive thermal mismatch stress between the Mo coating and the steel substrate.

In particular, the following information can be gained from this process map:

- Increasing kinetic energy of particles reduces coating porosity as indicated by an increase in the elastic modulus.
- Increasing kinetic energy of particles does not appear to affect the amount of residual stresses in the coating. However, a change of sign of the stress from compressive to tensile occurs at higher temperature when the thermal stress field superimposes on the quenching stress.
- Increasing kinetic energy of particles leads to increased coating adhesion as well as increased hardness and thermal conductivity.
- Increasing substrate temperature and decreasing kinetic energy of particles further formation of contiguous, disc-like splats as opposed to 'exploded' splats (see also Figure 10.3).

11.2
Stochastic Approaches

In contrast to the property-based approaches to coating design discussed above, stochastic treatment covers a much broader range of parameter interactions and their influence on coating properties and performance. The predictions gained from all results obtained from statistical data treatment, including statistical design of experiment (SDE) methodology, artificial neuronal networks (ANN), generic algorithms, and fuzzy logic control (FLC), require validation either by key experiments or first-principle calculations. Several options will be described below.

11.2.1
Statistical Design of Experiments (SDE)

11.2.1.1 The Experimental Environment and its Evolution

A brief survey of the techniques of statistical design of experiments most frequently applied to thermally sprayed coatings has been given by Bisgaard (1990), Heimann (1990) and Pierlot *et al.* (2008). Two-level factorial analysis applied to plasma spraying was reviewed by Lugscheider and Knepper (1991). The design methodologies are based on the ideas of Plackett and Burman (1946), Box (1960, 1978), Deming (1986) Taguchi and Konishi (1987). Software is available to perform the statistical calculations required with ease.

According to Tukey (1962), industrial experiments can be classified according to their depths of intellectual investment as (i) confirmatory experiments, (ii) exploratory experiments, and (iii) fundamental or 'stroke-of-genius' experiments. A second way of classification is based on the distance of their objective from the real world, i.e. from the market (Daniel, 1976). Finally, the continuity of factors provides a third

Table 11.1

	Classical	Statistical
Number of runs	many	few
Type of response	complex	simple
Synergism	absent	present
Error	small	large
Strategy	one-factor-at-a-time	factorial
Thought pattern	vertical	lateral (De Bono, 1970)

classification scheme. If the factors (parameters, variables) are continuous and controllable at preset levels, then the response surface methodology is the method of choice. If, however, some factors are orderable but not measurable, i.e. at discrete levels, the response surface analysis becomes less useful and should be replaced by nested or split-plot designs (Cochran and Cox, 1957). At a lower level of predicting power, screening designs like Plackett–Burman designs (Plackett and Burman, 1946) that can handle mixtures of continuous and discrete factors are particularly important as a statistical experimental design for the optimization of plasma sprayed coatings (see Sections 11.2.1.2, 11.2.2.1).

Every experiment attempts to approximate the 'real world' in some ways but must avoid, by a set of simplifying assumptions, the complex interactions occurring in real systems. There are, in principle, two ways to accomplish this: the *'classical' experimental strategy* that varies one parameter at a time but attempts to keep all others constant, and the *statistical strategy* that varies factors simultaneously to obtain a maximum of information with a minimum number of experiments. Thus the experimental economy becomes the overriding principle of the strategy. The classical experimental strategy yields accurate results but requires many experiments, and may give misleading conclusions to problems that have synergistic factor interactions, and also fails to elucidate the 'structure' of a system. Table 11.1 compares these two strategies (DuPont de Nemours, 1975).

11.2.1.2 Screening Designs

The evolution of the experimental environment usually starts with a screening design, for example a Plackett–Burman or Taguchi design with many independent (up to 40) variables. It yields a rather crude prediction of the relative magnitude and sign of the parameters and thus the ranking of importance of parameters through a first-order polynomial model. The experimenters should list and investigate carefully all possible parameters they can think of but should refrain from skipping some because of 'folklore', laboratory gossip, or preferences and hunches. The penalty for the tremendous reduction in the number of required experiments, however, will be the failure to detect synergistic interactions between parameters. On the other hand, an advantage of the screening designs is that they can accommodate a mix of continuous and discrete parameters. Plackett–Burman designs are *saturated*, i.e. they contain as many experimental runs as there are coefficients to determine in the first-order polynomial model. If the number of potential influencing factors is very large, *supersaturated* designs can be selected that contain less runs than coefficients to be estimated (Bandemer and Bellmann, 1994). An even more reduced design plan can be obtained using the principle of chance balance (Bandemer and Bellmann, 1994; Satterthwaite, 1959). More modern approaches consider evolutionary algorithms combined with fuzzy logic allowing to estimate the behavior of a complex system with only a few randomly chosen tests (see Section 11.2.4).

11.2.1.3 Response Surface (RSM) Designs

With the significant independent parameters (usually up to eight) identified by screening designs a 'limited response surface' experiment should be run such like a full two-level factorial, 2^p or a fractional three-level factorial (Box–Behnken) design (Box and Behnken, 1960) that yields higher quality predictions by allowing interpolation within the experimental space by a second-order polynomial model. Such a model determines nonlinear behavior, i.e., the curvature of the response surface and thus permits the estimation of synergistic parameter interactions.

11.2.1.4 Theoretical Models

The polynomial models approximate the 'true' response surface only in the necessarily narrow region of the investigated parameter space. Thus, any extrapolation beyond the proven validity of the predictions is dangerous and may lead to useless or even nonsensical results. In particular, there is a risk that local maxima or minima of the response surface are thought to be global ones. To avoid this, eventually theoretical models have to be built (Box *et al.*, 1978; Box and Draper, 1987) that yield the exact mathematical response surface, usually by the application of first-order differential equations.

11.2.1.5 Anatomy of Screening Designs

In a real experimental program the screening designs, in particular Plackett–Burman-type designs are the starting point for any investigation of a completely unknown system. They are designed to screen out the few really important, i.e. statistically significant variables from a large number of possible ones with a minimum of test runs. Plackett–Burman designs are fractions of an $N = 2^p$ factorial for which N is a multiple of 4. Although they allow a tremendous reduction in experimentation there is no estimate of synergistic nonlinear parameter interactions. In fact, only estimates of main effects clear of each other can be obtained. A saturated Plackett–Burman design is a useful and convenient tool for optimization of 11 plasma spray parameters in only 12 runs (Figure 11.4). The '+' signs are assigned to the parameters x_i at their maximum levels, the '−' signs to their minimum levels. The factor effects are calculated by adding in each column the '+' responses ($\Sigma+$) and subtracting the sum of the '−' responses ($\Sigma-$). The value $\Delta = (\Sigma+) - (\Sigma-)$, divided by the number of '+' (or '−') signs in the columns with assigned factors is the factor

Run	x_1	x_2	x_3	x_4	x_5	x_6	x_7	x_8	x_9	x_{10}	x_{11}
1	+	−	+	−	−	−	+	+	+	−	+
2	+	+	−	+	−	−	−	+	+	+	−
3	−	+	+	−	+	−	−	−	+	+	+
4	+	−	+	+	−	+	−	−	−	+	+
5	+	+	−	+	+	−	+	−	−	−	+
6	+	+	+	−	+	+	−	+	−	−	−
7	−	+	+	+	−	+	+	−	+	−	−
8	−	−	+	+	+	−	+	+	−	+	−
9	−	−	−	+	+	+	−	+	+	−	+
10	+	−	−	−	+	+	+	−	+	+	−
11	−	+	−	−	−	+	+	+	−	+	+
12	−	−	−	−	−	−	−	−	−	−	−

Figure 11.4 Saturated Plackett–Burman design for estimation by screening of 11 parameters in 12 runs.

effect for the parameter x_i (see Appendix, Table C.3). For the extra columns to which no factors have been assigned this 'factor effect' is an estimate of the experimental error. For example, assignment of only six variables, x_1 to x_6 leaves five degrees of freedom, i.e. unassigned factor effects x_7 to x_{11} that can be used to estimate the standard deviation of the factor effects

$$\sigma_{FE} = \{(1/q)\sum x_{iq}^2\}^{1/2} = \{(1/n)\sum E_i^2\}^{1/2}.$$

To determine which factors x_i are statistically significant the calculated factor effects are compared with the minimum factor significance, {min}. The minimum significant factor effect is

$$\{min\} = t_v^\alpha \, \sigma_{FE}, \tag{11.1}$$

with t_v^α = Student's t for v degrees of freedom, and α level of confidence for a double-sided t-distribution. All factors whose effects are larger than the (absolute) value of {min} are considered statistically significant. An example for the application of a Plackett–Burman screening test is given in Section 11.2.2.1. Although such designs are available for nearly any multiples of four trials, the most useful ones are for 12, 20 or 28 trial runs. Such designs nominally handle 11, 19 and 27 factors, respectively. It is good practice to conduct an operability review after having selected any p columns for factors. For increased precision of prediction, a larger design could be used and/or a reflected design could be added (Plackett and Burman, 1946). By using the next larger design partial replication is obtained. For example, with 6 factors, the recommended design has 12 runs. This leaves 5 degrees of freedom to estimate the experimental error. Alternatively, the 6 factors can be run in a 20 trial design, hence leaving

13 degrees of freedom to estimate the experimental error. Reflected designs are able to estimate main effects (almost) clear of two-factor interactions, and thus approach the predicting power of fractional factorial designs (see below).

The selection of the level of significance, α requires some discussion. When there is only a relatively small number of degrees of freedom, i.e. when a low number of runs has been performed, it is preferable to select a significance level lower than 0.95. In this case, the 'power of test' is greater, i.e. the likelihood to detect a significant factor effect if it exists. As a rule of thumb, for degrees of freedom ≤ 5, 5 to 30, and ≥ 30, respectively, significance levels of 0.90, 0.95, and 0.99, respectively should be selected.

If two-factor interactions are present, Plackett–Burman designs have the desirable property that significant factor effects stand out over a pool of background noise containing both experimental error and interactions when present. In the absence of interactions, the *precision ratio* achieved is

$$\sigma_{FE}/\sigma = 2/\sqrt{n}, \tag{11.2}$$

where σ_{FE} = standard deviation of a factor effect, σ = standard deviation of a single observation, and n = total number of observations (runs) in the selected design. The value of n, i.e. the number of experimental runs should be large enough to have a high probability of separating the significant signal from experimental noise, but small enough the optimize time and costs. The size of the factor effect to be detected is Ω, and the desired probability that a significant parameter can be detected when it has a true effect of size Ω is $(1-\beta) \geq 0.90$. This is satisfied if $\sigma_{FE} = \Omega/4$. From Equation 11.2 it follows that

$$n = (8\sigma/\Omega)^2 = [8/(\Omega/\sigma)]^2, \tag{11.3}$$

with Ω/σ = signal-to-noise ratio.

Equation 11.3 is called Wheeler's test. To detect effects twice as large as the experimental error ($\Omega = 2\sigma$), n must be 12 to 16. To detect effects the same size as the error ($\Omega = \sigma$), n become four times as large, i.e. 48 to 64. Weak factor effects thus require a large number of experiments.

11.2.1.6 Anatomy of Factorial Designs

Full Factorial Designs Factorial designs permit the estimation of the main (linear) effects of several factors, x_i simultaneously and clear of two-factor interactions, $\{x_i x_j\}$. The total number of experiments, N is obtained by running experiments at all combinations of the p factors with L levels per factor, i.e. $N = L^p$ for a full factorial design. In particular, two-level factorial designs (L = 2) are highly useful for a wide variety of problems. They are easy to plan and to analyze, and readily adaptable to both continuous and discrete factors. Such designs provide adequate prediction models for responses without a strong nonlinear behavior in the experimental region. However, for best results the responses Y should be continuous, and have uniform and independent errors. This means that (i) the experimental error must have approximately the same magnitude at all experimental points, and that (ii) the size and sign of error at any one experimental point must not be affected by the sizes and signs of errors that occur at other experimental points. If requirement (i) cannot be guaranteed a logarithmic transformation of the responses Y is often advisable. With factor coding as discussed above, using the so-called Yates order (Yates, 1937) and randomization of the experimental trials, factor effects for each main effect and two-factor interaction are calculated similar to the procedure mentioned in Section 11.2.1.2. The computation of factor effects and residuals can also be quickly and accurately accomplished using the IYA (inverse Yates algorithm; Hunter, 1966). The test of significance can be derived from an appropriate t-test:

$$\{\min\}_{\text{lin}} = t_v^\alpha \, \sigma_{\text{FE}} (2/mk)^{1/2}, \tag{11.4a}$$

$$\{\min\}_c = t_v^\alpha \, \sigma_{\text{FE}} (1/mk + 1/c)^{1/2}, \tag{11.4b}$$

where $\{\min\}_{\text{lin}}$, $\{\min\}_c$ are the minimum significant factor effects for linear effects and curvature, respectively, and $m = 2^{p-1}$ for factor effects and 2^p for the average, k = number of replicates, and c = number of center points. If a computed factor effect is larger (in absolute value) than $\{\min\}_{\text{lin}}$ than it can be safely concluded that the true effect Ω is non-zero. Also, when the curvature effect is larger than $\{\min\}_c$ then at least one factor has non-zero curvature associated with it. Applications of full factorial designs 2^3 to the design of (Ti, Mo)C–NiCo- and 88WC12Co-coatings on mild steel, and a 2^5 design applied to a self-fluxing NiCr coating on steel will be dealt with in Section 11.2.2.2.

Fractional Factorial Designs If the number of parameters p to be estimated becomes larger than 5 or 6 a full factorial may no longer be appropriate for reasons of experimental economy. Then fractions of a full factorial can be run, i.e. $N = 2^{p-q}$. In particular, half-fraction of full factorials, $N = 2^{p-1}$ estimate main effects and two-factor interactions clear of each other.

Fractional factorial designs are useful in the following situation (Box and Hunter, 1961):

> when some interactions can be reasonably assumed nonexistent from prior knowledge,
>
> in screening operations where it is expected that the effects of all but a few of the factors will be negligible, where blocks of experiments are run in sequence, and ambiguities remaining at an earlier stage of experimentation can be resolved by later blocks of experiments, and
>
> when some factors that may interact are to be studied simultaneously with others whose influence can only be described through main effects.

Fractional factorial designs can be divided into three types with regard to their *power of resolution*, i.e. their degree of fractionation. The higher the degree of fractionation the more comprehensive the assumptions required to arrive at unequivocal interpretation of the system. Designs of resolution III, for example 2^{3-1}, are those in which no main effect is confounded with any other main effect, but main effects are confounded with two-factor interactions, and two-factor interactions with each other. In designs of resolution IV, for example 2^{4-1} or 2^{8-4}, no main effect is confounded with any other main effect *or* two-factor interaction, but two-factor interactions are confounded with each other. Finally, in designs of resolution V, for example 2^{5-1}, no main effect or two-factor interaction is confounded with any other main effect or two-factor interaction, but two-factor interactions are confounded with three-factor interactions. Since in most real problems three- and higher-factor interactions can be safely neglected, a design of resolution V should be the design of choice. However, in order to reduce the number of runs required and thus keep within the economical bounds of most research and development projects, fractional factorial designs of resolution IV can be tolerated. The design 2^{8-4} discussed below is of resolution IV. The confounding of two-factor interactions in this design leads to composite two-factor interaction, i.e. the sum of four two-factor interactions. Examples of these fractional factorial designs applied to 88WC12Co, Ti, 97Al$_2$O$_3$-TiO$_2$, 85Fe15Si, stellite, and Cr$_2$O$_3$ coatings on mild and austenitic steel substrates will be shown in Section 11.2.2.3 below.

11.2.1.7 Box–Behnken Designs

Box–Behnken designs (Box and Behnken, 1960) are incomplete three-level factorial designs that allow an estimation of the coefficients in a second-degree graduating polynomial. They employ subsets of the corresponding full three-level factorial, 3^p. For example, the three-factor design uses only 13 of the 27 points of the full factorial 3^3 with two extra replicates of the center points added, for a total of 15 experimental points of a spherical space-filling[1] and rotatable design. Another desirable feature of such designs with p > 4 is the possibility to run it in separate blocks of points. Such *orthogonal blocking* permits to subtract out the effect of a shift of response between blocks, and thus to remove bias errors due to differences in extraneous variables not considered in the design. The 15 data points in the three-factor Box–Behnken design are five more than the minimum number of 10 required to estimate the coefficients of the design (3 linear main effects, 3 parabolic main effects, 3 two-factor effects, and 1 three-factor effect). Thus it provides five degrees of freedom for error. Orthogonal blocking is possible for designs from four up to ten factors. Rotatability is associated with the geometrical properties of the design, i.e. the arrangement of the array of data points, except for the center point, to be at the mid-points of the edges or faces of a hypercube whose dimensionality is given by the number of factors considered. Hence all points are situated on a single sphere and are thus equidistant from the center. This means that the design is balanced by the mathematical momentum condition. The replicated center point allows (i) to estimate the minimum factor significance, i.e. the inherent experimental error, and (ii) to give constant prediction variances as a function of distance from the center.

---

1) Note that the factorial designs are considered to have a 'cuboidal' or 'hypercube' factor space whereas the true response surface designs have a 'spherical' factor space.

It is good experimental strategy to employ such Box–Behnken-type designs at a rather advanced stage of experimentation when the number of potentially significant factors has been narrowed down to 3 to 6 continuous factors. As pointed out clearly by Bisgaard (1990), even considering nonlinearity of responses in plasma-sprayed designs does not warrant, for reasons of experimental economy, three-level factorial designs in the initial stage of experimentation. Thus second-order effects should be dealt with exclusively when they actually show up in the set of data, and not only because the experimenter suspects that the system under investigation may show some global nonlinearity! When switching from two- to three-level designs any discrete, i.e. noncontinuous factor effect must be considered constant. The response surface obtained through a Box–Behnken RSM design provides usually a high-quality prediction over a region where linear, parabolic (curvature) and two-factor interactions are needed to describe a response of the system, Y as a function of the coefficients of the independent input parameters X_j obtained by a full quadratic polynomial for p independent parameters:

$$Y = b_0 + \Sigma b_j X_j + \Sigma b_{jj'} X_j X_{j'} + \Sigma b_{jj} X_j^2, \tag{11.5}$$

with $j > j'$.

11.2.1.8 Designs of Higher Dimensionality

Three levels are the minimum number for each factor to accurately describe nonlinear (curvature) effects. To add additional power of prediction to a Box–Behnken design, it is advisable to use more than three factor levels (see for example Schuppert and Ohrenberg, 2005). One popular class of response surfaces are the *central composite* or Box–Wilson (Box and Wilson, 1951) designs that employ five levels for each factor. It is composed of a full two-level factorial 2^3 with added center points plus six star points outside the cube planes defined by the four points of the two-level factorial. The geometrical shape of the resulting five-level design is a tetrakishexahedron whose 14 design points surround a k-time replicated center point. This polyhedron is 'cuboidally' spacefilling and can be described as the dual polytope of a cuboctahedral Dirichlet domain (Voronoi polyhedron). An example of such a design used to determine the fractionality of Cr_2O_3 coatings (Zimmermann, 2000; Reisel and Heimann, 2004) has been described in Chapter 5.6.6.

11.2.1.9 Neyer D-optimal Designs

This test was designed to extract the maximum amount of statistical information from the test samples (Neyer, 1994). Unlike the other statistical test methods described above, this method requires detailed computer calculations to determine the test levels as it uses the results of all the previous tests interactively to compute the next test level.

There are three parts to this test. The first part is designed to 'close-in' on the region of interest to within a few standard deviations of the mean, as quickly and efficiently as possible. It thus resembles a Plackett–Burman screening test which indeed is D-optimal (Pumplün et al., 2005). The second part of the test is designed to determine efficiently the unique estimates of the parameters. The third part continuously refines the estimates once unique optimized estimates have been established.

This experimental design strategy has been applied to optimize thickness and porosity levels of plasma-sprayed tantalum oxide environmental barrier coatings (EBCs) to protect silicon nitride-based ceramics (Moldovan et al., 2004). For the first set of 20 runs, seven parameters based on historical knowledge were varied at two levels: injector angle, plasma power, total gas flow, percent hydrogen in plasma gas, spray distance, carrier gas flow rate, and presence or absence of air cooling, using a D-optimal based software program (Experimental Design Optimizer, Harold S. Haller & Co., Cleveland, OH). During the second round (N = 16), three parameters varied in the first set (plasma power, total plasma gas flow, percent hydrogen in the plasma gas) were held constant at their optimized value. To the remaining four parameters two more were added, powder feeder disk speed and robot scan rate, and varied at two and three levels, respectively. Multiple regression analyses were performed to determine the dependency of coating thickness and porosity on the parameters selected in the first and second set of trials. Since minimum porosity was deemed more important for EBCs than thickness considerations, the parameter levels resulting in the lowest porosity were selected to run a third confirmatory set of trials.

Since the estimate of the standard deviation is used only until overlap of the data occurs, the efficiency of the test is essentially independent of the number of parameters used in the test design.

11.2.2
Optimization of Coating Properties: Case Studies

11.2.2.1 Plackett–Burman (Taguchi) Screening Designs

Optimization of a novel (Ti, Mo)C–NiCo coating on mild steel (German steel number St38) was performed by vacuum plasma-spraying in an argon/hydrogen plasma using a 12-run 11-factor Plackett–Burman design (L_{12}-type according to Taguchi) (Thiele, 1994). The six factors varied at two levels (high and low) were **1** = powder (agglomerate) grain size,[2] **2** = plasma power, **3** = powder feed rate, **4** = plasmatron traverse speed, **5** = chamber pressure, **6** = spray (stand-off) distance. The plasma power **2** was not an independent parameter but obtained by an appropriate selection of the argon/hydrogen ratio and the current. The low value of **2** was 42 kW (argon: 48 L min^{-1}, hydrogen: 6 L min^{-1}, current: 800 A), the high value was 47 kW (argon: 48 L min^{-1}, hydrogen: 7 L min^{-1}, current: 900 A). The remaining factors and their levels were: **1** (−32 + 10 µm; −63 + 32 µm), **3** (0.5; 1 scale), **4** (4 m min^{-1}; 8 m min^{-1}), **5** (80 mbar; 100 mbar), and **6** (340 mm; 380 mm). Four responses Y_i were measured: surface roughness of the deposit (Y_1), microhardness (Y_2), porosity (Y_3) and fracture energy density (Y_4). The factors found to influence those responses at a level of confidence of 95% were **1** (positive effect for surface roughness and porosity,

---

[2] Here and in the following text the factors X_1, X_2 ... X_i will be denoted by **1, 2,... i**.

negative effect for fracture energy density) and **6** (positive effect for porosity, negative effect for fracture energy density). In addition, with the parameters **2**, **5** and **6** a full factorial design 2^3 was run for the fine powder ($-32 + 10\,\mu m$) with increased ranges of **2** (38 kW; 53 kW) and **5** (100 mbar; 180 mbar) and a reduced range of **6** (200 mm; 300 mm) in order to minimize the coating porosity. For the low levels of **2** (38 kW), **5** (100 mbar) and **6** (200 mm) a coating porosity around 2% could be achieved with reasonably low surface roughness and a fracture energy density around $30\,J\,mm^{-3}$. Problems occurred with a high heat transfer to the substrate that requires efficient substrate cooling.

Other optimization studies on plasma-sprayed coatings based on L_8 or L_{16} Taguchi designs relate to

- NiCrAl/bentonite abradable coatings (L_{16} design, 15 independent parameters, 3 dependent parameters: erosion resistance, tensile strength, hardness) (Novinski et al., 1990).
- HVOF-sprayed Al–Si/polyester abradable coatings (L_8 design, 5 independent parameters, 3 dependent parameters: hot erosion resistance, bond strength, hardness) (Chon et al., 1990).
- Thick *thermal barrier coatings* (TTBCs) (L_8 design, 7 independent parameters, 5 dependent parameters: erosion resistance, macrohardness, porosity, deposition efficiency, thermal shock resistance) (Nerz et al., 1990).

WC/Co-, Cr_3C_2/NiCr- and Al_2O_3/TiO_2 coatings (L_8 design, 7 independent parameters, 4 dependent parameters: microhardness, Rockwell macrohardness, tensile strength, composition) (Walter and Riggs, 1990).

11.2.2.2 Full Factorial Designs

Such designs are not very common in the literature since they are limited to a rather small subset of spray parameters. Only if enough knowledge of the system under investigation has been accumulated are such designs applied to fine-tune the parameters of a decreased factor space (see Section 11.4.1).

A two-level five-factor full factorial 2^5 was employed by Hurng et al. (1987) to evaluate the properties of a self-fluxing nickel-base alloy containing 15 to 18 mass% Cr and 5 mass% (Fe + C + B + Si) deposited by a hybrid APS/PTA (plasma transferred arc) process on mild steel. The five independent parameters and their ranges were: **1** plasma current (420 A; 460 A), **2** spray distance (15 mm; 17 mm), **3** powder feed rate (8; 12 wheel speed units), **4** plasmatron traverse speed ($25\,mm\,s^{-1}$; $35\,mm\,s^{-1}$) and **5** plasma transferred arc current (4; 6 scale units). Each parameter setting was repeated thus given two sets A and B of 32 runs each. In addition, 15 experiments were run at the 'working point, i.e. the midpoint between the low and high values selected for each parameter.

The optimized dependent response parameters were the microhardness (800 ± 40 $HV_{0.3}$), the porosity ($2.1 \pm 1.5\%$) and an oxide content as low as 0.13%. The significant parameters for optimizing the microhardness were the plasma current **1** with an effect per unit of $-0.83\,HV_{0.3}\,A^{-1}$, the two-factor interactions **12** (positive effect) and **45** (positive effect), and the spray distance **2** with an effect of about 11 $HV_{0.3}$ mm^{-1}. Thus **1** > **12** ≥ **45** > **2** (absolute). The significant parameters for optimizing the lumped together-values of coating porosity and percent oxide were the traverse speed **4** with 0.38% mm^{-1} s^{-1}, the two-factor interaction **23** (negative effect) and the plasma transferred arc current **5** with an effect per unit of -0.32% A^{-1}. Thus **4** > **23** > **5** (absolute).

These results clearly show the problems that such coating property predictions have when performed at an insufficient level of predicting power, i.e. only two-level design: the factor significances for microhardness and porosity are quite different and preclude unambiguous optimization treatment. Microhardness is presumably much affected by porosity but the parameter with the highest significance for the former, the plasma current **1**, does not show up at all in the response polynomial of the latter. Likewise, the parameter with the highest significance for the porosity, the traverse speed **4** is not part of the response polynomial of the microhardness. With the large number of experiments expended (79) a folded over, i.e. replicated three-level five-factor Box–Behnken design could have been executed with a slightly increased total number of runs of 92. This design had allowed for the estimation of nonlinear effects that are to be expected in the system. On the other hand, a result comparable to that described above would have been obtained with a mere 24 runs as a reflected five-factor 12-run Plackett–Burman or L_{12} Taguchi design that estimates main effects clear of two-factor interactions with 6 unassigned factors to determine the minimum factor significance. It should be emphasized again that in the interest of experimental economy it is good practice to always start with a simple screening to weed out the weakly significant parameters, and only then follow up with a full factorial design.

A simple two-level three-factor full factorial design 2^3 with 2 center points added was used by Troczynski and Plamondon (1992) to optimize the erosion rate, Rockwell A-macrohardness, density, thermal decomposition of WC, and surface roughness of 88WC12Co coatings deposited by APS at a plasma power of 24 ± 4 kW on mild steel. The independent parameters and their levels were hydrogen content in % in the argon/hydrogen plasma gas **1** {(2; 4) $\pm 0.3\%$}, spray distance **2** {(51; 127) ± 2.5 mm}, and powder feed rate **3** {(30; 60) ± 3 g min^{-1}}. The choice of the independent parameters was almost always a compromise, here between a limited experimental capability, a large number of variables potentially controlling the properties of the WC/Co coatings, and the practical relevance in industrial spraying operations. By plotting sections of the four-dimensional response hyperspaces [**1**, **2**, **3**, Y_i] it was possible to define a set of robust conditions that resulted in optimized coating properties (erosion rate <10 mg s^{-1}, hardness HRA >40, density >9 g cm^{-3}, and W_2C content <7% of initial amount of WC) insensitive to minor variations. These robust processing conditions were **1** = $2 \pm 0.5\%$ H_2, **2** = 76.2 ± 12.7 mm, and **3** = 60 ± 10 g min^{-1}. With this highly economical design a set of operating windows were obtained as shown in Figure 11.5.

Optimization of the porosity of APS-tungsten coatings on 6061 aluminum substrates was done using a combination of statistical designs (Varacalle *et al.*, 1995), e.g. a full factorial 2^3 with 4 center points, and a central composite design using a 2^3 full

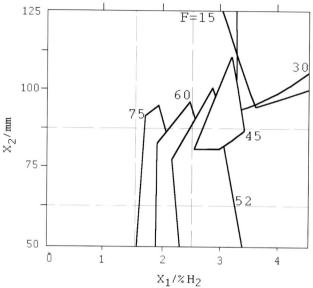

Figure 11.5 Response surface (RS) optimization of an 88WC12Co coating to achieve an erosion rate < 10 mg s^{-1}, Rockwell A hardness > 40, density > 9 g cm^{-3}. Shown are operating windows of the hydrogen content in the plasma gas in % (X_1) and the stand-off distance in mm (X_2) for powder feed rates F ranging from 15 to 75 g min^{-1} (Troczynski and Plamondon, 1992).

factorial plus 6 star points and 2 center points. The variables were **1** = total gas flow, **2** = secondary(H_2)-to-total gas flow ratio and **3** = stand-off distance. The average porosity measured ranged from 12.7 to 1.1%. The minimum porosity was obtained for **1** = 52 L min^{-1} (110 SCFH), **2** = 11%, and **3** = 70 mm (2.75') as shown on the response surface (Figure 11.6).

To determine the dependency of the phase composition and hence the photocatalytic efficiency of induction plasma sprayed (IPS) titania coatings on the processing parameters plasma power **1**, Ar carrier gas flow rate **2**, and powder feed rate **3** a full factorial 2^3 design was employed together with 3 center points to account for nonlinearity of the factors and their interactions (Burlacov et al., 2006). Responses were coating mass M, coating thickness d, surface roughness R_a, and several parameters obtained from Raman microprobe analyses of the coatings such as sum of intensities of the two characteristic Raman modes $\Sigma(I_A + I_R)$ for anatase (147 cm^{-1}) and rutile (447 cm^{-1}), the individual Raman intensities I_A and I_R, the intensity of the Raman inactive SRO phase associated with highly disordered, amorphous and/or molten splats of titania, I_{SRO}; and the ratio of the XRD intensities of the (110) and (101) interplanar spacings of rutile, $I_{(110)/(101)}$ as obtained by GIXRD. Table 11.2 shows the constant terms b_0 and the coefficients b_i of the factors **1–3** as well as the regression coefficients r^2 for the response functions $Y_i = b_0 + \Sigma b_j X_j + \Sigma b_{jj'} X_j X_{j'} + \Sigma b_{jj} X_j^2$ with j > j'.

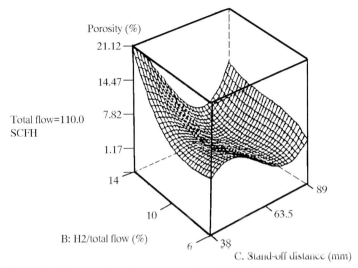

Figure 11.6 Optimization of the porosity of an APS WC/Co coating on 6061 aluminum. The response surface of the porosity is displayed as a function of the ratio of the hydrogen/total gas flow (B) and the stand-off distance (C) for a total gas flow of 110 SCFH (approx. 3.1 m³ h⁻¹) (Varacalle et al., 1995)

The crystallinity of the anatase and rutile phases as expressed by the intensities of the Raman signals depends in the approximation linearly on the carrier gas flow rate **2** and the powder feed rate **3** in a positive sense, i.e. increasing both factors improves the crystallinity of the anatase phase whereas the crystallinity of the rutile phase is only a function of **2**. It is noteworthy that neither the plasma power **1** nor any two-factor and higher interactions appear to influence the coating properties in a statistically significant way with the exception of the coating roughness. The fact that increasing the powder feed rate **3** increases the yield of anatase can be explained by a plasma cooling effect under dense loading conditions (Mawdsley et al., 2001,

Table 11.2

Y_i	b_0	b_1	b_2	b_3	r^2
M (mg)	17			+9	0.91
d (μm)	42			+21	0.91
$\Sigma(I_A + I_R)$ (a.u.)	241		+85	+30	0.83
I_A (a.u.)	131		+35	+24	0.80
I_R (a.u.)	109		+50		0.84
I_{SRO} (% of total area)	34		−3	−6	0.75
$I_{(110)/(101)}$ (a.u.)	0.96		+0.34	−0.09	0.89
R_a	5	−0.3	+0.8	+0.4	0.87

Lee et al., 2003; see also Section 6.3.1.2). The surface roughness of the coatings depends on all three factors, most importantly on the positive effect of the carrier gas flow rate **2**.

11.2.2.3 Fractional Factorial Designs

These designs permit a higher flexibility in terms of increasing the number of parameters under study. Although the number of parameters that can potentially influence the coating properties is very large (≥ 150) it is common practice to consider only 8 to 12 parameters in statistical designs (see Figure 11.7). Popular designs are 2^{8-4} fractional factorial designs of resolution IV.

Tungsten Carbide/Cobalt Coatings An example will be given of the optimization of 88WC12Co coatings deposited by APS (argon/helium plasma at Mach 2 velocities) onto low-carbon steel (Heimann et al., 1990a). The eight selected parameters and their ranges were the plasma current **1** (700; 900 A), argon gas pressure **2** (0.34; 1.36 MPa), helium gas pressure **3** (0.34; 1.36 MPa), powder gas pressure **4** (0.34; 0.68 MPa), powder feed rate **5** (0.5; 2 scale value), powder grain size **6** ($-45 + 5; -75 + 45$ μm), number of traverses **7** (5; 15 for set 1 and 20;30 for set 2) and spray distance **8** (250; 450 mm). If all possible experiments were to be executed, their total number would be 256 (2^8). In the present situation, however, only a fraction of the total, i.e. the 1/16 replicate, was selected (2^{8-4}).

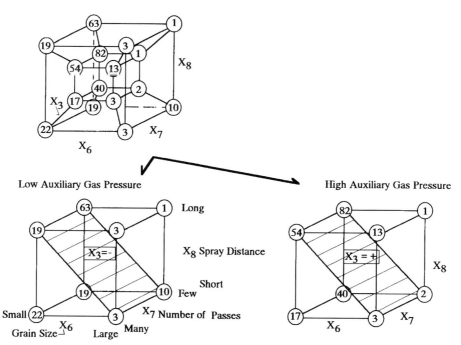

Figure 11.7 4D-hypercube of optimization of the thickness of 88WC12Co coatings as a function of the auxiliary (helium) gas pressure **3**, the powder grain size **6**, the number of traverses **7**, and the stand-off distance **8** (Heimann et al., 1990a).

Two sets of 16 spray runs each were performed. During the first set, the number of traverses was varied between 5 and 15; this produced relatively thin coatings with a maximum thickness of 80 µm. The data from this set were statistically evaluated to yield information on the main factor effects and synergistic two-factor interactions. The second set with the number of traverses varying between 20 and 30 resulted in coatings with a maximum thickness of 200 µm. The data from the second set were used to evaluate the microhardness and the cohesive strength of the coatings. The design allows the calculation of the eight main effects clear of the composite two-factor interactions E_i. As a general example the calculations are executed in Appendix C.

The statistical significance of the factor effects calculated analogous to the simple procedure outlined in Section 11.2.1.2 (see Appendix C) was checked against the minimum factor significance, $\{\min\} = \sigma_{FE} t_v^\alpha$, where $\sigma_{FE} = \sqrt{(1/n\Sigma E_i^2)}$, t_v^α is the Student t-value for a confidence level α of a double-sided significance test and $v =$ degrees of freedom. All absolute factor effects larger than or equal to $|\{\min\}|$ were considered to be significant.

The composite two-factor interactions E_i were calculated in a similar fashion (see Appendix C, Table C.4). They represent the unassigned factors that can be used to estimate the experimental error, i.e. the σ_{FE} value needed to calculate the minimum factor significance, $|\{\min\}|$. With the data obtained it follows that $\sigma_{FE} = \sqrt{(1/n\Sigma E_i^2)} = \sqrt{(3592/7)} = 22.6$, and $|\{\min\}| = 22.6\, t_{v=7}^{\alpha=0.90} = 22.6 \times 1.895 = 43$. Hence all main factor effects whose absolute values are larger than 43 should be considered significant at a confidence level of 90%. Checking with the data in Appendix C, this is the case for **5** = 54 and **6** = 55 for thicker coating, i.e. those obtained from the second set of experiments with the number of traverses ranging between 20 and 30. Both effects have positive signs, i.e. the thickness of the coating increases with increasing powder feed rate, **5** and increasing spray distance, **8**. Short spray distances lead to overheating of the alloy powder thus causing thermal decomposition and/or reaction of the WC with the cobalt metal matrix forming η-carbides Co_nW_mC (Figure 10.35; Hajmrle and Dorfman, 1985). Hence, the response polynomial of the thickness of plasma-sprayed 88WC12Co alloy coatings can be approximately expressed by the equation $d(\mu m) = 32 + 27X_5 + 28X_8$. For thinner coatings a different picture emerges. In this case the significant factors affecting coating thickness are **6** (negative effect) > **8** > E_4, E_6 (negative), E_5 > **7** > **3**.

Figure 11.7 shows the position of the experimental data points for the coating thickness in µm in a four-dimensional hypercube where the four axes of the cube are the factors **6**, **7**, **8** and **3**. The design consists of two nested cubes with **3** = −1 and **3** = +1. These 3D cubes are also shown in Figure 11.7 (bottom). Maximum thickness of the coatings can be obtained using fine powders, long spray distances, and high helium gas pressure. The complex statistical significance of the composite two-factor interactions $E_4 = 15 + 38 + 26 + 47$, $E_5 = 16 + 78 + 34 + 25$, and $E_6 = 17 + 23 + 68 + 45$ may be somewhat deconvoluted by assuming that there exists at least one large component interaction involving factors with significant main effect in each composite interaction. The most likely candidates are **38** for E_4, **78** for E_5, and **68** for E_6. Additional experiments with a larger set of runs at a higher power of prediction would be required to resolve these ambiguities.

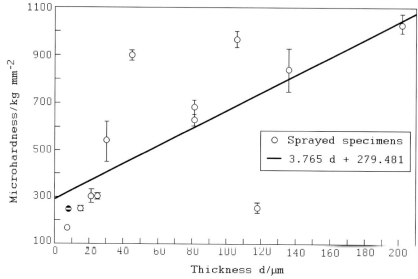

Figure 11.8 Quasi-linear relationship between microhardness ($HV_{0.2}$) and thickness for 88WC12Co coatings (Heimann et al., 1990a).

Only a weak significant factor, **8**, exists for the microhardness. There is a quasi-linear relationship between coating thickness and microhardness (Figure 11.8). The maximum microhardness obtained for coatings with thickness exceeding 200 µm is 1200 $HV_{0.2}$ (75 HRC). The cohesive strength of the coatings exceeded 60 ± 16 MPa and yielded a maximum value of 80 MPa. Additional SEM investigations were performed to obtain qualitative information on the porosity. The only significant parameter to describe the development of porosity is the powder grain size, **6**, i.e. fine powders produce higher coating porosity. Statistically non-significant results were obtained for the degree of densification; high plasma currents, **1** and a small powder grain size, **6** produced a higher degree of densification. These results are somewhat contradictory as increased densification should also produce a lower porosity. It may be, however, that overheating of the particles at short stand-off distances leads to vaporization, and the increased porosity simply reflects the eruption of gaseous decomposition products such as CO_2 generated by partial oxidation of tungsten carbide under formation of η-carbides (see Figure 10.35; Hajmrle and Dorfman, 1985).

In conclusion, the application of the statistical design matrix 2^{8-4} to 88WC/12Co coatings showed that the eight parameters selected as variables can be divided into highly significant ('soft'), and less or insignificant ('hard') parameters that can be varied widely (see Troczynski and Plamondon, 1992). 'Soft' parameters for optimizing coating thickness are powder grain size and spray distance, i.e. parameters that determine the degree of melting of the particles. Typical 'hard' parameters are argon gas pressure, powder gas pressure and powder feed rate.

Figure 11.9 shows the three-tiered hierarchy of the coating properties. Third-level parameters are plasma/particle velocity, degree of particle melting, and degree of

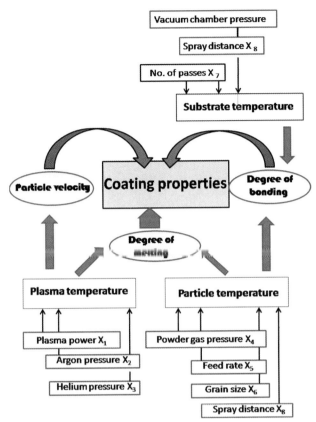

Figure 11.9 The development of coating properties shows a three-tiered hierarchy. 1st level: plasma spray parameters **1** to **8**, 2nd level: temperatures of plasma, particles and substrate, 3rd level: particle velocity, degree of melting, and degree of bonding (after Heimann et al., 1990a).

bonding of the coating to the substrate. These properties are influenced by the second-level parameters plasma temperature, particle temperature and substrate temperature. These in turn are determined by the eight first-level input parameters. This scheme illustrates once more that these parameters do not always act in the same directions (saddle points of the response surface) and thus a statistical multifactorial design is required to unravel the generally complex parameter interactions.

Ferrosilicon Coatings A second case study concerns the optimization in terms of coating thickness and microhardness of an 85Fe15Si (Valco 3603.0, $-22 + 5\,\mu\text{m}$) coating on carbon steel (Heimann, 1992). This material is known for its excellent corrosion resistance and may have applications in demanding environments such as those present in coal gasification where high hydrogen sulfide levels lead to severe stress corrosion cracking (SCC). Coupons were cut from St 37 steel plate and mounted on a specially designed sample holder that permits cooling by water flowing through a thin copper pipe attached by hard soldering to the backside of the sample holder. The plasma spray equipment used to deposit the coatings was a conventional METCO system with a vibrating hopper, and a METCO 9M plasmatron. Spraying was performed with an argon/hydrogen plasma gas. A fractional factorial design 2^{8-4} was selected whose factors and their levels were as follows: **1** plasma current (300; 500 A), **2** argon gas flow (100; 140 scale value), **3** hydrogen gas flow (2; 4 scale value), **4** substrate cooling[3] (no; yes), **5** ammonia shroud[4] (no; yes), **6** traverse speed (20; 30 m min^{-1}), **7** substrate roughness (0.45; 1.1 µm), and **8** stand-off distance (80; 120 mm). Constant factors were the anode type (GH 732), powder feed tube (A), powder spray nozzle (2), powder feed rate (60 g min^{-1}), number of preheating cycles (3), number of spray traverses (3) and overlap (5 mm). The main factor effects **1** to **8** as well as the composite two-factor interactions E_1 to E_7 were estimated for the coating thickness (Y_1) and the microhardness (Y_2). The significant main effects for coating thickness were **4** (negative) and **6** (positive), i.e. the coating thickness is maximized (175 µm) on a preheated substrate with increasing traverse speed. There is a conspicuously large negative composite two-factor interaction, $E_2 = 13 + 27 + 46 + 58$ (see Appendix C). This can be rather easily explained by the large value of the (negative) **46** two-factor interaction involving the two significant main effects **4** and **6**.

The significant main effects for the microhardness were **3** (positive), **4** (negative) and **8** (negative), i.e. the microhardness is maximized (325 kPa) at a short stand-off distance on a preheated substrate with increasing hydrogen flow rate (plasma enthalpy). Since it could be shown that for the optimization of the microhardness only three of the original eight selected parameters were significant, the 2^{8-4} fractional factorial design can be reduced to a replicated 2^3 full factorial design in variables **3**, **4** and **8**. Hence the assumption has been made that the remaining factors are essentially inert. However, the rather large standard deviation of the replicated microhardness values showed that this simplifying assumption cannot be upheld. For example, **1** is almost at the level of significance and may account for the variation of the replicated values. Thus the factor **1** behaves like a perturbation of the 3D response surface in **3**, **4** and **8**, and it must be concluded that larger plasma arc currents, i.e. higher plasma temperatures will also increase the microhardness of 85Fe15Si coatings.

Plotting the microhardness in a 4D hypercube design (Figure 11.10) in **1**, **3**, **4** and **8** shows clearly that the associated response surface has a saddle in the **34** plane for low and high level variations of **1**.

In conclusion, optimized microhardness values for 85Fe15Si coatings can be obtained by

- substrate preheating;
- increased flow rate of high-enthalpy gases, *e.g.* hydrogen;

---

3) 'No' substrate cooling means substrate preheating.
4) Ammonia was used as a shroud gas because the 85Fe15Si coating was developed as a bond coat for a plasma-sprayed silicon nitride-based HT erosion resistant coating. Ammonia was intended to counteract the thermal decomposition of silicon nitride during spraying (Heimann, 1992).

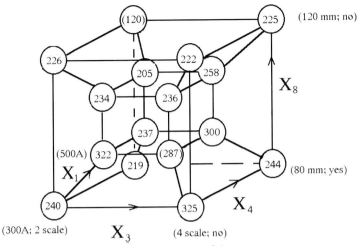

Figure 11.10 Geometric 4D-representation of a 2^{9-4} design in variables **1, 3, 4** and **8** to estimate the microhardness of a 85Fe15Si coating. The numbers refer to $HV_{0.2}$ values in kPa (Heimann, 1992).

- short stand-off distances;
- high plasma arc current.

Under these conditions the porosity of the coatings will also be minimized, and adhesion to the substrate will be maximized. A cross-sectional micrograph of a 85Fe15Si coating is shown in Figure 5.3.

Alumina/Titania Coatings Similar coating optimization has been performed for APS alumina–titania coatings applied to pump plungers operating at elevated temperature and as protective coatings against hot gas erosion of petrochemical processing equipment (Heimann et al., 1990b). As above, a 2^{8-4} fractional factorial design of resolution IV was selected with **1** plasma current (850; 950 A), **2** argon gas pressure (345; 517 kPa), **3** helium gas pressure (276; 414 kPa), **4** substrate preheating (23; 200 °C), **5** powder feed rate (4.5; 5.5 rpm), **6** roughening grit size (80; 40 mesh), **7** number of traverses (20; 50) and **8** stand-off distance (76; 127 mm). Coating thickness was found to be significantly influenced by **2, 7** and **8**. Figure 11.11 shows the 2^3 design cube with the results of the replicated coating thickness measurements. Figure 11.12 shows a different way to express parameter significance. The coefficients obtained by calculation identical to those shown in Appendix C have been plotted on a probability net. If the coefficients would only vary in a random fashion then their plot should give a straight line (Gaussian distribution). Deviation from this straight line signifies significant parameter effect. Figure 11.12 shows that the thickness of alumina–titania (97/3) coatings is significantly influenced by the argon gas pressure, **2** and the spray distance, **8** in a negative way, but positively influenced by the number of passes, **7**, the powder feed rate, **5** and a two-factor composite interaction, E_3 (positive effect) whose determining contributions are presumably

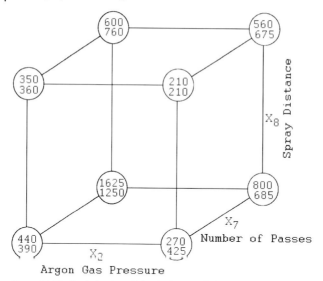

Figure 11.11 Optimization of the thickness of alumina/titania coatings as a function of argon gas pressure **2**, number of traverses **7**, and stand-off distance **8** (Heimann et al., 1990b).

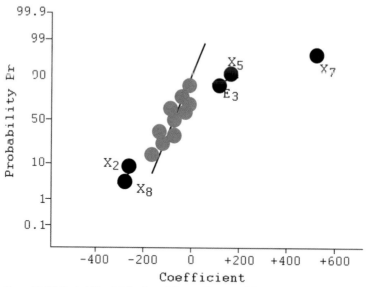

Figure 11.12 Probability distribution of the coefficients of the polynomial equation obtained for the thickness of alumina/titania coatings (Heimann et al., 1990b).

the two-factor interactions 28 and 57. In Figure 11.13 it is shown that the interaction 28 has a large difference in slope at low and high levels of 7 with a crossover at high 7. On the other hand, the probability plots of the coefficients of the second-order polynomial response equations for the microhardness (Vickers test, $HV_{0.3}$;

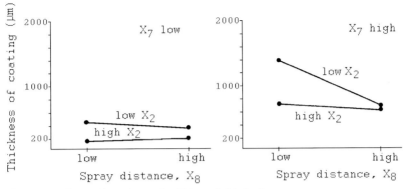

Figure 11.13 The two-factor interaction 28 is negligible for few traverses (7 < 20) but strong for many traverses (7 > 50) (Heimann et al., 1990b).

Figure 11.14) and the abrasion mass loss determined by the ASTM G 65 test (Figure 11.15) show random behavior thus indicating that in both cases the properties are not significantly influenced by the selected factors. The microhardness increases quasi-linearly (correlation factor 0.61) with coating thickness according to a limiting equation $HV_{0.3} = ad + b$, where $a = 120 \, kg \, mm^{-3}$, $b = 974 \, kg \, mm^{-2}$, and d = thickness in mm. The abrasion mass loss is inversely proportional (correlation factor -0.89) to the coating thickness and obeys the equation $\Delta m = c/d$ where $c = 45.3 \, mg \, mm^{-1}$ and d = thickness in mm. There is also an inverse relationship between abrasion mass loss and microhardness that can be expressed by the power

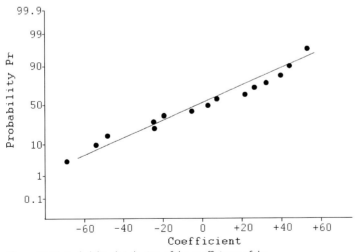

Figure 11.14 Probability distribution of the coefficients of the polynomial equation obtained for the microhardness of alumina/titania coatings, showing random i.e. statistically non-significant factor effects (Heimann et al., 1990b).

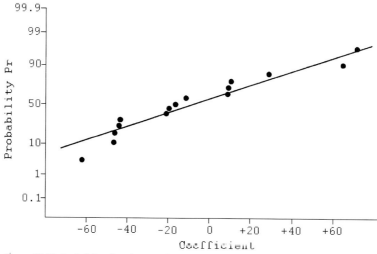

Figure 11.15 Probability distribution of the coefficients of the polynomial equation obtained for the abrasion mass loss (ASTM G65) of alumina/titania coatings, showing statistically non-significant factor effects (Heimann et al., 1990b).

function $\Delta m = A(HV_{0.3})^{-B}$ [kg] where $A = 4.3 \times 10^{24}$ and $B = 9.5$ (correlation factor −0.64).

In conclusion, application of a two level fractional factorial design 2^{8-4} to a set of 97Al$_2$O$_3$3TiO$_2$ coatings on low-carbon steel (A 569) surfaces showed that the eight parameters selected could be divided into highly significant or 'soft', and less or non-significant 'hard' parameters. 'Soft' parameters to describe the coating thickness are the stand-off distance **8**, the number of traverses **7**, and the argon gas pressure **2**. Typical 'hard' parameters are arc current, preheating temperature and grain size of the grit blasting material.

The flattening behavior of r.f. IPS Al$_2$O$_3$ particles was studied by a 2^4 factorial design by Fan et al. (1998). It was found that significant factors controlling the deformation of splats are (i) pressure in the spray chamber, (ii) spraying distance, and (iii) powder size. Interactions among these three factors are significant. Using plasma diagnostic techniques, a relation between the splat morphology and the surface temperature and velocity of particles prior to impingement on the substrate was established.

A Taguchi fractional factorial L$_8$ design was employed to compare the properties of APS and D-gun sprayed alumina coatings by evaluating the influence of plasma spray parameters on surface roughness, porosity, microhardness, abrasion resistance, and sliding wear resistance. D-gun sprayed coatings showed consistently denser and more homogeneous microstructures, higher hardness, and superior tribological performance (Saravanan et al., 2000).

Stellite Coatings Another example deals with stellite 6 (28Cr1C4W1Si, bal. Co) coating deposited by a plasma transferred arc (PTA) surfacing process on mild steel

(SIS 2172) bars by Herrström et al. (1993). A 2^{6-2} fractional factorial design was used in the parameters **1** plasma current (100; 130 A), **2** argon gas flow (1; 3 L min^{-1}), **3** powder gas flow (2; 4 L min^{-1}), **4** oscillation frequency (82; 90 min^{-1}), **5** weld speed (5; 7 cm min^{-1}) and **6** stand-off distance (6; 14 mm). Dependent parameters estimated were the dilution in %, the hardness HV$_{30}$, and the width and the height of the deposit in mm. The hardness of the PTA deposit depends significantly on **2 > 6 > 1 ≫ 16 45**, the width of the deposit shows **4 > 2**, and the height **1 > 2**. The dilution depends significantly at a confidence level < 95% on the parameters **2 > 1 ≫ 6**. It should be mentioned that the responses Y are not independent of each other. Thus the hardness depends not only on the dilution but also on the microstructure and the cooling rate that were not explicitly parts of the experimental design. At the working point, i.e. at an intermediate parameter level ('0') the following optimized responses were found: hardness 424.6 ± 3.78 HV$_{30}$, dilution $6.84 \pm 0.64\%$, width 15.24 ± 0.30 mm, and height 2.84 ± 0.29 mm.

Titanium Coatings Work by Lugscheider et al. (1987) on vacuum plasma-spraying of Ti coatings on 1.4571 and 1.4541 austenitic steel substrates employed an L$_8$ Taguchi matrix with four parameters varied. These parameters and their levels were the plasma current **1** (605; 655 A), argon gas flow **2** (33; 43 slpm), vacuum chamber pressure **4** (144; 164 mbar) and powder feed rate **6** (6.8; 10.8 g min^{-1}). The parameters **3** (hydrogen gas flow; 6.5 slpm), **5** (spray distance; 280 mm), **7** (plasma transferred arc current; 0 A) and **8** (sputter distance; 320 mm) were kept constant at the levels indicated. The eight runs performed actually constitute an 2^{8-5} matrix. Two additional runs with parameters **1** and **2** changed in the directions indicated by the fractional factorial design, and parameters **4** and **6** kept at their zero levels, i.e. midways between the upper and lower parameter levels led to the desired minimum porosity of 1.3% (Lugscheider and Lu, 1986). The other optimized dependent parameters were the microhardness (218 ± 37 HV$_{0.05}$), adhesion strength obtained by DIN 50 160 (75 ± 4 N mm^{-2}), and the traverse bending strength by DIN 50 111. The significant parameters for this optimization were **2** (negative effect) **≫ 14 > 1**.

Chromium Oxide Coatings Figure 11.16 shows response surfaces depicting the dependence of the thickness of APS Cr$_2$O$_3$ coatings on the plasma spray parameters arithmetic surface roughness of the substrate **1** (4; 20 μm), the plasma power **2** (29; 39 kW), the powder feed rate **3** (10; 40%), the powder grain size **4** (−45 + 25; −75 + 25 μm) and the spray distance **5** (80; 120 mm) varied according to a 2^{5-1} fractional factorial design of resolution V (Erne, 2004).

It could be shown that the coating thickness depends in a statistically significant way on the main effects **1, 2, 3, 4** and on the two-factor interactions **14, 15** and **34** yielding the regression polynomial

$$Y_{\text{thickness}} = 23 - 15X_1 + 18X_2 + 70X_3 - 27X_4 - 15X_1X_4 + 34X_1X_5 - 20X_3X_4. \quad (11.6)$$

Interpretation of the contour plots shown in Figure 11.16 provides deep insight into the interaction of the various plasma spray parameters and hence the anatomy of the

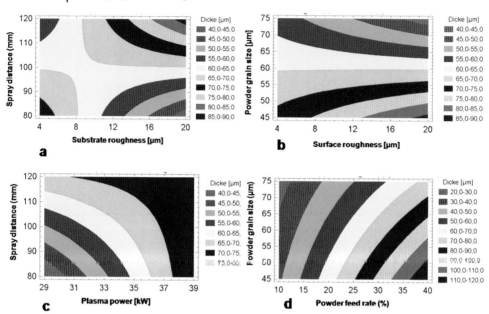

Figure 11.16 Response surface contours showing the dependence of the thickness of APS Cr_2O_3 coatings deposited on steel St37 on the arithmetic surface roughness of the substrate (X_1), the plasma power (X_2), the powder feed rate (X_3), the powder grain size (X_4) and the spray distance (X_5) (Courtesy: Dipl.-Min. Martin Erne, Leibniz-Universität Hannover, Germany).

system. The **15** plot (Figure 11.16a) yields a saddle (col) surface for intermediate parameter setting (**2** = 34 kW, **3** = 25%, **4** = 60 μm) resulting in average coating thicknesses between 60 and 65 μm (yellow field). Only with increasing substrate roughness and spray distance the thicknesses are varying appreciably caused by reduced heat dissipation of the rougher surfaces at short spraying distance. The **14** plot (Figure 11.16b) obtained for constant parameters **2** (34 kW), **3** (25%) and **5** (100 mm) shows that the coating thickness strongly increases with decreasing two-factor interaction **14**. The reason for the pronounced effect of the substrate roughness **1** relates to the reduced adhesion of the coarse powder particles at the very rough surface as well as a possible effect of surface activation and insufficient heat dissipation. On the other hand, during spraying of fine powder particles the surface activation provides excellent adhesion of the particles and thus large coating thickness. Although the two-factor interaction **25** was found to be non-significant during evaluation of the factor effects, its response surface (Figure 11.16c) obtained for intermediate parameter setting (**1** = 12 μm, **3** = 25%, **4** = 60 μm) shows that the coating thickness yields a high gradient at low plasma power values. The low coating thickness at low spray distance is likely related to insufficient melting of the powder particles at low enthalpies and the fact that a high proportion of unmelted particles will erode the already deposited material. Higher plasma powers result in a higher proportion of melted particles and consequently in higher coating thickness even at low spray distance. Finally, the **34** plot (**1** = 12 μm, **2** = 34 kW, **5** = 100 mm) reveals that the coating thickness increases steeply with decreasing two-factor interaction **34**. The maximum coating thickness will be realized for fine, well-melted particles that are deposited with high feed rates on a surface with low roughness.

Figure 11.17 shows the **24** response surfaces of the arithmetic coating roughness R_a (a) and the coating porosity (b) of APS chromium oxide coatings (Erne, 2004). The saddle in Figure 11.17a (**1** = 12 μm, **3** = 25%, **5** = 100 mm) is caused by the fact that the main parameters **2** and **4** have opposite signs in the reduced response polynomial (Equation 11.7) even though neither one is statistically significant:

$$Y_{Ra} = 11 + 9.9X_1 - 2.3X_5 - 1.9X_1X_3 - 1.7X_1X_4 - 1.6X_2X_4 - 2.3X_3X_5. \tag{11.7}$$

Figure 11.17b shows the contours of the response surface for the coating porosity at intermediate levels **1** = 12 μm; **3** = 25%, and **5** = 100 mm) in the **2** and **4** parameter plane. The corresponding reduced response polynomial was calculated to be

$$Y_P = 6.23 - 2.51X_2 - 2.54X_3 + 3.76X_4 - 2.54X_2X_4. \tag{11.8}$$

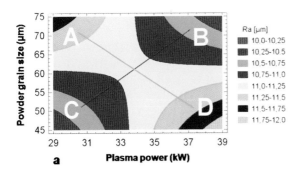

a

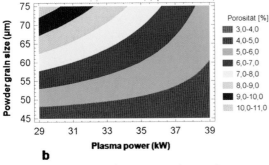

b

Figure 11.17 Response surface contours showing the dependence of the surface roughness R_a (a) and porosity (b) of APS Cr_2O_3 coatings deposited on steel St37 on the plasma power (X_2) and the powder grain size (X_4) (Courtesy: Dipl.-Min. Martin Erne, Leibniz-Universität Hannover, Germany).

The single significant effect is the powder grains size **4**, i.e. coarse powder produce coating with high porosity. Also, low plasma enthalpy and feed rates result in coatings with high porosity. The two-factor interaction **24** shown in Figure 11.17b has likewise a negative effect on the porosity, i.e. coarse powders spayed at low plasma power produce highly porous coatings whereas increased power and reduced grain size lead to denser coatings. Smoother coatings (areas B and C) were found for the parameter combinations: high power/coarse grains and low power/fine grains as opposed to rougher coatings (areas A and D) found for the parameter combinations: low power/coarse grains and high power/fine grains.

The examples of application of statistical design of experiment (SDE) methodology described above confirm amply the strong predictive power of these techniques. Even though parameter validation has to be performed and adapted to industrial requirements of total quality management (TQM; Chapter 10) in conjunction with theoretical modeling and numerical simulation (Sampath *et al.*, 2003; Wilden, 2007; see Chapter 6), SDE provides a relatively quick, easy and economical assessment of the dependence of desired coating properties on a potentially large number of independent plasma spray parameters.

11.2.3
Artificial Neuronal Network Analysis

The responses of coating properties to variations in input parameters are complex and in most cases strongly nonlinear (see above). Accordingly, to recognize parameter interactions, correlations and individual effect on coating properties a robust methodology is required (Guessasma *et al.*, 2004) that also enables response to parameter constraints based on economic and equipment considerations. Such a methodology has been found in artificial neuronal networks (ANN) based on database training implemented by a learning algorithm and supported by a set of experiments to predict property–parameter evolution. Artificial neuronal network analysis (Nelson and Illingworth, 1991; Tenenbaum *et al.*, 1996) offers several advantages over the classical statistical design of experiments (SDE) strategy. While treatment of nonlinear (quadratic) behavior of the response function Y for SDE requires a polynomial for p variables of the form (see also Equation 11.5)

$$Y = b_0 + \sum_{i=1}^{p} b_i x_i + \sum_{i=1, j=1}^{p} b_{ij} x_{ij} + \sum_{j=1}^{p} b_{ii} x_i^2 \qquad (11.9)$$

the response polynomial required for ANN optimization of one output parameter Y depending on 3 input parameters x_i is

$$Y = a_0 (1 + \exp - (a_i \times x_i + a_{ij} \times x_{ij} + a_{jj} \times x_j^2))^{-1}, \qquad (11.10)$$

where the four coefficients will be adjusted during the training procedure (Guessasma and Coddet, 2005). Accordingly, the number of experiments N_{SDE} required for a full factorial SDE is

$$N_{SDE} = (NL)^{NI}, \tag{11.11a}$$

where NL is the number of experimental levels and NI the number of input parameters.[5] The number of experiments required for ANN is smaller and yields

$$N_{ANN} = a \times NW = a \times (NI + NO)^{\alpha}, \tag{11.11b}$$

where NW is the number of weight parameter, NI and NO are the numbers of input and output parameters, respectively and α is the so-called layer number (dimensionality of the system) that determines the power of prediction.

Additional strengths of ANN analysis are its ability to disclose complex correlations with small structures, lack of need for prior assumptions about input/output correlations, and the advantage of using incomplete sets of experiments, i.e. missing data can be tolerated by ANN in contrast to SDE in which the database has to be complete. Constraints also exist: the database has to be representative, i.e. must show a good sampling of the input/output correlation, and the phenomena represented by the predicted relationships must have a physical relevance, hence forbidding the use of canonical variables (Guessasma and Coddet, 2005). Also the mathematical treatment is more complex and involved.

The mathematical concept of an artificial neuron was first introduced by McCulloch and Pitts (1943) to emulate the signal transmission by biological neurons in the brain. Neurons receive input from one or more dendrites and sum up these input signals to produce an output signal (synapse). The sums of each node are weighted. The function relating input and output is called the transfer function, introducing nonlinearity to the system. A frequently used canonical form of the transfer function is the sigmoid (logistic curve) but other nonlinear functions can also be used such a piecewise linear functions or step functions, e.g. the Heaviside function (antiderivative of the Dirac delta function).

A basic function underlying ANN is the expression

$$Y_k = \varphi\left(\sum_{j=0}^{m} w_{kj} \times x_j\right), \tag{11.12}$$

where $\varphi = \varphi(t) = 1/(1 + \exp(-t))$ is the sygmoid transfer function and m is the number of inputs with signals x_0 through x_m and weights w_0 through w_m. Figure 11.18 shows the simple scheme how input signals x_i (i = 0...m) are connected to the output signal Y_k (k = number of layers) by the transfer function φ. The output signal Y_k propagates to the next layer (k + 1) through a weighted synapse.

Guessasma and Coddet (2005) presented an instructive example of the power of ANN by predicting porosity levels of alumina–13% titania coatings deposited throughout execution of 19 sets of experiments with a total of 126 samples. The input parameters varied were the plasma arc current (I, 350–750 A, reference value: 530 A), the argon

---

[5] In the contribution by Guessasma and Coddet (2005), Equation (4), the NL and NI parameters were erroneously interchanged.

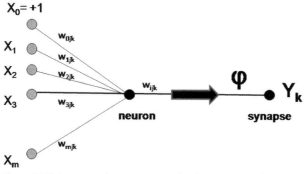

Figure 11.18 Input signals x_i are connected to the output signal Y_k through the transfer function φ.

primary gas flow rate (A, 40–70 slpm, reference value: 54 slpm), the hydrogen secondary gas flow rate (H, 0–50 slpm, reference value: 35 slpm), the carrier gas flow rate (CG, 2.2–4.2 slpm, reference value: 3.2 slpm) and the powder feed rate (D_m, 7–22 g min^{-1}, reference value: 22 g min^{-1}). Spray distance (125 mm), spray angle (90°), injector diameter (1.8 mm) and injection distance (6 mm) were kept constant throughout.

Background information on ANN, training and testing procedures, and ANN implementation and adaptation to databases were discussed by Guessasma and Coddet (2005) (see also Wang et al., 2007). In particular, the training rule applied was provided by a quick learning algorithm considering weight update as a function of the previous and current cycles until an optimum residual error occurs (Patterson, 1996). With the database obtained from the experiments described above, an optimized artificial neural network structure was developed as shown in Figure 11.19.

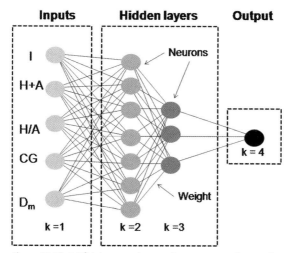

Figure 11.19 Artificial neuronal network structure used to predict the porosity (output) of plasma-sprayed alumina–13% titania coatings as a function of the input parameters I, H + A, H/A, CG and D_m (modified after Guessasma and Coddet, 2005).

The database was divided into three parts: 50% of the experiments were assigned to fine-tune the weight parameters NW in the training stage, 27% for the test of the network configuration during the test stage, and the remainder to test the generalization of the optimized network to predict porosity values in the input parameter range outside the experimentally applied values during the prediction stage. Hence calculations were performed in the following prediction ranges: arc current I → 300–800 A; total gas flow rate H + A → 20–80 slpm; ratio of hydrogen/argon gas flow rates H/A → 0–50 slpm; carrier gas flow rate CG → 1.5–5 slpm; powder feed rate D_m → 5–60 g min^{-1}.

As shown in Figure 11.19 the first layer of the ANN (k = 1) contains the five input parameters I, A, H, CG and D_m, the first hidden layer (k = 2) contains 7 neurons, the second hidden layer (k = 3) contains 3 neurons, and the output layer (k = 4) yields the predicted porosity. To compare predicted and experimental porosity values, each input parameter was varied individually, keeping the other at their reference values.

Historically available information on the possible anatomy of the system confirmed that with only two hidden layers a correct optimization could be obtained (Guessasma et al., 2002).

The experimental results were fitted to the following relationships:

1. arc current: P(%) = 6.26 + 3.28 × 10^{-5} I^2 − 5.14 × 10^{-8} I^3; r^2 = 0.83.
2. total gas flow rate: P(%) = 7.3 + 0.01 × (H + A); r^2 = 0.42.
3. hydrogen/argon ratio: P(%) = 10.13 − 0.09 × (H/A); r^2 = 0.93.
4. carrier gas flow rate: P(%) = 16.58 − 6.34 × CG + 1.10 × CG^2; r^2 = 0.38.
5. powder feed rate: P(%) = 5.45 + 0.1 × D_m; r^2 = 0.93.

The low regression coefficients of relationships 2 and 4 have been attributed to the scatter of the experimental results around the mean values.

The predicted parameter effects are 68% for input parameter I between 350 and 750 A (i.e. the predicted decrease of porosity from 9% to about 2.5% corresponds to 68%), 4% for H + A between 40 and 70 slpm, 34% for H/A between 13 and 50%, 9% for CG between 2.2 and 4.2 slpm, and 19% for D_m between 7 and 22 g min^{-1}. Hence the porosity of plasma-sprayed alumina–13% titania coatings decreases strongly with increasing arc current (nonlinearly) and hydrogen/argon ratio (linearly), i.e. plasma enthalpy, and decreases somewhat with decreasing powder feed rate (linearly), i.e. less cooling of the plasma jet under decreasing dense loading conditions. The contributions of the total plasma gas flow rate (linearly) and the carrier gas flow rate (parabolically) are weak. Based on these results the predicted individual control factors were arc current I, hydrogen fraction H/A, and powder feed rate D_m. Taking into account interaction of all parameters, the control factors were the arc current and the hydrogen fraction. Then porosity levels <2% could be achieved with the following parameter setting: I > 425 A, H/A > 37.4%. As expected, minimum porosities are generated by increased heat transfer in-flight from the plasma to the particles, reduction of viscosity of the melted materials and hence increase of the flattening ratio ξ during impact.

11.2.4
Fuzzy Logic Control

This approach is rarely used to optimize the properties of plasma-sprayed coatings. Fuzzy logic control (FLC) is a knowledge-based methodology that translates human linguistics into an optimized model of controllable reality using a set of fuzzy rules and membership functions. It is an organized and mathematical method of handling inherently imprecise concepts such as the nonlinear dependence of the thickness of plasma-sprayed coatings on a multitude of input parameters. This uncertainty originates from the chaotic nature of the coating process in which infinitesimally small changes in the input parameters cause large and in general nondeterministic changes in the output parameter, i.e. coating properties. As described in 4.4.4 nonlinearity is introduced by stochastic arc root fluctuation as well as entrainment of parcels of cold air by chaotic pulsation (see Figure 3.29) caused by the turbulent action of the plasma jet. As far as heat transfer is concerned nonequilibrium distributions of the phases of plasma waves cause electromagnetic and magneto hydrodynamic turbulences (see Section 3.1.1) that affect the local magnetic field strength, **B** and the electrical current density, **j**. Consequently their cross-product, the Lorentz force (**j** × **B**) fluctuates rapidly and with it the plasma compression ('z-pinch'). Finally, the local thermal equilibrium breaks down on a scale that is small compared with the overall volume of the plasma jet. Then the system enters the realm of a *'heat transfer catastrophy'*. This phenomenon can be tentatively described in terms of stability theory by an elementary cusp catastrophy of co-dimension two (Riemann–Hugoniot catastrophy) as schematically illustrated in Figure 4.14 above.

The fuzzy logic control system consists of several units including a 'fuzzificator', an inference engine containing a database and a rule base, and a 'defuzzificator' (Zadeh, 1965; Ross, 2004). Figure 11.20 shows a flow chart of a fuzzy logic controller coupled with Taguchi-structured control factors designed to optimize the thickness of plasma-sprayed ZrO_2–8%Y_2O_3 thermal barrier coatings (Jean et al., 2006). Coatings were deposited under statistical variation of seven input parameters at three levels and one parameter at two levels according to an orthogonal L18 ($2^1 \times 3^7$) Taguchi design. The operating parameters X_i were the number of traverses (A: 5; 8), the accelerating voltage (B: 65; 70; 75 V), the arc current (C: 550; 600; 650 A), the traverse speed of the plasmatron (D: 20; 25; 30 mm s^{-1}), the stand-off distance (E: 80; 100; 120 mm), the powder feed rate (F: 20; 25; 30 g min^{-1}), the carrier gas flow rate (G: 5; 6; 7 slpm) and the primary plasma gas flow rate (H: 50; 55; 60 slpm).

According to Jean et al. (2006) a fuzzy system is defined as

$$f : U = \bigcup_{i=1}^{n} U_i \subset R^n \Rightarrow V \subset R, \quad (11.13)$$

where U and V are input and outputs, and R refers to the rule base. Each fuzzy rule R can be expressed by a Boolean logic 'if-than' statement as follows:

Rule i : *If* input parameter X_1 is A_{i1} and X_2 is A_{i2} and .. X_n is A_{in} *then* Y_i is B_i,

$$(11.14)$$

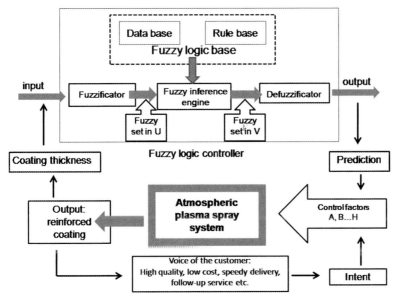

Figure 11.20 Flow chart of a fuzzy logic controller coupled with Taguchi-structured control factors (after Jean et al., 2006).

for i = 1,2, ... m. The terms A_{in} are linguistic sets[6] associated with the inputs X_n, the output Y_i is determined by the linguistic terms B_i. The range of inputs were partitioned into the three linguistic sets *S* (small), *M* (medium) and *L* (large), the range of outputs into nine graduating membership functions: *SS* (small small), *VS* (very small), *S* (small), *SM* (small medium), *M* (medium), *ML* (medium large), *L* (large), *VL* (very large) and *LL* (large large). The Mamdani centroid inference method was applied to defuzzify the output membership functions to obtain a valid prediction (see Figure 11.20).

From ANOVA the three factors accelerating voltage (B), stand-off distance (E), and carrier gas flow rate (G) were found to affect the coating thickness most, accounting for 78.6% of the total experimental variance whereby the individual percent contributions were about equal at 26%. Hence these highly significant factors were assigned to a fuzzy logic controller with 27 fuzzy rules to predict the outputs. From historical data it was concluded that a coating thickness of 50 μm was most desirable to guarantee maximum coating adhesion to the substrate.[7] Applying the logic rules along with the Mamdani centroid inference, the linguistic and membership values for the outputs can be determined. In the particular case described by Jean *et al.* (2006)

---

6) These linguistic terms are at least semantically akin to 'discrete' variables in the much less involved screening designs of Plackett–Burman or Taguchi type (see 11.2.1.2).

7) It should be mentioned that adhesion of plasma-sprayed coatings is not just a function of coating thickness which controls the residual stress level, but other factors such as splat cohesion, substrate roughness, temperature of the substrate, properties of the bond coats, level of impurities and degree of oxidation likewise play important roles.

out of the 27 fuzzy rules only four membership functions 'fired', i.e. fulfilled the requirements to accommodate the three significant factors to minimize the coating thickness to a value as close as possible to the desired 50 μm as indicated by a linguistic output value SS. These rules are as follows:

Rule 6: If (B is *S*, E is *M*, G is *L*) than coating thickness *SS*.

Rule 9: If (B is *S*, E is *L*, G is *L*) than coating thickness is *SS*.

Rule 15: If (B is *M*, E is *M*, G is *L*) than coating thickness is *SM*.

Rule 18: If (B is *S*, E is *L*, G is *L*) than coating thickness is *SS*.

The coating thickness for the input parameter set with A → *L*, B → *M*, C → *M*, D → *L*, E → *L*, F → *S*, G → *L*, H → *S* has been measured to be 51.67 μm, very close to the design value of 50 μm which compares favorably with the defuzzyfied values of 51.41. The fact that for this parameter combination the signal-to-noise ratio with −12.3 db shows the largest values of all tests confirms that the FLC design accurately predicted the target coating thickness.

11.3
Future Developments

Figure 11.21 shows the main aspects of plasma spraying that have to be considered in order to produce advanced metal, ceramic or composite coatings with high in-service performance, which have been dealt with in the preceding chapters.

Thermal spray R&D of advanced materials for high performance applications is increasing rapidly, and many developments are now being commercialized. As an indication of this rapid development it should be mentioned that over 90% of the advances that have been made over the last 90 years have been made in the last three decades! Equipment and process advances have typically led the technology in the past. Increasingly, on-line materials and intelligent process control (iSPC), development of new coating materials and technologies, and novel applications will lead in the next 10 years.

Future developments in the advanced materials coating field can be characterized and evaluated in terms of their economic feasibility by looking at different technology support and development strategies (SRI International, 1990). Figure 11.22 shows that wear and thermal barrier coatings are level I strategies that deploy current technologies to improve the competitive performance of small companies or create new companies. Bioceramic and diamond coatings are level II strategies that focus on the development of innovative applications of new technological discoveries. Finally, low friction coatings, high-temperature superconducting, and silicon nitride coatings are level III strategies whose efforts are still concentrated on basic R&D leading up to the discovery of new technologies. Those new technological breakthroughs expected for the future will predominantly assist large companies, and attract completely new industries.

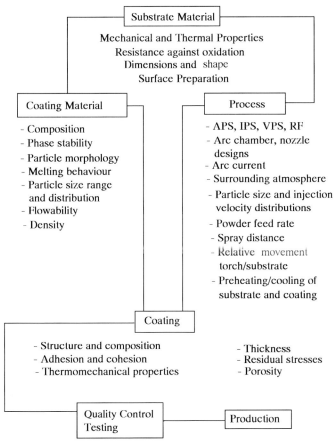

Figure 11.21 Main aspects of plasma spraying considering substrate–coating–process interactions.

The main areas of contemporary developments will be automotive coatings with high rate, low cost processing, while aerospace applications are triggering the advance of the technology for novel thermal barrier coatings, spray forming, near-net shape technologies, and composite materials processing.

Advances are being typically made in the following areas (see also Walser, 2004):

- new materials, processes, and equipment;
- control devices and automated robotic handling including motion control, gas pressure technology, real-time sensors and so on;
- use of iSPC with resulting close process monitoring.
- accumulation of data bases and development of expert systems;
- HVOF and CGDS processing, and their spin-off technologies;
- engineered powder production and quality control of feedstock;
- composites and intermetallic alloy spray forming through processing of free-standing near-net shape parts;

11 Design of Novel Coatings

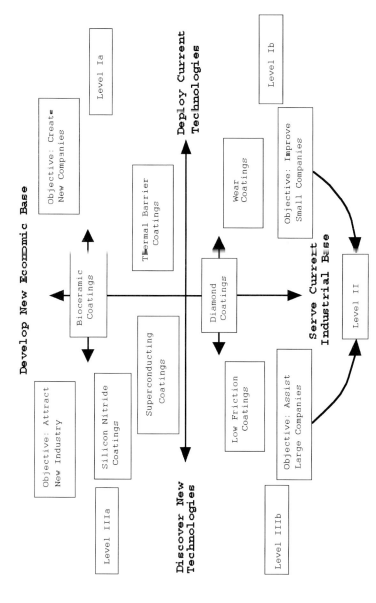

Figure 11.22 Technology support and development strategies applied to various types of coating (SRI International, 1990).

- improved methods for nondestructive testing and evaluation of coatings;
- thin film deposition through LPPS;
- computer-aided design/rapid prototyping techniques including stereolithography.

Education and training in thermal spray processing need to be implemented and managed on a broader base. Collaboration with industry in the resource and manufacturing sectors will lead increasingly to strategic alliances that enable industry to produce more competitively and environmentally compatible. Process control, including modeling of complex plasma–particle–substrate interactions, on-line process diagnostics, and development of novel coatings with improved performance are areas rich in research needs and opportunities.

Such areas can be predicted by application technology mapping (Sutliff, 1987; also Heimann and Lehmann, 2008). Marketers look for applications of materials, and then determine the performance needs for particular applications. These data are mapped against the value-in-use estimate and the customer's ability to pay (Figure 11.23).

Military clients can afford sophisticated materials and coatings, or products with a high value-in-use, for example piezoelectric, magnetic, ferroic and superconducting coatings for range finders and surveillance systems. Titanium nitride, titanium carbide, and diamond coatings for ceramic *cutting tools* are high-value-added but nevertheless cost-competitive because of their superior wear performance in numerically controlled high-speed machining of tough and hard steels and superalloys. On the other hand, *heat engine components* such as ceramic turbochargers for passenger cars, thermal barrier coatings (TBCs) for aerospace gas turbines, coatings for cylinder bores (see below), and a variety of automotive sensors based on functional ceramic coatings and thin films have a low value, and the ability of the car manufacturer to pay for them is also low in order to maximize profit. New developments are presently being considered such as thick thermal barrier coatings (TTBCs) for diesel engines to replace

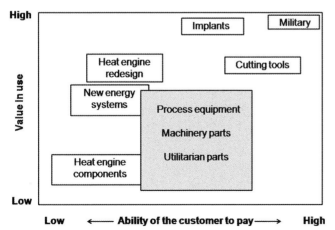

Figure 11.23 Application technology mapping for coatings (Sutliff, 1987; Heimann and Lehmann, 2008).

water cooling by air cooling systems. The middle ground of Figure 11.23 is occupied by wear-resistant parts for nonautomotive markets, i.e. *process equipment and machinery tools*. In this area a new strong driving force is evolving that is geared towards first-generation materials to improve process efficiency and overall productivity in the manufacturing and resource industries. Also, electrode and electrolyte coatings for solid oxide fuel cells are at the verge of a break-through to install new systems for environmentally conscious future energy generation concepts.

In the field of heat engine components much effort is being expended to develop light weight, low friction coatings for the internal bore of gasoline and diesel engine cylinders (Barbezat, 2005). In the ongoing quest to enhance fuel economy by reducing the weight of passenger cars hypoeutectic aluminum–silicon (Al8Si) cast alloys are widely used today for engine blocks. This alloy is cast by a low pressure die casting process and provides higher thermal conductivity and better corrosion resistance compared with more traditional engine block materials such as lamellar graphite grey cast iron (GGL) or vermicular graphite grey iron (GGV). However, its toughness and hardness, and hence its tribological performance are noticeably lower thus requiring coatings, in particular plasma-sprayed coatings based on carbon steel such as ferrite with fine inclusions of iron carbides, and solid lubricant wustite and magnetite as well as corrosion-resistant steel alloyed with Cr and Mo, or Ni, Cr and Cu (Uozato *et al.*, 2005). Such coating materials are deposited by rotating plasma spraying to a thickness of about 200 µm onto the inner surface of engine cylinders made from aluminum cast alloy. They significantly reduce the friction between cylinder wall and piston rings, reduce the weight of the engine block by reducing the pitch distance between the cylinders, reduce the oil consumption by providing a surface roughness after diamond honing as low as 0.3 µm, and as an overall result reduce the fuel consumption by as much as 2% with increasing tendency. However, the wear rates of plasma-sprayed coating are still significantly higher than those of cast iron, reaching values in critical piston ring areas as high as $5\,nm\,h^{-1}$ of service.

Several car manufacturers have equipped some of their high-end engines with plasma-spray coated cylinder liners, starting about in 2000. While the emphasis in case of gasoline engines has been in the field of racing car engines (V10 for Formula 1 and V8 for Formula 2 racing cars) with powers exceeding 400 kW and luxury cars such as the Bugatti W16 engine with a power exceeding 700 kW, coated diesel engine cylinder liners are available on the market for vehicles such as Volkswagen's high powered Touareg and Phaeton vehicles (V10 TDI, 230 kW) and the Van T5 (L5 EA 115, 130 kW) (Barbezat, 2005). It can be expected that such plasma-sprayed coatings will gain increasing acceptance, primarily in Europe and Japan in the years to come by spreading to mid-size cars. High-loaded diesel engines will particularly benefit from corrosion-resistant coatings to protect the Al8Si alloy from fuel containing sulfur, vanadium and formic acid impurities. Also scuffing- and abrasion-resistant coatings are required during recirculation of exhaust gases. This technology is increasingly used to boost engine efficiency, for example by installing a turbocharger rotor made from silicon nitride in the hot exhaust stream (Jack, 1994; Riley, 1996).

In conclusion, the future of advanced materials coatings applied by plasma spray technology looks very bright. There are, however, problems still to be solved.

Technical problems include optimization of plasmatron design, powder size and composition, rheological and flow properties of powders, overspray losses, and surface preparation. Quality control procedures must be developed or improved, and implemented for ceramic, metal and composite coatings to standardize impact testing, hardness testing, shear and bending testing, cavitation–erosion testing, slurry abrasion testing etc. Most of all, reliable and reproducible tests must be developed to measure adhesion strength and tribological performance of coatings. Finally, the development of computer codes that model the forces acting on the coating/substrate interface is necessary. The objective is to develop coatings that sustain in-film compressive loads during service, thus improving adherence to the substrate and, in turn, maximizing the service life of the coated equipment.

There is also the considerable challenge to improve the image of thermal spray coatings as a viable, reliable und immensely versatile option available to design engineers. Although it is widely recognized that plasma-sprayed coatings can provide many successful answers to engineering problems, its level of awareness in industry and government has to be raised (Heimann, 1991).

Thermal spray technology is now well on its way to becoming one of the most important tools of increasingly sophisticated surface engineering technology. Initially developed as a simple and relatively crude surfacing tool, thermal spraying is now considered a powerful and flexible materials processing method with a high potential of development. It appears that the return of investment (ROI) in this area of surface engineering is excellent, and that small and medium-sized enterprises (SMEs) can hugely benefit from entering a market segment that by many is considered the materials technology of the twentyfirst century.

At the end of this treatise on plasma-sprayed coatings the following final statement should be made. Strict quality control of well established coatings, and close attention to the design and testing of coating/substrate systems as a single synergistic entity, combined with the development of novel structural and functional coatings using improved automated equipment and comprehensive data bases, expert systems, and realistic modeling and simulation protocols will secure plasma spray technology a substantial market niche in the future.

References

ASM International, *Handbook of Thermal Spray Technology*, ASM International: Materials Park, OH. 2004.

H. Bandemer, A. Bellmann, *Statistische Versuchsplanung*, 4th edn, Teubner: Stuttgart, Leipzig. 1994, p. 133.

G. Barbezat, *Surf. Coat. Technol.*, 2005, **200**, 1990.

S. Bisgaard, in: Proc. 3rd NTSC, Long Beach, CA, May 20–25, 1990, p. 661.

G.E.P. Box, K.B. Wilson, J. Royal Stat. Soc., Series B, 1951, 1.

G.E.P. Box, D.W. Behnken, *Technometrics*, 1960, **2**, 455.

G.E.P. Box, J.S. Hunter, *Technometrics*, 1961, **3** (3), 311.

G.E.P. Box, W.G. Hunter, J.S. Hunter, *Statistics for Experimenters*, John Wiley & Sons: New York, 1978.

G.E.P. Box, N.R.D. Draper, *Empirical Model Building and Response Surface Methodology*, John Wiley & Sons: New York, 1987.

I. Burlacov, J. Jirkovský, M. Müller, R.B. Heimann, *Surf. Coat. Technol.*, 2006, **201**, 255.

T. Chon, A. Aly, B. Kushner, A. Rotolico, W.L. Riggs, in: Proc. 3rd NTSC, Long Beach, CA, May 20–25, 1990; p. 681.

X. Chu, S.A. Barnett, *J. Appl. Phys.*, 1995, **77**, 4403.

W.G. Cochran, G.M. Cox, *Experimental Designs*, John Wiley & Sons: New York, 1957.

C. Daniel, *Applications of Statistics to Industrial Experimentation*, John Wiley & Sons: New York, 1976.

E. De Bono, *Lateral Thinking*, Ward Lock Education, 1970.

W.E. Deming, *Out of the Crisis*, MIT Press: Cambridge, MA, 1986.

DuPont de Nemours & Co Strategy of Experimentation, 1975.

M. Erne, Optimierung der Dichte, Dicke und Oberflächenrauheit APS-gespritzter Chrom(III)oxid Korrosionsschutzschichten, Unpublished 4th Year Thesis, Technische Universität Bergakademie Freiberg, 2004. (http://www.wissen24.de/vorschau/25257.html).

X. Fan, F. Gitzhofer, M. Boulos, *J. Thermal Spray Technol.*, 1998, **7** (2), 197.

S. Guessasma, G. Montavon, P. Gougeon, C. Coddet, in: Proc. ITSC 2002, E. Lugscheider, P.A. Kammer (eds.), DVS-Verlag: Düsseldorf, Germany, 2002, 453.

S. Guessasma, G. Montavon, C. Coddet, *Comput. Mater. Sci.*, 2004, **29** (3), 315.

S. Guessasma, C. Coddet, *Surf. Coat. Technol.*, 2005, **197**, 85.

K. Hajmrle, M. Dorfman, *Modern Devel. Powder Metallurgy*, 1985, **15/17**, 409.

R.B. Heimann, in: Proc. First All-Alberta Appl. Stat. Biometr. Workshop. AECV91-P1, Edmonton, Alberta, Canada, October 18–19, 1990, 87.

R.B. Heimann, D. Lamy, T. Sopkow, *J. Can. Ceram. Soc.*, 1990a, **59** (3), 49.

R.B. Heimann, D. Lamy, T.N. Sopkow, in: Proc. 3rd NTSC, May 20–25, 1990b, Long Beach, CA., p. 491.

R.B. Heimann, *Process. Adv. Mater.*, 1991, **1**, 181.

R.B. Heimann, Development of Plasma-Sprayed Silicon Nitride-Based Coatings on Steel, Research Report to NSERC and EAITC, Canada, December 15, 1992.

R.H. Heimann, H.D. Lehmann, in: *Recent Patents on Mater. Sci.*, Bentham Science Publ., 2008, **1**, 41.

C. Herrström, H. Hallén, A. Ait-Mekideche, E. Lugscheider, in: Proc. TS'93 Aachen, 1993, DVS 152, p. 409.

H. Holleck, *Surf. Eng.*, 1991, **7**, 137.

J.S. Hunter, *Technometrics*, 1966, **8**, 177.

T.C.C. Hurng, M.B.C. Quigley, R.L. Apps, in: Proc. 2nd Int. Conf. on Surface Engineering, Stratford-upon-Avon, U.K., June 16–18, 1987, p. 413.

K.H. Jack, Prospects for Nitrogen Ceramics, in: *Silicon Nitride 93* (M.J. Hoffmann, P.F. Becher, G. Petzow, eds.), Key Eng. Mater., 1994, **89–91**, 345.

M.D. Jean, B.T. Lin, J.H. Chou, *Surf. Coat. Technol.*, 2006, **201**, 3129.

C.H. Lee, H. Choi, C. Lee, H. Kim, *Surf. Coat. Technol.*, 2003, **173**, 192.

E. Lugscheider, P. Lu, B. Hauser, D. Jäger, *Surf. Coat. Technol.*, 1987, **32**, 215.

E. Lugscheider, P. Lu, in: Proc. Int. Conf. Plasma Science Technol., Beijing, China, June 4–7, 1986; Science Press, Beijing, 1986, p. 250.

E. Lugscheider, M. Knepper, *DVS*, 1991, **136**, p. 88.

J.R. Mawdsley, Y.J. Su, K.T. Faber, T.F. Bernecki, *Mater. Sci. Eng. A*, 2001, **308**, 189.

D.B. Marshall, T. Noma, A.G. Evans, *Comm. Am. Ceram. Soc.*, 1982, **65**, C-175.

W. McCulloch, W. Pitts, *Bull. Math. Biophys.*, 1943, **7**, 115.

M. Moldovan, C.M. Weyant, D.L. Johnson, K.T. Faber, *J. Thermal Spray Technol.*, 2004, **13** (1), 51.

M.M. Nelson, W.T. Illingworth, *A Practical Guide to Neural Nets*. 3rd edn, Addison-Wesley: New York, NY, USA. 1991.

J.E. Nerz, B.A. Kushner, A.J. Rotolico, in: Proc. 3rd NTSC, Long Beach, CA, May 20–25, 1990, p. 669.

B.T. Neyer, *Technometrics*, 1994, **36** (1), 61.

E.R. Novinski, A.J. Rotolico, E.J. Cove, in: Proc. 3rd NTSC, Long Beach, CA, May 20–25, 1990; p. 151.

D. Patterson, *Artificial Neural Networks*, Prentice Hall: Singapore. 1996.

C. Pierlot, L. Pawlowski, M. Bigan, P. Chagnon, *Surf. Coat. Technol.*, in press.

R.L. Plackett, J.P. Burman, *Biometrika*, 1946, **33**, 305.

C. Pumplün, S. Rüping, K. Morik, C. Weihs, *D-optimal Plans in Observational Studies*, Report SFB 475 ('Reduction of Complexity in Multivariate Data Structures'), 2003, University Dortmund, Germany.

G. Reisel, R.B. Heimann, *Surf. Coat. Technol.*, 2004, **185**, 215.

F.L. Riley, Applications of Silicon Nitride Ceramics, in: *Advanced Ceramic Materials* (H. Mostaghaci, ed.), Key Eng. Mater., 1996, **122–124**, 479.

J.R. Rodgers, *AMPTIAC Newsl.*, 2001, **5**, 1.

T.J. Ross, *Fuzzy Logic with Engineering Applications*, John Wiley & Sons: Chichester, 2004.

S. Sampath, H. Herman, N. Shimoda, T. Saito, MRS Bulletin, January 1995, 27.

S. Sampath, X. Jiang, A. Kulkarni, J. Matejicek, D.L. Gilmore, R.A. Neiser, *Mater. Sci. Eng. A*, 2003, **348**, 54.

P. Saravanan, V. Selvarajan, M.P. Srivastava, D.S. Rao, S.V. Joshi, G. Sundararajan, *J. Thermal Spray Technol.*, 2000, **9** (4), 505.

S.I. Satterthwaite, *Technometrics*, 1959, **1**, 111.

A. Schuppert, A. Ohrenberg, Method and Computer for Experimental Design, EP1499930, 2005.

SRI International, Center for Economic Competitiveness. Assessment of Alberta Technology Centers, 1990.

S.K. Sutliff, in: Trans. 4th Workshop CUICAC, Toronto, ON, Canada, May 25–26, 1987.

G. Taguchi, S. Konishi, *Taguchi Methods, Orthogonal Arrays and Linear Graphics*, American Supplier Institute, 1987.

J.B. Tenenbaum, W.T. Freeman, M.C. Mozer, M.I. Jordan, T. Petsche (eds.), Proc. Advanc. Neural Process. Information Systems, Vol. 9. The MIT Press: Cambridge, MA, USA. 1996.

S. Thiele, Mikrohärte, Mikrostruktur und Haftung vakuum-plasmagespritzter TiC/Mo_2C/Ni, Co-Verbundschichten. Unpublished Master Thesis, Technische Universität Bergakademie Freiberg, June 1994.

T. Troczynski, M. Plamondon, *J. Thermal Spray Technol.*, 1992, **1** (4), 293.

J.W. Tukey, *Ann. Math. Stat.*, 1962, **33**, 1.

S. Uozato, K. Nakata, M. Ushio, *Surf. Coat. Technol.*, 2005, **200**, 2580.

D.J. Varacalle, L.B. Lundberg, B.G. Miller, W.L. Riggs, in: Proc. 14th ITSC'95, Kobe, May 22–26, 1995, 377.

S. Veprek, *J. Vac. Sci. Technol.*, 1999, **A17**, 2401.

B. Walser, *Spraytime*, 2004, **10** (4), 1.

L. Wang, J.C. Fang, Z.Y. Zhao, H.P. Zeng, *Surf. Coat. Technol.*, 2007, **201**, 5085.

J. Walter, W.L. Riggs, in: Proc. 3rd NTSC, Long Beach, CA, May 20–25, 1990; p. 729.

J. Wilden, *Mat. -wiss.u. Werkstofftech.*, 2007, **38** (2), 131.

F. Yates, Imperial Bureau Soil Sci., Techn. Comm., 1937, 35.

L.A. Zadeh, Fuzzy Sets. Inform. and Control, 1965, 8(3), 338.

L.R. Zhao, K. Chen, Q. Yang, J.R. Rodgers, S.H. Chiou, *Surf. Coat. Technol.*, 2005, **200**, 1595.

J. Zimmermann, Untersuchungen zur Haftfestigkeit und Rauheit plasmagespritzter Cr_2O_3-Schichten, Unpublished 4th Year Thesis, Technische Universität Bergakademie Freiberg, 2000.

Appendix A: Dimensionless Groups

The most important dimensionless groups used in equations to model mass, heat and momentum transfer in plasmas can conveniently be divided into groups as follows.

A.1
Momentum Transfer

Reynolds number	$\mathbf{Re} = VL/v$ (inertia force/viscous force)
Euler number	$\mathbf{Eu} = \Delta p/\rho V^2$ (fluid friction)
Grashof number	$\mathbf{Gr} = L^3 g\gamma \Delta T/v^2$ (buoyancy force/viscous force)
Weber number	$\mathbf{We} = \rho V^2/\gamma$ (inertia force/surface tension)

A.2
Heat Transfer

Fourier number	$\mathbf{Fo} = \kappa \Delta T/L^2$ (heat transfer by diffusion)
Péclet number	$\mathbf{Pe} = VL/\kappa = \mathbf{Re} \times \mathbf{Pr}$ (bulk heat transfer/conductive heat transfer)
Rayleigh number	$\mathbf{Ra} = L^3 g\gamma \Delta T/v\kappa = \mathbf{Gr} \times \mathbf{Pr}$ (free convection)
Nusselt number	$\mathbf{Nu} = hL/k_f = \mathbf{Re} \times \mathbf{St}$ (total heat transfer/conductive heat transfer; k_f = thermal conductivity of fluid)
Biot number	$\mathbf{Bi} = hL/k_s$ (heat transfer resistance inside/surface of a solid; k_s = thermal conductivity of solid)
Stefan number	$\mathbf{St} = \sigma L T^3/k$ (heat transfer by radiation)

Plasma Spray Coating: Principles and Applications. Robert B. Heimann
Copyright © 2008 WILEY-VCH Verlag GmbH & Co. KGaA, Weinheim
ISBN: 978-3-527-32050-9

A.3
Mass Transfer

Fourier number	$Fo^* = D\Delta T/L^2 = $ **Fo/Le** (unsteady state mass transfer)
Péclet number	$Pe^* = VL/D = $ **Re** × **Sc** (bulk mass transfer/diffusive mass transfer)
Grashof number	$Gr^* = L^3 g\beta' \Delta x/v^2$ (mass transfer at free convection)
Nusselt number	$Nu^* = K_c L/\rho D = $ **Sh** (Sherwood number; K_c = overall mass transfer coefficient)

A.4
Materials Constants

Prandtl number	$Pr = v/\kappa = $ **Pe/Re**
Schmidt number	$Sc = v/D = $ **Pe*/Re**
Lewis number	$Le = \kappa/D = $ **Sc/Pr**

Appendix B: Calculation of Temperature Profiles of Coatings

The following treatment has been adapted from the work by Houben 1988. It provides the calculations required to obtain the temperature profiles through Mo and AISI-316 coatings, respectively deposited onto a low carbon steel substrate as shown in Figure 5.14. The calculations are based on the Neumann-Schwartz equation (Kuijpers and Zaat, 1974).

B.1
Heat Conduction Equations

For the three regions identified in the coordinate system of Figure 5.12 the three heat conduction (Fourier) equations can be expressed by

$$\partial^2 \Theta_0 / \partial x^2 - (1/a_0) \partial \Theta_0 / \partial t = 0 \quad \text{for} \quad x \leq 0 \, \text{(solid substrate)} \tag{B.1a}$$

$$\partial^2 \Theta_1 / \partial x^2 - (1/a_1) \partial \Theta_1 / \partial t = 0 \quad \text{for} \quad 0 \leq x \leq X(t) \, \text{(solid deposit)} \tag{B.1b}$$

$$\partial^2 \Theta_2 / \partial x^2 - (1/a_2) \partial \Theta_2 / \partial t = 0 \quad \text{for} \quad x \geq X(t) \, \text{(liquid deposit)}, \tag{B.1c}$$

where Θ = temperature, a = thermal diffusivity, and the indices 0, 1 and 2 refer to substrate, solid deposit and liquid deposit, respectively.

The boundary conditions are ($T_{s0} = 20\,°C$)

$$\begin{aligned}
\Theta_0 &= T_{s0} \quad \text{as } x \to -\infty, \\
\Theta_0 &= \Theta_1 \quad \text{as } x = 0. \\
\Theta_1 &= \Theta_2 \quad \text{as } x = X(t)
\end{aligned} \tag{B.2}$$

For thermal equilibrium conditions at the interfaces it follows

$$k_0 (\partial \Theta_0 / \partial x) = k_1 (\partial \Theta_1 / \partial x) \quad \text{as } x = 0 \tag{B.3}$$

$$k_1 (\partial \Theta_1 / \partial x) - k_2 (\partial \Theta_2 / \partial x) = L\rho (dX/dt) \quad \text{as } x = X(t), \tag{B.4}$$

where L = latent heat of melting.

Plasma Spray Coating: Principles and Applications. Robert B. Heimann
Copyright © 2008 WILEY-VCH Verlag GmbH & Co. KGaA, Weinheim
ISBN: 978-3-527-32050-9

For the special case indicated in Figure 5.12 ($\Theta_2 = T_3 = T_m$ for $x \geq 0$) Equation B.4 simplifies to

$$k_1(\partial\Theta_1/\partial x) = L\rho(dX/dt) \quad \text{as } x = X(t). \tag{B.4a}$$

B.2
Solutions of the Equations

B.2.1
Substrate Temperature Profile

$$\text{Assumption}: \Theta_0 = T_{s0} + \alpha[1 + \text{erf}(x/\sqrt{4a_0 t})]. \tag{B.5}$$

The quantity α is the so-called contact conductivity at the interface substrate/solid deposit ($x=0$) and is defined as $\alpha = (k_1\rho_1 c_1)^{1/2}$, where k = thermal conductivity, ρ = density, and c = specific molar heat. It should be emphasized that the solution shown in Equation B.5 satisfies both Equation B.1a and, because of $\Theta_0 \to T_{s0}$ as $x \to -\infty$, also the first boundary condition (Equation B.2). The error function erf is as usually defined as $\text{erf } z = \int (2/\sqrt{\pi})\exp(-z^2)dz$.

B.2.2
Deposit Temperature Profile

$$\text{Assumption}: \Theta_1 = T_{s0} + \beta + \gamma\,\text{erf}(x/\sqrt{4a_1 t}). \tag{B.6}$$

The quantity β is the contact conductivity at the interface solid/liquid deposit ($x=X(t)$), and γ the "contact" diffusivity at the free interface receiving a constant stream of molten particles. Equation B.6 satisfies Equation B.1b, and also satisfies the second boundary condition if $\beta = \alpha$ ($\Theta_0 = \Theta_1$ as $x = 0$). Then Equation B.6 can be rewritten as

$$\Theta_1 = T_{s0} + \alpha + \gamma\,\text{erf}(x/\sqrt{4a_1 t}). \tag{B.6a}$$

To establish the required connection between α and γ the Equations B.5 and B.6 will be differentiated to yield

$$\partial\Theta_0/\partial x = \alpha[(\partial/\partial x)\text{erf}(x/\sqrt{4a_0 t})] = \alpha(2/\sqrt{\pi})\exp(-x^2/4a_0 t)(1/\sqrt{4a_0 t})$$
$$= [\alpha/\sqrt{\pi a_0 t}]\exp(-x^2/4a_0 t). \tag{B.7}$$

For $x = 0$ it follows

$$(\partial\Theta_0/\partial x)_{x=0} = \alpha/\sqrt{\pi a_0 t}. \tag{B.7a}$$

Analogously one obtains

$$(\partial\Theta_1/\partial x)_{x=0} = \gamma/\sqrt{\pi a_1 t}. \tag{B.7b}$$

Substituting Equations B.7a and B.7b into B.3 yields

$$k_0(\alpha/\sqrt{\pi a_0 t}) = k_1(\gamma/\sqrt{\pi a_1 t}).$$

With $\alpha/\gamma = B$, and $a = k/\rho c$ one obtains

$$B = (k_1\rho_1 c_1)^{1/2}/(k_0\rho_0 c_0)^{1/2}. \tag{B.8}$$

With Equation B.8, Equation B.6a can be expressed as

$$\begin{aligned}\Theta_1 &= T_{s0} + \gamma[(\alpha/\gamma) + \mathrm{erf}(x/\sqrt{4a_1 t})] \\ &= T_{s0} + \gamma[B + \mathrm{erf}(x/\sqrt{4a_1 t})].\end{aligned} \tag{B.9}$$

For $\Theta_1 = \Theta_2 = T_m$ (as $x \geq X(t)$) we obtain the explicit solution of the heat transfer equation at the interface liquid/solid deposit as

$$T_m - T_{s0} = \gamma[B + \mathrm{erf}(x/\sqrt{4a_1 t})]. \tag{B.10}$$

With the assumption

$$X = p(4a_1 t)^{1/2} \text{ or } t = X^2/4p^2 a_1 \tag{B.10a}$$

(solution of the first-order differential equation of heat transfer) it follows that

$$dX/dt = p(a_1/t)^{1/2}. \tag{B.11}$$

With Equation B.7 for $x = X(t)$, Equation B.4 can further be expressed by

$$k_1(\gamma/\sqrt{\pi a_1 t}) \exp(-X^2/4a_1 t) = L\rho(dX/dt) \tag{B.12}$$

and with Equation B.11 as

$$k_1(\gamma/\sqrt{\pi a_1 t}) \exp(-X^2/4a_1 t) = L\rho p(a_1/t)^{1/2}. \tag{B.13}$$

From Equations B.10 and B.10a it follows that

$$\gamma = (T_m - T_{s0})/[B + \mathrm{erf}(X/\sqrt{4a_1 t})] = (T_m - T_{s0})/(B + \mathrm{erf}\, p). \tag{B.14}$$

Inserting Equations B.14 and B.10a into Equation B.13 yields

$$(k_1/\sqrt{\pi a_1 t})[(T_m - T_{s0})/(B + \mathrm{erf}\, p)] \exp(-X^2/4a_1 t) = L\rho p(a_1/t)^{1/2} \tag{B.15}$$

or

$$(k_1/\sqrt{\pi a_1 t})[(T_m - T_{s0})/(B + \mathrm{erf}\, p)] \exp(-p^2) = L\rho p(a_1/t)^{1/2}. \tag{B.15a}$$

If the density of the molten particles equals the density of the solid material ($\rho = \rho_1$), Equation B.15a can be expressed by

$$(k_1/\sqrt{\pi a_1 t})[(T_m - T_{s0})/(B + \mathrm{erf}\, p)](1/\sqrt{a_1/t}) = (L\rho_1 p)\exp(p^2), \tag{B.16}$$

and finally, because $c_1 = k_1/\rho_1 a_1$,

$$(B + \mathrm{erf}\, p)p\exp(p^2) = c_1(T_m - T_{s0})/L\sqrt{\pi}. \tag{B.17}$$

Equation B.17 can be used to determine the constant p from the energy balance at the solidification front. It can also be taken from a nomogram provided by Kuijpers and Zaat (1974). With this one can also determine the *real temperature profiles*.

B.3
Real Temperature Profiles

From Equation B.8 it follows that $\alpha = \gamma B$, and with Equation B.14 one obtains

$$\alpha = [B(T_m - T_{s0})]/[(B + \operatorname{erf} p)]. \tag{B.18}$$

Inserting Equation B.18 into Equation B.5 yields the *real substrate temperature profile*

$$\Theta_0 = T_{s0} + \{[B(T_m - T_{s0})]/[B + \operatorname{erf} p]\}\{1 + \operatorname{erf}(X/\sqrt{4a_0 t})\}. \tag{B.19}$$

Likewise, the *real solid deposit temperature profile* becomes

$$\Theta_1 = T_{s0} + \{(T_m - T_{s0})/[B + \operatorname{erf} p]\}\{B + \operatorname{erf}(X/\sqrt{4a_0 t})\}. \tag{B.20}$$

On rewriting, Equations B.19 and B.20 yield their final expressions:

$$(\Theta_0 - T_{s0})/(T_m - T_{s0}) = [B/(B + \operatorname{erf} p)][1 + \operatorname{erf}(X/\sqrt{4a_0 t})] \tag{B.19a}$$

and

$$(\Theta_1 - T_{s0})/(T_m - T_{s0}) = [1/(B + \operatorname{erf} p)][B + \operatorname{erf}(X/\sqrt{4a_1 t})]. \tag{B.20a}$$

These equations should be compared with the approximate expression obtained for transient heat conduction in solid spheres as shown earlier:

$$(T_m - T_R)/(T_m - T_1) = f(Bi, Fo) = f[\alpha\theta/r^2] = f[a_1 t/r_0^2] \tag{B.16}$$

References

J.M. Houben, Relations of the adhesion of plasma sprayed coatings to the process parameters size, velocity and heat content of the spray particles. Ph.D. Thesis, Technical University Eindhoven, The Netherlands, 1988.

T.W. Kuijpers, J.H. Zaat, *Metals Technol.*, 1974, March, 142.

Appendix C: Calculation of Factor Effects for a Fractional Factorial Design 2^{8-4}

In this example, calculations will be shown for estimating the factor effects that describe the dependence of the thickness of a plasma-sprayed 88WC12Co coating on the eight selected plasma parameters and their ranges (Table C.1).

The fractional two level factorial design 2^{8-4} is shown in Table C.2.

The 2^{8-4} design chosen is a 1/16 replicate of a full 2^8 factorial of resolution IV that has the power to estimate the eight main factor effects X_i clear of each other, and clear of composite two-factor interactions E_i. The confounding pattern of the composite two-factor interactions E_i is as follows:

$$
\begin{aligned}
E_1 &= X_1X_2 + X_3X_7 + X_4X_8 + X_5X_6 & = 12 + 37 + 48 + 56 \\
E_2 &= & = 13 + 27 + 58 + 36 \\
E_3 &= & = 14 + 28 + 36 + 57 \\
E_4 &= & = 15 + 38 + 26 + 47 \\
E_5 &= & = 16 + 78 + 34 + 25 \\
E_6 &= & = 17 + 23 + 68 + 45 \\
E_7 &= & = 18 + 24 + 35 + 67
\end{aligned}
$$

The effects of higher-order interactions can usually be safely neglected.

Composite effects of the sum of four two-factor interactions, however, can be estimated from the unassigned factors. If only weak or no interactions exist, the effects of unassigned factors can be used to estimate the experimental (statistical) error, *i.e.* the minimum significant factor effect. The arrangement of the coded parameter levels in the design matrix (Table C.3) follows Yates' standard order.

Tables C.3 and C.4 show the numerical evaluation of the results for the main factor effects X_i (Table C.3) and the composite two-factor interactions E_i (Table C.4). First the sum $\Sigma(+)$ of all responses Y (thickness of coating) on the "+" level is calculated. Then the sum $\Sigma(-)$ of all responses Y on the "−" level is calculated. The factor effect is the difference Δ of the two sums, divided by the number of "+" (or "−") signs in each column. The coefficients C of the parameters in the polynomial response equation are obtained by dividing the factor effect by two. Factor significance is checked against the minimum factor significance, $\{min\} = \sigma_{FE} \times t_{\alpha,df}$ where $\sigma_{FE} = [(1/n)\Sigma E_i^2]^{1/2}$ with $t_{\alpha,df}$ = Student's t-value for a confidence level α of a double-sided significance test and df = degrees of freedom. All absolute factor effects larger than {min} are considered to be statistically significant.

Plasma Spray Coating: Principles and Applications. Robert B. Heimann
Copyright © 2008 WILEY-VCH Verlag GmbH & Co. KGaA, Weinheim
ISBN: 978-3-527-32050-9

Appendix C: Calculation of Factor Effects for a Fractional Factorial Design 2^{8-4}

Table C.1 Plasma spray parameters and their ranges used to estimate the factor effects on coating thickness and microhardness.

	Plasma spray parameter		Range
X_1	Plasma arc current		700–900 A
X_2	Argon gas pressure		0.34–1.36 MPa
X_3	Helium gas pressure		0.34–1.36 MPa
X_4	Powder gas pressure		0.34–0.68 MPa
X_5	Powder feed rate		0.50–2.00 (scale value)
X_6	Powder grain size		$(-45+5)$–$(-75+45)\mu m$
X_7	Number of passes	1st set	5–15
		2nd set	20–30
X_8	Spray distance		200–450 mm

Note that the second half of the main effect design matrix (Table C.3) is the mirror image of the first half, and that the two halves of the composite two-factor interaction design matrix are identical (Table C.4). From Table C.4 the minimum factor effect can be calculated as follows:

$$\sigma_{FE} = \sqrt{(1/n)\Sigma E_i^2} = \sqrt{(3592/7)} = 22.6 \tag{C.1a}$$

$$\{min\} = \sigma_{FE} \times t_{\alpha = 0.90; df = 7} = 22.6 \times 1.895 = 43. \tag{C.1b}$$

Table C.2 Fractional two-level factorial design in variables **1** to **8** and the responses Y (coating thickness) and their standard deviations σ.

Run #	x_1	x_2	x_3	x_4	x_5	x_6	x_7	x_8	Y (μm)	σ (μm)
1	700	0.34	0.34	0.68	2	coarse	20	450	118	79
2	900	0.34	0.34	0.34	0.5	coarse	30	450	16	8
3	700	1.36	0.34	0.34	2	fine	30	450	203	111
4	900	1.36	0.34	0.68	0.5	fine	20	450	57	25
5	700	0.34	1.36	0.68	0.5	fine	30	450	82	35
6	900	0.34	1.36	0.34	2	fine	20	450	138	87
7	700	1.36	1.36	0.34	0.5	coarse	20	450	30	12
8	900	1.36	1.36	0.68	2	coarse	30	450	82	44
9	900	1.36	1.36	0.34	0.5	fine	30	250	70	4
10	700	1.36	1.36	0.68	2	fine	20	250	108	104
11	900	0.34	1.36	0.68	0.5	coarse	20	250	65	30
12	700	0.34	1.36	0.34	2	coarse	30	250	16	10
13	900	1.36	0.34	0.34	2	coarse	20	250	26	21
14	700	1.36	0.34	0.68	0.5	coarse	30	250	9	12
15	900	0.34	0.34	0.68	2	fine	30	250	30	13
16	700	0.34	0.34	0.34	0.5	fine	20	250	22	19

Table C.3 Computing of main factor effects according to Yates' algorithm.

Run #	Mean	X_1	X_2	X_3	X_4	X_5	X_6	X_7	X_8	Y
1	+	−	−	−	+	+	+	−	+	118
2	+	+	−	−	−	−	+	+	+	16
3	+	−	+	−	−	+	−	+	+	203
4	+	+	+	−	+	−	−	−	+	57
5	+	−	−	+	+	−	−	+	+	82
6	+	+	−	+	−	+	−	−	+	138
7	+	−	+	+	−	−	+	−	+	30
8	+	+	+	+	+	+	+	+	+	82
9	+	+	+	+	−	−	−	+	−	7
10	+	−	+	+	+	+	−	−	−	108
11	+	+	−	+	+	−	+	−	−	65
12	+	−	−	+	−	+	+	+	−	16
13	+	+	+	−	−	+	+	−	−	26
14	+	−	+	−	+	−	+	+	−	9
15	+	+	−	−	+	+	−	+	−	30
16	+	−	−	−	−	−	−	−	−	22
Σ(+)	1009	421	522	528	551	721	362	445	726	
Σ(−)	0	588	487	481	458	288	647	564	286	
Δ	1009	−167	35	47	93	433	−285	−119	443	
Effect	63	−21	4	6	12	54	−36	−15	55	
Coeff.	b_0	−11	2	3	6	27	−18	−8	28	

Thus, all factor effects whose absolute values are larger than 43 are significant at a confidence level of 90%. From Table C.3 it follows that X_5 (powder feed rate) and X_8 (stand-off distance) are the only significant main factor effects. This holds true even when the confidence level is increased to 95%. There are no significant composite two-factor interactions (Table C.4). Both main factor effects have positive signs, *i.e.* the coating thickness increases with increasing powder feed rate and increasing spray distance. Hence the response polynomial can be roughly expressed by the equation

$$d[\mu m] = 63 + 27X_5 + 28X_8. \tag{C.2}$$

With the factors X_5 and X_8, and with addition of factor X_6 (powder grain size) whose effect is close to the minimum significance level a more detailed statistical design could be run, for example a Box-Behnken design with three variables at three levels (N = 15 runs) that allows to estimate nonlinear factor effects, or a Box-Wilson design (central composite design with star points) with three variables at five levels (N = 2^3 factorial points + 6 star points + 3 center points = 17 runs) to accurately describe curvature effects of the response surface. Considering the nonlinear effects of the three estimated factors the variability control factors (VCF) and target control factors (TCF) parameters within a Taguchi quality design methodology (Taguchi, 1987; Logothetis and Wynn, 1989; Eibl et al., 1992) can be selected and implemented in a statistical process control (SPC) protocol.

Table C.4 Computing of composite two-factor interactions.

Run #	E_1	E_2	E_3	E_4	E_5	E_6	E_7	Y
1	+	+	−	−	−	+	−	118
2	−	−	−	−	+	+	+	16
3	−	+	+	−	+	−	−	203
4	+	−	+	−	−	−	+	57
5	+	−	−	+	+	−	−	82
6	−	+	−	+	−	−	+	138
7	−	−	+	+	−	+	−	30
8	+	+	+	+	+	+	+	82
9	+	+	−	−	−	+	−	7
10	−	−	−	−	+	+	+	108
11	−	+	+	−	+	−	−	65
12	+	−	+	−	−	−	+	16
13	+	−	−	+	+	−	−	26
14	−	+	−	+	−	−	+	9
15	−	−	+	+	−	+	−	30
16	+	+	+	+	+	+	+	22
Σ(+)	410	644	505	419	604	413	448	
Σ(−)	599	365	504	590	405	596	561	
Δ	−189	279	1	−171	199	−183	−113	
Effect	−24	35	0	−21	25	−23	−14	
Coeff.	−12	18	0	−11	13	−12	−7	

References

S. Eibl, U. Kess, F. Puckelsheim, *J. Qual. Technol.*, 1992, **24** (1), 22.

N. Logothetis, H.P. Wynn, *Quality Through Design*, Clarendon Press: Oxford, 1989.

G. Taguchi, *Introduction to Quality Engineering*, Amer. Suppl. Inst. Center for Taguchi Methods, Dearborn, Michigan, 1987.

Index

2D scatter images 132
3-body block-on-ring sliding wear test 200, 209, 343
3D stress state 201
3D topographical data 140
4D hypercube design 384

a

Abbot's curve 348
Abel's inversion 67
abrasion-resistant coatings 402
abrasive wear 340
AC discharge mode 40
acoustic emission (AE) 239, 333
– inverse analysis 335
adaptive statistical process 23
adhesive wear friction 339
aerosol deposit (AD) 8
aerospace engines 214
– titanium-based components 214
agglomeration process 85
air-operated heating nozzle 327
alkaline phosphatase (ALP) activity 281
Almen test 151–152
alumina powder particles 107
alumina-based coatings 217
alumina-mullite coatings 251
alumina–titania coatings 218, 322, 385
alumina-yttria binder matrix 269
alumina-zirconia top layer 248
aluminum-based composite SiC coatings 213
amorphous calcium phosphate (ACP) 124, 125, 148, 281, 289–290
anode jet-dominated (AJD) 47
anode vapor jets 47
arc column 44
– structure 44
arc discharge generators 50

arc-fuse process. 84–85
Archimedes' technique 309, 338
area law of mixtures 330
area scaled fractal complexity (ASFC) method 140
argon-ion laser beam 346
artificial neural network (ANN) structure 367, 392, 394
– analysis 367, 392–393
– mathematical concept 393
a-scan diagram 324
Aston region 43
atmospheric argon jet 62
atmospheric plasma spraying (APS) 1, 3, 17–18, 134, 266, 284
auxiliary metal bath process 213
axial feed plasma torch 213
axial plasma velocity 74
axial temperature gradients 62
axial velocity isopleths 62

b

back wall echo (BE) 324
ballistic compressor 39
ballistic deposition process 115
ballistic model 136
band-gap materials 5
barium-strontium aluminum silicate (BSAS) 243
barium titanate (BT) 5
Basset–Boussinesq–Oseen equation 169–170
Bench-scale tests 353
Bennett equation 49
Berkovich indenters 329
Bernoullis theorem 71
bioactive calcium phosphate ceramics 278
– application 278
bioceramic HAp coatings 152

Plasma Spray Coating: Principles and Applications. Robert B. Heimann
Copyright © 2008 WILEY-VCH Verlag GmbH & Co. KGaA, Weinheim
ISBN: 978-3-527-32050-9

bioceramic materials 277
– coatings 3, 13, 278
  – properties 278
– development 277
– research 277
bioconductive monolithic ceramics 277
bioinert bond coats 289–290
biomaterials–cell responses 298
– *in vitro* test methods 298
Biot number 89, 91–92
bismuth-manganese alloy 5
blind hole test 145
block-on-ring abrasion test 201, 203
Boltzmann number 101
bond coat oxidation 253
bonding osteogenesis 277
bone morphogenetic protein (BMP) 298
bone-resorbing osteoclast 277
Boolean logic 306
boride diffusion coatings 210
Bouguet number 102
box counting method 137
Box–Behnken design 369, 373–374, 377
Box–Wilson design 374
Bragg angles 146, 150
Bragg equation 145
Brinell- indentation 335
Buckinghams rule 163
Bunsen burner 296
burner rig test 354

c

calcium oxide (CaO) 281
calcium phosphate coatings 123
– phase composition 123
calcium silicate 290
calcium titanate 281
– 2D-lattice match 281
calcium zirconate 292
Cameron–Plint tester 341–342
carbide-stabilizing additions 211
carbon-containing precursor gases 206
carrier gas flow rate 395
– contributions 395
cathode fall region 42, 44
Cause–Effect diagrams 304
cavitation–erosion testing 403
cemented carbides 193
centerline plasma velocity 74
central composite design 374
central powder injection 79
centro-symmetrical charge distribution 25
ceramic coatings 3
– market development 3

ceramic components 351
– hot corrosion 351
ceramic liquid droplets 311
– spreading process 311
ceramic–metal composite coatings 144
cermet coating clad 206
chaotic plasma spray process 112
charge coupled device (CCD) camera 76
chemical barrier coating (CBC) 212, 249, 251
chemical cladding 85
chemical corrosion resistance 351
– tests 351
chemical degradation 199
chemical reactions heating plasmas 38
chemical vapor deposition (CVD) coatings 2, 5, 215, 298, 361
Chicot–Lesage model 214
chipping mechanism 311
chromatic monitoring 69
chromium carbide coatings 209, 211–212
chromium-iron alloy 256
chromium-nickel-cobalt alloys 193
chromium oxide coatings 220, 389
classical experimental strategy 368
Clausius–Clapeyron equation 122
Clausius–Clapeyrons law 125
Cline–Anthony model 246
coating hardness values 328
coating optimization strategies 288
coating technology 4
– advantage 4
coating–substrate interface 282
coaxial powder injection 79
Co-based WC coatings 194, 203, 333
coefficient of thermal expansion (CTE) 224, 239
cohesion-type stripping 317
cold gas dynamic spraying (CGDS) 7, 20, 266, 298
collective interaction phenomena 27
collision path length 27
component tests 353
computer-assisted feedback loop 103
connected energy transmission 86
conservation equations 120, 166
contact osteogenesis 277
continuous detonation spraying (CDS)-HVOF 194
continuous process improvement (CPI) 303, 305
continuous wave photo-thermal radiometry (CW-PTR) 327
continuum-emission coefficients 47

convection–diffusion equation   94
convection-stabilized arcs   59
convective energy transfer   88
convective heat transfer mode   101
Coriolis forces   49
corrosion-resistant coatings   7, 191, 221, 349, 402
Coulomb logarithm   28
Coulter Counter   309
crack-arresting mechanism   245
crack-free coating   241
crack propagation   243
– management   243
cross-correlation function (CCF)   75
c-scan diagram   324
curvature monitoring technique   151
curvature-stress models   154
cylindrical plasmatron nozzles   65

d

DC discharge mode   40
DC plasma arc torch   56
DC plasma jet process   224
Debye screening length   27
deLaval nozzle   63, 65, 74
Delesse's principle   141
Deming's principle   305
dense vertically cracked (DVC) coatings   243
density correlation function   138
– principle   138
density functional theory (DFT)   365
D-gun techniques   208
diamond coatings   13, 224–225
diamond growth species   226
diamond jet hybrid (DJH)-HVOF   194
diamond-stylus surface roughness tester   140, 347
Dietzel equation   144
diffusion-controlled mechanism   237
dimensionless transfer equations   163, 407–408
disc-shaped splats   312
distance osteogenesis   277
double-torsion test   333
downstream injection   79
drag coefficient   170–171
– modeling   170
Draper Lin-Cube-Star design   141
drift waves   100
drug delivery systems (DDS)   298

e

E-beam coating   191
echo-impulse techniques   325

elastic–plastic anisotropic parameters   335
elastic–plastic finite element method   183
electric arc spraying operations   80
electric space charge effects   27
electric stabilization   59
electro-catalytically active coatings   262
electro-chemical machining (ECM)   256
electrode-less plasmas   54
electrode-stabilized arcs   60
electrode-supported plasmas   50
electromagnetically accelerated plasma spraying (EMAPS)   39, 192, 216, 267
electromagnetic interference (EMI)   14
electromagnetic pinch effect   56
electromagnetic shock tube   39
electron-beam physical vapor deposition (EB-PVD)   223
electron beam-plasma generator   40
electron energy loss spectroscopy (EELS)   193
electron–gas interactions   25
elementary cusp catastrophy   100, 396
Elenbaas–Heller approach   164
elutriation method   308
emission spectroscopy   69
empty plasma   68
energy economy   102
energy-efficient high temperature plasmatrons   62
energy transfer processes   22
enlightened participatory management   305
– team members   305
enthalpy   65
environmental barrier coating (EBC)   375
equilibrium plasmas   32–33
erosive wear   344
Euclidean distance mapping   337
Euler flow equation   49
evaporation constant   97
eximer laser ablation techniques   224
experimental turbocharger rotors   215

f

Fluorapatite (FAp)   284
feed materials   80
– characteristics   80
ferrosilicon coatings   383
filled wires   80
fine-grained multiphase structures   364
finger-like protrusions   178
finite element modeling (FEM)   154, 183
finite-size effect   264
first-order polynomial model   368
first-order process map   366

first sharp diffraction peak (FSDP)   132
flame assisted chemical vapor deposition (FACVD)   296
flame spraying (FS)   284
fluid flow   177
four-dimensional response hyperspaces   377
Fourier analysis   139
Fourier equation, *see* heat transfer equation
Fourier number   91–92, 118
Fourier's law   126
four-point bending test   333
fractal analysis method   313
fractal geometry   135, 137
fractional factorial design   372, 380, 384–385, 389
fracture profile analysis (FPA)   139
fracture toughness   332
free-standing coatings   147, 283
free stream velocity   71
fringe mode-measurements   73
FT–Raman (FTRS) investigations   290
fuel economy   235
full factorial designs   371, 376
full-potential linearized augmented plane-wave (FLAPW) method   365
full-potential linear muffin-tin orbital (FPLMTO) method   365
functionally gradient material (FGM)   363
fuzzificator   396
fuzzy logic control (FLC)   367, 396
fuzzy look-up model   166

g

gas-atomized particles   84
gas flow rate   172
gas-sheath stabilized plasma   60
Gauss function   135
Gaussian distribution   84, 246, 385
Gaussian laser beam   246
Gauss principle   337
Gerdien arc   54, 60
glancing-incidence X-ray diffraction (GIXRD)   132
glass-based coatings   5
glow discharge optical emission spectrometry (GD-OES)   43
gravity-driven counterflow   247
grey alumina   217
Grey probe   70–71
grit-blasting procedure   138, 240, 312

h

hard-phase titanium   7
heat engine components   401–402

heat transfer   87, 99, 111, 126, 166, 169, 177, 396
– calculations   98
– catastrophy   99–100, 102
– coefficient   88
– dense loading conditions   99
– exothermic chemical processes   37
– low loading conditions   87
– model   182
heat transfer equation   93, 115, 117
heat-treated coatings   291
helium plasma spray process   166
– modeling   166
He–Ne laser light pulse   105
high density plasmas   32
high electron temperature   40
high-energy erodent particle   344
– impact   344
high energy radiation   34
high field-strength hydrogen arc   60
high-frequency pulse detonation (HFPD)   215
high friction coefficient   267
high-intensity argon arc   47
high power plasma spraying (HPPS)   17
high pressure jets   60
high-pressure turbine blades   345
high-temperature erosion protection   210
high temperature protection coating system   353
high-temperature superconducting materials   3, 13, 254
high velocity oxyfuel (HVOF) spraying technique   17, 21, 194, 284
high-velocity plasma jet   118
high velocity pulsed plasma spraying (HVPPS)   20
Holdgren practice   210
hollow-spherical-powder (HOSP)   84–85
HT-superconducting coatings (HTSC)   253
hydrogen plasma   196
– jet   86
– use   196
hydroxyapatite (HAp)   287, 310
– coating properties   281, 287
– particle splats   310
HYPREPOC process   85

i

image analysis   312
image plate system   132
indentation size effect (ISE)   332
indium tin oxide (ITO)   5

induction plasma sprayed (IPS) titania coatings 378
– photo-catalytic efficiency 378
inductively coupled plasma (ICP) devices 50, 54
– spectrometry 307
inductively coupled torch 55
– flow fields 55
– temperature 55
industrial environment 2
inert gas plasma spraying (IGS) 17
in-flight particle temperature 76
infrared spectroscopy 286
in situ annealing 241
intelligent process control (iSPC) 9, 12, 398
– development 398
interfacial echo (IE) 325
inter-fringe spacing 73
inverse Yates algorithm (IYA) 372
ion arc device 52
ion beam coating 191
ionization frequency 40
isothermal plasmas 37

j

Jönsson and Hogmarks model 331
Jurans Quality Trilogy 305
just-in-time (JIT) concept 304

k

kinetic gas theory 28
Knoop indentation 328, 335
Knoop-type diamonds 329
Knudsen number 89
Kruskal–Willis tests 294

l

Lagrangian two-phase fluid model 168
lambda-type oxygen sensor probe 71
lamellar graphite grey cast iron (GGL) 402
Landau length 27
Langmuir plasma frequency 26
Langmuir probe 32
lanthanum hexaaluminate 249
lanthanum-strontium manganate (LSM) 263
Larmor radius 30–31
laser-assisted spraying 20
laser beam diffraction 70
laser-densified coatings 246
laser detection system 152
laser Doppler anemometer (LDA) 70, 72–73, 86, 103
laser-induced radial cracks 245

laser-melted coating 247
laser Raman spectrometers 155
laser surface remelting technique 248
laser treatment 244
linear isotropic elasticity theory 183
line-emission coefficients 47
line-of-sight limitation 210
liquid melt pool 111
liquid precursor plasma spraying 20
logistic curve 393
log-normal distribution 84
longitudinal cracks 246
long-term fatigue wear 343
long-term laboratory test 353
Lorentz force 39, 46, 49, 60, 100, 267
low cycle fatigue (LCF) 223
low density plasma 30–31
low earth orbit (LEO) 250
– spacecraft 6
low-energy plasmas 37
– generation 37
low-frequency ultrasonic energy 323
low plasma enthalpy 392
low pressure plasma coating system 53
low pressure-plasma jet 107
low pressure plasma spraying (LPPS) technology 17–18, 171, 176, 284, 287, 291, 312, 336
L_{12} Taguchi design 377

m

macrobonding 313
macroscopic plasma parameters 61
Madejski model 117
magnetic pinch 48
magnetically-stabilized arcs 61
magnetically-stabilized jet 60
magnetic plasma confinement 37
magneto-hydrodynamic (MHD) 26, 48, 61
– equations 26
– interactions 61
– Lorentz force 48
magnetron sputtering 6
main regime parameters (MRP) 117
Mamdani centroid inference 397
Mann–Whitney tests 294
Marker and Cell method 117
Marshall–Lawn indentation model 333
Marsh's relation 331
mass conservation equation 166
mass correlation function 139
mass transfer coefficient 93
Maxwell–Boltzmann distribution 28
Maxwell distribution 29

Maxwell electromagnetic field equations 99, 165
Maxwell model 246
mechanical anchorage 313
medium density plasma 30, 32
mercury pressure porosimetry 338
mesoscopic stresses 144
metal insulator semiconductor (MIS) memory devices 215
metallic coatings 221, 335
metallic conductors 145
– expansion-resistance effect 145
metallic hard materials 362
metallurgical bonds 314
metal–metal bonds 362
Mexican hat 112
Meyer's approach 329
micro-arc oxidation (MAO) 284, 293, 298
microbonding 313
microhardness tests 328–329
microhole drilling 183
microprocessor-controlled metering devices 99
microscopic stresses 143
microstructural modications 237
milling method 183
mining safety appliance (MSA) 309
mismatch-related failure 363
mixed-mode failures 316
modeling process 163
– principal aspects 163
mole-centered system 167
molecular beam epitaxy 6, 191
momentum conservation equation 49, 100
momentum transfer 86, 116, 169
monatomic gases 34
Monte Carlo approach 172–173
mullite coatings 250–251
multilayer concept 364
multi-layer holographic coatings 6
multiple plasmatron–independent feeding systems 364
multiple regression analyses 375
multi-wavelength pyrometry 103

n

nanostructured feedstock 243
– plasma-sprayed multilayer coatings 243
narrow-band interference 73
Navier–Stokes equation 49, 88, 164, 166, 177
Nelson–Riley extrapolation function plots 149
Nernst–Descartes law 150
neutron diffraction measurements 145, 157
Newton momentum equation 181
Newton's second law 119
Neyer D-optimal designs 374
Ni-based coatings 203
NiCrAlY bond coat 240
NiCr-based coatings 222
nitride coatings 213
noise-less plasma 37
nonautomotive markets 402
noncontact techniques 69
noncontinuum effects 171
nonequilibrium plasmas 32
nonisothermal plasma 37, 40
nonlinear hyperbolic differential equations 119
nontransferred electrode plasmatrons 50
nontransferred free plasma jet 52
novel advanced layered coatings 363
– design 363, 363
nuclear fission device 38
numerical simulation 163
numerical techniques 99
Nusselt number 88, 93–94, 101, 174

o

object-oriented finite (OOF) element method 238
Ohm's law 40
Ollard–Sharivker test 314, 317
one-dimensional radial equation 26
on-line process diagnostics 2
– modeling 2
Orowan–Petch relation 344
orthogonal blocking 373
output parameters 100
oxide coatings 216
oxide-cobalt-titanium anodes (OCTA) 262
oxygen-deficient coatings 254
oxyhydroxyapatite (OHAp) 281

p

parameter interactions 392
partially stabilized zirconia (PSZ) 212, 224, 233
particle dispersion 172
– modeling 172
particle injection 105
particle-loaded plasma gas stream 102
particle number density 114
– determination 107, 114
particle-source-in-cell (PSI) 166
particle–substrate interaction 1, 111, 143, 172
particle temperature determination 107

particle velocity determination   103
patchwork method   140
Peclet number   118
peel test   319
petrochemical industry   211
– problem   211
photo-catalytic process   264
photo-induced electron-hole transfer process   267
photoluminescence piezospectroscopy   155
physical vapor deposition (PVD) techniques   2, 6, 157, 214, 361, 363
– arc technique   214
physico-chemical processes   260
piecewise linear functions   393
pinch effect   39
pin-hole test   318
pin-on-disk test   269, 339
Pitot tube probe technique   70
Plackett–Burman designs   368–371, 374–375, 377
planar shock waves   118
plasma-arc wire-spraying operations   103
plasma-assisted CVD (PA-CVD)   224
plasma-assisted vapor deposition (PAVD)   33
plasma-confining Lorentz force   100
plasma diagnostic techniques   65, 388
plasma electrolytic oxidation (PEO)   297–298
plasma energies   33
– Maxwellian distribution   33
plasma enthalpy   395
plasma equations   166
plasma furnace   37
plasma gas flow rate   395
plasma generation   37, 39
plasma jet   46, 61, 142
– decarburization reactions   349
– temperature   61
– velocity distributions   61
plasma melted rapidly solidified (PMRS) method   206
plasma parameters   26, 32
plasma–particle interaction   1, 166, 168–169
– modeling   166
plasma–particle–substrate interactions   2, 401
plasma plume sensor   99
plasma-spray coated cylinder liners   402
plasma spray-deposited FAp coatings   284
plasma-sprayed alumina coatings   217, 333
plasma-sprayed calcium phosphate layers   280
– adhesion strength   280
plasma-sprayed ceramic nitride coatings   213

plasma-sprayed chromium oxide coating   140–141, 158, 252, 322
plasma-sprayed coating   99, 132, 147, 157, 176, 193, 289, 254, 306, 314, 328, 335, 337, 396, 402–403
– elastic moduli   335
– hydroxyapatite (Hap)   3, 283–284
– pore orientation   337
– properties   137
– success   306
plasma-sprayed oxide electrocatalytic materials   263
plasma-sprayed TBC systems   248
plasma-spray powder   307–308
– quality control measures   307
– testing procedures   307
plasma spray process   1–3, 20, 22, 86, 92, 102–103, 115, 166, 173, 180, 212, 284, 349, 363, 365, 398, 403
– features   22
– main aspects   398
– modeling   1
– operations   212
– operators   306
– osseoconductive hydroxyapatite   284
– parameters   20, 163, 279, 333, 389, 392
plasma transferred arc (PTA) surfacing process   388
plasmatron   48
– design   48
– input parameters   365
– models   57
plasma velocity   71, 108
plasma volume flow rate   172
plasma zone lengthening   57
point counting approach   337
Poisson's ratio   144, 146–147, 283, 342
positive column   44
positive electrode-electrolyte-negative (PEN) electrode   255
positive-ion emission   47
powder characterization   307
powder feed systems   364
powder injection velocity   172
powder particle diameters   81
Prandtl number   88
precipitation-strengthening mechanism   211
probability density function (PDF)   173
process mapping   365
process–property maps   172
propane–air pilot flame   296
property-based approaches   367
property–parameter evolution   392
protective ceramic layer   256

– function   256
pseudo-alloy coating   80
pseudo-Bragg reflections   135
pseudo-grazing incidence X-ray diffraction (GIXRD) method   150
pseudo-Voigt algorithm   156
pulsed laser deposition (PLD)   298
pulsed laser sealing   248

q

qualification procedures   306
quality assurance (QA)   306
quality control procedures   403
quality function deployment (QFD)   12, 83, 304–305
quality philosophy   304
quasi-elastic behavior   147
quasi-Gaussian distribution   105
quasi Knudsen effusion cells   6
quasi-laminar jets   63
quasi-linear relationship   382
quasi-metallurgical bond   118
quasi-neutral multiparticle systems   25
quasi one-dimensional quantum wells   6
quasi-steady state evaporation   97
– time   97
quenching stresses   282
quiet plasma, *see* noise-less plasma

r

radial temperature profile   61
radial velocity distribution   62
Radio frequency (r.f.)   54
rainbow sensor   346
Raman active modes   265
– spectroscopic characteristics   265
Raman inactive SRO phase   378
– intensity   378
Raman microprobe technique   264–265
– analyses   378
Raman microscopy   264
– advantage   264
Raman spectroscopic methods   264
random walk influence   108
Rankine–Hugoniot equation   118, 120
Ranz–Marshall expression   88
rare earth elements (REE)   346
rarefaction effect   89
Rayleigh number   101
reactive laser sealing   248
reactive laser treatment   248
reactive magnetron sputtering   191
reactive plasma spray (RPS) technique   20, 206, 214

– TiN coatings   214
reduced depression depth   348
reduced spike height   348
reflection high energy electron diffraction (RHEED)   6
refractory metal coatings   221
residual coating stresses   143–144, 180, 282
– modeling   180
– origin   144
residual tensile coating stresses   144, 242
resource industries   3
response surface (RSM) designs   369
return of investment (ROI)   403
Reynolds number   57, 70, 93, 116, 170–171
Richardson–Dushman equation   44
Richardson method   140
Riemann–Hugoniot catastrophy   100, 396
Rockwell 'C' diamond   321–322
Rockwell hardness   196
Rockwell superficial hardness tests   328
– procedure   328
Rokide process   80
room temperature impact consolidation (RTIC)   8
rubber wheel abrasion test   200–201, 339
Ryshkevich–Duckworth equation   344

s

Saha-Eggert equation   37
salt-shower test   354
scale-sensitive fractal analysis   140
scanning laser profilometer   140
Schmidt number   93
Schumann–Runge bands   67
scratch test   321–322
screening designs   368–369
– anatomy   369
screw-feed type metering system   344
second-degree graduating polynomial   373
second-generation bioceramics   3
second-order effects   374
second-order polynomial model   369
sedimentation techniques   309
self-affine geometries   136
self-fluxing nickel-base alloy   376
– properties   376
self-generated magnetic field   46
shear tests   318
Sherwood number   93
shock–compression process   121
shock wave   111, 120
short range order (SRO) structure   132, 281
– nature   265

short-term chemical corrosion tests   353
SiC-based heat exchanger tubes   251
– oxidation protection   251
signal-to-noise ratio   398
silicon nitride coatings   214–215, 267, 269
silver-coated steel substrates   254
single-particle trajectories calculations   99, 166
single plasmatron–multiple powder feeders   364
size–shape distribution   312
slit island analysis (SIA)   139
small and medium-sized enterprise (SME)   12, 403
small-angle neutron scattering (SANS)   238
smooth-rough crossover (SRC)   141
soft-vacuum conditions   105
sol-gel coatings   5
   processes   82
solidification   177
solid–liquid suspension   296
solid oxide fuel cells (SOFC)   3, 13–14, 254
– electrocatalytic coatings   254
solid particle erosion (SPE)   194–195, 209–210, 251, 345
– cavitation   194
solid-phase bond   314
solid wires   80
spallation technique   325, 327
spark plasma sintering   248
species conservation equation   167
specimen curvature technique   152
spectral double images   76
spectroscopic methods   66, 68
spherical cavity analysis   331
spherical semi-molten particle   125
splat configuration   310
splat morphology   177, 180
splat shapes   177
– modeling   177
spray parameter settings   286
spray powders   81, 307
– particle size distribution   307
spray pyrolysis   5
sputtering methods   6
stability theory   396
stationary diesel engines   220
statistical design matrix   382
– application   382
statistical design of experiment (SDE)   11–12, 20, 83, 304–306, 367, 392
statistical experimental strategy (SES)   1, 304
statistical process control (SPC) methods   1, 20, 83, 99, 304–305

statistical quality assurance (SQA)   304
statistical quality control (SQC)   304
Stefan problem   174
Stefan–Boltzmann coefficient   88–89
stellite coatings   388
stochastic approaches   367
Stoney equation   151, 154
strain to fracture (STF)   334
stress corrosion cracking (SCC)   383
stress development   239
stress–strain model   183
stress tensor   149
stroke-of-genius experiments   367
student t-value   381
substrate–coating interface   282, 290
superalloy coatings   223
supersaturated designs   369
surface coating techniques   1, 3
   survey   3
surface-sensitive analytical methods   264
suspension plasma spraying (SPS)   20, 218, 258–259, 266, 284, 296

t
Taguchi analysis   99
Taguchi design   368, 375–376, 388
Taguchi-structured control factors   396
tail-flame injection type plasmatrons   52
tape test   321
temperature-dependent parameters   133
temperature measuring techniques   66
tensile pull test   314
tensile stresses   283
tetracalcium phosphate (TetrCP)   124
thermal barrier coating (TBC)   7, 84, 182, 216, 224, 233, 312, 401
thermal decomposition products   281
thermal equilibrium   164
thermal expansion coefficient   252
thermally grown oxide (TGO)   233
thermally induced chemical changes   348
thermally sprayed coatings   3, 13, 312, 347
– quality   312
thermal pinch   48
thermal plasma chemical vapor deposition (TPCVD)   20, 258
thermal plasma cooperative processes   100
thermal protection systems (TPS)   250
thermal spray technology   1, 7, 9, 17, 20, 403
– development   9
– principles   17
– processing   76, 401
thermal stresses   81, 144, 182, 283
thermal wave interferometry   327

thermo-diffusive mass transport   102
thermonuclear fusion research   252
thermophoretic forces   104
thick thermal barrier coating (TTBC)   236
thin-shelled particles   85
third-level parameters   382
Thomas approach   329
three-electrode plasmatron   79
three-factor interactions   373
three-level factorial designs   373
three-phase alloys   198
three-point bend test   334
TiC-based coatings   206, 208
– wear properties   206
time-dependent mass flux   49
time-resolved infrared radiometry (TRIR)
    thickness variations   327
titania bond coats   292–294
titanium carbide based coatings   205
titanium dioxide   263
titanium nitride   149
– coatings   214
torch diagnostic system-spray plume trajectory (TDS-SPT) sensor   99
total quality management (TQM)   83, 303–306, 392
Townsend coefficient   40–41
Townsend discharge   41–42
traditional engine block materials   402
transferred electrode plasmatrons   51
transient heat conduction   91
tribological properties   338
tricalcium phosphate (TCP)   124, 281
triple torch plasma reactor (TTPR)   243
Tsui–Clyne models   154
tungsten carbide   192, 198
– coatings   194, 203, 210
– nozzle   344
tungsten coatings   377
turbine inlet temperature (TIT)   233
turbulent free plasma argon jet   108
turbulent Langmuir oscillations   40
two-beam model   154
two-color pyrometry   68, 76
two-factor interaction   372–373, 381, 384, 386, 389–391, 392
two-fluid interfacial flow   169
two level fractional factorial design   367, 388
– application   388
two-stage model   181
two-step electron transfer model   267
two-temperature plasmas   32
two-wavelengths pyrometry   68

u

ultimate tensile strength (UTS)   222
ultra large scale integration (ULSI) devices   6
ultrasonic c-scan technique   325
ultrasonic tests   323
ultrasonic waves
– holographic imaging   323
ultra supercritical (USC) power plants   222
underwater plasma spraying (UPS)   194
upstream injection   79

v

vacuum plasma spraying (VPS)
    techniques   17, 221
velocity measurements   65, 70
vermicular graphite grey iron (GGV)   402
vibrational spectroscopic methods   286
Vicker's diamond   329, 333, 340
Vicker's indentation test   320
VISAR velocimeter   325
Vlasov plasma   30
volume law of mixture model   332
volume-centered system   167
volumetric emission coefficient   67
Voronoi cells   337
vortex interactions   62
vortex-stabilized arc   60
VPS-sprayed chromium coatings   324

w

wall-stabilized arcs   59
water-atomized particles   84
water–cooled injection probe   108
water-cooled vitreous silica tube   56
water-stabilized plasma spraying (WSP)   193
water vortices   53
WC-based coatings   194–195, 200, 202
wear mechanism maps   339
wear model tests   339
wear-resistant coatings   335
Weibull analysis   334
Weibull modulus   182, 334
welding process   102, 111
Wheeler's test   371
wide-angle X-ray scattering (WAXS)   132
wipe test   122, 286, 310

x

X-ray diffraction (XRD)   132, 158, 237, 286
– analysis   353
– measurement   145, 147
X-ray fluorescence spectroscopy   307
XRD-based measurements   180

y

Yates order 372
Young's modulus 146, 151, 158, 210, 237, 243, 316, 337, 365
Y-PSZ coatings 244
– laser surface remelting 244

z

zirconia-based TBC 235
zirconia bond coats 291
zirconia coatings 182, 242
zirconia precursor powders 235
zirconia-reinforced HAp coatings 291
z-pinch configuration 39